NATO ASI Series

Advanced Science Institutes Series

A series presenting the results of activities sponsored by the NATO Science Committee, which aims at the dissemination of advanced scientific and technological knowledge, with a view to strengthening links between scientific communities.

The Series is published by an international board of publishers in conjunction with the NATO Scientific Affairs Division.

A	Life Sciences	Plenum Publishing Corporation
B	Physics	London and New York
C	Mathematical and Physical Sciences	Kluwer Academic Publishers
D	Behavioural and Social Sciences	Dordrecht, Boston and London
E	Applied Sciences	
F	Computer and Systems Sciences	Springer-Verlag
G	Ecological Sciences	Berlin Heidelberg New York Barcelona
H	Cell Biology	Budapest Hong Kong London Milan
I	Global Environmental Change	Paris Santa Clara Singapore Tokyo

Partnership Sub-Series

1.	Disarmament Technologies	Kluwer Academic Publishers
2.	Environment	Springer-Verlag / Kluwer Academic Publishers
3.	High Technology	Kluwer Academic Publishers
4.	Science and Technology Policy	Kluwer Academic Publishers
5.	Computer Networking	Kluwer Academic Publishers

The Partnership Sub-Series incorporates activities undertaken in collaboration with NATO's Cooperation Partners, the countries of the CIS and Central and Eastern Europe, in Priority Areas of concern to those countries.

NATO-PCO Database

The electronic index to the NATO ASI Series provides full bibliographical references (with keywords and/or abstracts) to about 50 000 contributions from international scientists published in all sections of the NATO ASI Series. Access to the NATO-PCO Database compiled by the NATO Publication Coordination Office is possible in two ways:

- via online FILE 128 (NATO-PCO DATABASE) hosted by ESRIN, Via Galileo Galilei, I-00044 Frascati, Italy.
- via CD-ROM "NATO Science & Technology Disk" with user-friendly retrieval software in English, French and German (© WTV GmbH and DATAWARE Technologies Inc. 1992).

The CD-ROM can be ordered through any member of the Board of Publishers or through NATO-PCO, B-3090 Overijse, Belgium.

Series F: Computer and Systems Sciences, Vol. 158

Springer
Berlin
Heidelberg
New York
Barcelona
Budapest
Hong Kong
London
Milan
Paris
Santa Clara
Singapore
Tokyo

Mathematical Methods in Program Development

Edited by

Manfred Broy

Birgit Schieder

Institut für Informatik
Technische Universität München
D-80290 München, Germany

Springer

Published in cooperation with NATO Scientific Affairs Division

Proceedings of the NATO Advanced Study Institute on Mathematical Methods in Program Development, held in Marktoberdorf, Germany, July 30 – August 11, 1996

Library of Congress Cataloging-in-Publication Data

Mathematical methods in program development/ edited by Manfred Broy, Birgit Schieder.
p. cm. -- (NATO ASI series. Series F, Computer and systems sciences ; vol. 158)
"Proceedings of the NATO Advances Study Institute on Mathematical Methods in Program Development, held in Marktoberdorf, Germany, July 30-August 11, 1996"--T.p. verso.
Includes bibliographical references.
ISBN 3-540-63195-X (hardcover : alk. paper)
1. Electronic digital computers--Programming. 2. Computer software--Development--Mathematical methods. I. Broy, M., 1949- . II. Schieder, Birgit. III. NATO Advances Study Institute on Mathematical Methods in Program Development (1996 : Marktoberdorf, Germany) IV. Series.
QA76.6.M3648 1997
005.1'01'51--dc21 97-22450
CIP

CR Subject Classification (1991):
D.2.1, D.2.2, D.2.4, F.3.1, F.4.1, I.2.2

ISBN 3-540-63195-X Springer-Verlag Berlin Heidelberg New York

Printed in Germany

Typesetting: Camera-ready by editor
Printed on acid-free paper
SPIN: 10552724 45/3142 – 5 4 3 2 1 0

Preface

Program development is one of the most fascinating challenges of modern technology. Still, the field is in an immature state. Only step by step is the scientific foundation being discovered and understood in more detail. The Summer School on Mathematical Methods in Program Development was held in Marktoberdorf from July 30 to August 11, 1996. More than 100 people, counting participants, lecturers, and staff, attended.

The Summer School centred around the techniques of the mathematics of program development, including programming paradigms treated by Kathleen Fisher, John C. Mitchell, and Gert Smolka, with their contributions on new programming paradigms, such as object-oriented programming and higher-order concurrent constraint programming. An excellent survey over the history of programming languages, including many interesting side-remarks explaining why the history went the way it did, is found in F.L. Bauer's contribution on the genesis of algorithmic languages. Trying to abstract away all the superficial differences in programming style are the contributions by C.A.R. Hoare on unified theories of programming and by Manfred Broy with respect to mathematical models used in early phases of program development.

The contributions by J.-R. Abrial and L. Mussat and by B. von Karger concentrated on case studies, demonstrating the development of protocols by applying well-founded mathematical techniques in the development process. The rest of the contributions are devoted to the aspect of proving programs correct. This includes questions of proof and program design and calculational logic, as in the contributions by Edsger W. Dijkstra, by W.H.J. Feijen and A.J.M. van Gasteren, and by David Gries. Finally the use of machine-based verification assistance is explained by Larry Paulson with the system Isabelle and by N. Shankar introducing PVS. The Summer School treated the topics in a very comprehensive, but nevertheless harmonious way. It turned out to be quite a surprise how closely all the aspects that were treated are related and how well the lectures fitted together.

With people from more than 20 countries all over the world together discussing, learning, and further developing this exciting topic of the mathematics of program development, the Summer School was a unique event both from a professional and from a social point of view. It is my privilege to thank all those people who helped to make the Summer School a big success. These are the participants, the lecturers, our hosts in Marktoberdorf, the people helping in the staff, and especially Dr. Ralf Steinbrüggen and Frau Ingrid Luhn, who took care perfectly of all the many questions of organization. Finally, I thank Dr. Hans Wössner from Springer-Verlag, who again was a very pleasant partner to work with when preparing these lecture notes.

Munich, April 1997 Manfred Broy

Table of Contents

Eliminating the Chaff —Again
Banquet Speech at Marktoberdorf 1996

David Gries
Computer Science, Cornell University
Ithaca, New York, 14853 USA

I was asked to speak at this banquet at a rather late date. In order to make preparation easy, I did what any self-respecting software engineer does: reuse an already-produced component —with changes to fit the new context. At Marktoberdorf 1978, the formal development of algorithms was a hot topic, discussed with religious fervor by those for and against. Yesterday, traces of that fervor cropped up in the discussion of operational reasoning versus formal proof development. At Marktoberdorf 1978, I delivered a sermon on the new religion of formal algorithm development, called "Eliminating the Chaff". This sermon is the component that I will reuse. "Chaff", by the way, is what you get when you thresh wheat to get the kernels —the chaff is the garbage that is thrown away.

Within this sermon, various authors are appropriated quoted (or misquoted, as the case may be). From time to time, I will give an aside (footnote) to give you historical information that you, at your age, may not know and to let you know when to laugh.

The reading for the day

The reading for this morning's sermon is taken from Knuth's first epistle[1] to the Structurians (page 6, pars. 2 and 3). This part of the epistle concerns Knuth's attempt to develop an algorithm discussed in the first book of the old testament[2] and the problems he faced:

> Whenever I'm trying to write a program without goto statements, I waste an inordinate amount of time deciding what type of iterative clause to use (**while** or **repeat**, etc.) ...I know in my head what I want to do, but I have to translate it painstakingly into a notation that often isn't well-suited to the mental concept ....
>
> [I wrote the program while] ...I was in bed with a pad of paper, ..., at about one AM; I expect I finished 15 or 20 minutes later. About 2 minutes were wasted trying to think of a suitable iteration statement.

Here endeth the reading for the day.

---

[1]Knuth, D.E. A review of *Structured Programming.* STAN-CS-73-371, Computer Science Department, Stanford University, June 1973.

[2]Dahl, O.-J., E.W. Dijkstra, and C.A.R. Hoare. *Structured Programming.* Academic Press, 1972.

The technical lesson

Keeping in mind the reading for the day, let us develop a divine little binary search algorithm in the calculational way and see what we can learn from it. We ardently seek an algorithm that, given x and an array segment $b[0..n-1]$ that satisfies Q, stores in variable i to truthify R:

$$Q: b[0] \leq x < b[n-1] \ \wedge 2 \leq n$$
$$R: 0 \leq i < n-1 \ \wedge \ b[i] \leq x < b[i+1]$$

We note with satisfaction that for the segment b shown below, the divine little algorithm will store in i the rightmost position of x —or if x is not in the array segment, the position after which x belongs.

	0	1	2	3	4	5	6
b	2	4	4	4	7	7	8

We reject with horror specifications that call for x to be found or that fail if x is not in the array. Such heathenistic suggestions, usually proposed by operational people who hide their shame in unreadable handwriting, must be suppressed! Nay, eradicated! Even if the ink is not tasteful!

We now seek simple ways to truthify R —or parts of it. We note that $i := 0$ truthifies all except the conjunct $x < b[i+1]$. Hence, using a method passed down to us by one of our leaders, we create a fresh variable j and weaken R to get a possible loop invariant P:

$$P: 0 \leq i < n-1 \ \wedge j < n \ \wedge \ b[i] \leq x < b[j] \quad ,$$

which is truthified by $i, j := 0, n-1$. In short order, we:

(a) Calculate the loop condition B using the fact that it must satisfy $P \wedge \neg B \Rightarrow R$.
(b) Divine the bound function $j-i-1$, using the fact that i starts out small, j starts out large, and termination occurs when $j = i+1$.
(c) Strengthen the loop invariant to include $i < j$.
(d) Develop the loop body —in the calculational way— to decrease the bound function and maintain invariant P, thus giving us the following algorithm:

```
i,j := 0,n-1;
{invariant P : 0 ≤ i < j ≤ n-1  ∧  b[i] ≤ x < b[j]}
{bound function : j-i-1}
do i+1 ≠ j → var e := (i+j) ÷ 2;
                {i < e < j}
                if b[e] ≤ x → i := e
                [] x < b[e] → j := e
                fi
od
```

Throughout the development, it is the words of our leaders that drive us to success:

> When the going gets tough, the tough [those who calculate] get going.
>
> Turn the rocks in your way [the various parts of a specification] into stepping stones.
>
> Let the formulas do the work!
>
> Conquer complexity with notation, enlightening formal properties, and manipulation!
>
> Balance formalism with common sense.

After this development, our concluding prayers are interrupted by an operational heathen, who would like the specification to be extended to:

$$Q' : 0 \leq n$$

$$R' : \begin{cases} x < b[0] & \rightarrow \quad i = -1 \\ b[0] \leq x < b[n-1] & \rightarrow \quad R \\ b[n-1] \leq x & \rightarrow \quad i = n-1 \end{cases}$$

We shudder at the specification because of all the case analysis, and the operational heathen's suggestion to implement by putting an **if** structure around *our* algorithm causes us to feel faint. However, repeating the leader's words, "when the going gets tough, the tough get going", gives us courage.

We warn the heathen that such ornate structures should not be constructed without serious consideration, that unnecessary case analysis is evil, that case analysis in specifications usually breeds the same in a program.

We then turn to eliminating the case analysis in R'. The case $i = -1$ cannot be included in the general case (the second case) because $b[-1] \leq x$ is undefined. We therefore invent a *virtuous array element* $b[-1] = -\infty$. This element does not exist and cannot be referenced by the algorithm, but it does remain in our thoughts. Now, $b[-1] \leq x$ can be assumed. So we change the 0 in the second case of R' to -1 and eliminate the first case.

Now, the introduction of $-\infty$ certainly changes the axioms underlying our theory of integers, but not in a way that disturbs us in this algorithm. We are simply applying an admonition given to us just a few days ago by that great Hoary unifier: apply whichever theory is most useful in a given context.

In the same way, we add a second virtuous element $b[n] = \infty$, and our virtuous reality is now complete. We can write the extended specification as follows.

$$Q' : 0 \leq n$$

$$R' : -1 \leq i < n \ \wedge \ b[i] \leq x < b[i+1]$$

(where virtuous elements $b[-1]$ and $b[n]$ satisfy $b[-1] = -\infty$ and $b[n] = \infty$)

It comes time to modify the algorithm to suit the extended specification. With a prayer for success, after some investigation we see the need to change only the invariant and the initialization —changing two 0's to -1's and two $n-1$'s to n's.

$i, j := -1, n;$
$\{\text{invariant } P' : -1 \leq i < j \leq n \ \wedge \ b[i] \leq x < b[j]\}$
$\{\text{bound function} : j - i - 1\}$
do $i + 1 \neq j \rightarrow$ **var** $e := (i + j) \div 2;$
$\{i < e < j\}$
if $b[e] \leq x \rightarrow i := e$
$[]\ x < b[e] \rightarrow j := e$
fi
od

We have a binary search algorithm that, if b is ordered and x is in b, finds the rightmost occurrence of x.

Moreover, if b is ordered and x is not in b, it finds the position after which x should be inserted. Moreover, the algorithm works even if b is empty.

Moreover, the array need not be ordered (in which case the algorithm finds a position i satisfying $b[i] \leq x < b[i+1]$), because nowhere did we use orderedness in the development!

Moreover, we know of situations where such a binary search in an unordered array is useful (including a version of quicksort that needs only constant extra space).

Moreover, one can show that this binary search is faster, on average, than a binary search that terminates as soon as x is found.

Moreover, the algorithm is correct because its proof of correctness was developed hand-in-hand with it.

Moreover, the presentation is such that its correctness can be ascertained easily and quickly with the given information and the theory of correctness handed down to us.

Our calculational methodology has indeed led us to a mathematically elegant, efficient, practical, useful, algorithm.

This week, we asked our audience of 90 to write a specification for a binary search, develop an algorithm, and argue about its correctness. Only 21 handed in answers to this exercise, and 13 of them had a bug! Backsliders all!

A moralistic discussion

Searchers for simplistic calculational beauty, we gather here today to discuss the fate of those who wallow in self-indulgent complexity, who revel in difficult solutions to simple problems, who insist on following outmoded, archaic religions, who stubbornly refuse to calculate.

The fate of Knuth, the author of today's reading, is a good example for all of us. An excellent mind, a creative thinker, the fastest pen in the West, Knuth still has his troubles. You see, he has taken upon himself the task of describing computer science in seven volumes. That is a never-ending task, for the faster he writes, the further behind he is. Even his attempts at literate programming are hampered by his web of operational reasoning. Knuth could be likened to Job; blameless and upright, he has enormous misfortune, being saddled with more page proofs than anyone in history. But unlike Job, it is Knuth's own fault. It is his own Job Control Language[3] that has caused his misfortune. His web of control and text causes programs and their descriptions to be twice —nay even thrice as long and complicated as necessary.

Just listen to the cry of a man who recognizes his problem but is incapable of syntactically transforming himself into an equivalent but less operational man.

> I wasted an inordinate amount of time deciding what type of iterative clause to use!
>
> About two minutes were wasted!
>
> I was in bed with a pad of paper!

Is he not like the heathen in the technical lesson, who breaks under the strain of unnecessary case analysis, inadequate control structures, and ineffective operational thoughts? Friends, Knuth's problem could be solved simply by casting out old habits and absorbing the guarded commandments given to us 20 years ago.

Yes, 20 years ago! Even then, in those old days, we were imploring people to guard their commands well and not listen to false prophets. Before tasting the fruits of their labors, we said, make sure it was Manna from heaven, and not from elsewhere.[4] And we warned against the God Bacchus, who had emerged notational orgies![5] But our imploring was of no avail, and today forces of operational thinking are more prevalent than before!

Others have had the audacity to misquote Shakespeare when they rail against iteration, "Oh, thou has damnable iteration and art indeed able to corrupt a saint." We all know, of course, that what Shakespeare really said was,

> Oh FORTRAN, thou has damnable iteration and art indeed able to corrupt a saint like Knuth!

This was said in a discussion with one of our old leaders about the guarded command loop, which Shakespeare praised so much. Part of that discussion found its way into "As You Like It", in which Shakespeare said of the leader's alternative construct,

> Your "if" is the only peacemaker; much virtue in "if".

Colleagues in clearness and conciseness, on my way here this morning I had occasion to walk through a vegetable garden with our leader, and he taught me a lesson I will never forget. The sun was shining brightly, a light breeze was blowing. We meandered through bushes bearing bright green beans, around vines of crisp cucumbers, every plant bore its

[3] A reference to the old Job control language in DOS.

[4] A reference to Zohar Manna, who had developed a theory of correctness in which an assertion in a program need not be true each time that place in the program was reached.

[5] A reference to John Backus, developer of Fortran and its first compiler, member of the Algol 60 committee, and developer of a functional/combinatorial programming notation around 1975–1978.

luscious fruit for the fulfillment of our stomachs. It was a nice indication that this wonderful town of Marktoberdorf was becoming more vegetarian.

But the poor tomato plant! This poor tomato plant was actually quite large and sprawling, with some of its branches hungrily lapping up the space between it and its neighbors, even flowing over and covering them. Yet, it had only a few, pitiful, tomatos.

Our leader saw its condition and began snapping off the sprawling branches and pinching off old shoots, and slowly the tomato plant was pruned to a manageable size. As he did this, our leader caressed the plant and gently said,

> Sucker the tomatoes to come unto me.

Ah, my friends, I saw the analogy at once and resolved then and there to spread the word far and wide. Our programs, our programming languages, nay, even our own *thoughts* should be purged of all life-sapping, useless growth. And I urge you to begin immediately to release those complexities, to burn those operational thoughts, to exorcise those deadly alleys, to prune those languages like C, and instead to concentrate on the simple, the elegant, the sweet. And soon, you will find yourself able to grow more luscious, tasteful fruit in the form of the heavenly algorithms and unified theories we so long for. As you eliminate the chaff, bit by bit you will find your memory expanding, your intuition intuiting, your real creative ability showing itself. And you will find the peace, the equanimity, the harmony, you so much desire.

Becoming more serious

Colleagues in CS —or Computational Simplicity— I would like to conclude on a serious note (if that is possible).

We in CS have to strive more to work together, to see each other's viewpoint, to learn from each other. As the Hoary Unifier said, the framework of complementarity must be understood before we can avoid gaps and duplication and achieve rational collaboration in place of unscientific competition and strife. In short, we must stand on each other's shoulders, not step on each other's toes.

We in CS have a harder time with this than most other disciplines, because our field is so close to reality and the real world is not rational —instead, it is driven by money. Two examples. I saw a report that the MacIntosh was 30% more effective (time wise) than the PC in a wide range of applications —word processing, spreadsheets, etc. Yet, people are moving to PC's. Another report gave names of executives in PC companies that used Macintoshes on their desks or used MacIntoshes for their own graphics and publications. These companies were using one product but selling their own inferior product.

In such a world, we must continue to remember that our job is to advance the state of computer science —in both research and teaching— and this requires the integrity and complementarity that our Unifier was talking about.

John Mitchell exhibited the right attitude in working to come to grips with object-oriented concepts in a scientific manner instead of engaging in the usual shouting matches about OOP. Abrial exhibited the right attitude in trying to reconcile the goals and concerns of theory with those of industry. And, although Paulson's goal of mechanical theorem proving are different from mine in human theorem proving, we were able to work together to arrive at a neat little proof, in a way that gained us both some understanding.

The person working on a different aspect of a problem that you are working on is not competing with you, they is complementing you.

Above, I mentioned teaching in connection with research. Many of you will become CS teachers, some of you are already. Teaching in CS is more difficult than in other scientific/engineering fields, for several reasons. Youth of the field, rapid change, influence of hardware, the closeness to the real world, the influence of money, too many students, etc.

You must also balance several conflicting issues: research versus teaching, teaching what industry needs versus what it thinks it wants, teaching what students need to survive later on versus what they think they want now, theory versus practice, and so on. It is in the teaching of programming —and programming is at the center of CS— that I see great difficulties. The world forces us —at least temporarily— into teaching C or C++, one of the most pedagogical abominations there is. For introductory teaching, it is a nightmare.

Many of you teach programming and do a lot of programming, and yet out of 21 binary searches turned in by you, 13 had a bug. This in spite of the fact (or perhaps because of it) that every introductory programming text contains a binary search! If you have any feeling of integrity, you should seriously investigate why so many people got the binary search algorithm wrong and work to effect a change. If you do, I think you will find that it is because of operational thinking that is not in balance with calculation. Operational and calculational thinking should not compete with each other, they should complement each other. The one without the other is ineffective. The other without the one is sterile and useless.

All those bugs may also be because of using inadequate notations. Why use C for expressing algorithms, when guarded commands and functional languages allow one to express thoughts so much more simply and economically. Why are we *teaching* C in introductory courses; why don't we start with Gofer, Scheme, MATLAB, or other notations that allow us to express some thoughts at a higher level?

You are, or will be, educating the programmers and computing scientists of the future. Are you giving them the right mental tools? Are you producing better educated people than we did twenty years ago? If not, learn better mental tools, as well as the supplemental tools (software packages) yourself and strive to teach them better. This calls for viewing others and what they do as complementary instead of competing.

A supplication

We all know what fools mortals are. But help us to break our bad habits, to put our houses in order, to learn to work together in pursing our common goal of the advancement of computer science. Teach us that every why needs a wherefore but not a watfor[6]. Help us to learn that knowledge without wisdom is useless —even dangerous. Help us to learn that correctness lies within a little and certain compass, but that error is immense. Teach us that the ideal is the union of the mathematician with the poet, of correctness with passion.

[6]A reference to the old WATFOR implementation of FORTRAN.

Envoi at Marktoberdorf 1996

C.A.R. Hoare

Congratulations! You are now graduates of one of the world's most prestigious and exclusive institutions of higher education, the Marktoberdorf Summer School in Computing Science. The School was founded in 1970, when the first two-week session was organised in Marktoberdorf.

The distinguished list of lecturers included Jim Horning, Gerhard Goos and John Kuck. Also, Ole-Johan Dahl lectured on his recent discoveries in object-orientated programming, and I talked about an axiomatic approach to data structuring. Together with a leading article by E.W. Dijkstra, these were afterwards published in a book entitled Structured Programming, which is referenced still by our lecturers in the last two weeks.

Since 1970, Marktoberdorf Summer Schools in Computing have taken place at roughly two-year intervals. Edsger Dijkstra, who much regrets his absence this year, has lectured at nearly all of them, and so have I. Our graduates by now must number over a thousand. Like you, they came from and returned to countries in all corners of the world. And I sincerely hope that you, like them, will retain through your life the strongest and most pleasant memories of what you have learnt, understood and accomplished in these last two weeks.

You surely will remember your lectures for the interest of their scientific discoveries, and their illuminating insights which will help you make further advances in research into your own area of speciality. You will remember also the interest and excitement of new ideas in areas adjacent to your own speciality. And in subjects further removed from your previous experience, you will remember the skill and care with which your lecturers have overcome your fear or reluctance to learn something completely new. As a result, you will take away with you a deeper understanding of the fundamentals of our whole subject, and the way in which its branches should be related, so that scientific advance can proceed on a broad front, with cooperative contributions from many differing researchers from varying traditions and cultures.

I hope that you will remember with enjoyment your many activities outside the lecture room: walking, riding, swimming, dancing, climbing, volleyball, football ... Those who have suffered injury from these activities, I hope you will soon discard your crutches and forget all the pain. But remember the concerts, the parties, the eating and the drinking, and above all the good companionship of the many friends you have made here, among the staff, the lecturers and among your fellow students.

I can assure you that your lecturers and directors of the school will retain the same happy memories of these two weeks. We have really enjoyed teaching you, answering your questions and engaging in conversation with an outstanding class of students. That is our greatest reward, and it is renewed whenever we encounter one of our former students again. So please do not hesitate to remind us when we meet again, because with passing years we cannot hope to maintain the achievement of David Gries and Netty van Gasteren to remember all your names individually.

Many of you applied to come to this School as a result of recommendations from contemporaries, colleagues, lecturers or supervisors, who were perhaps graduates of one of our previous Schools. If you have happy memories and wish to repay part of the benefit you have received, the best way to do this is to pass the benefit on to others who follow. Tell your friends about our School, and recommend it to those of your contemporaries, colleagues and students who could obtain the same or greater benefit. It is the quality and participation of the students that make these Schools a success. That is the reward that attracts new lecturers to join us, and keeps our old lecturers faithful.

In saying goodbye at the end of the 1996 summer school, I am confident that with your help that we will soon again be welcoming your worthy successors to future sessions of the Marktoberdorf Summer School in Computing Science.

Part 0

Designing Proofs and Programs

The professional design of proofs and programs requires a number of well-chosen ingredients. A clear and concise notation forms the basis. Simplicity of the logical rules and a handy proof format not only ease formula manipulation, but often let design steps arise merely from the shape of formulae. Strong heuristic principles and design methods guide the effective application of the rules. Together these ingredients yield a design discipline that is equally valuable in theorem proving, proof simplification and programming. An efficient design process, and elegant proofs and programs are the reward of systematic design.

Edsger W. Dijkstra

David Gries

Rina Feijen, Netty van Gasteren, Wim H.J. Feijen

Heuristics for a very simple Euclidean proof

(EWD1180)

Edsger W. Dijkstra

We consider 3 points on a given circle, A and B fixed and P movable along the clockwise circle segment from A to B. (See fig. 0) At my recent oral examinations "Mathematical Methodology", the students I asked did *not* know the theorem that $\angle APB$ is constant, i.e. independent of the position of P. So we designed the proof in as systematic a manner as possible.

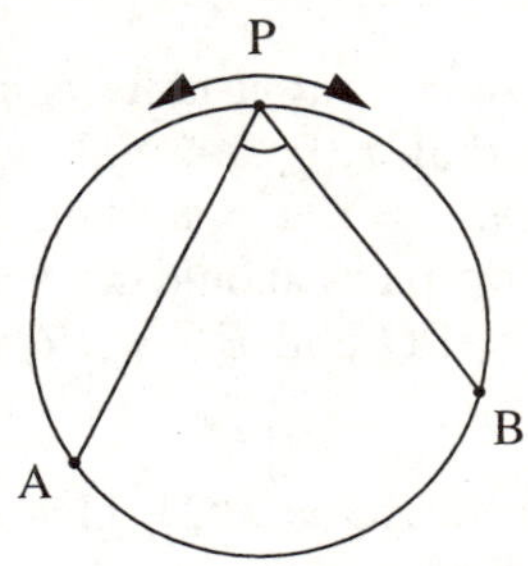

fig. 0

*   *

*

• As a matter of routine we checked the relevance of the data. Is it essential that A is fixed? Yes, obviously. Is it essential that P is confined to the circle? Yes. (Some let $\angle APB$ shrink by moving P outside the circle, some let it increase by moving P inside.) Moral: we *have* to take the circle into account.

• How is a circle defined? As a set of points equidistant to the centre. So we had better introduce the centre C and the three radii of equal length. (See fig. 1)

• Since it is only essential that A, B, and P lie on the circle and this fact has been captured, the circle is now erased. (See fig. 2)

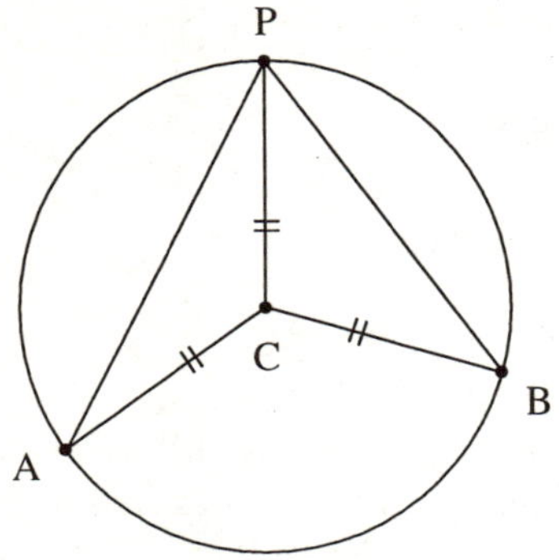

fig. 1

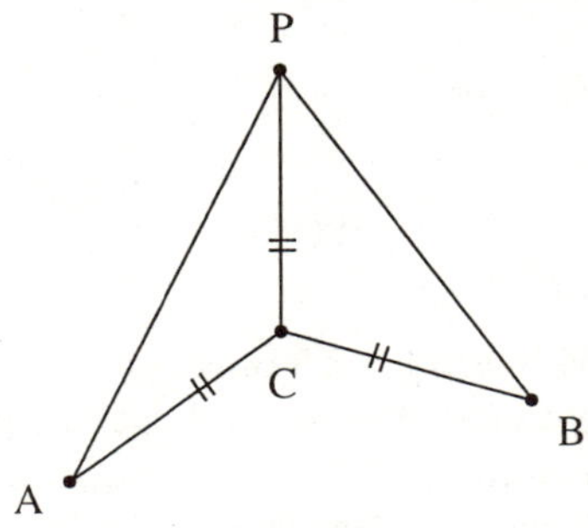

fig. 2

• *Three* distances are the same, but usually equality is considered to be a *binary* relation. If the fact that x, y, z are all equal has to be captured by the pairwise equalities $x = y$, $y = z$, $z = x$, then 2 of the 3 suffice (since = is transitive). Which of the two pairs should we choose in fig. 2? For reasons of symmetry these are $AC = PC$ and $BC = PC$. Let us concentrate on the first one. (See fig. 3)

• What can we conclude from $AC = PC$? This was invariably the moment some mathematical knowledge was introduced: an isosceles triangle has equal base angles. Together with the other case this leads to fig. 4, in which equal angles have been marked equally.

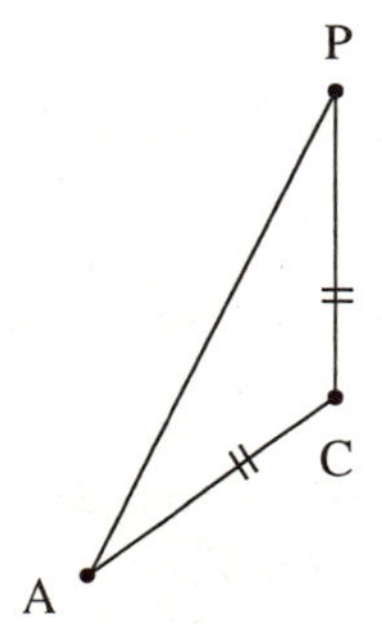

fig. 3

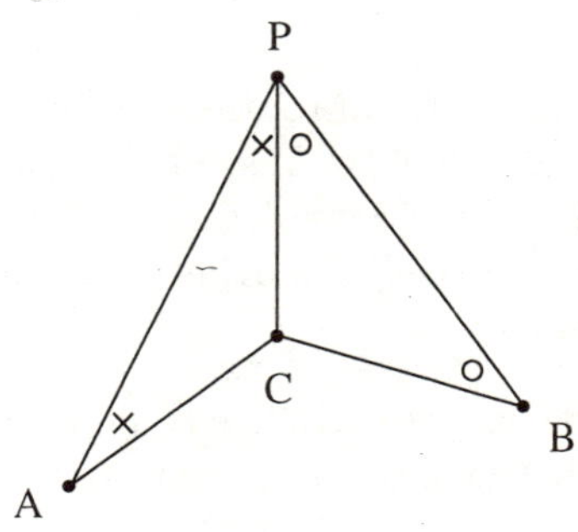

fig. 4

• At this stage, most students remarked without prompting that a vital step had been made: from a given in terms of lengths a conclusion about angles has to be drawn, and the last step is vital because it provides the transition from lengths to angles. This is the moment to decide to remove the markings ⫲ of equal lengths and to conduct the remainder of the argument in terms of angles.

• We have to show that $\angle APB$ is constant. We have not used yet that A, B, C are fixed, and that therefore $\angle ACB$ is constant. This observation draws our attention to the angles around C. (Had we not made this observation, we could have observed that the angles around C are in fig. 4 the only unmarked ones.) And this was the moment of the second introduction of some mathematical knowledge: most students chose to use that the angles of a triangle add up to π (which they called 180°). This leads to the marking as in fig. 5 and we conclude that $\angle APB = \frac{1}{2} \angle ACB$, which proves the theorem.

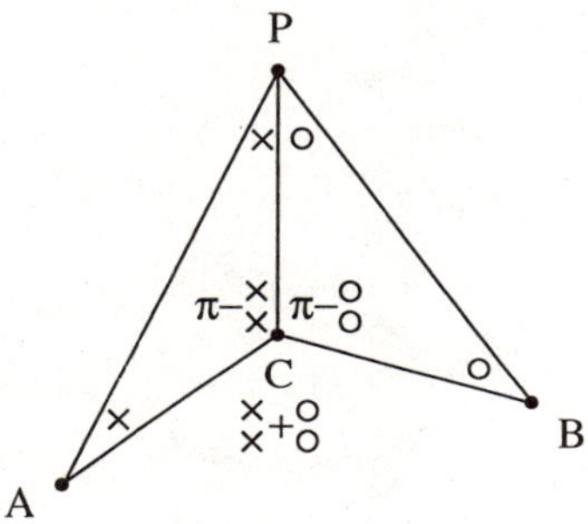

fig. 5

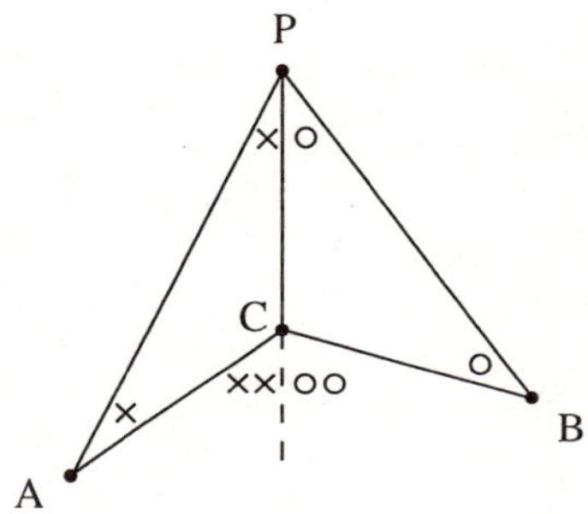

fig. 6

Note None of the students appealed to the theorem captured by

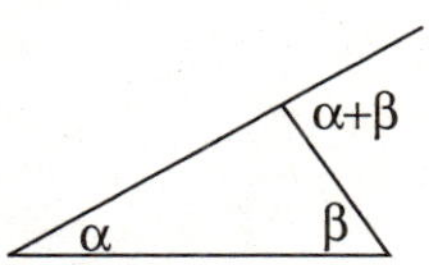

,

extended PC and produced the marking as in fig. 6. It has the advantage of not mentioning $\pi^{*)}$, and is suggested by the observation that, at P, we have a constant angle with a variable partitioning: by extending PC, we create another constant angle with variable partitioning.

$^{*)}$ The problem of introducing a constant (like π) is that then one has to use its properties. (End of Note.)

The problem was nice for warming up!

Winding Stairs Campground, 14 May 1994

Prof. Dr. Edsger W. Dijkstra
Department of Computer Sciences
The University of Texas at Austin
Austin, TX 78712–1188
USA

Complete DAGs

(EWD1204)

Edsger W. Dijkstra

[DAG is just another TLA I don't approve of; it stands for "Directed Acyclic Graph".]

In a recent manuscript—in the literal sense of the word—: "Some properties of the relative converse" (March 1995), Tony Hoare plays with tetrahedra of which the edges may be directed. (As by-product of his considerations he finds, for instance, that "any non-cyclic ascription of directions to any five of the edges can be non-cyclically extended to the sixth edge.") Here is a little bit of related theory.

* * *

Let K be the complete graph with n vertices, in which each edge has a direction. Then the following 3 statements are equivalent

(i) K contains no cyclic paths

(ii) K contains no cyclic paths of length 3

(iii) the outdegrees of the vertices of K are the values 0 through $n-1$

The equivalences hold for $n = 1$, since for $n = 1$ (i), (ii), and (iii), are all true. For general n, the proof proceeds by mathematical induction. The induction step itself is done by cyclic implication: we now consider the complete $(n+1)$-graph K' with directed edges, and observe for it

<u>(i) ⇒ (ii)</u>

This follows—independently of the induction hypothesis—from the fact that "of length 3" is a restriction.

<u>(ii) ⇒ (iii)</u>

Let A be of K' a vertex of maximum outdegree; let K be the n-graph that remains after removal of A and its n connecting edges. Because of (ii), also K contains no cycles of length 3 and, ex hypothese, the outdegrees of its vertices are the values 0 through $n-1$; consequently we are done when we show that A has outdegree n (note that then the vertices of K have the same outdegree in K as in K').

Let B be the vertex with outdegree $n-1$ in K; since, by construction, A is not a vertex of K, A and B are different vertices. We now first deal with the edge AB, and then with the remaining edges connecting A to K.

The direction of edge AB is $A \rightarrow B$ for the assumption $B \rightarrow A$ leads to a contradiction: then B would have in K' the (maximum) outdegree n, and so would A (by virtue of how it has been chosen: outdegree $A \geq$ outdegree B), but in the directed complete $(n+1)$-graph, at most 1 vertex has the maximum outdegree n.

Let C be a third vertex; then, because C is a vertex of K, in which B has the maximal outdegree $n-1$, the direction of edge BC is $B \rightarrow C$. From $A \rightarrow B$, $B \rightarrow C$ and the absence of cycles of length 3 in K', we conclude that the direction of AC is $A \rightarrow C$. (Note that, so far, we had only used the absence of cycles in K.) Hence, A has outgoing edges only, quod erat demonstrandum.

<u>(iii) ⇒ (i)</u>

Assume (iii) for K'; let A be the vertex of maximum outdegree n; let K be the n-graph that remains after removal of A and its n connecting edges. Possible cyclic paths in K' are then of two kinds, either they include A, or they lie in K.

Because A has outgoing edges only, no cyclic path leads through it; because of assumption (iii) for K' and of the fact that A has outgoing edges only, we conclude (iii) for K and, ex hypothese, that no cyclic paths lie in K. So, K' contains no cyclic paths, quod erat demonstrandum.

* *

*

The absence of cyclic paths in the complete graph tells us—with apologies for the picture!—that each triangle is of the form 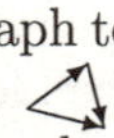 , i.e. $\rightarrow$ is a transitive relation. Hiding this fact so far, could be considered a conscious obfuscation on my part.

Austin, 29 March 1995

Prof. Dr. Edsger W. Dijkstra
Department of Computer Sciences
The University of Texas at Austin
Austin, TX 78712–1188
USA

A result of rabbit removal

(EWD1208)

Edsger W. Dijkstra

In this note, all variables are of the same type.

In "Introduction to Logic and to the Methodology of Deductive Sciences" by Alfred Tarski, I found the following theorem—my formulation—

Let $\cdot$ be an associative, symmetric infix operator such that

$$\text{equation } x : a = b \cdot x \text{ is solvable for all } a, b; \tag{0}$$

then

$$x \cdot y = x \cdot z \Rightarrow y = z \quad . \tag{1}$$

In comparison, Tarski calls his calculational proof "considerably more involved"; it was in any case too complicated to be remembered (at least: by me), it also seemed to me to be more complicated than necessary. (This last judgement was, of necessity, very tentative, because without detailed analysis it was impossible to see how much of the complexity belonged to the argument proper, and how much of it was the result of the style of presentation deemed acceptable in the 30s. Of course we cannot blame Tarski for having done most of his work before Wim Feijen developed his proof format.)

Under the circumstances there was only one action possible for me, viz. to ignore what I had seen and could not remember, and to try to design a proof myself. (Here, the ambitious reader is invited to stop reading and to try to prove the theorem himself.)

*　　　*

*

Since the consequent of our demonstrandum (1) is the equality $y = z$, the suggested structure of our proof could be

"We observe for any x, y, z

$$\begin{array}{ll} & y \\ = & \{\text{antecedent } x \cdot y = x \cdot z\} \\ & z \end{array}$$

",

but this sketch contains little more than a proposal how to let the proof structure reflect that our demonstrandum has the shape of an implication.

In the one and only step of the above proof sketch, an equality has to be justified by an equality, hence an application of Leibniz's principle is dictated! Our next proof sketch is therefore:

"We observe for any x, y, z

$$\begin{array}{ll} & y \\ = & \quad \{\text{definition of } f\} \\ & f.(x \cdot y) \\ = & \quad \{\text{antecedent } x \cdot y = x \cdot z\} \\ & f.(x \cdot z) \\ = & \quad \{\text{definition of } f\} \\ & z \quad \text{''}. \end{array}$$

Remark As far as the use of the antecedent via Leibniz's principle is concerned, one could consider the alternative

$$\begin{array}{ll} & y \\ = & \quad \{\text{definition of } f\} \\ & f.(x \cdot z) \\ = & \quad \{\text{antecedent } x \cdot y = x \cdot z\} \\ & f.(x \cdot y) \\ = & \quad \{\text{definition of } f\} \\ & z \quad , \end{array}$$

but that does not work, and we shall later see why. (End of Remark.)

Our next task is to design a suitable f, i.e. to propose how to rewrite y as an expression that contains $x \cdot y$ as subexpression. Since $\cdot$ is the only operator defined on our type, we have little freedom: (0) with $a, b := y, x \cdot y$ allows us to postulate that p satisfies

$$y = x \cdot y \cdot p \quad , \tag{2}$$

and now our proof can start as follows:

$$\begin{array}{ll} & y \\ = & \quad \{(2)\} \\ & x \cdot y \cdot p \\ = & \quad \{\text{antecedent } x \cdot y = x \cdot z\} \\ & x \cdot z \cdot p \end{array}$$

Our remaining problem is how to transform $x \cdot z \cdot p$ into z, and we have little choice but to use the same (2), which has just been used for (x, p)-introduction, a second time, but now for (x, p)-removal. But for this removal to work, we need a temporary reintroduction of y. For rewriting z so that it contains y as subexpression, the simplest solution, allowed by (0) with $a, b := z, y$, is to postulate that q satisfies

$$z = y \cdot q \quad . \tag{3}$$

And now we can give the complete proof. For any x, y, z, we postulate, as permitted by (0), that p satisfies (2) and q satisfies (3), and subsequently we observe

		Δy	Δz	Δp	Δq
	y				
$=$	$\{(2)\}$			$+1$	
	$x \cdot y \cdot p$				
$=$	$\{x \cdot y = x \cdot z\}$	-1	$+1$		
	$x \cdot z \cdot p$				
$=$	$\{(3)\}$	$+1$	-1		$+1$
	$x \cdot y \cdot q \cdot p$				
$=$	$\{(2)\}$			-1	
	$y \cdot q$				
$=$	$\{(3)\}$	-1	$+1$		-1
	z .				

To the right of the proof we have recorded the changes in the number of occurrences of variables y, z, p, q. Because the net effect of the appeals to (2) and (3) is $\Delta p = 0 \wedge \Delta q = 0$, it yields $\Delta y = 0 \wedge \Delta z = 0$ as well. The total effect of the proof (whose steps transform y into z) has to be $\Delta y = -1 \wedge \Delta z = +1$, and this is the change provided by the use of the antecedent via Leibniz's principle. This counting argument explains why the alternative proof sketch in the remark does not work: it works in the wrong direction, viz. $\Delta y = +1 \wedge \Delta z = -1$.

* *

*

To put the above proof into perspective, let me render—reduced to its essentials—Tarski's proof.

For any x, y, z, we postulate, as permitted by (0), that u, v, w satisfy,

$$y = y \cdot u \tag{4}$$

$$z = y \cdot v \tag{5}$$

$$u = x \cdot w \tag{6}$$

and subsequently observe

$$
\begin{array}{ll}
 & y \\
= & \quad \{(4)\} \\
 & y \cdot u \\
= & \quad \{(6)\} \\
 & y \cdot x \cdot w \\
= & \quad \{x \cdot y = x \cdot z\} \\
 & z \cdot x \cdot w \\
= & \quad \{(6)\} \\
 & z \cdot u \\
= & \quad \{(5)\} \\
 & y \cdot v \cdot u \\
= & \quad \{(4)\} \\
 & y \cdot v \\
= & \quad \{(5)\} \\
 & z \quad .
\end{array}
$$

Tarski's proof is 2 steps longer than ours because of the occurrence of u, which has to be introduced, and removed later. In fact, our proof can be obtained from Tarski's by eliminating u from (4) & (6). It is not quite clear why u was introduced in the first place; could it have happened because the heuristics remained completely implicit?

* *

*

The structure of Tarski's proof was further obscured by his insistence of fully parenthesizing his expressions: symmetry and associativity thus add another 7 steps to the proof, which I consider a high price for so little.

About 20 pages later, he proves another theorem—again my formulation—

Let $\cdot$ be an infix operator that is symmetric and associative wherever defined, and such that

equation $x : a = b \cdot x$ is solvable for all a, b; (0)

then

$x \cdot y$ is defined for all x, y. (7)

The plan of the proof is to construct, as permitted by (0), for given x, y, a quantity z which, by virtue of its construction, can be transformed into $x \cdot y$. For reasons of heuristics, we start at the other end and ask ourselves what we have to introduce that will enable us to transform $x \cdot y$ into z.

Starting from

$$x \cdot y \qquad (\alpha)$$

we have to do 3 things: remove x, remove y, and introduce z. We can achieve 2 of them by defining—as (0) allows us to do—z for suitably chosen w by

$$z : y = w \cdot z \quad ; \qquad (8)$$

then (α) can be rewritten as

$$x \cdot w \cdot z \quad . \qquad (\beta)$$

The remaining task consists of 2 things, viz. removal of x and removal of w. Remembering (i) that, as yet, w is undefined, and (ii) that (β) provides the environment in which to exploit that $\cdot$ is associative, we can effectuate both removals by defining—as (0) allows us to do—w for suitably chosen u by

$$w : u = x \cdot w \quad ; \qquad (9)$$

then (β) can be rewritten as

$$u \cdot z \quad . \qquad (\gamma)$$

The remaining task is to choose u in such a way that we can remove it. For *given* z, (0) would allow us to define u by

$$z = z \cdot u$$

and—thanks to the symmetry of $\cdot$ —such a u would do the job, but via (8) and (9), our z is functionally dependent on u. Having to define u independently of z, we generalize our last relation and define—as (0) allows us to do—u for suitably chosen c by

$$u : c = c \cdot u \quad , \qquad (10)$$

and hope that we can find a c that does the job.

Appealing to (10) requires in our rewriting exercise the temporary introduction of c into the intermediate result. Also, for the very last step of the transformation, we need an expression that can be equated to z. These two considerations suggest a next—and last—appeal to (0): we define v by

$$v : z = c \cdot v \quad . \qquad (11)$$

Now the transformation continues:

$$
\begin{array}{ll}
 & u \cdot z \qquad (\gamma) \\
= & \quad \{(11)\} \\
 & u \cdot c \cdot v \\
= & \quad \{(10)\} \\
 & c \cdot v \\
= & \quad \{(11)\} \\
 & z
\end{array}
$$

and we are done. For c, any element will do. In summary: choose a c, define *in order* u by (10), w by (9), z by (8), v by (11) and observe

$$
\begin{array}{ll}
 & z \\
= & \{(11)\} \\
 & c \cdot v \\
= & \{(10)\} \\
 & u \cdot c \cdot v \\
= & \{(11)\} \\
 & u \cdot z \\
= & \{(9)\} \\
 & x \cdot w \cdot z \\
= & \{(8)\} \\
 & x \cdot y \quad .
\end{array}
$$

Tarski's presentation of this proof is obscured by the fact that without any comment he has made the arbitrary choice $c = y$.

Ref. Tarski, Alfred, "Introduction to Logic and to the Methodology of Deductive Sciences", Dover Publications Inc., New York, 1995

Austin, 11 June 1995

Prof. Dr. Edsger W. Dijkstra
Department of Computer Sciences
The University of Texas at Austin
Austin, TX 78712–1188
USA

For the record: painting the squared plane

(EWD1212)

Edsger W. Dijkstra

Thanks to Richard S. Bird, the following problem circulated at the 3rd International Conference on the Mathematics of Program Construction, Kloster Irsee, Germany, 1995.

Starting with a sufficiently large piece of (white) squared paper, we play the following one-person game, that consists of two phases. In the first phase, we may select 9 squares and colour them red. In the second phase, we may repeatedly colour a white square red provided it already has at least 2 red neighbours (where being neighbours means sharing an edge). The question to be answered is "Is it possible to colour the 10 by 10 square?"

*   *

*

When, in the second phase, a white square may be coloured on account of exactly two "enabling" neighbours, we can draw the topological conclusion that there are two possibilities for the relative position of those two enabling neighbours: either around a corner or opposite to each other. When we demonstrate a possibility by showing how to do it, we can hardly avoid that each move in our witness is of some type. For instance: in order to show that the 9 by 9 square can be coloured, we can colour in the first phase the squares along a diagonal; in the second phase the enabling neighbours are then always situated around a corner. (For the 9 by 9 square there are also solutions in which opposite enabling neighbours occur as well.)

When demonstrating an impossibility, we consider no solutions because there aren't any, and in this type of argument it would be very nice to avoid the case analysis by not drawing the topological conclusion at all, i.e. by ignoring the cyclic arrangement of the four neighbours and of the four edges they share with the square they surround. In short, for an impossibility argument it is very tempting to try to get away with the simplicity of a counting argument. The only still open question is then what to count.

Counting red squares is too crude: their number increases by 1 at a time, which is only mildly interesting; worse is that the notion "red square" says nothing about the notion of being neighbours. In order to capture the latter, we have to turn to the edges.

How do we relate edges to colours of squares? Each edge is identified by—and identifies—the pair of squares that share it. This enables us to define three "edge colours":

an edge is red, when shared by two red squares,
an edge is white, when shared by two white squares, and
an edge is pink, when shared by a red square and a white square.

Accordingly, a white square has no red edges, and a red square has no white edges. More precisely: colouring a white square red, turns its pink edges into red ones, and its white edges into pink ones. The rule in the second phase of "at least two red neighbours" means that at least 2 pink edges turn red, and at most $4 - 2$ white edges turn pink. In other words: in the second phase, the number of pink edges does not increase. Since the first phase does create at most 36 pink edges, we never end up with the 40 pink edges that would occur in a red square of 10 by 10, and thus the question is settled.

* *
*

Faced with a problem, people usually react as follows. In their first explorations they try to colour the 10 by 10 square, and, finding they cannot do it, they change to trying to show the impossibility, but in that second effort they drag all the topological ballast into the argument, make pictures, talk about "occupied columns and rows" etc. In that second effort they are heading for a counting argument, but are insufficiently aware that avoiding case analysis is counting's major purpose.

* *
*

The final argument nicely illustrates the linguistic demands that the doing of mathematics makes: a term for what we called "pink edges" is absolutely essential.

Austin, 30 August 1995

Prof. Dr. Edsger W. Dijkstra
Department of Computer Sciences
The University of Texas at Austin
Austin, TX 78712–1188
USA

The marriage agency

(EWD1214)

Edsger W. Dijkstra

A marriage agency has in its files 25 men and 25 women and aims at acquainting the men with the women by organising dinner evenings, each one involving 12 tables, seating 2 men and 2 women. (Consequently, at each dinner evening, 1 man and 1 woman have a day off.) The question is how many of such dinner evenings the marriage agency can schedule without some man and woman sharing a table twice.

* *

*

Frans van der Sommen raised this problem at the end of the ETAC session of 8 August 1995. Rutger M. Dijkstra, who happened to attend that session, immediately showed how to schedule 12 dinner evenings: partition the men into a singleton and 12 pairs numbered [0..12), do the same with the women, and let on the nth evening the men of pair i share a table with the women of pair $(i+n)$ **mod** 12.

The demonstration that 12 is indeed the maximum took a little bit longer. Consider a sequence of 13 dinner evenings. Then at most 13 men have had an evening off, and therefore at least 12 (= 25 − 13) men dined all 13 evenings. Such a man enjoyed 26 (= 2 × 13) times the company of a woman; the number of women being 25, there is—pigeon-hole principle—at least one woman whose company he enjoyed at least twice.

[As said, this took a little bit longer. We first tried too a crude a counting argument in terms of man-woman combinations. Their total number is 625 $(= 25 \times 25)$, each evening 48 $(= 4 \times 12)$ combinations are getting acquainted. Since $625/48 = 13\frac{1}{48}$, this rules out a schedule of 14 or more evenings, but the feasibility of a schedule of 13 dinner evenings it leaves open.]

This is actually a sequel to EWD1212: feasibility is shown by means of a witness, infeasibility by a counting argument.

Austin, 5 September 1995

Prof. Dr. Edsger W. Dijkstra
Department of Computer Sciences
The University of Texas at Austin
Austin, TX 78712–1188
USA

Courtesy Dr. Birgit Schieder

(EWD1215)

Edsger W. Dijkstra

This note deals with the proof of the following

Theorem Given are N blue and N red points in the Euclidean plane such that no 3 of them are collinear; then blue and red points can be connected one-to-one by straight line segments such that no segments intersect.

The proof is by showing the termination of the following repetition, operating on an arbitrarily initialized variable of type "one-to-one correspondence between the given N blue and N red points":

do there are intersecting segments →
    flip two intersecting segments **od** ,

where the operation "flip" replaces as illustrated below two intersecting segments (the solid ones) by two non-intersecting segments (the dotted ones):

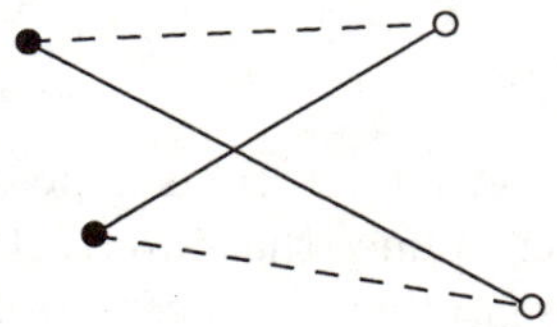

We can not conduct the proof of termination by stating that the flip reduces the number of segment/segment intersections, for it need not do so: the "dotted" segments may introduce new intersections with other segments.

The famous termination argument observes (i) that the number of possible states is finite (viz. $N!$), so that a real variant function will do, and (ii) that the flip decreases the sum of the lengths of the N connecting segments (and it is here that the noncollinearity is used to rule out all sorts of "pathological" cases). This argument is famous because it is also a little bit scandalous because of its illicit affaire with the notion of Euclidean distance. (It is most definitely *not* a metric theorem we are dealing with.) So, when I showed the above argument in 1994 at the Marktoberdorf Summer School, I challenged the audience to design a "decent" proof (i.e. using only concepts of affine geometry). The challenge was most elegantly met by Dr. Birgit Schieder (of the "Institut für Informatik der Technischen Universität München"), who constructed a counting argument. The challenge is, firstly, to choose what to count and, secondly, to design a crisp argument for the decrease.

The changing state of the computation is characterized by the one-to-one correspondence, i.e. by the set of N connecting segments. Below, we shall use the term "segment" to denote the so-called "open" segment, i.e. its two coloured endpoints are *not* included. We shall use the term "line" to denote any of the N^2 infinite lines through 1 blue and 1 red point; this set of N^2 lines is as fixed as the $2N$ coloured points. Finally, a line and a segment are said to "intersect", or to give rise to an "intersection", when they have exactly 1 point in common.

Lemma 0 For no line the number of segments it intersects is increased by a flip.

Proof

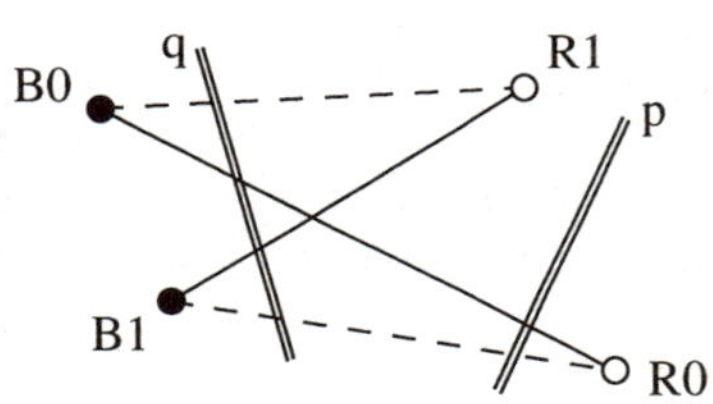

For a line whose number of intersections is possibly increased by the flip, we can confine our attention to lines that intersect at least 1 of the segments created by the flip. Representative instances are line p (intersects 1 dotted segment) and line q (intersects 2 dotted segments). But—not thanks to Euclid, but thanks to Hilbert—they each intersect as many solid segments, so the flip does not increase their number of intersections.

(End of Proof.)

Lemma 1 For two lines, the number of segments intersected is decreased by 1.

Proof Line $(B0, R0)$ loses its intersection with segment $(B1, R1)$, which disappears, and intersects neither of the new segments. Similarly for line $(B1, R1)$.

(End of Proof.)

From the lemmata we conclude that each flip decreases the number of line/segment intersections, and this concludes Birgit Schieder's counting argument.

*　　　　*
*

I posed the problem to the Tuesday Afternoon Clubs in Eindhoven and Austin; both solved the problem. At the ETAC, we then investigated how the case analysis could be reduced, and how the argument that it is exhaustive could be simplified. (The argument as given above, however, was designed later.) At the ATAC, we afterwards looked for heuristics that would have led to the idea of counting segment/line intersections. In retrospect we could make all sorts of explanatory and justifying remarks, but none of them could in honesty be dressed up as heuristic principle.

The counting argument annoys me because my "proof" of Lemma 0 relies so heavily on that part of geometry with whose axiomatization I am so unfamiliar: I know Euclid's axioms, but for a proof of Lemma 0 they don't suffice.

Austin, 17 September 1995

Prof. Dr. Edsger W. Dijkstra
Department of Computer Sciences
The University of Texas at Austin
Austin, TX 78712–1188
USA

WLOG, or the misery of the unordered pair

(EWD1223)

Edsger W. Dijkstra

Let p, q be predicates on unordered pairs of some type, i.e. (for all x, y)

$$p.(x,y) \equiv p.(y,x) \quad \text{and} \quad q.(x,y) \equiv q.(y,x) \quad , \tag{0}$$

and suppose we have to prove (for all x, y)

$$p.(x,y) \Rightarrow q.(x,y) \quad . \tag{1}$$

Suppose we can prove the weaker theorem

$$p.(x,y) \wedge r.x \Rightarrow q.(x,y) \quad . \tag{2}$$

By renaming in (2) with $x, y := y, x$ and appealing to (0), we immediately get

$$p.(x,y) \wedge r.y \Rightarrow q.(x,y) \quad , \tag{3}$$

and thus the proof of (2) also establishes the conjunction of (2) and (3):

$$p.(x,y) \wedge (r.x \vee r.y) \Rightarrow q.(x,y) \quad , \tag{4}$$

which is equivalent to our demonstrandum (1) provided

$$p.(x,y) \Rightarrow r.x \vee r.y \quad . \tag{5}$$

*　　　　*

*

The above describes what WLOG (= Without Loss Of Generality) is usually about. In principle—as shown above—the technique is perfectly sound, but the problem is that in practice it is only too often appealed to by the mere "WLOG we can assume $r.x$" without formulating, let alone proving (5).

Often, the underlying symmetry is not mentioned either, and such omissions make most "WLOG proofs" rather unsatisfactory.

* *
*

When I was introduced to the technique, (2) would be proved, and the proof of (3) would be omitted on the ground that mutatis mutandis that proof would be the same. At the time we were sufficiently vague about quantification and the role of dummies not to be able to see that (2) and (3) are the same theorem, the textual difference being introduced by our use of (x, y) to denote an unordered pair.

Austin, 12 November 1995

Prof. Dr. Edsger W. Dijkstra
Department of Computer Sciences
The University of Texas at Austin
Austin, TX 78712–1188
USA

My simplest theorem

(EWD1232)

Edsger W. Dijkstra

Theorem Any natural number that has a divisor greater than itself equals zero.

Proof We observe for any natural n, d, q

$$
\begin{array}{ll}
 & n = d \cdot q \wedge d > n \\
= & \quad \{\text{Leibniz}\} \\
 & n = d \cdot q \wedge d > d \cdot q \\
= & \quad \{d > 0\} \\
 & n = d \cdot q \wedge 1 > q \\
= & \quad \{q \text{ is natural}\} \\
 & n = d \cdot q \wedge q = 0 \\
\Rightarrow & \quad \{\text{Leibniz}\} \\
 & n = 0 \quad .
\end{array}
$$

(End of Proof.)

At least twice—EWD1088 & EWD1170—I had used that 0 is the only natural number with infinitely many divisors—e.g. 2^k for any k—, but I never took the trouble to prove it, and that probably explains why I missed the above.

Austin, 10 February 1996

Prof. Dr. Edsger W. Dijkstra
Department of Computer Sciences
The University of Texas at Austin
Austin, TX 78712–1188
USA

The transitive closure of a wellfounded relation

(EWD1241)

Edsger W. Dijkstra

Transitive closures can be defined in many ways, but today we define the nonreflexive transitive closure s of a relation r as the strongest s satisfying

$$[r \vee r\,;s \equiv s] \quad . \tag{0}$$

From (0) alone—i.e. not using that s is the strongest—we can derive

$$\langle \forall x :: [x \equiv s\,;x] \Rightarrow [x \equiv r\,;x] \rangle \quad . \tag{1}$$

Proof We observe for an x satisfying

$$[x \equiv s\,;x] \tag{2}$$

$$
\begin{array}{ll}
 & x \\
\equiv & \quad \{(2)\} \\
 & s\,;x \\
\equiv & \quad \{(0)\} \\
 & (r \vee r\,;s)\,;x \\
\equiv & \quad \{;\text{ over } \vee \text{ and associative}\} \\
 & r\,;x \vee r\,;s\,;x \\
\equiv & \quad \{(2)\} \\
 & r\,;x \vee r\,;x \\
\equiv & \quad \{\text{pred. calc.}\} \\
 & r\,;x \quad .
\end{array}
$$

(End of Proof.)

One of the formulations of "r is left-wellfounded" is

$$\langle \forall x :: [x \equiv r\,;x] \Rightarrow [\neg x] \rangle \quad . \tag{3}$$

Thanks to (1), (3) implies

$$\langle \forall x :: [x \equiv s\,;x] \Rightarrow [\neg x] \rangle \quad ,$$

in other words: if a relation is left-wellfounded, so is its nonreflexive transitive closure.

*　　　　*
*

From (0) alone—i.e. not using that s is the strongest—we can derive

$$\langle \forall x :: [x \Rightarrow r\,;x] \Rightarrow [x \Rightarrow s\,;x] \rangle \quad . \tag{4}$$

Proof We observe for any x

$$\begin{array}{ll} & [x \Rightarrow s\,;x] \\ = & \quad \{(0)\} \\ & [x \Rightarrow (r \vee r\,;s)\,;x] \\ \Leftarrow & \quad \{\text{pred. calc. and monotonicity of ;}\} \\ & [x \Rightarrow r\,;x] \quad . \end{array}$$

(End of Proof.)

Thanks to Knaster-Tarski, an alternative formulation of "s is left-wellfounded" is

$$\langle \forall x :: [x \Rightarrow s\,;x] \Rightarrow [\neg x] \rangle \quad . \tag{5}$$

From (4) and (5) we derive

$$\langle \forall x :: [x \Rightarrow r\,;x] \Rightarrow [\neg x] \rangle \quad ,$$

in other words: if the nonreflexive transitive closure of a relation is left-wellfounded, so is the relation itself.

Remark It is worth noting that the proofs of the crucial implications (1) and (4) use neither wellfoundedness nor the fact that s is the strongest s satisfying (0). (End of Remark.)

*　　　　*
*

For left-wellfounded r, (0) determines s uniquely, i.e. given

$$[r \vee r\,;s \equiv s] \tag{6}$$

$$[r \vee r\,;t \equiv t] \tag{7}$$

$$\langle \forall x :: [x \Rightarrow r\,;x] \Rightarrow [\neg x] \rangle \tag{8}$$

we have to prove $[s \equiv t]$.

Proof For reasons of symmetry, it suffices to prove $[t \Rightarrow s]$. We observe

$$
\begin{array}{ll}
 & [t \Rightarrow s] \\
\equiv & \quad \{\text{pred. calc.}\} \\
 & [\neg(t \wedge \neg s)] \\
\Leftarrow & \quad \{(8) \text{ with } x := t \wedge \neg s\} \\
 & [t \wedge \neg s \Rightarrow r\,;(t \wedge \neg s)] \\
\equiv & \quad \{\text{shunting and } (6)\} \\
 & [t \Rightarrow r \vee r\,; s \vee r\,;(t \wedge \neg s)] \\
\equiv & \quad \{\,;\text{ over } \vee \text{ and pred. calc.}\} \\
 & [t \Rightarrow r \vee r\,;(t \vee s)] \\
\Leftarrow & \quad \{\text{monotonicity of };\} \\
 & [t \Rightarrow r \vee r\,; t] \\
\equiv & \quad \{(7)\} \\
 & \mathit{true} \quad .
\end{array}
$$

(End of Proof.)

The above is a considerable streamlining of AvG88/EWD1079 dd 28 April 1990 (which made no use of the relational calculus).

I gratefully acknowledge the contribution of Rutger M. Dijkstra (viz. the isolation of the theorem (0) $\Rightarrow$ (1)) and that of Wim Feijen and Netty van Gasteren (viz. the final proof of unicity). I thank the ETAC in general for its willingness to discuss this old problem.

Nuenen, 28 June 1996

Prof. Dr. Edsger W. Dijkstra
Department of Computer Sciences
The University of Texas at Austin
Austin, TX 78712–1188
USA

The formula for $\sin.(\alpha + \beta)$

(EWD1244a)

Edsger W. Dijkstra

We consider a triangle with sides a, b, c and opposite angles α, β, γ respectively:

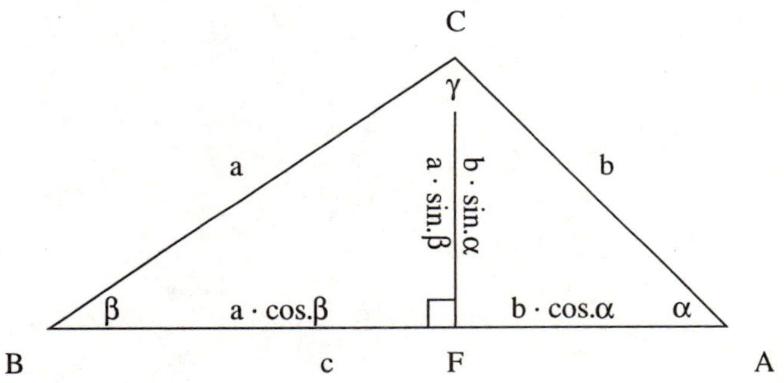

We have added the altitude CF; the additional annotation follows from the definitions of the sine and cosine functions. We observe

$$\begin{array}{ll} & true \\ \equiv & \quad \{\text{the two annotations for } CF\} \\ & a \cdot \sin.\beta = b \cdot \sin.\alpha \\ \equiv & \quad \{\text{algebra}\} \\ & a : b = \sin.\alpha : \sin.\beta \\ \equiv & \quad \{\text{symmetry}\} \\ & a : b : c = \sin.\alpha : \sin.\beta : \sin.\gamma \qquad (*) \end{array}$$

Next we observe

$$\begin{array}{ll} & true \\ \equiv & \quad \{\text{annotations for } BF \text{ and } FA\} \\ & c = a \cdot \cos.\beta + b \cdot \cos.\alpha \qquad (\dagger) \\ \equiv & \quad \{(*)\} \\ & \sin.\gamma = \sin.\alpha \cdot \cos.\beta + \sin.\beta \cdot \cos.\alpha \\ \equiv & \quad \{\alpha + \beta + \gamma = \pi\} \\ & \sin.(\alpha + \beta) = \sin.\alpha \cdot \cos.\beta + \sin.\beta \cdot \cos.\alpha \qquad (**) \end{array}$$

and so we have proved the addition formula (∗∗) for the sine function for $0 \le \alpha$, $0 \le \beta$ and $\alpha + \beta \le \pi$. (Note that F does not need to lie between A and B for (†) to be valid.)

Austin, 11 September 1996

Prof. Dr. Edsger W. Dijkstra
Department of Computer Sciences
The University of Texas at Austin
Austin, TX 78712–1188
USA

A kind of converse of Leibniz's Principle

(EWD1245)

Edsger W. Dijkstra

We owe to Gottfried W. Leibniz the principle that is informally known as "substituting equals for equals" and is formally expressed by

$$x = y \Rightarrow f.x = f.y \quad ,$$

which we can also read as "function application is equality-preserving". This note proves the

Theorem Besides the two constant relations, equality is the *only* relation preserved by function application.

In this note, function application is denoted by an infix dot (period); for the sake of simplicity we restrict ourselves to "endofunctions", i.e. functions for which argument and function value are of the same type, which is also the type of the variables (a, b, c, d, x, y) and the type on which the binary relations are defined. Boolean negation will be denoted by $\neg$, boolean equality by $\equiv$, which is given the lowest syntactic binding power; the dot of function application has the highest binding power.

For the rest of this note, R denotes—in the usual infix notation—a binary relation

- that differs from the constant relation T, given by $\langle \forall x, y :: x\, T\, y \rangle$, and
- that differs from the constant relation F, given by $\langle \forall x, y :: \neg(x\, F\, y) \rangle$, and
- that differs from equality.

To prove our theorem we have to show that, in general, R is *not* preserved by function application, i.e. that there exist an a, b and f such that

$$a\, R\, b \quad \text{and} \quad \neg(f.a\, R\, f.b) \quad .$$

We take care of the existential quantification over f by using the lemma—not proved here—

$$\langle \exists f :: f.a = c \wedge f.b = d \rangle \equiv a = b \Rightarrow c = d \quad ,$$

which allows us to rewrite our proof obligation as follows:

Show the existence of a quadruple a, b, c, d such that

(i) $a\,R\,b$

(ii) $\neg(c\,R\,d)$

(iii) $a = b \Rightarrow c = d$.

We now show this existence by constructing a witness.

That R differs from equality enables us to postulate (for the rest of this note) for some x, y

$$x\,R\,y \not\equiv x = y$$

and now we distinguish two cases.

Case 0: $x\,R\,y \wedge x \neq y$.

Choosing $a, b := x, y$ ensures that (i) and (iii) are satisfied; because R differs from T, c, d can be chosen so as to satisfy (ii). (End of Case 0)

Case 1: $\neg(x\,R\,y) \wedge x = y$.

Choosing $c, d := x, y$ ensures that (ii) and (iii) are satisfied; because R differs from F, a, b can be chosen so as to satisfy (i). (End of Case 1)

Since the case analysis was exhaustive, our proof obligation has been discharged.

Remark The unproved lemma can be demonstrated by mutual implication. The one direction—left $\Rightarrow$ right—follows directly from Leibniz's Principle; the other direction—the one needed in this note—requires a case analysis, say $a = b$ versus $a \neq b$. (End of Remark.)

Austin, 17 September 1996

Prof. Dr. Edsger W. Dijkstra
Department of Computer Sciences
The University of Texas at Austin
Austin, TX 78712–1188
USA

The strengths of the academic enterprise

(EWD1175)

Edsger W. Dijkstra

In the Western world, 66 institutions have enjoyed a continuously visible identity since 1530. Among those 66 are the Roman Catholic Church, the Lutheran Church and the Parliaments of Iceland and the Isle of Man. What makes these 66 so interesting—and I owe the knowledge of this fact to our President Dr. Berdahl—is that the remaining 62 are all universities! It strikingly demonstrates that universities have a potential for "longevity", but we should not make the mistake of concluding that they are "immortal" or invulnerable, for they are not: if they have existed for centuries, that is because successive generations of scholars and students have nurtured them well and with devotion. It is the obvious task of the current generation to hand over to the next what it got from the previous one, and in order to do so well, we had better understand how the strengths of the academic enterprise are maintained most effectively. Hence my title.

But before I can turn to my topic proper, I must make a few introductory remarks lest I be misunderstood.

The first one is that when we move from one society to another, all important words subtly change their meaning, and in connection with today's topic I must mention: university, education, training, teaching, scholar, scientist, engineer, theoretical, experimental, and applied. This was brought home to me in 1968 at a conference in Garmisch-Partenkirchen. I worked at the time at the Department of Mathematics of the Eindhoven University of Technology in the Netherlands, and told at that conference that the official academic title our graduates earned was "Mathematical Engineer", and most of the Americans began to laugh, because for them it sounded as a contradiction in terms, mathematics being sophisticated and unpractical, engineering being straightforward and practical. To give you another example, in the early 80s I learned that professors at Stanford University could use their grant money in the name of their research to pay someone else to do their teaching. When I heard that, I was shocked, when my wife heard it, she could not believe it, because we grew up with an academic culture in which teaching and research were considered warp and weft of the same fabric. In that view, a professor who does not get valuable inspiration from his own lectures and therefore does not regard his teaching as a precious experience, is just in the wrong business: he should teach at a vocational school or work at a research laboratory. So please, keep in mind that all important words I use may mean something different from what you are used to.

My second warning remark is that I shall refuse to discuss the academic enterprise in financial terms. The first reason is that the habit of trying to understand, explain, or justify in financial terms is unhealthy: it creates the ethics of the best-seller society in which saleability is confused with quality. The other day we had to discuss the professional quality of one of our colleagues, in whose favour it was then mentioned that one of his Ph.D.s had earned lots and lots of money in the computer business, and few people seemed to notice how ridiculous a recommendation this was. We also know that the financial success of a product can be totally independent of its quality (as everyone who remembers for instance the commercially successful IBM360 should know). The second reason for my refusal is that the value of money is a very fuzzy notion, so fuzzy in fact, that efforts to understand in financial terms always lead to greater confusion. [Remember this, for it is quite likely that this afternoon will give you the opportunity to observe the phenomenon. Note that money need not be mentioned explicitly for the nonsense to emerge, a reference to "the taxpayer" can do the job. The role of "the taxpayer" then invariably leads to the conclusion that of State Universities at least the undergraduate curriculum has to be second- or third-rate.] The final reason for my refusal is that the habit appeals to the quantitative mind and I come from a culture in which the primarily quantitative mind does not evoke admiration. [A major reason that we considered Roman Catholics to belong to a lower class was precisely their quantitative bent: they always counted, number of faithful, number of days in purgatory, you name it ....]

My third remark introduces you to the Buxton Index, so named after its inventor, Professor John Buxton, at the time at Warwick University. The Buxton Index of an entity, i.e. person or organization, is defined as the length of the period, measured in years, over which the entity makes its plans. For the little grocery shop around the corner it is about $\frac{1}{2}$, for the true Christian it is infinity, and for most other entities it is in between: about 4 for the average politician who aims at his re-election, slightly more for most industries, but much less for the managers who have to write quarterly reports. The Buxton Index is an important concept because close co-operation between entities with very different Buxton Indices invariably fails and leads to moral complaints about the partner. The party with the smaller Buxton Index is accused of being superficial and short-sighted, while the party with the larger Buxton Index is accused of neglect of duty, of backing out of its responsibility, of freewheeling, etc. . In addition, each party accuses the other one of being stupid. The great advantage of the Buxton Index is that, as a simple numerical notion, it is morally neutral and lifts the difference above the plane of moral concerns. The Buxton Index is important to bear in mind when considering academic/industrial co-operation.

My fourth and last introductory remark draws attention to a whole spectrum of techniques by which one generation transmits its insights and abilities to the next. At the one extreme we have the techniques of the guilds which treat their insights and abilities as valuable property, as a treasure to be kept secret. Their technique for protecting the secrecy is by keeping the secret knowledge unformulated; therefore, the apprentice has to join a master for seven meagre years, during which he can absorb the craft by osmosis, so to speak. The university is at the other end of the spectrum: it is the professor's task to bring the relevant insights and abilities into the public domain by explicit formulation. It is no accident that the universities as we know them started to flourish after the art of book printing had been established. There is more to be said about that spectrum of educational techniques, but I shall not do so now; I mentioned it to remind you why the absence of secrecy, or, more positively formulated, openness and honesty are characteristics that touch the heart of the academic enterprise: a university that hides or cheats can close its doors. The essential role of openness is something to remember when considering academic/industrial co-operation; it should also be remembered whenever a government invents reasons of national security or prosperity for the prevention of free publication of the results of academic research. Universities are not part of the nation's security organisation, they are not the nation's research laboratory either: they are the nation's universities.

In passing I would like to mention that in a rather different sense such openness is a precondition for academic survival. Just for being different and doing things the uneducated cannot understand, the academics are hated and feared, vide Socrates, executed in 399 BC, Archimedes, killed in 212 BC, and, more recently, Hypatia, AD 415 barbariously murdered by a Christian mob. The original Oxford Colleges were buildings fortified in order to protect the students against the rabble, and if you think that that is old hat, I refer you to the DDR or the People's Republic of China of only 25 years ago. It is a miracle whenever, these days, the academic world is tolerated at all; personally I am convinced that what tolerance there is would completely disappear, were the academic world to become secretive.

*   *

*

The University with its intellectual life on campus is undoubtedly a creation of the restless mind, but it is more than its creation: it is also its refuge. Regrettably, neither all professors nor all students are brilliant, but quite a few are and the unique thing is that, on campus, being brilliant is socially acceptable. Furthermore, the fabric of the academic world is such that it can absorb the most revolutionary ideas. And how essential that refuge is, we realize when we remember that many organizations impose a conformism that precludes even such deviant behaviour as wearing a moustache! (If you even wondered why I did not join Disneyland or IBM, you now know why.)

It is not only a refuge for the restless minds, it is also a reservation. It does not only protect the restless minds, it also protects the rest of the world, where they would create havoc if they were let loose. To put it in another way: the fence around campus is essential because it separates two worlds that otherwise would harm each other. The fence ensures that we have relatively little influence on the world "out there", but we would be foolish to complain, for our freedom to be as original or as radical as we like is based on the fact that industry and the world-at-large ignore our work anyhow. Currently there seems a world-wide tendency to try to lower the fence; the effort strikes me as ill-directed.

The unruly nature of academic life, of course, offends the orderly mind, and more than one regime has tried to deal with the problem by doing away with the restless minds, but the measure never had the effect the regime intended: destroy the campus, muzzle your intellectuals, and rapidly life deteriorates in all respects. The explanation is that, with all its aloofness, the university has an essential role to play, viz. to explain to the world the foolishness of its ways. Of course, all religions always try to do that, but religions being what they are, no pope, patriarch, ayatollah or dalai lama has enough authority to be taken seriously. Only the academic gadfly has so much authority that its sting really hurts.

President Reagan did not seem to see it that way, but even regimes of modest insight seem to understand that, as a corrective measure, the gadfly's sting is indispensable. The university has therefore the task to nurture the authority of the sting, both for its own protection and as a service to mankind. Aforementioned openness and honesty, though essential, are not enough; we should add a ruthless striving for perfection, ruthless in the sense that, on campus, there is no academically valid excuse for compromises.

The sting also defines the social responsibility of the universities. The question is: do we offer what society asks for, or do we offer what society needs? If the two co-incide, there is no problem, but often they don't, and in computing such co-incidence is extremely rare. In case of discrepancy, you must ignore what they ask for and give what they need, ignore what they would like and tell them what they don't want to hear but need to know. There are two compelling reasons for this uncompromising position.

The first one is that a *leading* university has no choice: to be leading means in this context showing new and better ways and possibilities no one else has dreamt of; if you give society what it asks for, you are not leading but led, viz. led by the demands of society as it sees them.

The second reason is that what society overwhelmingly asks for is snake oil. Of course, the snake oil has the most impressive names—otherwise you would be selling nothing—like "Structured Analysis and Design", "Software Engineering", "Maturity Models", "Management Information Systems", "Integrated Project Support Environments", "Object Orientation" and "Business Process Re-engineering" (the latter three being known as IPSE, OO and BPR, respectively). The external pressures to do the wrong thing are enormous, but yielding to them would be fatal for the academic enterprise, while resisting the pressure reinforces its strengths. The pressures are, in fact, so strong that I do not know a university where there is not some faculty or some department that has yielded, but there should be no mercy for snake oil pedlars on campus. [When a professor is no better than James Martin, he should start a business instead.]

* *

*

In the wake of the Cultural Revolution and now of the recession I observe a mounting pressure to co-operate and to promote "teamwork". For its anti-individualistic streak, such a drive is of course highly suspect; some people may not be so sensitive to it, but having seen the Hitlerjugend in action suffices for the rest of your life to be very wary of "team spirit". Very. I have even read one text that argued that university scientists should co-operate more in order to become more competitive .... Bureaucracies are in favour of teamwork because a few groups are easier to control than a large number of rugged individuals. Granting agencies are in favour of supporting large established organizations rather than individual researchers, because the support of the latter, though much cheaper, is felt to be more risky; it also requires more thinking per dollar funding. Teamwork is also promoted because it is supposed to be more efficient, though in general this hope is not justified. I have no first-hand experience with the ESPRIT projects of the European Community, as they started after I had left. Involvement of universities from different member states is, I believe, a conditio sine qua non, and here the purpose of the co-operation seems more to force the researchers to broaden their outlook than to increase the efficiency of the research. My impression is that regular contacts with academic colleagues from other countries are experienced as valuable, but that actual co-operation becomes extremely sticky each time industrial partners are included. And everybody complains about the amount of red tape and travel.

Interdisciplinary efforts on campus, that is co-operation between different departments of a university are almost always failures, and the reasons are clear. Why should a vigorous, flourishing department seek co-operation when it is doing just fine all by itself? It is the weak departments that are more tempted to seek each other's support and to believe that there is might in numbers. But such co-operation is of course based on the theory that, when you tie two stones together, the combination will float. Another reason is that the boundaries between our scientific disciplines are not arbitrary at all: the different disciplines represent a modularization of science that has been introduced for the sake of efficiency.

Co-operation between corresponding departments of different universities seems to work quite well, co-operation between the university and industry, however, is so much harder that it usually fails. We might even conclude that the effort is hopeless.

To begin with, there is the great difference in Buxton Index. For industry, the Buxton Index is less than 10, probably closer to 4 or 5, whereas for the academic scientist the Buxton Index is closer to, say, 50, for what you offer your students should last a lifetime, their lives, to be precise.

The second problem has to do with the openness, which is a hallmark of the university, whereas, like the guilds, industry tends to see its knowledge as trade secret. People have tried to find legal solutions for this dilemma, but I am afraid that such solutions only touch the surface: at a more profound level, either one of the parties forsakes its duty, or the co-operation collapses.

But the greatest limitation on the usefulness of co-operation between industry and academia is almost certainly that the two have completely different purposes. To quote Harvey Earl of GM: "General Motors is in business for only one reason. To make money. In order to do that we make cars. But if we could make money by making garbage cans, we would make garbage cans.". Some people might argue that they even tried to make money by making garbage. But the product is secondary; to quote Harvey Earl again: "Listen, I'd put smokestacks right in the middle of the sons of bitches if I thought I could sell more cars.". These quotations are from the fifties, but things have not changed that much. For instance, computing science has very convincingly shown that simplicity is a necessary precondition for reliability, but industry willfully complicates products so as to make them proprietary. The disgraceful state of affairs is fully revealed by the traditional disclaimer with which industrial software is sold.

Under current circumstances I would not even attempt to promote co-operation between the academic and the industrial worlds, because it seems pointless and dangerous. I have come to the conclusion that, industrial management being what it is, it is extremely unlikely that computing science can save the computing industry. Conversely, the computer industry can severely damage computing science; it does so quite regularly by the donation of equipment that had better be ignored. [To avoid misunderstanding, what I just said does not necessarily represent the official opinion of my employer!] So, the less contact we have, the better.

*　　　*
*

Academic computing science is doing fine, thank you, and unless I am totally mistaken, it will have a profound influence. I am not referring to the changes that result from computers in their capacity of tools. Okay, the equipment opens new opportunities for the entertainment industry, but who cares about that anyhow. The equipment has enabled the airline industry to make its rates so complicated and volatile that you need an expert to buy a ticket, and for this discouragement of air travel we can be grateful, but the true impact comes from the equipment in its capacity of intellectual challenge.

Thanks to the existence of computing equipment we have, for the first time in the intellectual history of mankind, an environment in which the large-scale application of formal techniques is feasible and necessary. Not too long ago, formal reasoning was regarded merely as a theoretically intriguing possibility, but so utterly unpractical that it was totally irrelevant for real mathematicians. Peano was ridiculed for his axiomatization of something as trivial as integer arithmetic! But it is precisely because of these "trivialities" that we can now do things of a power and a beauty, way beyond the wildest dreams I had as a youngster.

As a mathematician I enjoy the same type of excitement as the theoretical physicists enjoyed in the first decades of this century. The analogy is apt in more than one way. In either case the results were obtained not by mission-oriented research, but by trying to achieve the just feasible. If academic research is often astonishingly successful, it always is because the researchers had the wisdom and the opportunity to avoid both the trivial and the impossible, and to follow the very narrow path in between. It is that narrow path in between that defines the intellectual autonomy of successful scientific research.

The major strength of the academic enterprise is that in a very technical sense scientific progress is unique in a way that neither political nor commercial interests can change.

*　　　*
*

Let me end by quoting, by way of contrast, from the C^2E Report from the IC^2 Center for Commercialization and Enterprise, The University of Texas at Austin, Winter 1993–94. [I draw your attention to the "class-room theory": just "theory" was not bad enough!]

> "In order to supply businesses with the managers they need in an ever-changing world, it is critical that the University maintain direct ties with the business community. These ties give students real-world experience in which to apply class-room theory—to help them to be more effective on the job and to provide feedback to the University to ensure that its curriculum is meeting the needs of business."

Did the writer not know that the use of the term "the real world" is usually interpreted as a symptom of rabid anti-intellectualism, or did he not mind? It is not amazing that people wonder whether the School of Business Administration belongs on campus at all.

* *
*

The above was written for our "Industrial Forum" on Monday 7 February 1994. The quotations of Harvey Earl were taken from "The Fifties" by David Halberstam, Villard Books, New York 1993. Other recommended literature is "The Organization Man" by William H. Whyte, Simon & Schuster Inc., New York, 1956.

Austin, 9 February 1994

Prof. Dr. Edsger W. Dijkstra
Department of Computer Sciences
The University of Texas at Austin
Austin, TX 78712–1188
USA

On a Method for the Formal Design of Multiprograms

(AvG140/WF231)

W. H. J. Feijen* and A. J. M. van Gasteren†
Department of Mathematics and Computing Science
Eindhoven University of Technology
PO Box 513
5600 MB Eindhoven
The Netherlands

Keywords and Phrases

Program Design, Multiprogramming, Multibounds, Theory of Owicki and Gries, Distributed Phase Synchronization, "Simplicity is our Major Concern"

Abstract

We exemplify a method for the formal derivation of multiprograms, using the simple theory of Owicki and Gries as our main tool for coping with concurrency. In our first and simple example we derive a protocol for the problem of Concurrent Vector Writing, and in our second and more tricky example we design a distributed algorithm for the problem of Liberal Phase Synchronization.

0 Introduction

One of the greatest milestones in the history of computing science is, we believe, the emergence of C. A. R. Hoare's paper "An Axiomatic Basis for Computer Programming" [Hoa69]. At one fell swoop it turned the intuitively understood craft of programming into a scientific discipline of a mathematical nature. Although at first Hoare's proposal seemed to be meant for verification of existing programs only, it soon became an inspiration and a basis for yet another milestone in computing, namely the formal *development* of (sequential) programs [Dij76] [Gri81]. The great impact of Hoare's formalism—the Hoare-triple semantics as it was called—is that it treats program texts as mathematical objects in their own right, such without recourse to executions,

*wf@win.tue.nl
†netty@win.tue.nl

i.e. without reference to what happens when programs are run on particular machinery. This enormous—at the time revolutionary even—mental separation of concerns has been of decisive importance for the development of computing science.

Whereas operational reasoning has been identified as intractable in the world of sequential programming, it should be considered all but infeasible in the area of parallel programming. Or, to phrase it with an understatement like in [AO91]: For concurrent programs "The number of possible execution sequences is often forbiddingly large".

What Hoare did to sequential programming, was done—in principle at least—by Susan Owicki and David Gries to multiprogramming. The theory of Owicki and Gries [OG76] [Dij82] has become a well-received and established theory for verifying the partial correctness of existing parallel programs. By an unfortunate course of history, however, the theory has not become a basis for the formal *development* of parallel programs—multiprograms, as we call them—.

One possible reason why computing scientists have not explored how the theory of Owicki and Gries might assist in developing multiprograms is that, when the theory emerged (in the mid seventies), programmers were heavily involved in the science of sequential programming and in getting functional programming off the ground. Another—presumably more fundamental—reason could be that the theory can address partial correctness only, thus leaving aside the far more complicated problems of "liveness". And indeed, concurrency has brought about phenomena like deadlock, individual starvation, after-you-after-you blocking, which, in general, are much more complicated to handle than termination of ordinary sequential programs.

Nevertheless, it was the very simplicity of the Owicki-Gries theory (which is a slight extension of the Hoare-triple semantics) that for us formed the main incentive to investigate the possibilities and opportunities that the theory might offer for developing multiprograms. In fact, the prime purpose of this paper is to show how the theory of Owicki and Gries can be used quite effectively for the formal derivation of an even wide range of multiprograms, ranging from simple communication protocols to far more sophisticated distributed algorithms. Although it is true that the theory can only handle partial correctness, we shall see how in a number of cases this does not harm.

Our first example—the problem of Concurrent Vector Writing—is quite a simple one, and it mainly serves to convey several techniques and the programming style we have in mind. The second example—the problem of Liberal Phase Synchronization—shows the programming style at work for the development of a distributed algorithm that we ourselves could never have come up with by operational reasoning.

The technical skills needed for reading and understanding the rest of this text are familiarity with Hoare-triples and with the predicate calculus. In fact, fluency in the latter has proved to be indispensable for playing this game.

1 Preliminaries

1.0 On our computational model

A *multiprogram* is a set of ordinary sequential programs, called the multiprogram's *components*. Our computational model is an interleaving model: each computation of the multiprogram is an interleaving of the computations of the individual components.

The computations of an individual component are the usual ones known from sequential programming, but for one difference: guarded skip **if** $B \rightarrow$ skip **fi** is computationally equivalent to repetition **do** $\neg B \rightarrow$ skip **od** .

As for the granularity of the interleaving, this is, in principle, the designer's choice. Here, our atomic, indivisible statements—or grains of interleaving—are, first, the assignment statement, such as

$$x := x + 1 \quad \text{or} \quad x, b := x + 1, \neg b \quad ,$$

and, second, the guarded skip

if $B \rightarrow$ skip **fi** ,

in the sense that an evaluation of guard B to false is atomic, and so is an evaluation of B to true together with the execution of skip.

A final postulate on the permissible interleavings that constitute the computations of the multiprogram is that no component is neglected indefinitely. (This is usually called "weak fairness".)

1.1 The theory of Owicki and Gries, in brief

The components of a multiprogram can be annotated with assertions in the way familiar from sequential programming. Also, the multiprogram has a precondition, and in case all components terminate it has a postcondition as well. The theory of Owicki and Gries [OG76] [Dij82] tells us that such an annotation is correct if each assertion of it is both *locally* and *globally* correct:

- An assertion is locally correct, i.e. correct with respect to the component in which it occurs, if it is established by the (dynamically) preceding atomic statement in that component—or, if there is no such preceding statement, if it is implied by the precondition of the multiprogram as a whole—.

- An assertion is globally correct, i.e. correct with respect to the rest of the system, if it is not falsified by any atomic statement of any other component. (This is usually called "interference freedom".)

Finally, for the postcondition—if any—of the multiprogram to be correct, it should be implied by the conjunction of the postassertions of the individual components.

This is the *Core* of the Owicki-Gries theory. It can be succinctly summarized and memorized by

> the annotation is correct
> $\equiv$
> each assertion is established by the component in which it occurs, and it is maintained by the rest of the system.

* * *

For a number of reasons, it is useful at this point to reveal the intended operational interpretation of correct annotation. The intention is that when the execution of a component is "at" an assertion P, then the state of the system as a whole satisfies P, no matter how this component has arrived at P—P's local correctness—and no matter what interleaving is exhibited by the rest of the system while this component resides at P—P's global correctness—. This interpretation is useful to know because

- it makes the link between the Owicki-Gries theory and the computational model "sweetly reasonable"
- it is needed for the formal specification of multiprograms, e.g.
  when a consumer is about to consume a portion from a buffer, the precondition of that action ought to be: "buffer nonempty"
- it emphasizes our earlier claim that the theory of Owicki and Gries does away with operational reasoning in a highly rigorous manner
  —the only thing that matters about an assertion is that it *is* correct, not via which computational process it *became* correct—
- it tells us that if the state of the system satisfies a condition P, we must be willing to accept that it also satisfies conditions weaker than P.

It is this last observation which calls for an extra rule to be added to the Core of the Owicki-Gries theory, namely the rule of

Weakening the Annotation
The annotation remains correct if an assertion—P, say—is replaced by a weaker one, that is by an assertion that is implied by P.
End .

As we shall see later, this additional rule has severe technical and methodological consequences. In fact, it formed a first stimulus for switching from a posteriori verification to development of multiprograms [Fe87].

* *
*

The care for the global correctness of assertions creates a quadratic proof obligation: for each individual assertion in each of the components we have to check that it is not falsified by any atomic statement of any other component. A prime means for bridling this huge proof obligation is the use of *system invariants*. A system invariant is a relation that holds initially—i.e. is implied by the precondition of the multiprogram—and is not falsified by any atomic statement of the multiprogram. As a result, a system invariant can be added as a conjunct to each assertion, and therefore we can afford the freedom of not writing it anywhere in our annotation, but just once in a central position. (This clerical rule is of great importance for clarity and economy of exposition.)

Finally we note that, in what follows, the guarded skip is our only means for achieving synchronization. As mentioned before, guarded skip **if** $B \rightarrow$ skip **fi** is computationally equivalent to **do** $\neg B \rightarrow$ skip **od** . Thus, in Hoare-triple semantics (for partial correctness), its proof rule is

$$\{B \Rightarrow R\} \ \textbf{if}\ B \rightarrow \text{skip}\ \textbf{fi}\ \ \{R\} \quad .$$

More often than not, we shall use the special case

$$\textbf{if}\ B \rightarrow \text{skip}\ \textbf{fi}\ \ \{B\} \quad ,$$

i.e. we use a guarded skip **if** $B \rightarrow$ skip **fi** so as to establish the local correctness of assertion B.

1.2 Progress

Guarded skips are our only means for achieving synchronization. Operationally speaking, they serve to temporarily block a component—make it "wait"—in order to preclude violation of some synchronization condition. As indispensable as such blocking statements are for the purpose of synchronization, they bring about the danger of infinite blocking as well. Proving "individual progress", i.e. proving that the blocking as caused by a guarded skip is temporary, is, in principle outside the scope of the Owicki-Gries theory. It is here where other formalisms—like Temporal Logic—tend to creep in.

However, there are some special circumstances in which individual progress can be demonstrated quite easily. Consider, for example, the following rather canonical situation of a multiprogram. Suppose that, projected on the variables of sequence x, each component has the form

$$\text{Comp}.i: \quad *[\, x.i := 1 + x.i \,]$$ [0]

Also, suppose that the multiprogram maintains system invariant

$$MB: \quad x.i \le K.i.j + x.j \quad (\forall i, j) \quad ,$$

where $K.i.j$ is some constant which may depend on i and j. From MB and the structure of the programs we then conclude that, if for one reason or another, one component gets stuck forever, so will all the others. So, as far as progress is concerned, the multiprogram as a whole can offer just two scenarios, viz.

either *all* components get stuck forever
—this we call "total deadlock"—,
or *each* individual component makes progress.

We conclude that in the presence of a *multibound* like MB, individual progress can be proved by proving the absence of the danger of total deadlock. And the nice thing is that the latter is within the scope of the Owicki-Gries theory. We next explain how.

Since the guarded skip is our only mechanism for blocking a component, total deadlock refers to a state in which all components are engaged in the execution of a guarded skip while all these guards are false. With this in mind, absence of total deadlock can be demonstrated as follows. Choose one guarded skip in each component, provide it with a correct preassertion, and prove that the conjunction of these preassertions implies the disjunction of the corresponding guards. Do so for all possible combinations of guarded skips.

*　　　*
*

So much for progress. In one of the subsequent examples we shall see how to effectively exploit this "multibound-technique".

1.3 Three theorems and an example

The three theorems to be presented here each deal with lessening the quadratic burden of proving the global correctness of assertions. The first two theorems—the Rule of Orthogonality and the Rule of Widening—are absolutely trivial, but nevertheless useful in practice. The third theorem—the Guard Conjunction Lemma—requires a tough proof, which we will not give here.

[0] $*[S]$ is short for **do** true $\rightarrow S$ **od**

Rule of Orthogonality

Let P be a predicate in which variable x does not occur. Then P is maintained by whatever assignment to x.

Proof $\{P\}\ x := E\ \{P\}$ is a correct Hoare-triple.

End .

In what follows we will never mention the Rule of Orthogonality again, but we will extensively exploit it when dealing with the global correctness of assertions P by focussing on just those assignments in the other components that may affect the values of the variables occurring in P.

Rule of Widening

Let Δx and Δy be such that $\Delta x \leq 0$ and $0 \leq \Delta y$. Then neither $x := x + \Delta x$ nor $y := y + \Delta y$ falsifies $x \leq y$.

Proof $\{x \leq y\}\ x := x + \Delta x\ \{x \leq y\}$ and
$\{x \leq y\}\ y := y + \Delta y\ \{x \leq y\}$

are correct Hoare-triples, by the transitivity of $\leq$.

End .

The Rule of Widening says that "increments" of y or "decrements" of x maintain $x \leq y$. It goes without saying that the Rule comes in many variations. Whenever we appeal to it, we do so by just a single word: Widening. Experience has shown that it is useful to have a (catchy) name for a standard phenomenon.

The Guard Conjunction Lemma

For globally correct B, the coarse-grained guarded skip **if** $B \wedge C \rightarrow$ skip **fi** can be replaced by the succession

if $B \rightarrow$ skip **fi** ; **if** $C \rightarrow$ skip **fi**

of these two finer grained guarded skips, without impairing the correctness of the multiprogram, i.e. without invalidating the partial correctness and without endangering absence of total deadlock or individual progress.

Proof We are grateful to Dr. Jozef Hooman for having supplied a proof of this theorem [Hoom93].

End .

The more fine-grained the constituent atomic statements of a multiprogram are, the more cumbersome the required proof obligations. Theorems that tell us about opportunities to refine programs with impunity are very welcome. The Guard Conjunction Lemma is such a theorem, and it is even a simple one. It tends to advocate a programming style, in the sense that one seeks to construct the programs such that the theorem becomes applicable. Unfortunately, we are not aware of many more theorems of this kind.

An example

The example we deal with is the traditional Producer-Consumer problem. It serves to illustrate some of the issues we touched upon so far.

We are given the following two-component multiprogram

Pre: $in = 0 \wedge out = 0$
Prod: $* [\, in := in + 1 \,]$
Cons: $* [\, out := out + 1 \,]$.

Variable *in* records the number of portions produced and added to a buffer, say, by Prod. Variable *out* records the number of portions consumed and taken from the buffer by Cons. Because the buffer has finite capacity C, $1 \leq C$, the synchronization task ahead of us is to maintain as a system invariant

$$0 \leq in - out \leq C$$

or, equivalently,

$$P: \quad out \leq in \ \wedge \ in \leq C + out \quad .$$

Before proceeding, we observe that, in combination with the structure of the components, P is a perfect multibound. Hence our care for individual progress in the ultimate solution can be replaced by the care for absence of total deadlock, *no matter* what our ultimate solution will look like.

The problem is highly symmetric in Prod and Cons. Let us focus on Prod, say. Its statement $in := in + 1$ does not violate the first conjunct of P—widening!—. In order that it maintain P's second conjunct, we demand it to have precondition $in < C + out$. Thus, our first approximation of Prod becomes

```
Prod:  * [ {? in < C + out}
           in := in + 1
         ]
```

Now, the only thing that remains is to take care of the queried assertion in Prod. (The ? is a bookkeeping device to remind us of what remains to be done.)

Assertion $in < C + out$ is globally correct—widening!—. We ensure its local correctness by prefixing it with guarded skip **if** $in < C + out \rightarrow$ skip **fi** , which completes this little derivation. The final multiprogram reads

Pre: $in = 0 \wedge out = 0$
Inv: $P:\ \ out \le in \wedge in \le C + out$
Prod: $*$[**if** $in < C + out \rightarrow$ skip **fi**
; $\{in < C + out\}$
$in := in + 1$
]
Cons: $*$[**if** $out < in \rightarrow$ skip **fi**
; $\{out < in\}$
$out := out + 1$
]

As for the absence of total deadlock we investigate a state in which both guards are false:

$$
\begin{array}{ll}
 & C + out \le in \wedge in \le out \\
\equiv & \quad \{\text{invariant } P\} \\
 & C + out = in \wedge in = out \\
\equiv & \quad \{\text{invariant } P\} \\
 & C = 0 \\
\equiv & \quad \{1 \le C\} \\
 & \text{false}
\end{array}
$$

And as a result, there is no danger of total deadlock.

*   *
*

Herewith, we conclude this section on preliminaries.

2 The problem of Concurrent Vector Writing

The prime purpose of this section is to explain the programming style we have in mind for the formal design of multiprograms. For the current example, we do so in all relevant detail and in very small steps. We also intersperse the technicalities with remarks concerning notation, nomenclature, and heuristics.

*   *
*

We consider an array $x[0 \cdot\cdot N)$, $0 \le N$, and two components, A and B. Component A writes zeroes into x, and component B ones. Both components terminate, and we want the final state of the multiprogram to be $\vec{x} = \vec{0}$.

A more precise specification of this problem is given by the figure below

Pre:	$i = 0 \land j = 0$
A:	**do** $i \neq N \rightarrow x.i := 0 \; ; \; i := i + 1$ **od**
B:	**do** $j \neq N \rightarrow x.j := 1 \; ; \; j := j + 1$ **od**
Post:	? $\langle \forall k : 0 \leq k < N : x.k = 0 \rangle$

It is more precise in that it reveals how the components write into x. We shall refer to the combination of the precondition and the program texts for A and B as the *Computation Proper.* The figure also contains a notational device that we shall use all through, namely a query. Such queries indicate that we still have to take care of something—here: the correctness of the postcondition—. As we go along, we will also encounter queried assertions and queried system invariants.

The way we shall always take care of a queried item is by extending the program text—with code, assertions, or invariants—so as to accomplish the correctness of the item. However, the extensions are constrained by the rule that the computation proper should *not* be changed.

The last clause prohibits, for instance, that we replace the parallel composition of A and B with the sequential program B ; A, which would vacuously establish the required postcondition. The last clause also prohibits further changes to x, i, or j. The code to be added may, however, contain inspections of the variables of the computation proper, although in many cases it will be phrased in terms of fresh (synchronization) variables.

*　　　*

*

The simplest solution to our current problem is one that mimics the behaviour of the sequential program B ; A. This can be achieved at the expense of one fresh boolean, f say. A solution is

Pre: $i = 0 \land j = 0 \land \neg f$

```
A:      if f → skip fi              B:    do j ≠ N →
      ; do i ≠ N →                           x.j := 1
            x.i := 0                       ; j := j + 1
          ; i := i + 1                    od
        od                              ; f := true
```

It is a—nice—exercise to prove that this, indeed, establishes the required postcondition.

Although there is no logical objection to this solution, there is a strategical one: it kills all the potential parallelism. Component A is denied progress until B has terminated, whereas there seems to be no logical need for that. In constructing multiprograms, we prefer to adhere to the rule that progress of a component ought not to be hampered unnecessarily, unless there is a good reason for doing so.

Sometimes there are good reasons for hampering progress, and we will encounter one near the end of this example.

* *

*

After these preliminaries, we now proceed with our problem by making a first design decision, viz. restricting our solution to a multiprogram of the form

Pre: $i = 0 \land j = 0$	
A: **do** $i \neq N$	B: **do** $j \neq N$
$\rightarrow$ $S0$	$\rightarrow$ $T0$
; $x.i := 0$	; $x.j := 1$
; $S1$	; $T1$
; $i := i + 1$	; $j := j + 1$
; $S2$	; $T2$
od	**od**
Post: ? $\langle \forall k : 0 \leq k < N : x.k = 0 \rangle$	

Version 0

The S's and the T's serve as placeholders for additional code: they are the *only* places where we allow code to be inserted for the purpose of synchronization. Observe that by offering no placeholders before or after the repetitions, we have ruled out the "sequential" solution discussed before.

It is a good habit to be always explicit about the placeholders, and in our current example we will be until the very end. Drawing from experience, however, we know that they are mostly well-understood, so that we can afford to leave them implicit. But when in doubt, don't leave them out!

* *

*

Version 0 contains one query, viz. the postcondition. From the Owicki-Gries theory we know that its correctness is guaranteed whenever it is implied by the conjunction of the postconditions of the individual components. Now observe

$$\langle \forall k : 0 \leq k < N : x.k = 0 \rangle$$

$$\Leftarrow$$

$$\langle \forall k : 0 \leq k < i : x.k = 0 \rangle \ \land \ i = N \quad .$$

From the structure of component A, we see that it is reasonable that we venture $i = N$ as a correct postcondition of A. We enforce the correctness of $\langle \forall k : 0 \le k < i : x.k = 0 \rangle$ by requiring it to be a system invariant. With these choices it suffices to take true as a postcondition of B. And because true is a correct assertion in each multiprogram, we never write it down.

Thus, we arrive at our next version

Pre: $i = 0 \wedge j = 0$	
Inv: ? P : $\langle \forall k : 0 \le k < i : x.k = 0 \rangle$, Note 1	
A: **do** $i \neq N$	B: **do** $j \neq N$
$\rightarrow$ $S0$	$\rightarrow$ $T0$
; $x.i := 0$	; $x.j := 1$
; $S1$	; $T1$
; $i := i + 1$	; $j := j + 1$
; $S2$	; $T2$
od	**od**
{? $i = N$, Note 0}	
Post: $\langle \forall k : 0 \le k < N : x.k = 0 \rangle$	

Version 1

Observe that the postcondition has lost its query, because in passing from Version 0 to Version 1 we turned it into a correct one, be it at the expense of two new queried items, viz. assertion $i = N$ in component A and system invariant P. Please ignore the references to Note 0 and Note 1 for a moment.

Here is the place to point out what we think is an important methodological issue. Version 1 tells us, in a nutshell, what has been achieved—viz. the correctness of the postcondition—, and what remains to be achieved—viz. the correctness of the queried items—. Thus the figure that we named Version 1 acts as the *precise* interface between the past and the future of the development process. As a result Version 1 is the specification of the programming problem ahead of us, and it is absolutely irrelevant how we arrived at it. This is of importance because, as the development process evolves from the one version to the next, we, at any time, only need to pay attention to how to transform the current version into the next one. Thus, the procedure is very similar to the method of stepwise refinement (and, in a way, very similar to calculation where, at any time, we are only concerned with how to transform the current expression into the next one).

Here, too, is the place to argue why such a transformation is valid according to the Owicki-Gries theory. In transforming Version 0 into Version 1, we clearly strengthened the annotation: the latter requires the invariance of P and the correctness of assertion $i = N$, both absent in the former.

Now suppose we succeed in removing the queries from Version 1 by appropriate choices for the S's and the T's, then Version 1 turns into a correctly annotated multiprogram; but thanks to the postulate of *Weakening the Annotation*, Version 0 then turns into a correctly annotated multiprogram as well, and that is what we were after!

*　　　*

*

After these intermediate remarks time has come to continue our development. In moving to a next version, we have to remove one or more queries. In general, the choice is completely free. We indicate the queries chosen by supplying them with a reference to a Note. Thus, the two Notes in Version 1 express our intention to remove both queries in moving to the next version.

Note 0 $i = N$

L: correct, from the preceding guard

G: correct, because i is a private variable of A, i.e. is not changed by B.

End Note 0.

Because Note 0 deals with an assertion, we have to address local and global correctness. Therefore, such a note will, in principle, always contain two entries, viz. L and G.

Note 1 Inv P: $\langle \forall k : 0 \leq k < i : x.k = 0 \rangle$

Init: correct, from $i = 0$ in Pre

- $\{\text{true}\}$ $x.i := 0$
- $\{?\ x.i = 0\}$ $i := i + 1$
- $\{?\ i \leq j\}$ $x.j := 1$

End Note 1.

Because Note 1 deals with a system invariant we have to ensure that this invariant holds initially—entry Init—, and that it is maintained by all atomic statements of the multiprogram. For this latter purpose we investigate all statements that might affect the invariant—here $x.i := 0$, $i := i + 1$, and $x.j := 1$—and we calculate under what additional precondition these statements maintain the invariant—here true, $x.i = 0$, and $i \leq j$ respectively—. Here, we did these calculations by heart and gave the results at once. Also, observe that we (tacitly) used that the S's and the T's do not change the computation proper, i.e. do not change x, i, or j.

Thus we arrive at

Pre: $i = 0 \wedge j = 0$ Inv: $P: \langle \forall k : 0 \leq k < i : x.k = 0 \rangle$	
A: **do** $i \neq N$ $\rightarrow$ $S0$; $x.i := 0$; $S1$; $\{?\ x.i = 0\}$ $i := i + 1$; $S2$ **od** $\{i = N\}$	B: **do** $j \neq N$ $\rightarrow$ $T0$; $\{?\ i \leq j$, Note 0$\}$ $x.j := 1$; $T1$; $j := j + 1$; $T2$ **od**
Post: $\langle \forall k : 0 \leq k < N : x.k = 0 \rangle$	

Version 2

Next, we tackle just assertion $i \leq j$, leaving $x.i = 0$ for later. The reason is that $i \leq j$ will get a different treatment that is worth our special attention.

Note 0 $i \leq j$
We ensure its correctness by requiring it to be a system invariant.
End Note 0.

The reason for this design decision is a very general one. The alternative would have been to ensure the local correctness of $i \leq j$ in B via guarded skip **if** $i \leq j \rightarrow$ skip **fi** . The trouble with this choice is that component A cannot truthify guard $i \leq j$—it only increments i—. In general, it is not a good idea to install a guarded skip whose guard cannot be truthified by the rest of the system, because this could severely endanger progress. The alternative, viz. turning such a condition into a system invariant, has proven to be a reasonably successful strategy.

Remark In the current example, adoption of guarded skip **if** $i \leq j \rightarrow$ skip **fi** in component B would nevertheless lead to a happy end, but that is just a stroke of good luck. We would like to encourage the reader to trace this alternative derivation.
End Remark.

In order to shorten our treatment, we deal with the invariance of $i \leq j$ right away

Re inv $i \leq j$

Init: correct from Pre

• $\{?\ i < j\}\ i := i+1$

End .

The next version thus becomes—juxtaposition of assertions denoting their conjunction—

Pre: $i = 0 \wedge j = 0$ Inv: P: $\langle \forall k : 0 \leq k < i : x.k = 0 \rangle$, $i \leq j$
A: **do** $i \neq N$ $\rightarrow$ $S0$; $x.i := 0$; $S1$; $\{?\ x.i = 0$, Note 0$\}\{?\ i < j$, Note 1$\}$ $i := i+1$; $S2$ **od** $\{i = N\}$ — — — — — — — — — — — — — — B: **do** $j \neq N$ $\rightarrow$ $T0$; $\{i \leq j\}$ $x.j := 1$; $T1$; $j := j+1$; $T2$ **od**
Post: $\langle \forall k : 0 \leq k < N : x.k = 0 \rangle$

Version 3

We continue.

Note 0 $x.i = 0$

L: Choose $S1$ = skip. Then $x.i = 0$ follows from the preceding $x.i := 0$.

G: Only $x.j := 1$ in B can violate $x.i = 0$. We calculate

$$(x.i = 0)(x.j := 1)$$
$$\equiv \quad \{\text{axiom of assignment}\}$$
$$x.i = 0 \wedge i \neq j$$

$\equiv$ {invariant $i \leq j$}
$x.i = 0 \wedge i < j$.

Hence, $x.i = 0$ is not violated because it has $i < j$ as a co-assertion—see remarks below—.

End Note 0.

Remarks

(i) Notice that co-assertion $i < j$ is absolutely vital for the correctness of $x.i = 0$.

(ii) In showing the global correctness of $F \wedge G$ under a statement S, we have to show the correctness of Hoare-triple

$$\{F \wedge G\}\ S\ \{F \wedge G\} \quad .$$

But since this is equivalent to the conjunction of

$$\{F \wedge G\}\ S\ \{F\} \quad \text{and} \quad \{F \wedge G\}\ S\ \{G\} \quad ,$$

the proof of F's global correctness may use co-assertion G of F (and vice versa!). And that is what we did above: $i < j$ is a co-assertion of $x.i = 0$.

(iii) It looks as if we played a mean trick by using an assertion—$i < j$—that still carries a query. But we should not forget that such a query will sooner or later be removed, here turning $i < j$ into a correct assertion.

End Remarks.

Note 1 $i < j$

L: With skip as our choice for $S1$, we ensure the local correctness of $i < j$ by demanding
{? $i < j$, Note 2}
$x.i := 0$.

G: Widening.

End Note 1.

Note 2 $i < j$

L: Choose guarded skip **if** $i < j \rightarrow$ skip **fi** for $S0$.

G: Widening.

End Note 2.

In the above we slightly accelerated the design process by cascading, viz. by handling the consequences of assertion $i < j$ (in Note 1) right away. This is standard practice when the situation has become simple and transparent.

Now we are done: all queries have been removed! Before we write down our solution we would like to inform the reader that we played a tricky game with the order in which we handled the assertions $x.i = 0$ and $i < j$ in Version 3. Had we first tackled $i < j$ and chosen to establish its local correctness by choosing **if** $i < j \rightarrow$ skip **fi** for $S1$, the whole design would have failed. We invite the reader to try this and to pay attention to remark (i) given earlier.

The final solution—placeholders omitted—is

Pre: $i = 0 \land j = 0$ Inv: P : $\langle \forall k : 0 \leq k < i : x.k = 0 \rangle$, $i \leq j$	
A: **do** $i \neq N$ $\rightarrow$ **if** $i < j \rightarrow$ skip **fi** ; $\{i < j\}$ $x.i := 0$; $\{x.i = 0\}\{i < j\}$ $i := i + 1$ **od** $\{i = N\}$	B: **do** $j \neq N$ $\rightarrow$ $\{i \leq j\}$ $x.j := 1$; $j := j + 1$ **od**
Post: $\langle \forall k : 0 \leq k < N : x.k = 0 \rangle$	

Version 4

Our final task is to show termination of the multiprogram. Component B certainly terminates, in a state satisfying $j = N$. Because $i < N$ is a correct precondition for A's guarded skip, this guard is true for $j = N$. Hence, A terminates as well.

*　　　　　　*

*

Herewith we conclude the formal development of an—admittedly—very simple multiprogram. Our main purpose was

- to show, in very small steps, how such a development process evolves.
- to show how the validity of such a process is carried by the theory of Owicki and Gries.
- to exhibit the similarity with "stepwise refinement" and the induced benefits of stimulating separation of concerns.

- to introduce the notational and clerical aids in organizing such a development.

A few final remarks are in order.

- Version 4 not only presents the final solution, but it also does so in a fully documented way, in the sense that someone who embarks on an a posteriori proof of correctness with the rules from the Core of the Owicki-Gries theory, can successfully complete that with just the supplied annotation. The traditional way of giving a posteriori proofs is that one starts with the "raw code" and then *invents* the proper annotation. This is in general quite a nasty task and usually requires firm operational understanding of the multiprogram. Or to quote Leslie Lamport: "Finding the proper annotation to prove a property of a concurrent program is a difficult art; anything that makes the task easier should be welcome." [Lam88]. Our answer to Lamport's quest is to ignore this problem altogether, by not accepting the very idea of a posteriori verification, and by focussing on the construction of (correct) programs instead. Because program construction is a goal directed activity, a development will usually only call for what is needed for reaching the goal, and for nothing else. Version 4 exhibits just that. Typically, a posteriori correctness proofs of multiprograms are far too lavishly annotated, often with the aid of control predicates *at*, thus adding complication. And besides all this, one always wonders where the programs come from in the first place, if they are not developed.

- In the presentation given here, we proceeded from the one version to the next in *very* small steps. We did so in order to explain the rules of the game. Once one has grown a little more familiar with the game, one is likely to combine several refinement steps, as we, for instance, did in moving from Version 3 to Version 4. Combining refinement steps surely shortens the presentation (on paper), but it may evoke oversights or mistakes as well. It is always a matter of good taste how coarse- or how fine-grained one chooses one's steps. There are a number of good rules-of-thumb for this, which we do not elaborate on here.

Postscript (for the circuit designer)

In our solution—Version 4—the synchronization runs under control of the originally given integers i and j. In particular—see the guarded skip in component A—only the difference $j - i$ matters. In the algorithm as is this difference is an $(N + 1)$-valued entity—$0 \leq j - i \leq N$ is a system invariant—. For circuit design, it could be advantageous if $j - i$ were just 2-valued, because that would enable a transformation of the algorithm into the boolean domain. For this reason we shall now constrain the computation proper by strengthening system invariant $i \leq j$ into

$$Q: \quad i \leq j \ \wedge \ j \leq i+1 \quad .$$

This has an impact on the code of component B, which increments j. We give the adjusted B at once, leaving to the reader to check that relation Q has become a system invariant.

Pre: $i=0 \ \wedge \ j=0$	
Inv: $Q: \ i \leq j \ \wedge \ j \leq i+1$	
A: **do** $i \neq N$	B: **do** $j \neq N$
$\rightarrow$ **if** $i < j \rightarrow$ skip **fi**	$\rightarrow$ **if** $j \leq i \rightarrow$ skip **fi**
; $x.i := 0$	; $x.j := 1$
; $\{i < j\}$	; $\{j \leq i\}$
$i := i+1$	$j := j+1$
od	**od**

For reasons to become clear shortly, B has been given the same syntactic structure as A has.

By inserting a guarded skip in B, we hamper progress. Does the multiprogram still terminate? It does, because

- at least one of the guards of the two guarded skips is true, so that
- $i+j$ increases, and since
- $i \leq j$ is a system invariant, B will terminate in a state satisfying $j = N$, so that,
- as before, A will terminate as well.

Now we transform the program according to the coordinate transformation

$$c \ \equiv \ (i = j)$$

$$\neg c \ \equiv \ (i+1 = j) \quad .$$

Then we can eliminate the guards thanks to

$$i < j \ \equiv \ \neg c \quad \text{and} \quad j \leq i \ \equiv \ c \quad .$$

And statement $\{i < j\} \ i := i+1$ gets replaced by

$$\{\neg c\} \ \ c, i := \neg c, i+1$$

or—using preassertion $\neg c$—by

$$c, i := \text{true}, i+1 \quad .$$

The (raw) program code thus becomes

Pre: $i = 0 \wedge j = 0 \wedge c$	
A: **do** $i \neq N$ $\rightarrow$ **if** $\neg c \rightarrow$ skip **fi** ; $x.i := 0$; $c, i :=$ true$, i + 1$ **od**	B: **do** $j \neq N$ $\rightarrow$ **if** $c \rightarrow$ skip **fi** ; $x.j := 1$; $c, j :=$ false$, j + 1$ **od**
Post: $\vec{x} = \vec{0}$	

Now the situation has become so symmetric that if we wish the above machinery to establish postcondition $\vec{x} = \vec{1}$, we only need to flip c's initial value. And postcondition $\vec{x} = \vec{0} \vee \vec{x} = \vec{1}$ is even guaranteed if c's initial value is left unspecified.
End Postscript.

* *
*

3 The problem of Liberal Phase Synchronization

The problem of Liberal Phase Synchronization is a—not too minor—variation of a problem that we learnt from J. Misra [Mis91]. It is as follows. We consider a multiprogram consisting of a nonempty set of components of the form

$$\text{Comp}.p: \quad *[\ S.p\ ] \quad .$$

We assume that all successive invocations of $S.p$—the *phases* of Comp.p—terminate, for all p. The problem is to synchronize the components in such a way that no component is "too far ahead" of the others in the number of phases completed. Moreover, we will develop a solution in which the components are arranged in the nodes of a given rooted tree, in which the communication facilities are restricted to communication with neighbouring nodes only.

3.0 A formal specification

In order to make the specification more precise, we introduce *fresh* variables—one per component—that keep track of the number of phases completed, i.e. we have

$$\begin{array}{ll} \text{Pre}: & \langle \forall q :: x.q = 0 \rangle \\ \text{Comp}.p: & *[\ S.p\ ;\ x.p := 1 + x.p\ ] \quad . \end{array}$$

To express what we mean by "not too far ahead", we introduce natural constant M, $0 \leq M$, in terms of which the precise specification of the problem reads

Pre:	$\langle \forall q :: x.q = 0 \rangle$
Comp.p:	$*[\ \{?\ Q.p : \langle \forall q :: x.p \leq M + x.q \rangle\}$ $S.p$ $;\ x.p := 1 + x.p$ $]$

Version 0

Our task is to add synchronization code to the components so as to establish the correctness of the assertions $Q.p$.

Remarks

- The variables $x.p$ have been introduced for the purpose of specification and, as such, they belong to the computation proper. Therefore the synchronization code to be added should not change any of these variables.
- The specification does not express that in our ultimate solution the components are located in the nodes of the given tree. We do justice to this constraint as we proceed.

End Remarks.

3.1 Developing a solution

We assume the components to be arranged in a network, in which only neighbours can communicate with each other, i.e. can inspect each other's variables. If every pair of components were directly connected, our task would be simple: in order to make assertion $Q.p$ in Comp.p locally correct, we would prefix it with guarded skip **if** $Q.p \rightarrow$ skip **fi** ; by Widening, global correctness is for free.

We want to make do, however, with a sparser network, viz. a (rooted) tree. To that end we note that

$$Q.p \equiv x.p \leq M + \langle \downarrow q :: x.q \rangle \text{ [1]} \quad ,$$

and we decide to choose one fixed component R—located in the root of the tree—to always hold the minimal x-value, i.e. we require the system invariance of

$$P0: \quad x.R = \langle \downarrow q :: x.q \rangle \quad .$$

[1] $\downarrow$ denotes the minimum

Then target assertion $Q.p$ can be rewritten as

$$Q.p \equiv x.p \le M + x.R \quad .$$

Summarizing, the next version of our multiprogram becomes

Pre:	$\langle \forall q :: x.q = 0 \rangle$
Inv:	$?\ P0: \quad x.R = \langle \downarrow q :: x.q \rangle$
Comp.p:	$*[\ \{?\ x.p \le M + x.R\}$ $\quad S.p$ $;\ x.p := 1 + x.p$ $]$

Version 1

*　　　　*
*

If component R were directly connected to every other component—like, e.g., in a bus-like architecture—, our task would again be simple: the target assertion in Comp.p would be established by guarded skip **if** $x.p \le M + x.R$ $\rightarrow$ skip **fi** , and invariant $P0$—which equivales $x.R \le \langle \downarrow q :: x.q \rangle$—would be maintained by prefixing $x.R := 1 + x.R$ in Comp.R with guarded skip **if** $1 + x.R \le \langle \downarrow q : q \ne R : x.q \rangle \rightarrow$ skip **fi** .

We want to make do, however, with an arbitrary rooted tree, in which not all components are directly connected to root R.

In order to deal with target assertion $x.p \le M + x.R$, we somehow have to "separate" the "local" $x.p$ from the "remote" $x.R$. We do so by means of a set of fresh variables $y.q$—one for each component—, by splitting M into naturals K and L, i.e. $M = K + L$, and by strengthening the target assertion as follows:

$$x.p \le M + x.R$$
$$\Leftarrow$$
$$x.p \le K + y.p \ \wedge\ y.p \le y.R \ \wedge\ y.R \le L + x.R \quad .$$

Now, the first conjunct will replace the target assertion and the other two conjuncts we decide to be system invariants, i.e. we shall maintain

$P1: \quad y.R = \langle \uparrow q :: y.q \rangle$ and
$P2: \quad y.R \le L + x.R$.

Remark Our only reason for splitting "slack" M at Comp.p into K and L is that we do not have a good argument for choosing, in the strengthening above, between $K = 0$ and $L = 0$.
End Remark.

Before we summarize these design decisions in our next version, there are two more things to be dealt with. First, the above strengthening is superfluous for Comp.R, because for $p = R$ the original target assertion $x.p \leq M + x.R$ is vacuously true—$0 \leq M$—. Second, lest the new target assertions $x.p \leq K + y.p$ will become stably false, y-values will have to be increased somewhere. We now decide to equip each component with a rather independent co-component to take care of the orderly increments of the y's. Thus, our next version is

```
Pre:  ⟨∀q :: x.q = 0⟩
Inv:  ? P0 :  x.R = ⟨↓ q :: x.q⟩ ,  Note 2
      ? P1 :  y.R = ⟨↑ q :: y.q⟩ ,  Note 3
      ? P2 :  y.R ≤ L + x.R ,  Note 1
-----------------------------------------------------------
Node.p  (p ≠ R)
  Comp.p:  *[ {? x.p ≤ K + y.p ,  Note 0}{x.p ≤ M + x.R}
              S.p
            ; x.p := 1 + x.p
            ]
  Co.p:    *[ y.p := 1 + y.p]
- - - - - - - - - - - - - - - - - - - - - - - - - - - - - -
Node.R
  Comp.R:  *[ S.R
            ; x.R := 1 + x.R
            ]
  Co.R:    *[ y.R := 1 + y.R]
```

Version 2

*   *
*

And now we can complete the development. We do so according to the scheme explained in the previous section on Concurrent Vector Writing.

Note 0 $x.p \leq K + y.p$

L: Prefix the assertion with guarded skip
if $x.p \leq K + y.p \rightarrow$ skip **fi**

G: Widening

End Note 0.

Note 1 Inv $P2$: $y.R \leq L + x.R$

Init: Add $\langle \forall q :: y.q = 0 \rangle$ to Pre.

- Demand $\{?\ 1 + y.R \leq L + x.R\}$ $y.R := 1 + y.R$, and we handle this new assertion by

  L: Prefix $y.R := 1 + y.R$ with
  if $1 + y.R \leq L + x.R \rightarrow$ skip **fi**

  G: Widening

End Note 1.

Note 2 Inv $P0$: $x.R = \langle \downarrow q :: x.q \rangle$

We satisfy $P0$ by seeing to it that each path from a node to the root exhibits a descending sequence of x-values. Thus the x-values in the tree form a "downheap". Formally, with $f.q$ denoting the father of node q (for $q \neq R$) we satisfy $P0$ by demanding

Inv: ? $P3$: $\langle \forall q : q \neq R : x.(f.q) \leq x.q \rangle$, Note 4

End Note 2.

Note 3 Inv $P1$: $y.R = \langle \uparrow q :: y.q \rangle$

We satisfy $P1$ by seeing to it that the y-values in the tree form an "upheap", i.e. by demanding

Inv: ? $P4$: $\langle \forall q : q \neq R : y.q \leq y.(f.q) \rangle$, Note 5

End Note 3.

Note 4 Inv $P3$: $\langle \forall q : q \neq R : x.(f.q) \leq x.q \rangle$

Init: From Pre

- $P3$ can be falsified by increments of $x.p$ (both for $p \neq R$ and $p = R$). The weakest additional precondition for $x.p := 1 + x.p$ not to falsify $P3$ is
  $\langle \forall q : q \neq R \ \wedge \ p = f.q : 1 + x.(f.q) \leq x.q \rangle$.
  Since $p = f.q \ \Rightarrow \ q \neq R$, this can be simplified to
  $\langle \forall q : p = f.q : 1 + x.p \leq x.q \rangle$.
  We can now establish this demanded precondition of $x.p := 1 + x.p$ by a guarded skip, since only inspections of neighbours are involved. Global correctness of this precondition follows from Widening.

End Note 4.

Note 5 Inv $P4$: $\langle \forall q : q \neq R : y.q \leq y.(f.q) \rangle$

Init: We added $\langle \forall q :: y.q = 0 \rangle$ to Pre

- $P4$ can be falsified by increments of $y.p$ in Co.p, for $p \neq R$. The weakest additional precondition for $y.p := 1 + y.p$ not to falsify $P4$ is
$$\langle \forall q : q \neq R \ \wedge \ q = p : 1 + y.q \leq y.(f.q) \rangle \quad ,$$
which by $p \neq R$ equivales
$$1 + y.p \leq y.(f.p) \quad ;$$
again this condition—being globally correct—is dealt with by a guarded skip.

End Note 5.

Summarizing, our next and final version of the multiprogram is

```
Pre:  ⟨∀q :: x.q = 0⟩ ∧ ⟨∀q :: y.q = 0⟩   .
Inv:  P0 :  x.R = ⟨↓ q :: x.q⟩   ,
      P1 :  y.R = ⟨↑ q :: y.q⟩   ,
      P2 :  y.R ≤ L + x.R   ,
      P3 :  ⟨∀q : q ≠ R : x.(f.q) ≤ x.q⟩   ,
      P4 :  ⟨∀q : q ≠ R : y.q ≤ y.(f.q)⟩   .
------------------------------------------------------------
Node.p   (p ≠ R)
  Comp.p:   * [ if x.p ≤ K + y.p → skip fi
              ; {x.p ≤ K + y.p}{x.p ≤ M + x.R}
                S.p
              ; if ⟨∀q : p = f.q : 1 + x.p ≤ x.q⟩ → skip fi
              ; x.p := 1 + x.p
              ]
  Co.p:     * [ if 1 + y.p ≤ y.(f.p) → skip fi
              ; y.p := 1 + y.p
              ]
- - - - - - - - - - - - - - - - - - - - - - - - - - - - - -
Node.R
  Comp.R:   * [ S.R
              ; if ⟨∀q : R = f.q : 1 + x.R ≤ x.q⟩ → skip fi
              ; x.R := 1 + x.R
              ]
  Co.R:     * [ if 1 + y.R ≤ L + x.R → skip fi
              ; y.R := 1 + y.R
              ]
```

Version 3

3.2 Progress

We shall prove progress by, first, proving absence of total deadlock, and, second, proving that each individual component engaged in the computation proper makes progress.

3.2.0 No total deadlock

We prove the absence of the danger of total deadlock by assuming that all components are blocked in a guarded skip and then proving the validity of false. Because most of the components Comp.p have two guarded skips, we have to distinguish between being blocked in the one and being blocked in the other. By definition, Comp.p is

- white: if blocked in **if** $x.p \leq K + y.p \rightarrow$ skip **fi**
  (absent in root R)
- black: if blocked in **if** $\langle \forall q : p = f.q : 1 + x.p \leq x.q \rightarrow$ skip **fi**
  (impossible in the leaves, because the range of the universal quantification is empty)

Thus, if all components are blocked, Comp.R is black and the leaf Comp.p's are white. Then there exists a p such that Comp.p is black and each of its children is white. Let Comp.p be such a component. Now, if all components are blocked in a guarded skip, all the blocking guards are (stably) false. In particular, we have (the stable validity of)

(i) $\langle \exists q : p = f.q : 1 + x.p > x.q \rangle$ —Comp.p (black) is blocked—

(ii) $\langle \forall q : p = f.q : x.q > K + y.q \rangle$ —p's (white) children are blocked—

(iii) $\langle \forall q : q \neq R : 1 + y.q > y.(f.q) \rangle$ —all Co.q ($q \neq R$) are blocked—

(iv) $1 + y.R > L + x.R$ —Co.R is blocked—

Now, let q be a witness of (i), i.e. let

(v) $p = f.q$, and

(vi) $x.p \geq x.q$.

We then observe:

$$
\begin{array}{ll}
 & x.p \\
\geq & \{\text{(vi)}\} \\
 & x.q \\
\geq & \{\text{(ii), using (v)}\} \\
 & 1 + K + y.q \\
\geq & \{\text{(iii), using that (v) implies } q \neq R\} \\
 & 1 + K + y.(f.q)
\end{array}
$$

$$
\begin{array}{ll}
= & \{(\mathrm{v})\} \\
 & 1 + K + y.p \quad .
\end{array}
$$

Now we have to distinguish $p = R$ and $p \neq R$:

$\underline{p = R}$

$$
\begin{array}{ll}
 & 1 + K + y.p \\
\geq & \{(\mathrm{iv}),\ \text{using } p = R\} \\
 & 1 + K + L + x.p \\
\geq & \{0 \leq K + L\} \\
 & 1 + x.p
\end{array}
$$

$\underline{p \neq R}$

$$
\begin{array}{ll}
 & 1 + K + y.p \\
\geq & \{\text{precondition of blocking statement, using } p \neq R\text{—see Version 3—}\} \\
 & 1 + x.p \quad .
\end{array}
$$

Thus we have established $x.p \geq 1 + x.p$, thereby establishing the absence of the danger of total deadlock.

3.2.1 A multibound

From the very specification of our problem—Version 0—we already see that each solution will inevitably induce system invariant

$$MB: \quad \langle \forall p, q :: x.p \leq 1 + M + x.q \rangle \quad .$$

This is a perfect multibound for the Comp.p's, i.e. for precisely those components that increase the $x.p$'s. Because it is also these components that make up the computation proper—the $S.p$'s—it suffices to show that each of them makes progress. To that end we observe

$$
\begin{array}{lll}
 & \text{some Comp.}p \text{ gets stuck} & \\
\Rightarrow & \{MB\} & \\
 & \text{each Comp.}p \text{ gets stuck} & (*) \\
\Rightarrow & \{\text{instantiate } p := R\} & \\
 & \text{Comp.}R \text{ gets stuck} & \\
\Rightarrow & \{\text{inv} \quad P2: \quad y.R \leq L + x.R\} & \\
 & \text{Co.}R \text{ gets stuck} & \\
\Rightarrow & \{\text{inv} \quad P1: \quad y.R = \langle \uparrow q :: y.q \rangle\} & \\
 & \text{each Co.}p \text{ gets stuck} & (*)
\end{array}
$$

and we conclude—see the lines marked $(*)$—

$$
\begin{array}{l}
\text{some Comp.}p \text{ gets stuck} \\
\Rightarrow \\
\text{total deadlock.}
\end{array}
$$

Since we have shown the absence of the danger of total deadlock, we conclude that no Comp.p gets stuck, as demanded.

* *

*

And this completes the development of the algorithm.

3.3 Final remarks

- Our final solution—Version 3—contains the rather coarse-grained guarded skip **if** $\langle \forall q : p = f.q : 1 + x.p \leq x.q \rangle \to$ skip **fi** . Because each of the individual conjuncts of its guard is globally correct (Widening), the Guard Conjunction Lemma tells us that these conjuncts may be evaluated one after the other, and in any order, thus yielding a much finer-grained solution.
- The algorithm uses ever-growing integers x and y, but, as can be seen from the guards, only differences between these variables matter. It can be shown that all differences are bounded from above and can thus be replaced by finite-state variables. We did not want to carry out this coordinate transformation here.
- Both the problem and the derivation are a variation on [BF97], where $M = 0$, in which case the co-components need not see the light. It is worth observing, however, that the introduction of co-components has led to a smoother development of the algorithm, but at the same time to a slightly more complicated progress argument.
- We are convinced that all possible behaviours of the computation proper that satisfy the specification, can be exhibited by the algorithm. In other words, we are convinced that our synchronization code does not constrain the potential parallelism of the computation proper.

* *

*

4 Acknowledgements

We are very much indebted to Prof. Dr. Edsger W. Dijkstra for giving us the opportunity to lecture on Multiprogramming during the 1996 International Summer School at Marktoberdorf, by granting us his slots.

We are also indebted to the Board of Directors for letting us in, and to Dr. Birgit Schieder for her gracious offer to have this text LaTeX-ed for us.

On the scientific level we owe a lot to the (other) members of the Eindhoven Tuesday Afternoon Club, who have always been willing to confront us with their comments and criticisms. But most of all we are grateful to our former student Daniela Buhăceanu. Without her, we would never have written the section on Liberal Phase Synchronization.

Eindhoven, 20 December 1996

W. H. J. Feijen and A. J. M. van Gasteren

References

[AO91] Krzysztof R. Apt and Ernst-Rüdiger Olderog. *Verification of Sequential and Concurrent Programs.* Springer-Verlag, Berlin, 1991.

[BF97] Daniela Buhăceanu and W. H. J. Feijen. Formal Derivation of an Algorithm for Distributed Phase Synchronization. *Information Processing Letters.* To appear.

[Dij76] Edsger W. Dijkstra. *A Discipline of Programming.* Prentice-Hall, Englewood Cliffs, N. J., 1976.

[Dij82] Edsger W. Dijkstra. A Personal Summary of the Gries-Owicki Theory. In *Selected Writings on Computing: A Personal Perspective.* Springer-Verlag, New York, 1982.

[Fe87] W. H. J. Feijen. Some correctness proofs for the Safe Sluice. Internal Memorandum, Department of Mathematics and Computing Science, Eindhoven University of Technology, September 1987.

[Gri81] David Gries. *The Science of Programming.* Springer-Verlag, New York, 1981.

[Hoa69] C. A. R. Hoare. An Axiomatic Basis for Computer Programming. *Communications of the ACM,* 12(10): 576–580 and 583, October 1969.

[Hoom93] J. Hooman. Properties of Program Transformations. Technical Note, Department of Mathematics and Computing Science, Eindhoven University of Technology, 1993.

[Lam88] Leslie Lamport. Control Predicates are Better than Dummy Variables for Reasoning about Program Control. *ACM Transactions on Programming Languages and Systems,* 10(2): 267–281, April 1988.

[Mis91] Jayadev Misra. Phase Synchronization. *Information Processing Letters,* 38: 101–105, 1991.

[OG76] S. Owicki and D. Gries. An Axiomatic Proof Technique for Parallel Programs I. *Acta Informatica,* 6: 319–340, 1976.

Foundations for Calculational Logic [1]

David Gries
Computer Science, Cornell University
Ithaca, New York, 14853 USA

With help from Fred B. Schneider

Abstract: For the past 20 years, researchers in the formal development of algorithms —programming methodology— have been developing a *calculational approach* to logic, in which a calculational format using substitution of equals for equals, weakening, and strengthening are the major tools. This work is giving us a perspective on logic that is different from that of others: for us, logic is a tool instead of just an area of academic study. These lectures summarize our present understanding of this calculational logic. Covered are: propositional logic, predicate logic, treating undefined terms, the use of weakening or strengthening steps, a useful theorem on monotonicity, metatheorem Witness, and the everywhere operator. Some interesting proofs are developed.

Contents

[1] This work was supported by the NSF under grants CDA-9406743 and CCR-9503319.

Acknowledgements

These notes are culled from various places:

(a) Text [14] by myself and F.B. Schneider, as well as the second edition, which is under preparation.
(b) Articles by myself and Schneider, most notably [15] and [16].
(c) A few technical reports by myself.
(d) Article [11] by Rutger Dijkstra.

My debt to Schneider is obvious. I am also indebted to Edsger W. Dijkstra, Wim Feijen, Netty van Gasteren, and other colleagues for their work in this area and their support over many, many years.

1 Propositional logic

1.1 Basic concepts of propositional logic

The language of propositional logic

Let VP be a set of propositional variables. Throughout, lower-case letters $p, q, r, \ldots$ are any variables in VP, and upper-case letters $P, Q, R, \ldots$ are *metavariables*, which stand for formulas. A formula that contains a metavariable is called a formula *schema*; a formula schema stands for the set of formulas constructed by replacing its metavariables by formulas:

(1) **Definition of uniform substitution.** $E[V := G]$ denotes the uniform replacement of all occurrences of metavariable V in formula schema E by formula (schema) G. $E[V := G]$ is called *uniform substitution*, and the resulting formula (schema) is called an *instance of* E.[2].

[2] We extend uniform substitution in the usual manner to simultaneous uniform substitution by letting V be a list of metavariables and G be a list of corresponding formulas (schemas).

A (propositional) *formula* has one of the following forms.

$$(2)\qquad \begin{array}{cccc} p & \mathit{false} & \mathit{true} & \\ (P) & \neg P & (P \vee Q) & (P \wedge Q) \\ (P \equiv Q) & (P = Q) & (P \Rightarrow Q) & (P \Leftarrow Q) \\ (P \not\equiv Q) & (P \neq Q) & (P \not\Rightarrow Q) & (P \not\Leftarrow Q) \end{array}$$

Formally, all binary operations are fully parenthesized, which ensures unique parsing and simplifies some later definitions and analyses. However, as usual, precedence rules allow the elimination of many parentheses. Prefix operator $\neg$ binds tightest, then $=$ and $\neq$, then $\vee$ and $\wedge$, then $\Rightarrow$ and $\Leftarrow$ and $\not\Rightarrow$ and $\not\Leftarrow$, and finally $\equiv$ and $\not\equiv$. Operators $\vee$, $\wedge$, $\equiv$, $\not\equiv$, $\Leftarrow$, and $\not\Leftarrow$ are left associative. Operators $\neg$, $\Rightarrow$, and $\not\Rightarrow$ are right associative.

We use two symbols for equality, each having different calculational attributes: $=$ and $\equiv$. We regard $=$ as *conjunctional*: $p = q = r$ is shorthand for $p = q \ \wedge \ q = r$. Operation $\equiv$, on the other hand, is used associatively [3]: $p \equiv q \equiv r$ is equivalent to $(p \equiv q) \equiv r$ and to $p \equiv (q \equiv r)$. The implicit use of associativity (and symmetry) of $\equiv$ can simplify manipulations quite a bit, just as the implicit use of associativity and symmetry of $+$ simplifies numerical calculations.

Schematic logic P_1

A *logic* consists of *axioms* and *inference rules*. An axiom is simply a formula that is stated to be a *theorem*, without proof; another word for axiom is *postulate*. An inference rule has the following form.

$$\vdash P_1, \ldots, \vdash P_n \quad \longrightarrow \ \vdash Q$$

Formulas P_i are called the *premises* of the inference rule; Q, its *conclusion.* [4] The notation $\vdash P$ denotes "P is a theorem", and the inference rule presented above has the meaning: If P_1, ..., P_n are theorems, then Q is a theorem. Sometimes, when several different logics are being discussed, we indicate the particular logic being used with a subscript, as in $\vdash_{P_1} P$.

A *theorem* of a logic is either : (i) an axiom or (ii) the conclusion of an inference rule whose premises are theorems. Note that this is a recursive definition.

Table 1 gives a schematic presentation of propositional logic P_1, due to Church [5]. P_1 consists of one inference-rule schema and three axiom schemas. The inference-rule schema denotes the infinite set of inference rules obtained by replacing metavariables P and Q by formulas, using uniform substitution. (Similarly for the axiom schemas.)

P_1 defines a logic with formulas that contain only variables, *false*, and implication $\Rightarrow$. Other propositional operators are viewed as abbreviations (given in Table 2). This view of

[3] The fact that $\equiv$ is associative was known in the early 1920's. In his doctoral thesis, Alfred Tarski says that he learned of it from Jan Lukasiewicz —see [25]. However, this associativity has not been used much in the past. For example, Rosser [23] uses equivalence only conjunctionally.

[4] Metasymbol $\vdash$ has lowest precedence; it applies to the longest formula that follows it.

Table 1: Schematic logic P_1

Modus Ponens: $\vdash P, \vdash P \Rightarrow Q \quad \longrightarrow \vdash Q$
Axiom schema A0: $P \Rightarrow (Q \Rightarrow P)$
Axiom schema A1: $(R \Rightarrow (P \Rightarrow Q)) \Rightarrow ((R \Rightarrow P) \Rightarrow (R \Rightarrow Q))$
Axiom schema A2: $((P \Rightarrow \mathit{false}) \Rightarrow \mathit{false}) \Rightarrow P$

Table 2: Abbreviations used in Schematic Logic P_1

$\mathbf{P_1}$ **Def. of** *true*:	$\mathit{false} \Rightarrow \mathit{false}$
$\mathbf{P_1}$ **Def. of** $\neg P$:	$P \Rightarrow \mathit{false}$
$\mathbf{P_1}$ **Def. of** $P \not\Leftarrow Q$:	$\neg(Q \Rightarrow P)$
$\mathbf{P_1}$ **Def. of** $P \vee Q$:	$(P \Rightarrow Q) \Rightarrow Q$
$\mathbf{P_1}$ **Def. of** $P \wedge Q$:	$(P \not\Leftarrow Q) \not\Leftarrow Q$
$\mathbf{P_1}$ **Def. of** $P \equiv Q$:	$(P \Rightarrow Q) \wedge (Q \Rightarrow P)$
$\mathbf{P_1}$ **Def. of** $P \not\equiv Q$:	$(P \not\Leftarrow Q) \vee (Q \not\Leftarrow P)$
$\mathbf{P_1}$ **Def. of** $P \Leftarrow Q$:	$Q \Rightarrow P$
$\mathbf{P_1}$ **Def. of** $P \not\Rightarrow Q$:	$Q \not\Leftarrow P$

logic, in which the number of inference rules and axioms are kept to a minimum, reduces the amount of work needed in analyzing various aspects of the language and logic.

Given the definition of a theorem, a proof of P (i.e. a proof that formula P is a theorem) must either (i) indicate which axiom it is or (ii) indicate an inference rule that has P as its conclusion and premises that are theorems. This leads naturally to the "Hilbert style of proof". Such a proof consists of a sequence of lines, each of which contains a formula and an indication of why it is a theorem. The last line contains the theorem to be proved. For example, here is a proof of $p \Rightarrow p$.

1.	$p \Rightarrow ((q \Rightarrow p) \Rightarrow p)) \Rightarrow ((p \Rightarrow (q \Rightarrow p)) \Rightarrow (p \Rightarrow p))$	
		A1, with $P, Q, R := q \Rightarrow p, p, p$
2.	$p \Rightarrow ((q \Rightarrow p) \Rightarrow p)$	A0, with $P, Q := p, q \Rightarrow p$
3.	$(p \Rightarrow (q \Rightarrow p)) \Rightarrow (p \Rightarrow p)$	Modus Ponens, 2, 1
4.	$p \Rightarrow (q \Rightarrow p)$	A0, with $P, Q := p, q$
5.	$p \Rightarrow p$	Modus Ponens, 4, 3

In the proof, the terminology "P, with $Q, R := S, T$" is longhand for the uniform substitution $P[Q, R := S, T]$. As an example, line 4 of the proof is Axiom A0, $P \Rightarrow (Q \Rightarrow P)$, with every occurrence of P and Q replaced (simultaneously) by p and q, i.e. $p \Rightarrow (q \Rightarrow p)$.

The proof of $p \Rightarrow p$ given above is not obvious. Few people could reconstruct it without studying it for a while. This proof style appears to be inherently opaque. No one has developed strategies for developing such proofs, and few people actually write such proofs —it is just too hard to use for practical inference. The proof style is (almost) of academic interest only. This is perhaps why logicians have never really used logic, but just studied it. Later, we will present a more useful, calculational, proof style. Nevertheless, this logic and proof are useful as examples to explain the basic concepts of logic.

Schematic proofs

The proof given above proves one formula, $p \Rightarrow p$. If we want to know that $s \Rightarrow s$ is a theorem as well, we could carry out another, almost identical, proof. Instead, we can give the same proof but in a schematic style, in which all variables used are metavariables. Below, we prove $P \Rightarrow P$. It is easy to see that in the schematic proof one could substitute any formula for P, so this proof establishes $p \Rightarrow p$, $s \Rightarrow s$, $(p \Rightarrow s) \Rightarrow (p \Rightarrow s)$, etc.

1. $P \Rightarrow ((Q \Rightarrow P) \Rightarrow P)) \Rightarrow ((P \Rightarrow (Q \Rightarrow P)) \Rightarrow (P \Rightarrow P))$
   A1, with $P, Q, R := Q \Rightarrow P, P, P$
2. $P \Rightarrow ((Q \Rightarrow P) \Rightarrow P)$ — A0, with $P, Q := P, Q \Rightarrow P$
3. $(P \Rightarrow (Q \Rightarrow P)) \Rightarrow (P \Rightarrow P)$ — Modus Ponens, 2, 1
4. $P \Rightarrow (Q \Rightarrow P)$ — A0
5. $P \Rightarrow P$ — Modus Ponens, 4, 3

In general, then, we can use schematic proofs of schematic formulas. We prove formulas with metavariables in them, for we know that we can instantiate the metavariables of the proof with concrete formulas of our choice to arrive at a theorem that is an instantiation of the schematic theorem.

Formula $p \Rightarrow p$ is a theorem. Strictly speaking, $P \Rightarrow P$ is not a theorem, but a "theorem schema". But any formula obtained from $P \Rightarrow P$ by uniform substitution, in other words any *instantiation* of $P \Rightarrow P$, is a theorem. We therefore relax our terminology and call $P \Rightarrow P$ a theorem, just as we call axiom A0, $P \Rightarrow (Q \Rightarrow P)$, an axiom instead of an axiom schema.

Logic P_1 with a finite number of axioms

Schematic logic P_1 of Table 2 has an infinite number of axioms, which are given by a finite number of axiom schemas. It is possible to axiomatize the same logic with a finite number of axioms. This new axiomatization, given in Table 3, can be compared to Schematic P_1 as follows. The axioms of Table 3 are those of Schematic P_1, but with all metavariables replaced by propositional variables. Inference rule Modus ponens appears, without change. Finally, inference rule Uniform Substitution is introduced; this rule indicates that if formula P is a theorem, then so is any instance of P.

For example, here is a proof of $p \Rightarrow p$, using this new axiomatization.

1. $p \Rightarrow ((q \Rightarrow p) \Rightarrow p)) \Rightarrow ((p \Rightarrow (q \Rightarrow p)) \Rightarrow (p \Rightarrow p))$
   Uniform Subs., A1 (with $p, q, r := q \Rightarrow p, p, p$)
2. $p \Rightarrow ((q \Rightarrow p) \Rightarrow p)$ — Uniform Subs., A0 (with $q := q \Rightarrow p$)
3. $(p \Rightarrow (q \Rightarrow p)) \Rightarrow (p \Rightarrow p)$ — Modus Ponens, 2, 1

Table 3: Logic P_1

Modus Ponens: $\vdash P, \vdash P \Rightarrow Q \longrightarrow \vdash Q$
Uniform Substitution: $\vdash P \longrightarrow \vdash P[\beta := Q]$
(β a list of formula variables)
Axiom schema A0: $p \Rightarrow (q \Rightarrow p)$
Axiom schema A1: $(r \Rightarrow (p \Rightarrow q)) \Rightarrow ((r \Rightarrow p) \Rightarrow (r \Rightarrow q))$
Axiom schema A2: $((p \Rightarrow \mathit{false}) \Rightarrow \mathit{false}) \Rightarrow p$

4. $p \Rightarrow (q \Rightarrow p)$	A0
5. $p \Rightarrow p$	Modus Ponens, 4, 3

At first sight, it may seem that the two methods of axiomatizing a logic are equivalent —one should always be able to change a schematic presentation into one with a finite number of axioms by changing the metavariables of the axioms to propositional variables and adding inference rule Uniform Substitution. However, in Sec. 5.4 we will see that the two methods are not equivalent in power. There is a useful logic with an infinite number of axioms for which this approach does not work.

It is rarely made clear in texts, especially computer science texts that teach logic and discrete math, which kind of axiomatization is being used. For us, the distinction between the two kinds has become clear only in the past several years. For the rest of these notes, we use the schematic approach, for two reasons: (i) It is almost mandatory when one introduces predicate logic (quantification), and (ii) when dealing with the everywhere operator (which we do later), the finite-number-of-axioms approach becomes unwieldy.

The model for propositional calculus

The standard model of propositional logic consists of the set of all total functions (called *states*) $VP \rightarrow \{\mathbf{t}, \mathbf{f}\}$, along with rules for evaluating a formula in a state. For a state s, the value $s[\![P]\!]$ of an expression P in state s is given by:[5]

(3) $s[\![\mathit{false}]\!] = \mathbf{f}$

(4) $s[\![x]\!] = s.x$ (for variable x, $s.x$ denotes the value of x in state s)

(5) $s[\![P \Rightarrow Q]\!] = \mathbf{if}\ s[\![P]\!] = \mathbf{f}\ \mathbf{then}\ \mathbf{t}\ \mathbf{else}\ [\![Q]\!]$

A formula P is *satisfiable* iff it evaluates to $\mathbf{t}$ in at least one state (i.e. there is a state s in which $s[\![P]\!] = \mathbf{t}$); otherwise, it is *unsatisfiable*. A formula P is *valid*, written $\models P$, iff it evaluates to $\mathbf{t}$ in all states.

A logic is *sound* with respect to a model iff every theorem (of the logic) is valid (in the model). A logic is *complete* with respect to a model iff every valid formula (of the model) is a theorem (of the logic).

[5] Since P_1 formally has only operator $\Rightarrow$ and constant *false*, we need not define evaluation of the other operators and of *true*; these can be determined from the abbreviations defined in Table 2.

Table 4: Propositional logic **E**: inference rules

(6)	**Leibniz:**	$\vdash P = Q \longrightarrow \vdash E[V := P] = E[V := Q]$ (V a list of metavariables)
(7)	**Transitivity:**	$\vdash P = Q, \vdash Q = R \longrightarrow \vdash P = R$
(8)	**Equanimity:**	$\vdash P, \vdash P \equiv Q \longrightarrow \vdash Q$

Logic P_1 is known to be sound and complete with respect to standard model (3)–(5) of evaluation.

1.2 Calculational propositional logic E

We have already seen two different but equivalent propositional logics. We now give a third, logic **E**. **E** has a different set of inference rules and axioms, but it has the same set of theorems. **E** is designed to make proofs easier to develop, present, and understand. It is based on work performed in the field of formal programming methodology over the past 15 years, most notably Dijkstra, Feiejen, Scholten, and others —see for example [10, 14].

The inference rules of Logic **E** are given in Table 4. We now discuss them.

Inference rule Leibniz is also called "substitution of equals for equals". It is named *Leibniz* because Gottfried Leibniz (1646–1716) first mentioned it (not precisely in these terms) as a way to help characterize equality. Here is an example of the use of Leibniz. Suppose it is known that $\vdash y \wedge y \equiv y$. Then, Leibniz tells us that $\vdash p \Rightarrow y \wedge y \equiv p \Rightarrow y$, because we can instantiate the inference rule as follows:

P:	$y \wedge y$	E:	$p \wedge V$
Q:	y	$E[V := P]$:	$p \Rightarrow y \wedge y$
		$E[V := Q]$:	$p \Rightarrow y$

In [10], function application is used to present substitution of equals for equals. In our notation, the Dijkstra-Scholten view of it is given by

(9) $\vdash P = Q \longrightarrow \vdash f.P = f.Q$

While this rule may appear to be a beautiful link between substitution of equals for equals and function application, it is too weak to be of use when function bodies contain quantifications or other functions, as we will see later. Therefore, we do not use it.

Inference rule Transitivity should need no explanation.

Here is an example of the use of inference rule Equanimity. Suppose $\vdash$ *true* (i.e. *true* is a theorem) and also $\vdash (P \equiv P) \equiv true$. Then, by Equanimity, $\vdash P \equiv P$.

Inference rules Leibniz and Transitivity are written in terms of equality $=$ because, later on, we use them also for types other than boolean. Leibniz, formulated as in Table 4, holds whether the types of P and Q are both *int*, both *bool*, even both *graph*. Inference

Table 5: Propositional logic **E**: axiom schema

Associativity of $\equiv$: $((P \equiv Q) \equiv R) \equiv (P \equiv (Q \equiv R))$
Symmetry of $\equiv$: $P \equiv Q \equiv Q \equiv P$
Identity of $\equiv$: $true \equiv Q \equiv Q$

Definition of *false*: $false \equiv \neg true$
Distributivity of $\neg$ over $\equiv$: $\neg(P \equiv Q) \equiv \neg P \equiv Q$
Definition of $\not\equiv$: $(P \not\equiv Q) \equiv \neg(P \equiv Q)$

Associativity of $\vee$: $(P \vee Q) \vee R \equiv P \vee (Q \vee R)$
Symmetry of $\vee$: $P \vee Q \equiv Q \vee P$
Idempotency of $\vee$: $P \vee P \equiv P$
Distributivity of $\vee$ over $\equiv$: $P \vee (Q \equiv R) \equiv P \vee Q \equiv P \vee R$
Excluded Middle: $p \vee \neg p$

Golden rule: $P \wedge Q \equiv P \equiv Q \equiv P \vee Q$

Implication: $P \Rightarrow Q \equiv P \vee Q \equiv Q$
Consequence: $P \Leftarrow Q \equiv Q \Rightarrow P$

Anti-implication: $P \not\Rightarrow Q \equiv \neg(P \Rightarrow Q)$
Anti-consequence: $P \not\Leftarrow Q \equiv \neg(P \Leftarrow Q)$

rule Equanimity is written in terms of $\equiv$ because P and Q must be of type *bool* for the inference rule to make sense.

The axiom schemas of logic **E** are given in Table 5. The order of introduction of the axioms has been chosen to make the presentation of the logic as simple, enlightening, and useful as possible (see the text [14]). The operators are defined (by the axioms) in a nontraditional order: first $\equiv$, *true*; then $\neg$, *false*, $\not\equiv$; then $\vee$, then $\wedge$, and finally $\Rightarrow$, $\Leftarrow$, $\not\Rightarrow$, $\not\Leftarrow$. Thus, equivalence is given the place of honor, while in most propositional logics it is a second-class citizen —usually, $P \equiv Q$ is simply an abbreviation for $(P \Rightarrow Q) \wedge (Q \Rightarrow P)$ and is introduced last.

The first axiom states that $\equiv$ is associative. Thus, we can (and do) dispense with parentheses in sequences of equivalences. For example, the second axiom, Symmetry of $\equiv$, would usually be written as $(P \equiv Q) \equiv (Q \equiv P)$. But writing it as $P \equiv Q \equiv Q \equiv P$ allows us to parse it in many ways, e.g. $P \equiv ((Q \equiv Q) \equiv P)$. The use of associativity of $\equiv$ reduces tremendously the number of different formulas we have to write down and manipulate; the same economy is produced that associativity of integer addition yields.

In fact, associativity and symmetry of various operators are often used transparently in proofs, without explicit mention. This provides a tremendous savings in manipulation, helping to make this way of proving theorems quite effective.

It is shown in [15] that P_1 and **E** have the same theorems. Thus, **E** is also a sound and complete propositional logic.

The axioms and many theorems of propositional logic, as presented in the text [14], are given at the end of this article. Within these notes, we will refer to the theorems by number, in `typewriter font`. For example, "Associativity of $\equiv$ `(3.1)` refers to the first axiom of logic **E**.

Proofs in logic E

It is possible to write Hilbert-style proofs in **E**. However, the inference rules have been chosen to enable a proof style that is more in tune with the way people normally prove many mathematical facts: the calculational proof. The foundation of a calculational proof is substitution of equals for equals, written in the form shown to the left below. To the right, below, we display it as an instance of inference rule Leibniz.

$$\begin{array}{ll} & E[V:=P] \\ = & \quad \langle P=Q \rangle \\ & E[V:=Q] \end{array} \qquad\qquad \vdash P=Q \ \longrightarrow\ \vdash E[V:=P]=E[V:=Q]$$

Thus, the hint (the text within angle brackets to the right of $=$) is the premise of a Leibniz inference rule, and the formulas on the first and third lines are the LHS and RHS of the conclusion of that inference rule.

As an example, we prove Reflexivity of $\equiv$ `(3.5)`, $P \equiv P$.

$$\begin{array}{ll} & P \equiv P \\ = & \quad \langle \text{Symmetry of} \equiv \texttt{(3.2)}, \text{ with } Q:=P \rangle \\ & P \equiv P \equiv P \equiv P \quad \text{—Symmetry of} \equiv \texttt{(3.2)}, \text{ with } Q:=P \end{array}$$

This single Leibniz step of this proof shows that the following formula is a theorem: $(P \equiv P) \equiv (P \equiv P \equiv P \equiv P)$.

Inference rule Equanimity is used as follows in this proof. To the right of the last formula appears an em dash followed by an indication that the second formula is a theorem, an instance of Symmetry of $\equiv$. Since the second formula equivales the first, Equanimity tells us that the first formula, $P \equiv P$, is also a theorem.

Below, we prove `(3.11)`, $\neg P \equiv Q \equiv P \equiv \neg Q$, in order to show a use of inference rule Transitivity. By Leibniz, the first formula equivales the second; by Leibniz, the second formula equivales the third; and by Transitivity, the first formula equivales the third. Since the third is marked as a known theorem, by Equanimity, the first formula is a theorem.

$$\begin{array}{ll} & \neg P \equiv Q \equiv P \equiv \neg Q \\ = & \quad \langle \texttt{(3.9)},\ \neg(P \equiv Q) \equiv \neg P \equiv Q \rangle \\ & \neg(P \equiv Q) \equiv P \equiv \neg Q \\ = & \quad \langle \texttt{(3.9)}, \text{ with } P,Q := Q,P \text{ —i.e. } \neg(Q \equiv P) \equiv \neg Q \equiv P \rangle \\ & \neg(P \equiv Q) \equiv \neg(P \equiv Q) \quad \text{—Reflexivity of} \equiv \texttt{(3.5)} \end{array}$$

The hint on the last formula says that the last formula is theorem $P \equiv P$. Actually, it is $P \equiv P$ with P replaced by $\neg(P \equiv Q)$. We often don't state the instance of a theorem that is being used when it is easy to see what that instance is. This is another way to remove unnecessary detail in order to end up with palatable proofs.

We have explained an effective calculational proof format that is supported by logic **E**. Compared to the Hilbert-style proof format, the calculational format offers several advantages. First, the inference rules don't have to be mentioned, because the context indicates where each is used. Second, the Hilbert style often forces the same formula to be written twice, in different places. To see this, as an exercise, translate the proof given above into the Hilbert style. Third, as demonstrated in the next subsection, the calculational proof format offers us the chance to develop proof strategies and principles, which can be taught, learned, and put into practice.

Proof principles and strategies

Here is a list of proof principles and strategies that can be taught with the calculational proof format. They are not obvious to mathematical novices —and perhaps even to seasoned mathematicians. Conscious application of these principles can be very effective. Strategy fold-unfold (more properly called unfold-fold) is used by mathematicians all the time, but they never tell students about it. The name comes from its use in work on functional programming. In text [14], the use of each of these principles and strategies is demonstrated on many examples.

- **Principle.** Structure proofs to avoid repeating the same subexpression on many lines.
- **Principle.** Structure proofs to minimize the number of "rabbits pulled out of a hat" —make each step obvious, based on the structure of the expression and the goal of the manipulation.
- **Principle.** Lemmas may provide structure, bring to light interesting facts, and ultimately shorten a proof.
- **Strategy (fold-unfold).** To prove something about an operator, eliminate it (using its definition), manipulate the result, and then, if necessary, reintroduce the operator.
- **Strategy.** The shape of an expression can focus the choice of theorems to be used in manipulating it. Thus, identify applicable theorems by matching the structure of the expression and its subexpressions with the theorems.
- **Strategy.** To prove $P \equiv Q$, transform the term with the most structure (either P or Q) into the other. The idea here is that the one with the most structure provides the most insight into the next Leibniz transformation.

Propositional logic: the foundation of informal proof methods

In informal proofs, various "proof methods" are used. Table 6 lists four of the main ones, along with the propositional theorem on which each is based. This provides a sound underpinning to these proof methods.

Table 6: Informal Proof Techniques

Informal proof technique	Basis for the informal technique
Case analysis	$(P \vee Q) \wedge (P \Rightarrow R) \wedge (Q \Rightarrow R) \Rightarrow R$
Mutual implication	$P = Q \equiv (P \Rightarrow Q) \wedge (Q \Rightarrow P)$
Contradiction	$\neg P \Rightarrow \mathit{false} \equiv P$
Contrapositive	$P \Rightarrow Q \equiv \neg Q \Rightarrow \neg P$
Assuming the antecedent	Deduction Theorem

2 Handling the undefined using underspecification

> Everything should be made as simple as possible, but not any simpler.
> —Albert Einstein

One of the principles guiding research by those studying the formal development of programs has been *simplicity and uniformity of concept and notation.* Tools should be as simple as possible, if they are to be used reliably. Unnecessary distinctions and details should be avoided. For example, a logic should be usable for developing proofs without formal detail and complexity overwhelming. Engineering and style are legitimate concerns when constructing proofs, just as they are in other creative activities.

We illustrate taking simplicity as a goal by discussing the treatment of undefined terms logic. Partial functions are ubiquitous in programming —some basic mathematical operations are partial (e.g. division), some basic programming operations are partial (e.g. array subscripting $b[n]$), and many functions that arise through recursive definitions are partial. Therefore, a logic for reasoning about programs must handle partial functions.

Avoiding undefined through underspecification

We avoid the undefined by using *underspecification*:

> All operations and functions are assumed to be defined for all values of their operands —they are *total* operations and functions. However, the value assigned to an expression need not be uniquely specified in all cases.

Thus, the value of every (type-correct) expression is defined. We don't have partial functions, we have *underspecified* functions. Therefore, our logics don't have to deal with undefined values. The propositional and pure predicate logics need not be changed.

In this approach, division $/ : \mathbb{R} \times \mathbb{R} \to \mathbb{R}$ has to be total, so what should be the value of $x/0$? According to the above principle, we leave it unspecified. The value of $x/0$ could be 2.4 or $20 \cdot x$ or any value in $\mathbb{R}$; we simply don't say what it is.

Theorems mentioning an underspecified term $b \circ c$ usually involve an implication whose antecedent describes the set of states in which $b \circ c$ is uniquely specified.[6] In states in

[6] In some formal systems, a predicate *Dom*.'E' is introduced to represent the set of states in which expression E is uniquely specified. For example, *Dom*.'x/y' denotes $y \neq 0$. Note that *Dom* is not

which the antecedent is *false*, the expression evaluates to *true*, even though $b \circ c$ evaluates to some (unknown) value. For example, the law $y/y = 1$ would be written as

(10) $y \neq 0 \;\Rightarrow\; y/y = 1$

because y/y is uniquely specified iff $y \neq 0$.

We can prove theorems that rely on such laws in the standard manner. For example, to prove $y \neq 0 \Rightarrow x \cdot y/y = x$, we assume the antecedent $y \neq 0$ and prove the consequent:

$$
\begin{array}{ll}
 & x \cdot y/y \\
= & \quad \langle \text{Modus ponens with assumption } y \neq 0 \text{ and (10) yields } y/y = 1 \rangle \\
 & x \cdot 1 = x \\
= & \quad \langle \text{Identity of } \cdot \rangle \\
 & x
\end{array}
$$

Using underspecification requires a slight change to the definition of validity.

(11) **Modified definition of validity.** A formula E is *valid* iff, for every combination of values that can be assigned to an underspecified term of E, E evaluates to *true* in all states.

For example, for $x, y{:}\mathbb{R}$, $x/x = y/y$ is not valid, because one possible value for x/x is 2, one possible value for y/y is 3, and $2 \neq 3$. However, $x/x = x/x$ is valid.

The handling of recursively defined functions has been a concern of several articles that explore the undefined, notably [4] and [22]. Article [22] introduces the following function *subp* (see Sec. 3 for a definition of the notation for quantification).

$$
\begin{array}{l}
subp : \mathbb{Z} \times \mathbb{Z} \to \mathbb{Z} \\
(\forall i{:}\mathbb{Z} \mid : subp(i,i) = 0) \\
(\forall i,j{:}\mathbb{Z} \mid i > j : subp(i,j) = subp(i, j+1) + 1)
\end{array}
$$

and uses proofs of the following property as a leitmotif.

(12) $(\forall i,j{:}\mathbb{Z} \mid : i \geq j \;\Rightarrow\; subp(i,j) = i - j)$.

Function application $subp(i,j)$ is normally considered to be defined only if $i \geq j$. However, with the approach of underspecification, $subp(i,j)$ is always defined but its value is unspecified if $i < j$. Property (12) can then be proved by induction in a straightforward manner.

a function symbol of the logic. Instead, *Dom*.'E' is shorthand for a predicate. To understand the need for quoting the argument of *Dom*, recall that a function application like $abs.E$ applies function *abs* to the value of E. For example, $abs(2-5)$ evaluates to $abs(-3)$, which evaluates to 3. Thus, $Dom(x/0)$ makes no sense, since it would result in $x/0$ being evaluated. Rather, the argument of *Dom* is some representation of an expression, and the result is a predicate that describes the set of states in which the expression is uniquely specified.

Jones [21] warns that handling undefined through underspecification can lead to difficulties if recursive definitions inadvertently overspecify in such a way that unwanted properties can be deduced. For example, suppose $fact.i : \mathbb{Z} \to \mathbb{Z}$ is defined by

(13) $fact.i = \textbf{if } i = 0 \textbf{ then } 1 \textbf{ else } i \cdot fact(i-1) \textbf{ fi}$.

Then, we can use recursive definition (13) to deduce

$$\begin{array}{lcl} fact(-1) & = & -1 \cdot fact(-2) \\ fact(-2) & = & -2 \cdot fact(-3) \\ & \ldots & \end{array}$$

In this case, *fact* has indeed been overspecified, since $fact(-1) = -fact(-2)$ can be deduced. But the fault lies in the recursive definition rather than in handling undefined using underspecification. The problem can be solved by defining *fact* so that it indicates nothing about $fact.i$ for negative i:

$$fact.0 = 1$$
$$(\forall i{:}\mathbb{Z} \mid i > 0 : fact.i = i \cdot fact(i-1))$$

In conclusion, underspecification allows us to continue using simple two-valued logic, with all its nice properties, and still be formal, rigorous, and clear. Only when dealing with a partial function does it become necessary to exert extra effort, which takes the form of using antecedents that describe the set of arguments for which the partial function is specified.

The mistake of adding a new constant for undefined

Keeping in mind the following perspective, we now outline other approaches to dealing with partial functions. In calculational logic, Leibniz is the main inference rule, and equivalence of $\equiv$ plays a more prominent role than in other propositional logics. Further, the properties of equivalence are so useful that it is unwise to sacrifice them in an extension to handle the undefined. Equivalence should remain an equivalence relation and should have the identity *true*, and inference rule Leibniz should remain sound.

Consider changing just the model of evaluation of expressions by introducing a third constant $\bot$ to represent an undefined value. Thus, the value of a propositional variable or term in a state is one of *true*, *false*, and $\bot$. Since $p \equiv p \equiv true$ is a theorem of logic **E**, and since $\equiv$ is associative and symmetric, the following must hold:

(a) $\bot \equiv \bot$ evaluates to *true* (since $(\bot \equiv \bot) \equiv true$ is a theorem).
(b) $\bot \equiv true$ evaluates to $\bot$ (since $(\bot \equiv true) \equiv \bot$ is a theorem).
(c) $true \equiv \bot$ evaluates to $\bot$ (since $(true \equiv \bot) \equiv \bot$ is a theorem).

Now, what should be the value of $\bot \equiv false$?

(a) Suppose $\bot \equiv false$ evaluates to *false*. Then $(\bot \equiv false) \equiv false$ evaluates to *true* and $\bot \equiv (false \equiv false)$ evaluates to $\bot$, so $\equiv$ would not be associative.

(b) Suppose $\bot \equiv \mathit{false}$ evaluates to *true*. Then $(\bot \equiv \mathit{false}) \equiv \mathit{false}$ evaluates to *false* and $\bot \equiv (\mathit{false} \equiv \mathit{false})$ evaluates to $\bot$, so $\equiv$ would not be associative.

(c) Suppose $\bot \equiv \mathit{false}$ evaluates to $\bot$. Then $(\bot \equiv \bot) \equiv \mathit{false}$ evaluates to *false* and $\bot \equiv (\bot \equiv \mathit{false})$ evaluates to *true*, so $\equiv$ would not be associative.

With every choice of value for $\bot \equiv \mathit{false}$, associativity of equivalence is sacrificed.

Other standard logical properties no longer hold with this scheme. For example, suppose we take the so-called "strict" approach: the value of an expression is $\bot$ if any of its constituents are $\bot$. Then, the law of the Excluded Middle (3.28), $p \vee \neg p$, and Zero of $\vee$ (3.29), $\mathit{true} \vee p \equiv \mathit{true}$, are no longer valid.

Any way we turn, the underlying propositional logic has to be changed. We are not willing to accept such radical changes to propositional logic when a more suitable alternative for dealing with undefined terms exists.

As the field has explored handling undefined terms, various proposals have been made to extend two-valued logics. Jones and Middleburg [22], for example, developed a typed version of logic LPF of partial functions, which allows them to define and analyze "undefined". Their extension requires an operator that denotes "definedness" and two kinds of equality (weak and strong). Many useful properties in classical logic (eg. excluded middle and associativity of $\equiv$) do not hold. For us, their logic is far too complicated for use. We have no need to define and analyze "undefined"; we need only to deal with partial functions.

A model of evaluation for underspecification

In [1], Bijlsma gives a model of evaluation for propositional logic that seems to violate our conclusion that the basic logic has to be changed when the undefined, $\bot$, is introduced.

> Suppose expression E contains some variables v whose values are $\bot$ in a state s. E evaluates to $\mathbf{t}$ in s iff E evaluates to $\mathbf{t}$ with all possible combinations of the values $\mathbf{t}$ and $\mathbf{f}$ for the values of variables v. Similarly, E evaluates to $\mathbf{f}$ in s iff E evaluates to $\mathbf{f}$ with all possible combinations of the values $\mathbf{t}$ and $\mathbf{f}$ for the values of variables v. Otherwise, E evaluates to $\bot$ in s.

For example $v \vee \neg v$ and $\bot \equiv \bot$ both evaluate to $\mathbf{t}$ in all states, but $v \vee \neg w$ does not.

Think of $\bot$ as denoting the set $\{\mathbf{t}, \mathbf{f}\}$, meaning that $\bot$ is one of $\mathbf{t}$ and $\mathbf{f}$, but we don't know which. Then, $\bot$ denotes underspecification instead of the undefined, and this model of evaluation suitably expresses our notion of underspecification.

However, this is not a compositional model of evaluation: the value of an operation does not depend entirely on the value of its operands. For example, consider a state in which v has the value $\bot$. Thus, v and $\neg v$ both have the value $\bot$, so we might assume that $v \vee \neg v$ evaluates to $\bot \vee \bot$, which evaluates to $\bot$. However, $v \vee \neg v$ evaluates to $\mathbf{t}$.

We prefer to the use the standard compositional model of evaluation (see Sec 1.1) and to use the slightly different definition (11) of validity.

Abortive approaches that avoid undefined

Some approaches to avoiding partial functions assume a typed predicate logic. In a typed logic, every expression has a type, which can be determined syntactically from the expression itself and the types of its constituents (constants, variables, and function symbols).

One approach is to turn partial functions into total functions by restricting the types of their arguments. For example, instead of giving division the type $\mathbb{R} \times \mathbb{R} \to \mathbb{R}$, we give it the type $\mathbb{R} \times (\mathbb{R} - \{0\}) \to \mathbb{R}$. Unfortunately, with this approach, any set can be a type. There is no hope of having type be a syntactic property that can be checked mechanically, since the type system is undecidable. This approach is discussed further in [4].

A second approach (which can be used with a typed or an untyped logic) to avoiding the undefined is to view a function $f : B \to C$ as a relation on $B \times C$. Thus, $f.b = c$ is written as $(b, c) \in f$. Function application $f.b$ cannot be written, and just about the only way to refer to a value of a function is in an expression $(b, c) \in f$. To use this form, one must have an expression c for the value of $f.b$, which is awkward if not impossible in many situations. This approach is discussed briefly in [4].

Yet a third approach (which can be used with a typed or an untyped logic) to avoiding the undefined was suggested by Scott [24]: classify all atomic formulas that contain an undefined term as *false*. This means that the law of the excluded middle holds. However, with almost any partial function, we will be able to find conventional laws that no longer hold. For example in arithmetic, the law of trichotomy,

$$x = 0 \vee x > 0 \vee x < 0 \quad ,$$

no longer holds, since $x/0 = 0 \vee x/0 > 0 \vee x/0 < 0$ evaluates to *false* (terms $x/0 = 0$, $x/0 > 0$ and $x/0 > 0$ are all undefined). This approach therefore seems unworkable.

In conclusion, underspecification is the only approach to dealing with partial functions that allows us to use simple two-valued logic, with all its nice properties, and still be formal, rigorous, and clear.

3 Predicate calculus

Logicians typically define universal quantification $\forall x.P$ and existential quantification $\exists x.P$ using four axioms —to introduce and eliminate each of the kinds of quantification.

We take a different approach. We first define a uniform notation that includes universal quantification, existential quantification, summation, etc. This notation allows the "acumulation" of a set of values using any symmetic and associative operator. We then give general axioms that can be used to manipulate all quantifications. Then, we particularize to operations $\wedge$ and $\vee$ to get universal and existential quantification.

In this section, we present just the basics of this formulation, pointing to the axioms and theorems that appear at the end of this article.

The conventional summation $\Sigma_{i=1}^{n}\ e$ is written in our notation as

$$(\Sigma i \mid 1 \leq i \leq n : e) \quad \text{or} \quad (+i \mid 1 \leq i \leq n : e)$$

Note that we may specify the type of i, as in $(\Sigma i{:}int \mid 1 \leq i \leq n : e)$. Ours is a typed logic, where type is a syntactic property of a constant, variables, or any expression. Lack of time and space preclude a full discussion of types.

More generally, let $\star$ be any binary operator that is symmetric, is associative, and has an identity u (say):

Symmetry: $b \star c \;=\; c \star b$
Associativity: $(b \star c) \star d \;=\; b \star (c \star d)$
Identity u**:** $u \star b \;=\; b \;=\; b \star u$

For example, for $\star$ and u, we could choose $+$ and 0, $\cdot$ and 1, $\wedge$ and *true*, or $\vee$ and *false*. The general form of a *quantification* over $\star$ is exemplified by

(14) $(\star x{:}t1, y{:}t2 \mid R : P)$

where:

- Variables x and y are distinct. They are called the *bound variables* or *dummies* of the quantification. There may be one or more dummies.
- $t1$ and $t2$ are the types of dummies x and y. If $t1$ and $t2$ are the same type, we may write $(\star x, y{:}t1 \mid R : P)$. In the interest of brevity, we usually omit the type when it is obvious from the context, writing simply $(\star x, y \mid R : P)$.
- R, a predicate, is the *range* of the quantification —values assumed by x and y satisfy R. R may refer to dummies x and y. If the range is omitted, as in $(\star x \mid : P)$, then the range *true* is meant.
- Expression P is the *body* of the quantification. P may refer to dummies x and y.
- The type of the result of the quantification is the type of P.

Here are examples of quantifications, assuming, that i has type *int*.

$$\begin{aligned}
(+i \mid 0 \leq i < 4 : i \cdot 8) &\;=\; 0 \cdot 8 \;+\; 1 \cdot 8 \;+\; 2 \cdot 8 \;+\; 3 \cdot 8 \\
(\cdot i \mid 0 \leq i < 2 : i + (i+6)) &\;=\; (0 + (0+6)) \cdot (1 + (1+6)) \\
(\wedge i \mid 0 \leq i < 2 : i \cdot d \neq 6) &\;\equiv\; 0 \cdot d \neq 6 \;\wedge\; 1 \cdot d \neq 6 \\
(\vee i \mid 0 \leq i < 21 : b[i] = 0) &\;\equiv\; b[0] = 0 \;\vee\; \cdots \;\vee\; b[20] = 0
\end{aligned}$$

We use the linear notation $(\star x \mid R : P)$ throughout, for all quantifications, but we bow to convention and use a different symbol for $\star$ in some cases. In particular, we may write

$$\begin{aligned}
&(+x \mid R : P) \quad \text{as} \quad (\Sigma x \mid R : P) \\
&(\cdot x \mid R : P) \quad \text{as} \quad (\Pi x \mid R : P) \\
&(\vee x \mid R : P) \quad \text{as} \quad (\exists x \mid R : P) \\
&(\wedge x \mid R : P) \quad \text{as} \quad (\forall x \mid R : P)
\end{aligned}$$

We can now define the scope of a dummy variable, bound variable, and bound and free occurrences of variables, as usual. We leave this to the reader.

Earlier, we defined uniform substitution $E[V := G]$ as a copy of E in which each occurrence of metavariable V is replaced by G. We now define recursively the notion of *textual substitution* E^v_G as the replacement of all *free* occurrences of variable v in formula schema E by formula G.[7] Note that G must be a formula, and not a formula schema, so $E^v_{x \vee y}$ is not defined. This definition avoids "capture" of a free variable in G.

(15) **Definition of textual substitution.** In the following, G must be a formula, and not a formula schema.

c^v_G	$=$	c	(c a constant)
v^v_G	$=$	G	
w^v_G	$=$	w	(w a variable different from v)
$(\sim E)^v_G$	$=$	$\sim (E^v_G)$	($\sim$ a prefix operator)
$(E1 \circ E2)^v_G$	$=$	$E1^v_G \circ E2^v_G$	($\circ$ a binary operation)
$(\star x \mid R : P)^v_G$	$=$	$(\star y \mid (R^x_y)^v_G : (P^x_y)^v_G)$	(y a variable that does not occur free in R, P, v, or G)

Here are examples of textual substitution.

- $(true \vee x)^x_{y \wedge z} \;\equiv\; true \vee (y \wedge z)$
- $\neg(x \Rightarrow Y)^x_z \;\equiv\; \neg(z \Rightarrow Y)$
- $(+i \mid 1 \le i < n : 2 \cdot i)^i_x = (+j \mid 1 \le j < n : 2 \cdot j)$
- $(+i \mid 1 \le i < n : 2 \cdot i)^n_{x+y} = (+j \mid 1 \le j < x + y : 2 \cdot i)$
- $(+i \mid 1 \le i < n : i + n)^n_{x+i} = (+j \mid 1 \le j < x + i : j + x + i)$
- $\neg(x \Rightarrow Y)^Y_z$ (not defined)
- $\neg(x \Rightarrow Y)^x_{W+V}$ (not defined)

Note that the definition calls for the replacement of the dummy by a fresh variable, always. However, using (8.21) Dummy renaming, the dummy can be changed back. For example, we may write $(\exists i \mid : i > k)^k_m = (\exists i \mid : i > m)$.

Axioms (8.13)-(8.21) at the end of this article provide the basic theorems for manipulating all quantifications, for any operator $\star$ that is associative, is symmetric, and has an identity. (For an operator that is only associative and symmetric but has no identity, all axioms can be used as long as the range of all quantifications is not *false*).

These axioms (and all theorems derived from them) are valid as long as the quantifications in them are specified. For example, the infinite sum of all positive integers, $(+i \mid 0 < i : i)$, is unspecified. The treatment of unspecified quantifications falls outside the scope of these notes. For example, convergence of integer summations is a deep topic of

[7] The extension to simultaneous substitution, e.g. $E^{v,w}_{a+b,c}$, is left to the reader.

its own. Here, we state the following requirement concerning the uses of axioms and inference rules dealing with quantification, but we will not give rules for determining whether a quantification is defined.

(16) **Requirement.** The axioms and theorems for quantification hold only in cases where all the individual quantifications are specified.

Note that all quantifications specified if (i) the operator is one of $\wedge$, $\vee$ or the range is finite, and (ii) the range and body are specified.

We give an example of a proof of a theorem in predicate logic, theorem Change of Dummy (8.22). To motivate it, consider the expression

$$(+x \mid 2 \leq x \leq 10 : x^2) \quad .$$

Suppose we want an equal quantification whose range starts at 0 instead of 2. We can write:

$$(+y \mid 0 \leq y \leq 8 : (y+2)^2) \quad .$$

This can be rewritten as $(+y \mid 2 \leq y+2 \leq 10 : (y+2)^2)$. Using $f.y = y+2$ and $f^{-1}.x = x-2$, we see that the equality of these two summations is an instance of Change of dummy (8.22), with $+$ for $\star$.

The proof of Change of Dummy is interesting because it contains no rabbits. It is an "opportunistic proof", in that at each step, the current formula (together with the goal) gives insight into and provides the opportunity for that step. Thus, it kakes use of the principles and strategies for proof development outlined earlier. It is a *memorable* proof, because once studied, it can be reconstructed easily. And, in a lecture, it can be presented without notes. Not because you memorize it character for character but because you know the basic idea and can simply develop it. The proof starts with the RHS of (8.22), because it has more structure than the LHS.

$(\star y \mid R[x := f.y] : P[x := f.y])$

$=$ ⟨One-point rule (8.14) —Quantification over x has to be introduced. The One-point rule is the *only* theorem that can be applied at first.⟩

$(\star y \mid R[x := f.y] : (\star x \mid x = f.y : P))$

$=$ ⟨Nesting (8.20) —Moving dummy x to the outside gets us closer to the final form.⟩

$(\star x, y \mid R[x := f.y] \wedge x = f.y : P)$

$=$ ⟨Substitution (3.84a) —$R[x := f.y]$ must be removed at some point. This substitution makes it possible.⟩

$(\star x, y \mid R[x := x] \wedge x = f.y : P)$

$=$ ⟨ $R[x := x] \equiv R$; Nesting (8.20), $\neg occurs(\text{'}y\text{'}, \text{'}R\text{'})$ —Now we can get a quantification in x alone.⟩

$(\star x \mid R : (\star y \mid x = f.y : P))$

$=$ ⟨ $x = f.y \equiv y = f^{-1}.x$ —This step prepares for the elimination of y using the One-point rule.⟩

$$
\begin{array}{ll}
& (\star x \mid R : (\star y \mid y = f^{-1}.x : P)) \\
= & \quad \langle \text{One-point rule (8.14)} \rangle \\
& (\star x \mid R : P[y := f^{-1}.x]) \\
= & \quad \langle \text{Textual substitution} \text{ — } \neg occurs(\text{`}y\text{'}, \text{`}P\text{'}) \rangle \\
& (\star x \mid R : P)
\end{array}
$$

Now, when we define predicate calculus as an extension of propositional calculus, we use the symbols $\forall$ and $\exists$ in quantifications to denote $\wedge$ and $\vee$, respectively. Then, we may use the general axioms and theorems just discussed. In addition, we introduce a few axioms to characterize the relationship between $\forall$ and $\exists$, at the end of this article.

That is all we have to say about predicate logic in this section. Further sections will show the use of predicate logic.

4 Three important theorems about predicate calculus

4.1 Strengthening/weakening and Metatheorem Monotonicity

Thus far, Leibniz has been the basic step of calculational proofs. We now extend the proof format to allow weakening and strengthening steps. Suppose $P \equiv Q$ and $Q \Rightarrow R$ are theorems; we show how we may prove $\vdash P \Rightarrow R$:

$$
\begin{array}{ll}
& P \\
= & \quad \langle \text{Why } P \equiv Q \rangle \\
& Q \\
\Rightarrow & \quad \langle \text{Why } Q \Rightarrow R \rangle \\
& R
\end{array}
$$

This proof relies on a "transitivity" theorem (3.82b), $(P \equiv Q) \wedge (Q \Rightarrow R) \Rightarrow (P \Rightarrow R)$.

Formally, in order to accept a proof in this format, we have to show that we can translate it into a proof of $P \Rightarrow R$ that does not use the extension. Here is the proof (note that it uses the same two theorems).

$$
\begin{array}{ll}
& (P \equiv Q) \wedge (Q \Rightarrow R) \Rightarrow (P \Rightarrow R) \quad \text{— Transitivity (3.82b)} \\
= & \quad \langle \text{Redundant } \mathit{true}\ ^{8}\text{: Why } P \equiv Q \rangle \\
& \mathit{true} \wedge (Q \Rightarrow R) \Rightarrow (P \Rightarrow R) \\
= & \quad \langle \text{Redundant } \mathit{true}\text{: Why } Q \Rightarrow R \rangle \\
& \mathit{true} \wedge \mathit{true} \Rightarrow (P \Rightarrow R) \\
= & \quad \langle \text{Idempotency of } \wedge \text{ (3.38); Left identity of } \Rightarrow \text{ (3.73)} \rangle \\
& P \Rightarrow R
\end{array}
$$

Generalizing, we allow any number of $\equiv$ steps and $\Rightarrow$ steps to be used in the extended proof format. Similarly, from a sequence of $\Leftarrow$ and $\equiv$ steps we conclude that the first expression is a consequence of the last.

[8] Derived Inference rule Redundant *true* is $\vdash X \longrightarrow \vdash X \equiv \mathit{true}$.

Using inference rule Modus Ponens

Suppose we have a proof of $P \Rightarrow Q$ as shown to the left below.

P	P —reference to theorem P
$\Rightarrow$ ⟨Why $P \Rightarrow P1$⟩	$\Rightarrow$ ⟨Why $P \Rightarrow P1$⟩
$P1$	$P1$
$\ldots$	$\ldots$
Pn	Pn
$\Rightarrow$ ⟨Why $Pn \Rightarrow Q$⟩	$\Rightarrow$ ⟨Why $Pn \Rightarrow Q$⟩
Q	Q

Suppose in addition that P is a theorem. Then, by inference rule Modus Ponens,

$$\textbf{Modus Ponens:} \quad \vdash P,\ P \Rightarrow Q \quad \longrightarrow \vdash Q \quad ,$$

we can conclude that Q is a theorem as well. In such situations, we indicate a use of Modus Ponens by placing a reference to theorem P to its right, as shown to the right above.

Modus Ponens has been shown to be a derived inference rule of calculational logic, so it can be used.

Basic monotonicity theorems

In using the extended proof format, the basic monotonicity theorems of Table 7 become important. Their names come from the fact that a boolean function f is called *monotonic* if $\vdash (X \Rightarrow Y) \Rightarrow (f.X \Rightarrow f.Y)$ and *antimonotonic* if $\vdash (X \Rightarrow Y) \Rightarrow (f.X \Leftarrow f.Y)$.

On Metatheorem Monotonicity

Which of the following two formulas is valid, if either?

$$(\forall x \mid \neg P : S) \Rightarrow (\forall x \mid \neg(P \vee Q) : S)$$
$$(\forall x \mid \neg P : S) \Leftarrow (\forall x \mid \neg(P \vee Q) : S)$$

The answer is given by the following definition and theorem.

(26) **Definition.** Let metavariable Z occur once in a formula E, and not within an operand of an equivalence (or inequivalence). The position of Z within E has *even parity* if it is nested within an even number of negations, antecedents, and ranges of universal quantifications; otherwise, it has *odd parity.*

(27) **Metatheorem Monotonicity.** Suppose $P \Rightarrow Q$ is a theorem. Then:

(a) $\vdash E[Z := P] \Rightarrow E[Z := Q]$ (provided the parity of Z in E is even)
(b) $\vdash E[Z := P] \Leftarrow E[Z := Q]$ (provided the parity of Z in E is odd)

Table 7: Basic monotonicity theorems

(17) **Monotonic $\vee$:** $(P \Rightarrow Q) \Rightarrow (P \vee R \Rightarrow Q \vee R)$

(18) **Monotonic $\wedge$:** $(P \Rightarrow Q) \Rightarrow (P \wedge R \Rightarrow Q \wedge R)$

(19) **Antimonotonic $\neg$:** $(P \Rightarrow Q) \Rightarrow (\neg P \Leftarrow \neg Q)$

(20) **Monotonic consequent:** $(P \Rightarrow Q) \Rightarrow ((R \Rightarrow P) \Rightarrow (R \Rightarrow Q))$

(21) **Antimonotonic antecedent:** $(P \Rightarrow Q) \Rightarrow ((P \Rightarrow R) \Leftarrow (Q \Rightarrow R))$

(22) **Monotonic $\forall$-body:**
$(\forall x \mid R : P \Rightarrow Q) \Rightarrow ((\forall x \mid R : P) \Rightarrow (\forall x \mid R : Q))$

(23) **Antimonotonic $\forall$-range:**
$(\forall x \mid \neg R : P \Rightarrow Q) \Rightarrow ((\forall x \mid P : R) \Leftarrow (\forall x \mid Q : R))$

(24) **Monotonic $\exists$-body:**
$(\forall x \mid R : P \Rightarrow Q) \Rightarrow ((\exists x \mid R : P) \Rightarrow (\exists x \mid R : Q))$

(25) **Monotonic $\exists$-range:**
$(\forall x \mid R : P \Rightarrow Q) \Rightarrow ((\exists x \mid P : R) \Rightarrow (\exists x \mid Q : R))$

Sketch of proof. The proof is by induction on the structure of expression E. Reduce the case analysis by first manipulating E so that one has only to deal with formulas that contain variables, constants, negations, disjunctions with Z in the first operand, and existential quantifications with *true* ranges. Thus, make the following changes (in order).

(a) Replace $(\forall x \mid F1 : F2)$ by $\neg(\exists x \mid F1 : \neg F2)$.
(b) Replace $(\exists x \mid F1 : F2)$ by $(\exists x \mid : F1 \wedge F2)$.
(c) Replace $F1 \not\Leftarrow F2$ by $\neg(F1 \Leftarrow F2)$.
(d) Replace $F1 \Leftarrow F2$ by $F2 \Rightarrow F1$.
(e) Replace $F1 \not\Rightarrow F2$ by $\neg(F1 \Rightarrow F2)$.
(f) Replace $F1 \Rightarrow F2$ by $\neg F1 \vee F2$.
(g) Replace $F1 \wedge F2$ by $\neg(\neg F1 \vee \neg F2)$.
(h) If Z is in the second operand $F2$ of $F1 \vee F2$, replace $F1 \vee F2$ by $F2 \vee F1$.

These manipulations do not change the parity of the position of Z. Now comes a straightforward proof by induction on the structure of the more restricted expressions E, which will rely on monotonic/antimonotonic properties (17), (19), and (24). □

Using Metatheorem Monotonicity

In a weakening/strengthening step of a calculation, the hint should explain why the step is sound. Here is a simple example.

$$
\begin{array}{ll}
 & P \\
\Rightarrow & \langle \text{Weakening } (3.76a),\ P \Rightarrow P \vee Q \rangle \\
 & P \vee Q
\end{array}
$$

But in the following example, the hint is not precise because the soundness of the step depends not only on Weakening but also on Monotonic $\wedge$ (18).

$$
\begin{array}{ll}
 & P \wedge R \\
\Rightarrow & \langle \text{Weakening } (3.76a),\ P \Rightarrow P \vee Q \text{ —NOT A PRECISE HINT} \rangle \\
 & (P \vee Q) \wedge R
\end{array}
$$

We seek a uniform way of substantiating weakening/strengthening steps. Rather than rely directly on all the individual monotonicity properties (17)–(25), it is easier to rely on Metatheorem Monotonicity, which can be used to substantiate all such steps.

We use "Monotonicity:" and "Antimonotonicity:" to show reliance on this metatheorem, as shown below. "Monotonicity" suggests that the parity of the position of the subexpression involved in the replacement is even, so the step is weakening. "Antimonotonicity" suggests that the parity of the position involved in the replacement is odd, so the step is strengthening. In the examples given below, to the right we have shown how the formulas can be rewritten in terms of variable z, so that the use of the metatheorem is more easily seen.

$$
\begin{array}{lll}
 & (\forall x \mid : P \wedge R) & \text{i.e. } (\forall x \mid : Z \wedge R)[Z := P] \\
\Rightarrow & \langle \text{Monotonicity: Weakening } P \Rightarrow P \vee Q \rangle & \\
 & (\forall x \mid : (P \vee Q) \wedge R) & \text{i.e. } (\forall x \mid : Z \wedge R)[Z := P \vee Q]
\end{array}
$$

$$
\begin{array}{lll}
 & \neg(\forall x \mid \neg P \wedge R : S) & \text{i.e. } \neg(\forall x \mid \neg Z \wedge R : S)[Z := P] \\
\Leftarrow & \langle \text{Antimonotonicity: Weakening } P \Rightarrow P \vee Q \rangle & \\
 & \neg(\forall x \mid \neg(P \vee Q) \wedge R : S) & \text{i.e. } \neg(\forall x \mid \neg Z \wedge R : S)[Z := P \vee Q]
\end{array}
$$

Discussion

Monotonicity properties (17)–(25), as well as metatheorem Monotonicity, are well-known. They can be found, in one guise or another, in several texts on logic. But the two major books that deal with the calculational approach do a bad job of explaining how monotonicity/antimonotonicity is to be used. On page 61 of [10], Dijkstra and Scholten discuss the monotonic properties of negation and implication, but they don't state general theorem (27) and they don't give a good method of explaining when it is being used. On page 93 of [10], a hint explicitly states the use of monotonicity of $\wedge$ and $\exists$ in a weakening step, but on pages 73 and 77, monotonicity of $\forall$-body is used without mention. Perhaps the authors didn't realize that monotonicity would be a problem for many people.

Gries and Schneider [14] also do not treat monotonicity well, and this has resulted in confusion among students about monotonicity and its use. The next edition of [14] will use the approach of this note in order to eliminate the confusion.

4.2 A theorem concerning universal quantification

A boolean expression that has free occurrences of variables, like $b \vee x < y$, is called *open*, and its value may differ from state to state. The expression is *closed* if it has no free variables, as in $(\forall b, x, y \mid: b \vee x < y)$. The value of a closed expression does not depend on the state in which it is evaluated, since it has no free variables. Therefore, a closed expression is equivalent either to *true* or to *false*. The following metatheorem characterizes (at least partially) when quantifying over a variable does not change the value of a boolean expression.

(28) **Metatheorem.** P is a theorem iff $(\forall x \mid: P)$ is a theorem.

Proof. The proof is by mutual implication.

LHS $\Rightarrow$ RHS. We assume P is a theorem and prove that $(\forall x \mid: P)$ is a theorem.

$$\begin{array}{ll} & (\forall x \mid: P) \\ = & \quad \langle \text{Redundant } \mathit{true}: \text{Assumption } P \rangle \\ & (\forall x \mid: \mathit{true}) \\ = & \quad \langle (9.8),\ (\forall x \mid R: \mathit{true}) \equiv \mathit{true} \rangle \\ & \mathit{true} \end{array}$$

RHS $\Rightarrow$ LHS. Assume $(\forall x \mid: P)$ is a theorem. Using Instantiation (9.13) with x for E, we conclude that P is a theorem. □

Here are some applications of Metatheorem (28). Since $p \vee q \equiv q \vee p$ is a theorem, then so are $(\forall p \mid: p \vee q \equiv q \vee p)$, $(\forall q \mid: p \vee q \equiv q \vee p)$, and $(\forall p, q \mid: p \vee q \equiv q \vee p)$.

Here is the standard terminology for proving a universal quantification using Metatheorem (28):

To prove $(\forall x \mid R: P)$, we prove P for arbitrary x in range R.

Thus, $(\forall x \mid R: P)$ can be proved by proving $R \Rightarrow P$ (and $R \Rightarrow P$ itself can be proved by assuming antecedent R and proving consequent P). This method of proof is often used informally in mathematics.

4.3 Metatheorem Witness

We now use a new method of proof to prove Interchange of quantifications (9.29) (with all ranges being *true*),

$$(\exists x \mid: (\forall y \mid: P)) \Rightarrow (\forall y \mid: (\exists x \mid: P)) \quad ,$$

and then prove the metatheorem that substantiates it.

The antecedent indicates that there exist values for dummy x for which $(\forall y \mid: P)$ holds. Let us give a name, say $\hat{x}$, to one of these values, and use instead the antecedent $(\forall y \mid: P^x_{\hat{x}})$. Thus, we prove the theorem by proving $(\forall y \mid: P^x_{\hat{x}}) \Rightarrow (\forall y \mid: (\exists x \mid: P))$:

$$\begin{array}{ll} & (\forall y \mid: (\exists x \mid: P)) \\ \Leftarrow & \quad \langle \text{Monotonicity: } \exists\text{-Introduction } (9.28),\ P^x_{\hat{x}} \Rightarrow (\exists x \mid: P) \rangle \\ & (\forall y \mid: P^x_{\hat{x}}) \end{array}$$

The name $\hat{x}$ in the above proof is called a *witness* for the existential quantification. The use of a witness has simplified the formula to be manipulated, as well as the manipulations themselves.

We now prove a metatheorem that substantiates such a use of a witness.

(29) **Metatheorem Witness.** Suppose $\neg occurs(`\hat{x}', `P,Q,R')$. Then

$$(\exists x \mid R : P) \Rightarrow Q \text{ is a theorem iff } (R \wedge P)^x_{\hat{x}} \Rightarrow Q \text{ is a theorem.}$$

Identifier $\hat{x}$ is called a *witness* for the existential quantification.[9]

Proof. $\quad (\exists x \mid R : P) \Rightarrow Q$
$=\quad$ ⟨Trading (9.19)⟩
$\quad (\exists x \mid : R \wedge P) \Rightarrow Q$
$=\quad$ ⟨Implication (3.59); De Morgan (9.18b)⟩
$\quad (\forall x \mid : \neg(R \wedge P)) \vee Q$
$=\quad$ ⟨Dummy renaming (8.21) — $\neg occurs(`\hat{x}', `P,R')$ ⟩
$\quad (\forall \hat{x} \mid : \neg(R \wedge P)^x_{\hat{x}}) \vee Q$
$=\quad$ ⟨Distributivity of $\vee$ over $\forall$ (9.5) — $\neg occurs(`\hat{x}', `Q')$ ⟩
$\quad (\forall \hat{x} \mid : \neg(R \wedge P)^x_{\hat{x}} \vee Q)$
$=\quad$ ⟨Implication (3.59)⟩
$\quad (\forall \hat{x} \mid : (R \wedge P)^x_{\hat{x}} \Rightarrow Q)$

By Metatheorem (28), the last line is a theorem iff $(R \wedge P)^x_{\hat{x}} \Rightarrow Q$ is a theorem. □

Example of the use of Metatheorem Witness.

We prove $a+b=a+c \Rightarrow b=c$. The proof relies on a property of the integers called "additive inverse":

(30) $(\exists x{:}int \mid : x+a=0)$.

Thus, we are really proving (30) $\Rightarrow (a+b=a+c \Rightarrow b=c)$, where the antecedent is assumed.

Axiom (30) says that, for any integer a, there exists another integer x such that $x+a=0$. Let us use the name $\hat{a}$ for the integer x that satisfies $x+a=0$, so that we can use $(x+a=0)^x_{\hat{a}}$, i.e. $\hat{a}+a=0$.

To prove $a+b=a+c \Rightarrow b=c$, we assume the antecedent $a+b=a+c$ and prove the consequent:

$\quad b$
$=\quad$ ⟨ $0+b=b$ ⟩
$\quad 0+b$
$=\quad$ ⟨(30), $(\exists x{:}int \mid : x+a=0)$, with witness $\hat{a}$ ⟩

[9] Identifier x itself can be used for $\hat{x}$ if x does not occur free in Q.

$$
\begin{array}{ll}
 & \hat{a}+a+b \\
= & \quad \langle \text{Assumption } a+b=a+c \rangle \\
 & \hat{a}+a+c \\
= & \quad \langle\, \hat{a}+a=0 \text{ —again} \rangle \\
 & 0+c \\
= & \quad \langle\, 0+c=c \,\rangle \\
 & c
\end{array}
$$

If two (or more) conjuncts of an implication are existential quantifications, then each may be replaced using a witness, as long as the witnesses are distinct. We illustrate the inconsistency that may arise if two witnesses are given the same name. Consider again (30). Use Substitution to replace a by $a+5$, yielding theorem

$$(\exists x{:}int \mid{:}\ x+a+5=0) \quad .$$

From this theorem, derive using Witness (29) the assumption $\hat{a}+a+5=0$, where $\hat{a}$ is the witness. From theorem (30), derive the assumption $\hat{a}+a=0$, where the mistake is made of using the same witness $\hat{a}$. Together, these two assumptions yield the contradiction $5=0$.

4.4 Examples of the use of these metatheorems

4.4.1 Proving EA $\Rightarrow$ AE

Carroll Morgan [2] derives a nice proof of

(31) $(\exists x \mid{:}\ (\forall y \mid{:}\ P.x.y)) \Rightarrow (\forall y \mid{:}\ (\exists x \mid{:}\ P.x.y))$

and Wim Feijen [12] presents convincing heuristics for the development of this proof:

$$
\begin{array}{ll}
 & (\exists x \mid{:}\ (\forall y \mid{:}\ P.x.y)) \\
\Rightarrow & \quad \langle R \Rightarrow (\forall y \mid{:}\ R) \text{, provided } y \text{ does not occur free in } R \text{ —a previous} \\
 & \quad\ \ \text{theorem. Universal quantification over } y \text{ must appear at the outer} \\
 & \quad\ \ \text{level; this gets it there.} \rangle \\
 & (\forall y \mid{:}\ (\exists x \mid{:}\ (\forall y \mid{:}\ P.x.y))) \\
\Rightarrow & \quad \langle \text{Monotonicity: Instantiation } \texttt{(9.13)} \text{ —The inner quantification} \\
 & \quad\ \ \text{over } y \text{ has to be removed, and this does it.} \rangle \\
 & (\forall y \mid{:}\ (\exists x \mid{:}\ P.x.y))
\end{array}
$$

Dijkstra [9] discusses the same theorem but with non- *true* ranges:

(32) $(\exists x \mid R.x : (\forall y \mid S.y : P.x.y)) \Rightarrow (\forall y \mid S.y : (\exists x \mid R.x : P.x.y))$.

But now the proof gets messier, because of the non- *true* ranges. The purpose of Dijkstra's note [9] was to present a way of dealing with such non- *true* ranges, by introducing the notion of "Range diffusion" for punctual functions and predicates.

Here, we present an alternative proof of (32), which rests on Metatheorem Witness (29). The use of Witness strips away an existential quantification, making dealing with ranges easier and simplifying the proof in other ways as well.

Proof. By Metatheorem Witness (29), (32) is a theorem precisely when the following is:

$$R.x \land (\forall y \mid S.y : P.x.y) \Rightarrow (\forall y \mid S.y : (\exists x \mid R.x : P.x.y)) \quad .$$

So we prove this formula.

$$\begin{array}{ll} & R.x \land (\forall y \mid S.y : P.x.y) \\ \Rightarrow & \quad \langle \land \text{ over } \forall \text{ (9.7)} \rangle \\ & (\forall y \mid S.y : R.x \land P.x.y) \\ \Rightarrow & \quad \langle \text{Monotonicity: } \exists\text{-Introduction (9.28)} \rangle \\ & (\forall y \mid S.y : (\exists x \mid : R.x \land P.x.y)) \\ = & \quad \langle \text{Trading (9.19)} \rangle \\ & (\forall y \mid S.y : (\exists x \mid R.x : P.x.y)) \end{array}$$

□

4.4.2 A Proof of a characterization of induction

Let U be a nonempty set and $<$ be an transitive relation over U. We say that the pair $\langle U, < \rangle$ *admits induction*, i.e. mathematical induction can be performed over $\langle U, < \rangle$, iff for all predicates $P : U \to bool$,

(33) $(\forall x \mid : P.x) \equiv (\forall x \mid : (\forall y \mid y < x : P.y) \Rightarrow P.x) \quad .$

The pair $\langle U, < \rangle$ is called *FT* iff for every function H (of appropriate type) there exists a unique function F such that

$$F.x = H(x, \text{restriction of } F \text{ to } y \text{ such that } y < x) \quad .$$

Franzen Torkel suggested proving that $\langle U, < \rangle$ admits induction iff $\langle U, < \rangle$ is FT, using the calculational approach. The purpose of this section is to prove one direction of this theorem in order to show how the shape of formulas can be used to provide insight into the development of a proof.

Our first task is to formalize the "restriction of F to y such that $y < x$". To this end, think of a function F as a set of pairs $\{\langle y, F.y \rangle \mid y \in \text{domain of } F\}$. Then, for a function $F : U \to bool$, the restriction of F to y such that $y < x$ is $\{\langle y, F.y \rangle \mid y < x\}$. Thus, the type of function H is $U \times (U \times bool) \to bool$.

We say that $\langle U, < \rangle$ is FT iff, for all such H, there exists a unique function F that satisfies

(34) $F.x = H(x, \{\langle y, F.y \rangle \mid y < x\}) \quad .$

(35) **Theorem.** If $\langle U, < \rangle$ does not admit induction, then it is not FT.

Proof. We assume that $\langle U, < \rangle$ does not admit induction, and we exhibit a function H for which an F satisfying (34) does not exist. Since $\langle U, < \rangle$ does not admit induction, there is a predicate P for which

$(\forall x \mid : P.x)$ is *false* and
$(\forall x \mid : (\forall y \mid y < x : P.y) \Rightarrow P.x))$ is *true*

Hence, there is an x for which $P.x$ is *false*. Further, for any x satisfying $\neg P.x$, $(\exists y \mid y < x : \neg P.y)$ holds. Use $\hat{x}$ for the witness of this existential quantification, so that we have:

(36) $\neg P.x \Rightarrow \hat{x} < x \land \neg P.\hat{x}$.

Let function H be defined by the following.

(37) $H(x, S) = (\forall y, z \mid \langle y, z \rangle \in S : z) \Rightarrow P.x$.

This choice of H is inspired by the form of the RHS of the principal of mathematical induction, (33), and definition (34) of F —in this case, (34) particularizes to

(38) $F.x \equiv (\forall y \mid y < x : F.y) \Rightarrow P.x$.

We show that no function F satisfies (38). We begin with two corollaries of (38).

(39) $P.x \Rightarrow F.x$
(40) $\neg P.x \Rightarrow (F.x \equiv \neg(\forall y \mid y < x : F.y))$

We now prove $\neg P.x \Rightarrow F.x$, by assuming $\neg P.x$ and proving $\neg F.x \Rightarrow$ *false* —we prove that $\neg F.x$ implies both $F.\hat{x}$ and $\neg F.\hat{x}$. First, we have four simple corollaries of $\neg P.x$, (40), and (36):

(41) $F.x \equiv \neg(\forall y \mid y < x : F.y)$
(42) $\hat{x} < x$
(43) $\neg P.\hat{x}$
(44) $F.\hat{x} \equiv \neg(\forall y \mid y < \hat{x} : F.y)$

We now calculate:

$\neg F.x$
= ⟨(41); Double negation (3.12)⟩
$(\forall y \mid y < x : F.y)$
= ⟨Idempotency of $\land$ (3.38); Instantiation (9.13)—$\hat{x} < x$⟩
$F.\hat{x} \land (\forall y \mid y < x : F.y)$
$\Rightarrow$ ⟨Antimonotonicity: $<$ is transitive and $\hat{x} < x$⟩
$F.\hat{x} \land (\forall y \mid y < \hat{x} : F.y)$
= ⟨Double negation (3.12), $\neg\neg p \equiv p$; (44)⟩
$F.\hat{x} \land \neg F.\hat{x}$
= ⟨Contradiction (3.42) —$\neg p \land p \equiv$ *false*⟩
false

We have proved $\neg P.x \Rightarrow F.x$; together with (39), this means that the only possible solution to (38) is

$$F.x \equiv true \quad (\text{for all } x).$$

But this is not a solution, since for x for which $\neg P.x$ holds —and there is at least one such value— $(\forall x \mid: F.x \equiv true)$ violates (38). □

4.4.3 Andrews's challenge

At the Marktoberdorf summer school in August 1996, Larry Paulson lectured on his mechanical theorem prover, Isabelle, Natarajan Shankar lectured on his mechanical theorem prover, PVS, and Gries lectured on calculational logic. Both Paulson and Shankar suggested trying the calculational approach on Andrews's challenge, which is to prove theorem (45), given below (from F.J. Pelletier's collection of problems in first-order logic).

(45) $((\exists x \forall y \mid: p.x \equiv p.y) \equiv ((\exists x \mid: q.x) \equiv (\forall y \mid: p.y))) \equiv$
$((\exists x \forall y \mid: q.x \equiv q.y) \equiv ((\exists x \mid: p.x) \equiv (\forall y \mid: q.y)))$

Now, $\equiv$ is both associative and symmetric, so we can rewrite Andrews's challenge as

$$P \equiv Q$$

where P and Q are defined by the following.

$P:\ (\exists x \forall y \mid: p.x \equiv p.y) \equiv (\exists x \mid: p.x) \equiv (\forall y \mid: p.y)$
$Q:\ (\exists x \forall y \mid: q.x \equiv q.y) \equiv (\exists x \mid: q.x) \equiv (\forall y \mid: q.y)$

This form gives us the impression that perhaps P is valid (or invalid), regardless of p. If this is the case, then Q is also valid (or invalid). Hence, we need only prove P. In fact, we give two proofs: our first proof and a second proof developed independently by Rajeev Joshi and Kedar Sharad Namjoshi.

We prove P by mutual implication, by proving

(46) $((\exists x \forall y \mid: p.x \equiv p.y) \equiv (\exists x \mid: p.x)) \Leftarrow (\forall y \mid: p.y)$ and
(47) $((\exists x \forall y \mid: p.x \equiv p.y) \equiv (\exists x \mid: p.x)) \Rightarrow (\forall y \mid: p.y)$

Proof of (46):

Assume $(\forall y \mid: p.y)$, so also $p.y$ and $p.x$
$(\exists x \forall y \mid: p.x \equiv p.y) \equiv (\exists x \mid: p.x)$
= ⟨Redundant *true*: Assumptions $p.x$ and $p.y$⟩
$(\exists x \forall y \mid: true \equiv true) \equiv (\exists x \mid: true)$
= ⟨Identity of $\equiv$ (3.3); $(\forall y \mid: true) \equiv true$⟩
$(\exists x \mid: true) \equiv (\exists x \mid: true)$ —Reflexivity of $\equiv$ (3.5)

Proof of (47):

$$
\begin{array}{cl}
 & (47) \\
= & \langle \text{Contrapositive, } X \Rightarrow Y \equiv \neg Y \Rightarrow \neg X \rangle \\
 & \neg(\forall y \mid : p.y) \;\Rightarrow\; \neg((\exists x \forall y \mid : p.x \equiv p.y) \;\equiv\; (\exists x \mid : p.x)) \\
= & \langle \text{De Morgan (9.17) on antecedent;} \\
 & \quad \neg(X \equiv Y) \;\equiv\; X \;\equiv\; \neg Y \text{ and De Morgan (9.18b) on the consequent} \rangle \\
 & (\exists y \mid : \neg p.y) \;\Rightarrow\; ((\exists x \forall y \mid : p.x \equiv p.y) \;\equiv\; (\forall x \mid : \neg p.x))
\end{array}
$$

By Metatheorem Witness (29), the last formula is a theorem iff the following one is.

$$\neg p.\hat{y} \;\Rightarrow\; ((\exists x \forall y \mid : p.x \equiv p.y) \;\equiv\; (\forall x \mid : \neg p.x))$$

We calculate:

$$
\begin{array}{cl}
\multicolumn{2}{l}{\text{Assume } \neg p.\hat{y}\text{, so also } p.\hat{y} \;\equiv\; \mathit{false}} \\
 & (\exists x \forall y \mid : p.x \equiv p.y) \\
= & \langle \text{Zero of } \vee \text{ (3.29) on range of } \forall y \text{ —heading to change } p.x \text{ to } p.\hat{y} \rangle \\
 & (\exists x \mid : (\forall y \mid \mathit{true} \vee (y \equiv \hat{y}) : p.x \equiv p.y)) \\
= & \langle \text{Range split (8.18); One-point rule (8.14)} \rangle \\
 & (\exists x \mid : (\forall y \mid : p.x \equiv p.y) \wedge p.x \equiv p.\hat{y}) \\
= & \langle \text{Substitution (3.84)} \rangle \\
 & (\exists x \mid : (\forall y \mid : p.\hat{y} \equiv p.y) \wedge p.x \equiv p.\hat{y}) \\
= & \langle \text{One-point rule; Range split; Zero of } \vee \text{ —eliminate the conjunct } p.x \equiv p.\hat{y} \rangle \\
 & (\exists x \forall y \mid : p.\hat{y} \equiv p.y) \\
= & \langle \text{Assumption } p.\hat{y} \equiv \mathit{false}\text{; } \mathit{false} \;\equiv\; X \;\equiv\; \neg X \rangle \\
 & (\exists x \forall y \mid : \neg p.y) \\
= & \langle \text{Provided } x \text{ doesn't occur free in } X\text{, } (\exists x \mid : X) \;\equiv\; X \rangle \\
 & (\forall y \mid : \neg p.y)
\end{array}
$$

We now turn to the Joshi/Namjoshi proof of P. This proof rests on theorem (3.52), which we repeat here as (48); (49), which is derived easily from Exclusive or (3.53); and (50), whose proof we leave to the reader.

(48) $R \;\equiv\; S \;\equiv\; (R \wedge S) \vee (\neg R \wedge \neg S)$

(49) $R \;\equiv\; S \;\equiv\; (\neg R \vee S) \wedge (R \vee \neg S)$

(50) $(\neg R \vee X) \wedge (R \vee Y) \;\equiv\; (R \wedge X) \vee (\neg R \wedge Y)$

The theorem to be proved is $(\exists x \forall y \mid : p.x \equiv p.y) \;\equiv\; (\exists x \mid : p.x) \;\equiv\; (\forall y \mid : p.y)$.

$$
\begin{array}{cl}
 & (\exists x \forall y \mid : p.x \equiv p.y) \\
= & \langle (49) \rangle \\
 & (\exists x \forall y \mid : (\neg p.x \vee p.y) \wedge (p.x \vee \neg p.y)) \\
= & \langle \text{Distributivity of } \forall \text{ over } \wedge \text{ (8.16))} \rangle \\
 & (\exists x \mid : (\forall y \mid : \neg p.x \vee p.y) \wedge (\forall y \mid : p.x \vee \neg p.y))
\end{array}
$$

$$
\begin{array}{ll}
= & \langle \text{Distributivity of } \vee \text{ over } \forall \text{ (9.5)), twice} \rangle \\
& (\exists x \mid: ((\forall y \mid: \neg p.x) \vee (\forall y \mid: p.y)) \;\wedge\; ((\forall y \mid: p.x) \vee (\forall y \mid: \neg p.y))) \\
= & \langle \text{Quantified freshy (9.31)), twice} \rangle \\
& (\exists x \mid: (\neg p.x \vee (\forall y \mid: p.y)) \;\wedge\; (p.x \vee (\forall y \mid: \neg p.y))) \\
= & \langle (50), \text{ with } P, X, Y := p.x, (\forall y \mid: p.y), (\forall y \mid: \neg p.y) \rangle \\
& (\exists x \mid: (p.x \wedge (\forall y \mid: p.y)) \;\vee\; (\neg p.x \wedge (\forall y \mid: \neg p.y))) \\
= & \langle \exists \text{ over } \vee \text{ (8.15))} \rangle \\
& (\exists x \mid: p.x \wedge (\forall y \mid: p.y)) \;\vee\; (\exists x \mid: \neg p.x \wedge (\forall y \mid: \neg p.y)) \\
= & \langle \wedge \text{ over } \exists \text{ (9.21)), twice} \rangle \\
& ((\exists x \mid: p.x) \wedge (\forall y \mid: p.y)) \;\vee\; ((\exists x \mid: \neg p.x) \wedge (\forall y \mid: \neg p.y)) \\
= & \langle \text{De Morgan (9.17)) and (9.18b)} \rangle \\
& ((\exists x \mid: p.x) \wedge (\forall y \mid: p.y)) \;\vee\; (\neg(\forall x \mid: p.x) \wedge \neg(\exists y \mid: p.y)) \\
= & \langle (48) \rangle \\
& (\exists x \mid: p.x) \;\equiv\; (\forall y \mid: p.y)
\end{array}
$$

5 Propositional logic and the everywhere operator

5.1 Introduction

We now explore the "everywhere" operator $[P]$, with intended meaning "P is true in all states", the use of which was introduced to researchers in the formal development of programs by Dijkstra (see e.g. [10, 8]). We restrict our attention to propositional logic extended with the everywhere operator, and in this setting $[\,]$ is the modal operator $\Box$ of modal logic S5. (See [20] for an extensive treatment of modal logics.)

It is known that S5 is not complete with regard to the model that interprets $\Box P$ as "P is true in all states" (called the "Carnapian model" model C, because it is due essentially to Carnap). Extensions to S5 to make the logic complete with respect to this model are known [26, 6, 3, 18, 17], dating from as early as 1973 —see Gottlob's survey [13].

We provide a new sound and complete extension of S5 with respect to the Carnapian model. This extension allows seamless integration of the propositional and modal parts of the logic, giving a practical system for writing formal proofs. We also use this extension to show that the two following approaches to axiomatizing a logic are not equivalent: (i) give axiom schemes that denote an infinite number of axioms and (ii) write a finite number of axioms in terms of propositional variables and introduce inference rule Uniform Substitution.

This section of these notes is a condensation of a paper [16] by Gries and Schneider. It also uses material of Rutger Dijkstra [11].

A logic extended with $\Box$ can be used for proving theorems that could otherwise be handled only at the metalevel, and most likely informally. For example, the statement

(51) P is valid iff $(\forall x \mid: P)$ is valid

is formalized as $\Box P \equiv \Box(\forall x \,|{:}\, P)$. In contrast, formalizing (51) as the two inference rules

$$\vdash P \quad\longrightarrow\quad \vdash (\forall x \,|{:}\, P) \qquad \text{and} \qquad \vdash (\forall x \,|{:}\, P) \quad\longrightarrow\quad \vdash P$$

demotes it to a meta-logical notion. When the equivalence of P and $(\forall x \,|{:}\, P)$ is not expressible by a formula of the logic, it is not directly available for use in calculational reasoning.

As another example, the following fact about set theory,

$$\{x \mid Q\} = \{x \mid R\} \text{ is valid} \quad \text{iff} \quad Q \equiv R \text{ is valid} \quad ,$$

is formalized as $\Box(\{x \mid Q\} = \{x \mid R\}) \equiv \Box(Q \equiv R)$.

The Hoare triple $\{P\}\ S\ \{Q\}$, with the meaning "execution of statement S begun in a state in which P is *true* is guaranteed to terminate in a state in which Q is *true*" [19], can also be formalized using $\Box$. Using weakest-precondition predicate transformer *wp* [7], we define:

$$\{P\}\ S\ \{Q\} : \ \Box(P \Rightarrow wp(S,Q)) \quad .$$

5.2 Logics S5 and C

Consider the language of propositional logic P_1, extended with two operators $\Box$ and $\Diamond$. Operation $\Box P$ has the meaning "necessarily P" (this will be formalized later). Operation $\Diamond P$, meaning "possibly P", can be viewed simply as an abbreviation for $\neg\Box\neg P$ —i.e. "not necessarily not P". We call this new language S5. A formula of S5 that contains neither $\Box$ nor $\Diamond$ is called a *propositional formula.*

We consider models that are triples $\langle W, R, V\rangle$ in which:

- W is a nonempty set of *worlds.*
- R is an *accessibility relation,* a binary relation over W: $w\ R\ u$ signifies that *world u is accessible from world w*.
- Valuation $V : \mathit{formulas} \longrightarrow \mathit{set}(W)$, which delivers the set of worlds in which a formula evaluates to *true* [10], satisfies [11]

$$V.\mathit{false} = \{\,\}$$
$$V(P \Rightarrow Q) = -V.P \cup V.Q$$
$$V.\Box P = \textbf{if } V.P = W \textbf{ then } W \textbf{ else } \{\,\}$$

An S5-*model* is a model $\langle W, R, V\rangle$ in which R is an equivalence relation —reflexive, transitive, and symmetric. An S5-formula P is S5-*valid*, written $\models_{S5} P$, iff for every S5-model $\langle W, R, V\rangle$, $V.P = W$.

[10] Often, V is defined be a function of a formula and a state, and $V(P, s)$ is the value of formula P in state s. The definition used here simplifies notation and some calculations, since V has only one argument.

[11] We restrict attention to constant *false* and operator $\Rightarrow$ since *true* and the other operators are considered to be abbreviations in P_1.

Table 8: Schematic S5 and C

P_1:	The schematic propositional logic of Table 1
S5:	**Necessitation:** $\vdash P \longrightarrow \vdash \Box P$ **Axiom scheme $\Box$-Instantiation:** $\Box P \Rightarrow P$ **Axiom scheme Monotonicity:** $\Box(P \Rightarrow Q) \Rightarrow (\Box P \Rightarrow \Box Q)$ **Axiom scheme Necessarily Possible:** $\Diamond P \Rightarrow \Box\Diamond P$
C:	**Textual Substitution:** $\vdash P \longrightarrow \vdash P^v_Q$ (v a list of propositional variables)

The first part of Table 8 is Schematic propositional logic P_1 of Table 1. The second part of Table 8 extends P_1 to modal logic S5, by adding one inference rule and three axioms. S5 is sound and complete with respect to S5-validity [20].

We now define a single model in which a state associates **t** or **f** with each propositional variable and in which each state is accessible from all the others. In this model, $\Box P$ has the interpretation " P is true in all states".

Let $\widehat{W}$ be the set of all total functions $w : VP \rightarrow \{\mathbf{t},\mathbf{f}\}$. Let $\widehat{R}$ be the universal relation over $\widehat{W}$, i.e. $w\ \widehat{R}\ u$ holds for all w, u in $\widehat{W}$. Let $\widehat{V}$ be the valuation defined by $\widehat{V}.p = \{w{:}\widehat{W} \mid w.p = \mathbf{t}\}$. Then, model C is $\langle \widehat{W}, \widehat{R}, \widehat{V} \rangle$, and a formula P is C-valid iff for every w in $\widehat{W}$, $\widehat{V}.w = \widehat{W}$. Model C is also an S5-model, since $\widehat{R}$ is an equivalence relation. Consequently, $\models_{S5} P$ implies $\models_C P$.

Logic S5 is sound but not complete with respect to C-validity. To see this, consider the formula $\neg\Box p$. It is not S5-valid, since $V(\neg\Box p) = \{\,\}$ in the S5-model $(\{w\}, I, V)$, where $V.p = \{w\}$ and I is the identity relation. Since S5 is sound, $\neg\Box p$ is not a theorem of S5. However, $\models_C \neg\Box p$ holds, because $\widehat{V}.p \neq \widehat{W}$.

According to definition (15), textual substitution for a bound variable has no effect, e.g. $(\forall x \mid: P.x)^x_G = (\forall x \mid: P.x)$. We consider all variables within an operand of $\Box$ or $\Diamond$ to be bound, so that we extend the definition of textual substitution by the following:

(52) **Extension of definition (15) of textual substitution.**

$$(\Box P)^v_Q = \Box P$$
$$(\Diamond P)^v_Q = \Diamond P$$

The third part of Table 8 extends logic S5 with inference-rule scheme Textual Substitution to yield logic C. It is easy to show that Textual Substitution preserves C-validity.

Textual Substitution and all the inference rules of S5 preserve C-validity. Also, the axioms of S5 are C-valid. Therefore, logic C is sound with respect to C-validity.

We now prove that $\neg\Box p$ is a theorem of logic C.

$(\Box p \Rightarrow p)^p_{false}$ —Textual Substitution in Axiom $\Box$-Instantiation
$=$ ⟨Definition of textual substitution for propositional variable p⟩

$$
\begin{array}{ll}
& \Box p \ \Rightarrow\ \mathit{false} \\
= & \langle (3.74),\ P \Rightarrow \mathit{false} \equiv \neg P,\ \text{with } P := \Box p \rangle \\
& \neg\Box p
\end{array}
$$

5.3 Proving completeness of logic C with respect to C-validity

We prove that C is complete with respect to C-validity. Since C is an extension of S5, in this proof, we can use S5-theorems presented in Hughes and Cresswell [20] as C-theorems. Also, we rely on the following lemma, which follows directly from the definition of $\widehat{V}.\Box P$.

(53) **Lemma.** For any formula P, either $\models_C \Box P$ or $\models_C \neg\Box P$.

Hughes and Cresswell define *ordered modal conjunctive normal form* (ordered MCNF). A formula is in ordered MCNF if it has the form $C_1 \wedge \ldots \wedge C_n$ and each C_i has the form

(54) $Q \vee \Box S_1 \vee \ldots \vee \Box S_m \vee \Diamond T$,

where Q, the S_i, and T are propositional formulas (i.e. they don't contain $\Box$ or $\Diamond$).

We prove three lemmas, leading up to a proof that $\models_C$ (54) implies $\vdash_C$ (54).

(55) **Lemma.** For propositional formula Q, $\models_C Q$ implies $\vdash_C Q$.

Proof. Suppose $\models_C Q$. Since Q is a propositional formula, and since C contains complete propositional logic PM, $\vdash_C Q$. □

(56) **Lemma.** For propositional formula T, $\models_C \Diamond T$ implies $\vdash_C \Diamond T$.

Proof. If $\models_C \Diamond T$, some world w is in $\widehat{V}(\Diamond T)$. Thus, there is an assignment of constants $\overline{c}$ (each constant being *true* or *false*) to the propositional variables $\overline{p}$ of T such that $w \in \widehat{V}(T^{\overline{p}}_{\overline{c}})$. Since $T^{\overline{p}}_{\overline{c}}$ contains no variables, it evaluates to **t** in all worlds and is valid. By Lemma (55), $\vdash_C T^{\overline{p}}_{\overline{c}}$.

The following calculational proof shows that $(\Box\neg T \Rightarrow \neg T)^{\overline{p}}_{\overline{c}}$ is equivalent to $\Diamond T$. Further, since the first formula is a theorem (it is an instance of axiom $\Box$-Instantiation on which Textual Substitution is performed), the last formula is also a theorem. This establishes $\vdash_C \Diamond T$.

$$
\begin{array}{ll}
& (\Box\neg T \Rightarrow \neg T)^{\overline{p}}_{\overline{c}} \quad \text{—Textual Substitution in } \Box\text{-Instantiation} \\
= & \langle \text{Contrapositive } (3.61);\ \text{Double negation } (3.12) \rangle \\
& (T \Rightarrow \neg\Box\neg T)^{\overline{p}}_{\overline{c}} \\
= & \langle \text{Definition of textual substitution} \rangle \\
& T^{\overline{p}}_{\overline{c}} \Rightarrow \neg\Box\neg T \\
= & \langle T^{\overline{p}}_{\overline{c}} \equiv \mathit{true}\ (\text{since } \vdash_C T^{\overline{p}}_{\overline{c}});\ \text{Abbreviation } \Diamond \rangle \\
& \mathit{true} \Rightarrow \Diamond T \\
= & \langle \text{Left identity of } \Rightarrow 3.73) \rangle \\
& \Diamond T
\end{array}
$$

□

(57) **Lemma.** For propositional formula S, $\models_C \Box S$ implies $\vdash_C \Box S$.

Proof. If $\models_C \Box S$, then $\widehat{V}.S = \widehat{W}$. Hence, $\models_C S$. By Lemma (55), $\vdash_C S$. By inference rule Necessitation, $\vdash_C \Box S$. □

(58) **Theorem.** $\models_C$ (54) implies $\vdash_C$ (54).

Proof. Suppose $\models_C$ (54). By Lemma (53), either $\widehat{V}(\Box S_i) = \{\}$ or $\widehat{V}(\Box S_i) = \widehat{W}$, for each i, and similarly for $\diamond T$. The proof uses a three-case analysis: (i) $\widehat{V}(\diamond T) = \widehat{W}$, (ii) $\widehat{V}(\Box S_i) = \widehat{W}$ for some S_i, and (iii) $\widehat{V}(\diamond T) = \widehat{V}(\Box S_i) = \{\}$ for all i.

Case $\widehat{V}(\diamond T) = \widehat{W}$. Then $\models_C \diamond T$ and, by Lemma (56), $\vdash_C \diamond T$. Note that $\diamond T \Rightarrow$ (54) is of the form $P \Rightarrow P \vee Q$, which is a theorem of propositional logic PM, so $\vdash_C \diamond T \Rightarrow$ (54). By Modus ponens, $\vdash_C$ (54).

Case $\widehat{V}(\Box S_i) = \widehat{W}$. The proof is similar to the proof of the previous case, using Lemma (57) instead of (56).

Case $\widehat{V}(\diamond T) = \widehat{V}(\Box S_i) = \{\}$ **for all** i. Since (54), i.e. $Q \vee \Box S_1 \vee \ldots \vee \Box S_m \vee \diamond T$, is C-valid, $V.Q = \widehat{W}$, so $\models_C Q$. The rest of this proof is similar to the proof of the first case, using Lemma (55) instead of (56). □

Hughes and Cresswell [20, p. 117] prove the following theorem.

(59) **Ordered MCNF Theorem.** For any formula P, there exists an ordered MCNF formula $mcnf.P$ such that $\vdash_{S5} P \equiv mcnf.P$.

(60) **Corollary.** $\vdash_C P \equiv mcnf.P$

(61) **Corollary.** $\vdash_C P$ iff $\vdash_C mcnf.P$

(62) **Corollary.** $\models_C P$ iff $\models_C mcnf.P$.

Corollary (60) holds because C is an extension of S5. Corollary (61) follows from (60), the definition of abbreviation $\equiv$, and Modus Ponens. For Corollary (62), note that (60) together with the soundness of C yields $\models_C P \equiv mcnf.P$ and use properties of $\widehat{V}$.

To prove completeness of C, we use the following properties of propositional logic (which is included in logic C) and model C.

(63) $\vdash_C P \wedge Q$ iff $(\vdash_C P)$ and $(\vdash_C Q)$

(64) $\models_C P \wedge Q$ iff $(\models_C P)$ and $(\models_C Q)$

(65) **Theorem.** For any formula P, $\models_C P$ implies $\vdash_C P$.

Proof. $mcnf.P$ has the form $C_1 \wedge \ldots \wedge C_n$ where each C_i has form (54). We have,

$\models_C P$

iff ⟨(62) —where $mcnf.P$ is $C_1 \wedge \ldots \wedge C_n$⟩

$\models_C C_1 \wedge \ldots \wedge C_n$

iff $\langle$(64), $n-1$ times$\rangle$
$(\models_C C_1)$ and ... and $(\models_C C_n)$
implies $\langle$Monotonicity: Theorem (58) ($n-1$ times)$\rangle$
$(\vdash_C C_1)$ and ... and $(\vdash_C C_n)$
iff $\langle$(63), $n-1$ times$\rangle$
$\vdash_C C_1 \wedge \ldots \wedge C_n$
iff $\langle$(61) —where $mcnf.P$ is $C_1 \wedge \ldots \wedge C_n$$\rangle$
$\vdash_C P$ □

Comparison with earlier complete axiomatizations

As mentioned in Sec. 5.1, a number of complete axiomatizations of C have been given [26, 6, 3, 18, 17]. All of them are similar in nature to the following one, which we take from [17]. Begin with Schematic S5 (see Table 8). Instead of adding inference rule Textual Substitution, add as axioms

All formulas of the form $\diamond T$ for T a satisfiable propositional formula.

(i.e. all propositional formulas P for which $\widehat{V}.P \neq \{\}$. Lemma (56) now holds trivially, and we can prove completeness with respect to C-validity in the same way that we proved completeness of C.

This axiomatization is unsatisfactory to us because it refers to the semantic notion of satisfiability. However, this semantic notion can be eliminated, leading to a complete syntactic axiomatization. A propositional formula is satisfiable iff its disjunctive normal form contains a disjunct that does not contain some literal together with its negation. Hence, to discover whether $\diamond T$ (for T a propositional formula) is a theorem, convert T to disjunctive normal form and determine whether one of its disjuncts contains a literal and its negation. (Private communications with Rob Goldblatt and Joe Halpern).

The resulting axiomatization is still unsatisfactory to us, because of the need to reformulate T of a conjectured theorem $\diamond T$ in disjunctive normal form. This reformulation is not in keeping with our usual way of proving theorems (using a calculational approach, where suitable [14, 15, 10]). For example, to prove $\diamond T \vee \diamond S$, we would be forced to prove that one of T and S were satisfiable, rather than simply performing syntactic manipulations to obtain $\diamond T \vee \diamond S$, as is our preference. Inference rule Textual Substitution provides an alternative that is more in tune with the way we prove theorems calculationally; it allows for a more seamless integration of proofs of the various kinds of theorems of C.

5.4 C with a finite number of axioms

An axiomatization with a finite number of axioms is usually derived from one with axiom schemes by (i) replacing the metavariables in the axiom schemes with propositional variables and (ii) introducing inference rule uniform substitution to substitute formulas for propositional variables.

Table 9: P_1', S5$'$, and C$'$

P_1':	**Modus Ponens:** $\vdash P, \vdash P \Rightarrow Q \;\longrightarrow\; \vdash Q$
	Uniform Substitution: $\vdash P \;\longrightarrow\; \vdash P[V := Q]$ (V a list of formula variables)
	Axiom A0: $\mathtt{P} \Rightarrow (\mathtt{Q} \Rightarrow \mathtt{P})$
	Axiom A1: $(\mathtt{R} \Rightarrow (\mathtt{P} \Rightarrow \mathtt{Q})) \Rightarrow ((\mathtt{R} \Rightarrow \mathtt{P}) \Rightarrow (\mathtt{R} \Rightarrow \mathtt{Q}))$
	Axiom A2: $((\mathtt{P} \Rightarrow \mathit{false}) \Rightarrow \mathit{false}) \Rightarrow \mathtt{P}$
S5$'$:	**Necessitation:** $\vdash P \;\longrightarrow\; \vdash \Box P$
	Axiom $\Box$-Instantiation: $\Box P \Rightarrow P$
	Axiom Monotonicity: $\Box(P \Rightarrow Q) \Rightarrow (\Box P \Rightarrow \Box Q)$
	Axiom Necessarily Possible: $\Diamond P \Rightarrow \Box\Diamond P$
C$'$:	**Textual Substitution:** $\vdash P \;\longrightarrow\; \vdash P^v_Q$ (P, Q concrete; v a list of propositional variables)

This method for eliminating axiom schemes does not work in the case of Schematic C of Table 8, because uniform substitution for propositional variables does not preserve C-validity. For example, $\neg\Box p$ is C-valid (as proven earlier), but $(\neg\Box p)^p_{true}$, which is $\neg\Box \mathit{true}$, is not C-valid.

Instead, we obtain a sound axiomatization of C that has a finite number of axioms as follows. Let VF be a new set of *formula variables.* We use typewriter-font letters $\mathtt{P}, \mathtt{Q}, \mathtt{R}, \ldots$ for formula variables. The formulas of C$'$ are the formulas of C except that formula variables may also appear in them. For example, $p \vee q$, $\mathtt{P} \vee \mathtt{Q}$, and $p \vee \mathtt{Q}$ are formulas of C$'$. A formula of C$'$ is concrete if it does not contain a formula variable. For example, $p \vee q$ is concrete, but $\mathtt{P} \vee \mathtt{Q}$ and $p \vee \mathtt{Q}$ are not concrete. Language C contains exactly the concrete formulas of C$'$.

An axiomatization for C$'$ is given in Table 9. Its axioms are those of C, except that metavariables have been replaced by formula variables. The inference rules of C$'$ include those of C (even to requiring that, in Textual Substitution, P and Q be concrete). Inference rule Uniform Substitution is used only for replacing formula variables.

One may view logic C$'$ as simulating metavariables by formula variables. Note that neither Uniform Substitution nor Textual Substitution can be used to derive a non-concrete theorem from a concrete theorem.

In [16] it is proved that C and C$'$ have the same concrete theorems: For concrete formula P, $\vdash_C P$ iff $\vdash_{C'} P$.

5.5 Three different interpretations for S5

Rutger Dijkstra [11] has shown that there are three classes of models for S5, one of them being essentially the set of all complete boolean lattices. This provides evidence that S5 is indeed the proper axiomatization for propositional logic with the everywhere operator, except for the fact that $\neg\Box p$ (for p a propositional variable) is not a theorem. In some situations, it is desirable *not* to have such formulas be theorems; such logic has been called a *pointless* logic, because the propositional variables are the points or coordinates of the state space.

Call the S5 models defined earlier the *standard models*, and let **Std** denote this class.

A *simple model* is a pair (W, V) where W is a set and V is a function from formulas to subsets of W such that, for all formulas P and Q:

$$\begin{array}{lcl} V.false & = & \{\} \\ V.(P \Rightarrow Q) & = & -V.P \cup V.Q \\ V.(\Box P) & = & \mathbf{if}\ V.P = W\ \mathbf{then}\ W\ \mathbf{else}\ \{\}\ \mathbf{fi} \end{array}$$

Let **Spl** denote the class of simple models.

An *abstract model* is a pair (L, V) where L is a complete boolean lattice and V is a function from formulas to L such that, for all formulas P and Q:

$$\begin{array}{lcl} V.false & = & \bot \\ V.(P \Rightarrow Q) & = & \sim V.P \sqcup V.Q \\ V.(\Box P) & = & \mathbf{if}\ V.P = \top\ \mathbf{then}\ \top\ \mathbf{else}\ \bot\ \mathbf{fi} \end{array}$$

where $\sim, \sqcup, \top$ and $\bot$ denote 'complement', 'supremum', 'maximum' and 'minimum' respectively. Let **Abt** denote the class of abstract models.

Since we have three different classes of models, we also get three (potentially) different notions of validity. For every formula P:

$$\begin{array}{lcl} \models_{\mathbf{Std}} P & = & (\forall W, R, V \mid (W, R, V) \in \mathbf{Std} : V.P = W) \\ \models_{\mathbf{Spl}} P & = & (\forall W, V \mid (W, V) \in \mathbf{Spl} : V.P = W) \\ \models_{\mathbf{Abt}} P & = & (\forall L, V \mid (L, V) \in \mathbf{Abt} : V.P = \top) \end{array}$$

We now outline a proof that these three classes of models, as well as theoremhood, are equivalent, i.e. $\vdash_{S5} P$, $\models_{\mathbf{Abt}} P$, $\models_{\mathbf{Spl}} P$, and $\models_{\mathbf{Std}} P$ are equivalent. We do this by cyclic implication, in the order just given.

We leave to the reader to verify that S5 is sound for **Abt**: $\vdash_{S5} P$ implies $\models_{\mathbf{Abt}} P$.

Since **Spl** is a subclass of **Abt**, we immediately get $\models_{\mathbf{Abt}} P$ implies $\models_{\mathbf{Spl}} P$.

We now prove that $\models_{\mathbf{Spl}} P$ implies $\models_{\mathbf{Std}} P$. Assume $\models_{\mathbf{Spl}} P$ and let $(W, R, V) \in \mathbf{Std}$. We have to show that $V.P = W$ or, equivalently, $w \in V.P$ for all $w \in W$. For any (fixed) w in W define the set W' and the function V' as follows

$$\begin{array}{lcll} W' & = & \text{the equivalence class of } w & \\ V'.P & = & V.P \cap W' & \text{(for all formulas } P\text{)} \end{array}$$

One easily checks that $(W', V') \in \mathbf{Spl}$, so $V'.P = W'$, so $w \in V.P$.

Finally, since S5 is complete for **Std**, we have: $\models_{\mathbf{Std}} P$ implies $\vdash_{S5} P$.

References

[1] Bijlsma, A. Semantics of quasi-boolean expressions. In Feijen, W.H.J., et al (eds.) *Beauty is Our Business.* Springer Verlag, New York, 1990, 27–35.

[2] Carroll M. Email communications in January-February 1995.

[3] Carroll, M.J. An axiomatization of S13. *Philosophia, Philosophical Quarterly of Israel 8*, 2–3 (November 1978), 381–382.

[4] Cheng, J.H., and C.B. Jones. On the usability of logics which handle partial functions. In: C. Morgan and J.C.P. Woodcock (eds.). Third Refinement Workshop, pp. 51–69. Workshops in Computing Series, Heidelberg, 1991.

[5] Church, A. *Introduction to Mathematical Logic.* Princeton University Press, Princeton, 1956.

[6] Cocciarella, N.B. Logical atomism and modal logic. *Philosophia, Philosophical Quarterly of Israel 4*, 1 (1974), 41–66.

[7] Dijkstra, E.W. *A Discipline of Programming.* Prentice Hall, Englewood Cliffs, New Jersey, 1976.

[8] Dijkstra, E.W. The everywhere operator once more. EWD1086, 26 November 1990.

[9] Edsger W. Dijsktra. Triggered by Wim Feijen's treatment of " $\exists\forall \Rightarrow \forall\exists$ ". EWD1201, 27 February 1995.

[10] Dijkstra, E.W., and C.S. Scholten. *Predicate Calculus and Program Semantics.* Springer Verlag, New York 1990.

[11] Dijkstra, R. "Everywhere" in predicate algebra and modal logic. Accepted for publication in IPL, April 1996.

[12] Wim H.J. Feijen. Title unknown at the moment. WHJ189.

[13] Gottlob, G. Survey on a Carnapian extension of S5. Technical report, Institut für Informationssysteme, Technische Universität Wien, 1996.

[14] Gries, D., and F.B. Schneider. *A Logical Approach to Discrete Math.* Springer-Verlag, New York, 1993.

[15] Gries, D., and F.B. Schneider. Equational propositional logic. *IPL 53* (1995), 145-152.

[16] Gries, D., and F.B. Schneider. Adding the everywhere operator to propositional logic. Submitted for publication, May 1996.

[17] Halpern, Y.J., and Kapron, B. Zero-one laws for modal logic. *Annals of Pure and Applied Logic 69* (1994), 157-193.

[18] Hendry, H.E., and M.L. Pokriefka. Carnapian extensions of S5. *J. Philosophical Logic 14* (1985), 111-128.

[19] Hoare, C.A.R. An axiomatic basis for computer programming. *Comm. ACM 12* (October 1969), 576-580, 583.

[20] Hughes, G.E., and M.J. Cresswell. *An Introduction to Modal Logic.* Mehuen and Co., New York, 1968.

[21] Jones, C.B. Partial functions and logics: a warning. *IPL 54* (1995), 65–68.

[22] Jones, C.B., and C.A. Middelburg. A typed logic of partial functions reconstructed classically. *Acta Informatica 31* (1994), 399–430.

[23] Rosser, B. *Logic for Mathematicians.* McGraw-Hill, New York, 1953.

[24] Scott, D.S. Existence and description in formal logic. In R. Schoenman (ed.) *Bertrand Russell, Philosopher of the Century.* St. Leonards: Allen and Unwin, 1967, 181–200.

[25] Tarski, Alfred. *Logic, Semantics, Metamathematics.* (A collection of papers by Tarski, translated by J.H. Woodger.) Oxford University Press, London, 1956.

[26] Thomason, S.K. A new representation of S5. *Notre Dame J. Formal Logic XIV*, 2 (April 1973), 281–284.

6 Theorems of propositional and predicate logic

We present here the theorems of propositional and predicate logic, as given in the text [14]. Within the paper, references in `typewriter font` refer to these theroems.

6.1 Theorems of the propositional calculus

EQUIVALENCE AND TRUE

(3.1) **Axiom, Associativity of $\equiv$:** $((P \equiv Q) \equiv R) \equiv (P \equiv (Q \equiv R))$

(3.2) **Axiom, Symmetry of $\equiv$:** $P \equiv Q \equiv Q \equiv P$

(3.3) **Axiom, Identity of $\equiv$:** $true \equiv Q \equiv Q$

(3.4) *true*

(3.5) **Reflexivity of $\equiv$:** $P \equiv P$

NEGATION, INEQUIVALENCE, AND FALSE

(3.8) **Axiom, Definition of *false*:** $false \equiv \neg true$

(3.9) **Axiom, Distributivity of $\neg$ over $\equiv$:** $\neg(P \equiv Q) \equiv \neg P \equiv Q$

(3.10) **Axiom, Definition of $\not\equiv$:** $(P \not\equiv Q) \equiv \neg(P \equiv Q)$

(3.11) $\neg P \equiv Q \equiv P \equiv \neg Q$

(3.12) **Double negation:** $\neg\neg P \equiv P$

(3.13) **Negation of *false*:** $\neg false \equiv true$

(3.14) $(P \not\equiv Q) \equiv \neg P \equiv Q$

(3.15) $\neg P \equiv P \equiv \mathit{false}$

(3.16) **Symmetry of $\not\equiv$:** $(P \not\equiv Q) \equiv (Q \not\equiv P)$

(3.17) **Associativity of $\not\equiv$:** $((P \not\equiv Q) \not\equiv R) \equiv (P \not\equiv (Q \not\equiv R))$

(3.18) **Mutual associativity:** $((P \not\equiv Q) \equiv R) \equiv (P \not\equiv (Q \equiv R))$

(3.19) **Mutual interchangeability:** $P \not\equiv Q \equiv R \;\equiv\; P \equiv Q \not\equiv R$

DISJUNCTION

(3.24) **Axiom, Symmetry of $\vee$:** $P \vee Q \equiv Q \vee P$

(3.25) **Axiom, Associativity of $\vee$:** $(P \vee Q) \vee R \equiv P \vee (Q \vee R)$

(3.26) **Axiom, Idempotency of $\vee$:** $P \vee P \equiv P$

(3.27) **Axiom, Distributivity of $\vee$ over $\equiv$:** $P \vee (Q \equiv R) \equiv P \vee Q \equiv P \vee R$

(3.28) **Axiom, Excluded Middle:** $P \vee \neg P$

(3.29) **Zero of $\vee$:** $P \vee \mathit{true} \equiv \mathit{true}$

(3.30) **Identity of $\vee$:** $P \vee \mathit{false} \equiv P$

(3.31) **Distributivity of $\vee$ over $\vee$:** $P \vee (Q \vee R) \equiv (P \vee Q) \vee (P \vee R)$

(3.32) $P \vee Q \equiv P \vee \neg Q \equiv P$

CONJUNCTION

(3.35) **Axiom, Golden rule** : $P \wedge Q \equiv P \equiv Q \equiv P \vee Q$

(3.36) **Symmetry of $\wedge$:** $P \wedge Q \equiv Q \wedge P$

(3.37) **Associativity of $\wedge$:** $(P \wedge Q) \wedge R \equiv P \wedge (Q \wedge R)$

(3.38) **Idempotency of $\wedge$:** $P \wedge P \equiv P$

(3.39) **Identity of $\wedge$:** $P \wedge \mathit{true} \equiv P$

(3.40) **Zero of $\wedge$:** $P \wedge \mathit{false} \equiv \mathit{false}$

(3.41) **Distributivity of $\wedge$ over $\wedge$:** $P \wedge (Q \wedge R) \equiv (P \wedge Q) \wedge (P \wedge R)$

(3.42) **Contradiction:** $P \wedge \neg P \equiv \mathit{false}$

(3.43) **Absorption:** (a) $P \wedge (P \vee Q) \equiv P$

(b) $P \vee (P \wedge Q) \equiv P$

(3.44) **Absorption:** (a) $P \wedge (\neg P \vee Q) \equiv P \wedge Q$

(b) $P \vee (\neg P \wedge Q) \equiv P \vee Q$

(3.45) **Distributivity of $\vee$ over $\wedge$:** $P \vee (Q \wedge R) \equiv (P \vee Q) \wedge (P \vee R)$

(3.46) **Distributivity of $\wedge$ over $\vee$:** $P \wedge (Q \vee R) \equiv (P \wedge Q) \vee (P \wedge R)$

(3.47) **De Morgan:** (a) $\neg(P \wedge Q) \equiv \neg P \vee \neg Q$

(b) $\neg(P \vee Q) \equiv \neg P \wedge \neg Q$

(3.48) $P \wedge Q \equiv P \wedge \neg Q \equiv \neg P$

(3.49) $P \wedge (Q \equiv R) \equiv P \wedge Q \equiv P \wedge R \equiv P$

(3.50) $P \wedge (Q \equiv P) \equiv P \wedge Q$

(3.51) **Replacement:** $(P \equiv Q) \wedge (R \equiv P) \equiv (P \equiv Q) \wedge (R \equiv Q)$

(3.52) **Definition of $\equiv$:** $P \equiv Q \equiv (P \wedge Q) \vee (\neg P \wedge \neg Q)$

(3.53) **Exclusive or:** $P \not\equiv Q \equiv (\neg P \wedge Q) \vee (P \wedge \neg Q)$

(3.55) $(P \wedge Q) \wedge R \equiv \quad P \equiv Q \equiv R \equiv P \vee Q \equiv Q \vee R \equiv R \vee P \equiv P \vee Q \vee R$

IMPLICATION

(3.57) **Axiom, Definition of Implication:** $P \Rightarrow Q \equiv P \vee Q \equiv Q$

(3.58) **Axiom, Consequence:** $P \Leftarrow Q \equiv Q \Rightarrow P$

(3.59) **Definition of implication:** $P \Rightarrow Q \equiv \neg P \vee Q$

(3.60) **Definition of implication:** $P \Rightarrow Q \equiv P \wedge Q \equiv P$

(3.61) **Contrapositive:** $P \Rightarrow Q \equiv \neg Q \Rightarrow \neg P$

(3.62) $P \Rightarrow (Q \equiv R) \equiv P \wedge Q \equiv P \wedge R$

(3.63) **Distributivity of $\Rightarrow$ over $\equiv$:** $P \Rightarrow (Q \equiv R) \equiv P \Rightarrow Q \equiv P \Rightarrow R$

(3.64) $P \Rightarrow (Q \Rightarrow R) \equiv (P \Rightarrow Q) \Rightarrow (P \Rightarrow R)$

(3.65) **Shunting:** $P \wedge Q \Rightarrow R \equiv P \Rightarrow (Q \Rightarrow R)$

(3.66) $P \wedge (P \Rightarrow Q) \equiv P \wedge Q$

(3.67) $P \wedge (Q \Rightarrow P) \equiv P$

(3.68) $P \vee (P \Rightarrow Q) \equiv \mathit{true}$

(3.69) $P \vee (Q \Rightarrow P) \equiv Q \Rightarrow P$

(3.70) $P \vee Q \Rightarrow P \wedge Q \equiv P \equiv Q$

(3.71) **Reflexivity of $\Rightarrow$:** $P \Rightarrow P \equiv \mathit{true}$

(3.72) **Right zero of $\Rightarrow$:** $P \Rightarrow \mathit{true} \equiv \mathit{true}$

(3.73) **Left identity of $\Rightarrow$:** $\mathit{true} \Rightarrow P \equiv P$

(3.74) $P \Rightarrow \mathit{false} \equiv \neg P$

(3.75) $\mathit{false} \Rightarrow P \equiv \mathit{true}$

(3.76) **Weakening/strengthening:** (a) $P \Rightarrow P \vee Q$

(b) $P \wedge Q \Rightarrow P$

(c) $P \wedge Q \Rightarrow P \vee Q$

(d) $P \vee (Q \wedge R) \Rightarrow P \vee Q$

(e) $P \wedge Q \Rightarrow P \wedge (Q \vee R)$

(3.77) **Modus ponens:** $P \wedge (P \Rightarrow Q) \Rightarrow Q$

(3.78) $(P \Rightarrow R) \wedge (Q \Rightarrow R) \equiv (P \vee Q \Rightarrow R)$

(3.79) $(P \Rightarrow R) \wedge (\neg P \Rightarrow R) \equiv R$

(3.80) **Mutual implication:** $(P \Rightarrow Q) \wedge (Q \Rightarrow P) \equiv (P \equiv Q)$

(3.81) **Antisymmetry:** $(P \Rightarrow Q) \wedge (Q \Rightarrow P) \Rightarrow (P \equiv Q)$

(3.82) **Transitivity:** (a) $(P \Rightarrow Q) \wedge (Q \Rightarrow R) \Rightarrow (P \Rightarrow R)$

(b) $(P \equiv Q) \wedge (Q \Rightarrow R) \Rightarrow (P \Rightarrow R)$

(c) $(P \Rightarrow Q) \wedge (Q \equiv R) \Rightarrow (P \Rightarrow R)$

LEIBNIZ AS AN AXIOM

(3.83) **Axiom, Leibniz:** $F = G \;\Rightarrow\; E^Z_F = E^Z_G$

(3.84) **Substitution:** (a) $F = G) \wedge E^Z_F \;\equiv\; (F = G) \wedge E^F_G$

(b) $F = G) \Rightarrow E^Z_F \;\equiv\; (F = G) \Rightarrow E^Z_G$

(c) $Q \wedge F = G) \Rightarrow E^Z_F \;\equiv\; Q \wedge (F = G) \Rightarrow E^Z_G$

(3.85) **Replace by** *true* **:** (a) $P \Rightarrow E^Z_P \;\equiv\; P \Rightarrow E^Z_{true}$

(b) $Q \wedge P \Rightarrow E^Z_P \;\equiv\; Q \wedge P \Rightarrow E^Z_{true}$

(3.86) **Replace by** *false* **:** (a) $E^Z_P \Rightarrow P \;\equiv\; E^Z_{false} \Rightarrow P$

(b) $E^Z_P \Rightarrow P \vee Q \;\equiv\; E^Z_{false} \Rightarrow P \vee Q$

(3.87) **Replace by** *true* **:** $P \wedge E^Z_P \;\equiv\; P \wedge E^Z_{true}$

(3.88) **Replace by** *false* **:** $P \vee E^Z_P \;\equiv\; P \vee E^Z_{false}$

(3.89) **Shannon:** $E^Z_P \;\equiv\; (P \wedge E^Z_{true}) \vee (\neg P \wedge E^Z_{false})$

(4.1) $P \Rightarrow (Q \Rightarrow P)$

(4.2) **Monotonicity of** $\vee$ **:** $(P \Rightarrow Q) \Rightarrow (P \vee R \Rightarrow Q \vee R)$

(4.3) **Monotonicity of** $\wedge$ **:** $(P \Rightarrow Q) \Rightarrow (P \wedge R \Rightarrow Q \wedge R)$

PROOF TECHNIQUES

(4.4)**Deduction:** To prove $P \Rightarrow Q$, assume P and prove Q.

(4.5)**Case analysis:** If E^Z_{true}, E^Z_{false} are theorems, then so is E^Z_P.

(4.6)**Case analysis:** $(P \vee Q \vee R) \wedge (P \Rightarrow S) \wedge (Q \Rightarrow S) \wedge (R \Rightarrow S) \;\Rightarrow\; S$

(4.7)**Mutual implication:** To prove $P \equiv Q$, prove $P \Rightarrow Q$ and $Q \Rightarrow P$.

(4.9)**Proof by contradiction:** To prove P, prove $\neg P \Rightarrow$ *false*.

(4.11)**Proof by contrapositive:** To prove $P \Rightarrow Q$, prove $\neg Q \Rightarrow \neg P$

6.2 General laws of quantification

For symmetric and associative binary operator $\star$ with identity u.

(8.13) **Axiom, Empty range:** $(\star x \mid \mathit{false} : P) \;=\; u$

(8.14) **Axiom, One-point rule:** Provided $\neg occurs(\text{'}x\text{'}, \text{'}E\text{'})$,
$(\star x \mid x = E : P) \;=\; P[x := E]$

(8.15) **Axiom, Distributivity:** Provided $P, Q{:}\mathbb{B}$ or R is finite,
$(\star x \mid R : P) \star (\star x \mid R : Q) \;=\; (\star x \mid R : P \star Q)$

(8.16) **Axiom, Range split:** Provided $R \wedge S \equiv \mathit{false}$ and
$P{:}\mathbb{B}$ or R and S are finite,
$(\star x \mid R \vee S : P) \;=\; (\star x \mid R : P) \star (\star x \mid S : P)$

(8.17) **Axiom, Range split:** Provided $P{:}\mathbb{B}$ or R and S finite,
$(\star x \mid R \vee S : P) \star (\star x \mid R \wedge S : P) \;=\; (\star x \mid R : P) \star (\star x \mid S : P)$

(8.18) **Axiom, Range split for idempotent** $\star$ **:**
$(\star x \mid R \vee S : P) \;=\; (\star x \mid R : P) \star (\star x \mid S : P)$

(8.19) **Axiom, Interchange of dummies:** Provided $\star$ is idempotent or
R, Q are finite, $\neg occurs(\text{'}y\text{'}, \text{'}R\text{'})$, and $\neg occurs(\text{'}x\text{'}, \text{'}Q\text{'})$,
$(\star x \mid R : (\star y \mid Q : P)) = (\star y \mid Q : (\star x \mid R : P))$

(8.20) **Axiom, Nesting:** Provided $\neg occurs(`y', `R')$,
$(\star x, y \mid R \wedge Q : P) = (\star x \mid R : (\star y \mid Q : P))$

(8.21) **Axiom, Dummy renaming:** Provided $\neg occurs(`y', `R, P')$,
$(\star x \mid R : P) = (\star y \mid R[x := y] : P[x := y])$

(8.22) **Change of dummy:** Provided $\neg occurs(`y', `R, P')$, and f has an inverse, $(\star x \mid R : P) = (\star y \mid R[x := f.y] : P[x := f.y])$

(8.23) **Split off term:** $(\star i \mid 0 \leq i < n+1 : P) = (\star i \mid 0 \leq i < n : P) \star P^i_n$

6.3 Theorems of the predicate calculus

UNIVERSAL QUANTIFICATION

(9.2) **Axiom, Trading:** $(\forall x \mid R : P) \equiv (\forall x \mid : R \Rightarrow P)$

(9.3) **Trading:** (a) $(\forall x \mid R : P) \equiv (\forall x \mid : \neg R \vee P)$
(b) $(\forall x \mid R : P) \equiv (\forall x \mid : R \wedge P \equiv R)$
(c) $(\forall x \mid R : P) \equiv (\forall x \mid : R \vee P \equiv P)$

(9.4) **Trading:** (a) $(\forall x \mid Q \wedge R : P) \equiv (\forall x \mid Q : R \Rightarrow P)$
(b) $(\forall x \mid Q \wedge R : P) \equiv (\forall x \mid Q : \neg R \vee P)$
(c) $(\forall x \mid Q \wedge R : P) \equiv (\forall x \mid Q : R \wedge P \equiv R)$
(d) $(\forall x \mid Q \wedge R : P) \equiv (\forall x \mid Q : R \vee P \equiv P)$

(9.5) **Axiom, Distributivity of $\vee$ over $\forall$:** Prov. $\neg occurs(`x', `P')$,
$P \vee (\forall x \mid R : Q) \equiv (\forall x \mid R : P \vee Q)$

(9.6) Provided $\neg occurs(`x', `P')$, $(\forall x \mid R : P) \equiv P \vee (\forall x \mid : \neg R)$

(9.7) **Distributivity of $\wedge$ over $\forall$:** Provided $\neg occurs(`x', `P')$,
$\neg(\forall x \mid : \neg R) \Rightarrow ((\forall x \mid R : P \wedge Q) \equiv P \wedge (\forall x \mid R : Q))$

(9.8) $(\forall x \mid R : true) \equiv true$

(9.9) $(\forall x \mid R : P \equiv Q) \Rightarrow ((\forall x \mid R : P) \equiv (\forall x \mid R : Q))$

(9.10) **Range weakening/strengthening:** $(\forall x \mid Q \vee R : P) \Rightarrow (\forall x \mid Q : P)$

(9.11) **Body weakening/strengthening:** $(\forall x \mid R : P \wedge Q) \Rightarrow (\forall x \mid R : P)$

(9.12) **Monotonicity of $\forall$:**
$(\forall x \mid R : Q \Rightarrow P) \Rightarrow ((\forall x \mid R : Q) \Rightarrow (\forall x \mid R : P))$

(9.13) **Instantiation:** $(\forall x \mid : P) \Rightarrow P[x := e]$

(9.16) P is a theorem iff $(\forall x \mid : P)$ is a theorem.

EXISTENTIAL QUANTIFICATION

(9.17) **Axiom, Generalized De Morgan:**
$(\exists x \mid R : P) \equiv \neg(\forall x \mid R : \neg P)$

(9.18) **Generalized De Morgan:** (a) $\neg(\exists x \mid R : \neg P) \equiv (\forall x \mid R : P)$
(b) $\neg(\exists x \mid R : P) \equiv (\forall x \mid R : \neg P)$
(c) $(\exists x \mid R : \neg P) \equiv \neg(\forall x \mid R : P)$

(9.19) **Trading:** $(\exists x \mid R : P) \equiv (\exists x \mid : R \wedge P)$

(9.20) **Trading:** $(\exists x \mid Q \wedge R : P) \equiv (\exists x \mid Q : R \wedge P)$

(9.21) **Distributivity of $\wedge$ over $\exists$:** Provided $\neg occurs(`x', `P')$,
$P \wedge (\exists x \mid R : Q) \equiv (\exists x \mid R : P \wedge Q)$

(9.22) Provided $\neg occurs(`x', `P')$, $(\exists x \mid R : P) \equiv P \wedge (\exists x \mid : R)$

(9.23) **Distributivity of $\vee$ over $\exists$:** Provided $\neg occurs(`x', `P')$,
$(\exists x \mid : R) \Rightarrow ((\exists x \mid R : P \vee Q) \equiv P \vee (\exists x \mid R : Q))$

(9.24) $(\exists x \mid R : \mathit{false}) \equiv \mathit{false}$

(9.25) **Range weakening/strengthening:** $(\exists x \mid R : P) \Rightarrow (\exists x \mid Q \vee R : P)$

(9.26) **Body weakening/strengthening:** $(\exists x \mid R : P) \Rightarrow (\exists x \mid R : P \vee Q)$

(9.27) **Monotonicity of $\exists$:**
$(\forall x \mid R : Q \Rightarrow P) \Rightarrow ((\exists x \mid R : Q) \Rightarrow (\exists x \mid R : P))$

(9.28) **$\exists$-Introduction:** $P[x := E] \Rightarrow (\exists x \mid : P)$

(9.29) **Interchange of quantifications:**
Provided $\neg occurs(`y', `R')$ and $\neg occurs(`x', `Q')$,
$(\exists x \mid R : (\forall y \mid Q : P)) \Rightarrow (\forall y \mid Q : (\exists x \mid R : P))$

(9.30) Provided $\neg occurs(`\hat{x}', `Q')$,
$(\exists x \mid R : P) \Rightarrow Q$ is a theorem iff $(R \wedge P)[x := \hat{x}] \Rightarrow Q$ is a theorem

(9.31) **Quantified freshy:** [12] Provided $\neg occurs(`x', `P')$,
$P \Rightarrow (\forall x \mid R : P)$

[12] This theorem appears in the future second edition of *A Logical Approach to Discrete Math*, not the first edition. *Freshy* is slang for a freshman in a college or university. Here, we use *freshy* to denote the fresh variable x. It is fresh because it does not appear in P, but is introduced in the quantification.

Part 1

Examples of Protocol Design

The need for communication protocols arises if communication is restricted in some way: communication channels may be unreliable and lose messages, or the communication network may be confined to a structure that does not connect all nodes directly. In order to make communication reliable in spite of such hindrances, it is common to work with communication protocols. Being difficult in general, the design of protocols nicely illustrates principles and methods of program design. The following two examples show how different formalisms and methods can be used for the stepwise design of protocols.

Jean-Raymond Abrial

Wim H.J. Feijen,

Burghard von Karger

Specification and Design of a Transmission Protocol by Successive Refinements Using B[1]

by J.-R. Abrial[2] and L. Mussat[3]

1 Introduction

In this text, we propose a formal technique of protocol specification and design by *successive refinements*. The key idea is that the protocol is first specified in a time-less fashion by *concentrating* its intended behavior on *one step only*, all other steps being just void: in other words, time does not exist yet and, of course, no distribution can take place between the various phases of the protocol. This is the reason why the desired properties of the protocol are expressed so easily.

Then the protocol is refined by considering the other steps in turn. By doing so, the implicit *time dimension is refined* (stretched) gradually till it reaches its final extent where the various phases of the protocol might be executed concurrently. Of course, each refinement step has to be proved to satisfy the properties expressed in the preceding abstraction. As refinement is transitive, the final refinement is thus indeed a correct implementation of the initial specification.

In what follows, we shall experiment this approach on a well known protocol called the Bounded Retransmission Protocol (for short BRP) [1], [2], [3], [4]. We shall first give an informal description of the protocol (section 2). Then we shall make a little more precise what has been presented informally by giving a pseudo-code description of the various events composing the protocol (section 3). In the next section, we give a brief tutorial on B [5] and explain how it can be used to tackle the problem of the construction of distributed programs. In section 5, we present our plan for the systematic construction of the protocol, which we undertake in the remaining sections.

2 An Informal Presentation of the BRP

The effect of the execution of the BRP is to (partially) copy a file from one site to another. Hereafter, these sites are called respectively the Sender and the Receiver.

[1]This work has been supported by a grant from DRA and also by RATP, SNCF, and INRETS.
[2]Consultant.
[3]Service Central de la Sécurité des Systèmes d'Information.

The file is supposed to be transmitted piece by piece. The Sender sends (**snd_s**) a certain data item on the, so-called, Data Channel connecting the Sender to the Receiver. As soon as the Receiver receives (**rcv_r**) that data item it stores it in its own file and sends back an acknowledgment to the Sender on the, so-called, Acknowledgment Channel connecting the Receiver to the Sender. As soon as the Sender receives (**rcv_s**) this acknowledgment, it sends the next data item, and so on. We suppose that the *last* data item sent by the Sender contains a special information so that the Receiver is able to know when the file transmission is completed. Notice that it nevertheless has to send a final acknowledgment.

All this can be represented in the following diagram where the events (**snd_s**, **rcv_r**, and **rcv_s**) are supposed to represent the various phases of the protocol together with their synchronization:

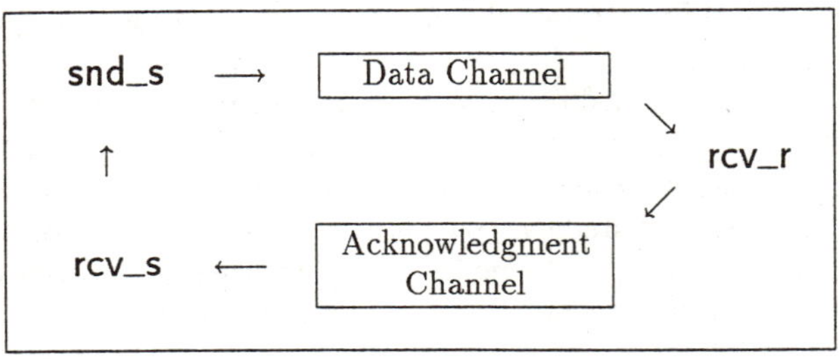

What we have just described is the "normal" behaviour of the protocol, where an entire file is transmitted from the Sender to the Receiver. We shall also describe below a "degraded" behaviour, where the Sender's file is transmitted only partially to the Receiver due to some problems on the transmission channels.

Unreliability of the Communications

The transmission channels (Data and Acknowledgement) situated between the Sender and the Receiver might be faulty: that is, some data items sent by the Sender or some acknowledgments sent by the Receiver might be lost. In order to cope with the unreliability of these channels, the Sender starts a timer when it sends a data item. This device is adjusted so that it wakes up the Sender (provided, of course, it has not received an acknowledgment in the meantime) after a certain delay. This delay is *guaranteed* to be greater than the *maximum* delay D_{max} that is required to first send a data item and subsequently send back the corresponding acknowledgment. In other words, the Sender can conclude that a message has necessarily been lost when the time is over, that is if a delay D_{max} has passed since the last data item has been sent to the Receiver without receiving a corresponding acknowledgment.

But, of course, when the timer wakes it up, the Sender does not know whether the lost message corresponds to the data item that it has just sent or to the corresponding acknowledgment supposed to be sent back by the Receiver. In any case, the Sender re-transmitts the previous data item and waits for the corresponding acknowledgment.

Protocol Abortion

In case of successive losses of messages, the process of data re-transmission can be repeated a number of times. When this number reaches a certain pre-defined limit N_{max}, the Sender decides that the transmission is definitely broken and aborts the protocol (from its own point of view). The question that arises immediately is then, of course, that of the synchronization with the Receiver. In other words, how does the Receiver know that the protocol has aborted ? Clearly, the Sender cannot communicate any longer with the Receiver in order to send it this abortion information because the communication is now broken.

This problem is solved by means of a second timer situated in the Receiver's site. This timer is activated by the Receiver when it receives a new data item (that is, not a re-transmitted one). This timer is adjusted so that it wakes up the Receiver (provided, of course, it has not received a new data item in the meantime) after a certain delay that is *guaranteed* to be such that the Receiver can be certain that the Sender has already given up (aborted the protocol). Clearly this delay has to be greater than or equal to the quantity $N_{max} \times D_{max}$, since after that delay the Sender must have given up as we have seen above. When the second timer wakes it up, the Receiver aborts the protocol (from its own point of view). As can be seen, in case of problems, the two participants are nevertheless indirectly synchronized by means of these timers.

Alternating Bit

As we have seen above, the Sender may re-transmitt the same data item several times. But, it may also transmitt two (or more) successive data items, which might happen to have the *same value*. Of course, this is annoying, since the Receiver may confuse a re-transmitted data item with a new one that is identical to its predecessor. In order to solve this problem, each data item is accompanied by a bit whose value is alternating from one item to the next. When the Receiver receives two successive items accompanied by the same bit, it can thus be certain that the latter is a re-transmission of the former.

Final Situation of the Protocol

At the end of the protocol execution, we might be in one of the following three situations: (1) either, the protocol has successfully been able to transfer the entire file from the Sender to the Receiver and the Sender has indeed received the last acknowledgment from the Receiver, (2) or the protocol has successfully been able to transfer the entire file from the Sender to the Receiver but the Sender has never received the last acknowledgment (in spite of successive re-transmissions, this message is definitely lost in the Acknowledgment Channel) so that the Sender aborts the protocols whereas the Receiver doesn't, (3) or else the protocol has aborted on both sites. Notice that the fourth possibility, where the Receiver would have aborted the protocol whereas the Sender wouldn't, is not possible.

3 A Pseudo-code Description of the BRP

In this section, we present a pseudo-code version of our protocol. The rôle of this description is to make a little more precise the completely informal presentation of the preceding section. Each event of the protocol (that is, snd_s, rcv_r, and rcv_s) and the two extra events corresponding to the timers (which we call, out_s and abt_r) are described in terms of an *enabling condition*, introduced by the key-word select, followed by an *action part*, introduced by the key-word then. The former contains the condition under which the event *may* be "fired", whereas the latter contains a description of what the event is supposed to do once it is enabled.

Our first event, snd_s, is enabled by a condition expressing that this event is indeed waken up (we shall see below that this is done either by the event rcv_s or by the event out_s). The action of snd_s consists in acquiring the next data item from the Sender's file, storing it on the Data Channel together with the corresponding alternating bit, starting the Sender's timer, and finally activating the Data Channel (effectively sending the data and the bit). Here is the pseudo-code of this event:

```
snd_s  ≙  select
              snd_s is waken up
          then
              Acquire data from Sender's file;
              Store acquired data on Data Channel;
              Store Sender's bit on Data Channel;
              Start Sender's timer;
              Activate Data Channel
          end
```

The next event, rcv_r, is enabled by the interrupt of the Data Channel on the Receiver's site. The action consists first in testing whether the alternating bit sent by the Sender is identical to the alternating bit previously stored by the Receiver. If this is the case, then this means, by convention, that we have a new data item: this item is extracted from the Data Channel, it is subsequently stored on the Receiver's file, the Receiver's alternating bit is modified, and, finally, the Receiver's timer is started if the received item is not the last one. In any case, the interrupt of the Data Channel is de-activated whereas the Acknowledgment Channel is itself activated. Here is the pseudo-code of this event:

```
rcv_r  ≙  select
              Data Channel interrupt occurs
          then
              Acquire Sender's bit from Data Channel;
              if Sender's bit = Receiver's bit then
                  Acquire data from Data Channel;
                  Store data on Receiver's file;
                  Modify Receiver's bit;
                  if data is not the last one then
                      Start Receiver's timer
                  end
              end;
              Reset Data Channel interrupt;
              Activate Acknowledgment Channel
          end
```

The next event, rcv_s, is enabled by the interrupt of the Acknowledgment Channel on the Sender's site. The action consists in removing the previously sent item from the Sender's file (although that data item has already been sent, it was nevertheless kept in the file in case of a re-transmission; now it can be definitely removed since we have just received the acknowledgment telling us that the Receiver has indeed received it). The Sender's alternating bit can now be modified for the next data item, the event snd_s is waken up, and, finally, the Acknowledgment Channel is de-activated.

```
rcv_s  ≙  select
              Acknowledgment Channel interrupt occurs
          then
              Remove data from Sender's file;
              Reset re-try counter;
              Modify Sender's bit;
              Wake up event snd_s;
              Reset Acknowledgment Channel interrupt
          end
```

The event **out_s** is enabled when the Sender's timer reaches its specified delay. The action consists in testing whether the re-try counter has reached its maximum value, in which case the protocol is aborted (from the point of view of the Sender). When this is not the case, then the re-try counter is incremented and, of course, the event **snd_s** is waken up for a re-transmission.

```
out_s  ≙  select
              Sender's timer interrupt occurs
          then
              if re-try counter is too big then
                  Abort protocol on Sender's site
              else
                  Increment re-try counter;
                  Wake up event snd_s
              end
          end
```

The event **abt_r** is enabled when the Receiver's timer reaches its specified delay. The action consists in aborting the protocol (from the point of view of the Receiver).

```
abt_r  ≙  select
              Receiver's timer interrupt occurs
          then
              Abort protocol on Receiver's site
          end
```

The definition of our protocol by means of this pseudo-code (or by means of any other similar descriptive notation) raises a number of questions. Are we sure that such a description is "correct" in the sense that it effectively corresponds to a *file transfer* protocol? Are we sure that the described protocol does terminate (no infinite loop, no deadlock)? What kind of properties should this protocol maintain?

It is our opinion that these questions cannot be answered on the basis of such an informal description. We nevertheless believe that it is quite useful to have such a description at one's disposal, since it may act as a *goal* to our future protocol construction. In the sequel, and as said above, we shall formally construct our protocol starting from a mathematical specification of its main properties, and ending up in a formal description of its components, which we might then fruitfully *compare* to their informal pseudo-code counterparts.

4 About B

In the sequel, we shall explore how B [5] can be used to specify, construct, and prove such a protocol, where the distribution of computation plays an important rôle. Clearly, such a system do not belong to the main target that was considered when B was developed over its long period of gestation. However, the usage of B for distributed system is clearly within reach. In this section, we recall a number of principles underlying B and we see how they can be applied to our problem. This utilization of B for developing distributed systems has already been quickly presented by one of us [10]. Also notice that our approach is very close to that already proposed by the *Action System* formalism of R. Back and others [9].

Abstract Machine

The main structuring device of B is the, so-called *Abstract Machine.* The basic idea behind this concept has been borrowed from those ideas that are behind some well-known programming features such as modules, packages, abstract data types or classes. In all these cases, the main principle at work is one of *encapsulation* of some (real or abstract) variables, protected by some "entry points" that constitute the only means for modifying or accessing the encapsulated variables in question. Everybody knows that using such devices has some nice structuring effects.

In an abstract machine, the encapsulated *variables* are declared and given some properties by means of a predicate called the *invariant* of the machine. And the "entry points" are called its *operations.*

Proof Obligation for an Abstract Machine

Proving that such a machine is *coherent* consists then in proving that the invariant is indeed left invariant after the "execution" of each operation. For a machine with invariant I and operation S, this takes the following form:

$$I \Rightarrow [S]\,I$$

In this formula, the notation $[S]\,I$ denotes the transformation of the predicate I by the *generalized substitution* S [5], considered to be a predicate transformer [6].

Operations with Pre-conditions

More precisely, the operation of an abstract machine is defined by a generalized substitution made of a *pre-condition* followed by an *action*. This takes the following form:

$$\boxed{\textsf{pre}\ P\ \textsf{then}\ S\ \textsf{end}}$$

The notion of pre-condition has the following meaning: this is simply a predicate that one has to prove for using the corresponding operation. In fact, any attempt at using the operation (at run time) without being sure that the proof of its pre-condition has been performed successfully may have the catastrophic effect of breaking the invariant and thus of possibly destroying the abstract machine.

The proof obligation corresponding to the preservation of an invariant I by an operation of the above form is the following:

$$I \wedge P \Rightarrow [\textsf{pre}\ P\ \textsf{then}\ S\ \textsf{end}]\,I$$

Notice that the occurrence of P in the antecedent of this implication is due to the fact that there is no point in proving the re-establishment of the invariant I by **pre** P **then** S **end** in the case where P does not hold, since the operation **pre** P **then** S **end** cannot be used outside its pre-condition (provided this is proved, of course). The above proof obligation can be expanded as follows (since, by definition, the predicate [**pre** P **then** S **end**] I is equivalent to $P \wedge [S]\,I$):

$$I \wedge P \Rightarrow P \wedge [S]\,I$$

reducing to

$$I \wedge P \Rightarrow [S]\, I$$

Abstract Machine Refinement

We now consider the refinement of an abstract machine. The refinement R of an abstract machine M consists of a formal text containing some variables (all distinct from those of M), an invariant J and a number of operations exactly corresponding to the operations of M (although their contents might be different). The well known idea is that one goes from a machine to its refinement by means of a change of variable that is expressed in the invariant J, which thus depends on *both* the variables of M and those of R. This is the reason why J is called a *gluing invariant.*

In the case of a machine M with an invariant I and an operation S refined by a refinement R with gluing invariant J and a corresponding operation T, the statement to be proved for the correctness of the refinement of S by T is the following [5]:

$$I \wedge J \wedge P \Rightarrow [T] \neg [S] \neg J$$

Notice that the predicate $[T] \neg [S] \neg J$ has to be parsed as follows:

$$[T]\,(\neg\,(\,[S]\,(\neg J\,)\,)\,)$$

In the case of a pre-conditioned operation **pre** P **then** S **end** refined by **pre** Q **then** T **end**, under the extra gluing invariant J, the proof obligation is:

$$I \wedge J \wedge P \Rightarrow Q \wedge [T] \neg [S] \neg J$$

reducing to the two independant statements

$$I \wedge J \wedge P \Rightarrow Q$$

$$I \wedge J \wedge P \Rightarrow [T] \neg [S] \neg J$$

The first of the above statements is interesting: it shows that the concrete pre-condition Q is necessarily *weaker* than its more abstract counterpart P.

Abstract System

In the case where the computation to be described can be performed in a distributed way (be it a real distribution or a simulated one), the situation is quite different from the one described above in the sense that we potentially have *several* computations taking place simultaneously. In other words, when a certain computation C is suspended for some reason, then this is not necessarily catastrophic (as far as the entire system is concerned) because some other computations can proceed instead and possibly invalidate the reason why computation C has been forced to stop its course.

The notion of abstract machine can also be used for developing distributed systems. The only difference is that we are going to give it *another interpretation.* In order to make a clear distinction between the two, let us call an abstract machine, an *abstract system* when using it for constructing distributed systems.

The first question we can ask ourselves is concerned with the status of the data that are "encapsulated" in an abstract system. Since, potentially, an abstract system can be used to formalize the behaviour of an *entire* distributed system, it is no longer the case that these data correspond to a single part that is to be encapsulated within a module.

For instance, when formalizing the distributed execution of a communication protocol between two partners, the data that are formalized within the corresponding abstract system are not the ones corresponding to a *single* partner, they rather correspond to the two of them put together in the same box. As can be seen, the vision is rather global.

An abstract system is no longer a "machine" protecting some data, it rather represents the mathematical model of an entire system as seen from a unique global abstract point of view. This has the immense advantage of allowing us to express, in the invariant of such an abstract system, some *global properties* that would otherwise be impossible to write down, should our system be decomposed, to begin with, in several independent sub-systems.

Guarded Events

Our next question is concerned with the status of the operations of such an abstract system. Such operations will now be called *events.* An event is not any more made of a pre-condition and an action as is the operation of an abstract machine. We still have an action part, but the pre-condition is now replaced by a *guard*. The guard is a predicate that represents the condition allowing the event to be *enabled*. Notice that it is not necessary any more to have a (static) proof of the guard of an event before using it. The only thing that counts is that the action of the event is now performed *under the assumption of the guard*.

The events of an abstract system are used to formalize what happens dynamically and spontaneously (guards permitting) in our system "as seen from the sky". The mathematical model we would like to build is that of the *entire* dynamic behaviour of our system.

Each operation of an abstract system is thus defined by means of a *guard* and an *action*. For example, in an abstract system, an event takes the following form:

$$\textsf{select } P \textsf{ then } S \textsf{ end}$$

Proof Obligation for an Abstract System

And the proof obligation corresponding to the preservation of an invariant I by an event like the one above is the following:

$$I \;\Rightarrow\; [\textsf{select } P \textsf{ then } S \textsf{ end}]\, I$$

that is (since, by definition, the predicate $[\textsf{select } P \textsf{ then } S \textsf{ end}]\, I$ is equivalent to $P \Rightarrow [S]\, I$)

$$I \;\Rightarrow\; (P \Rightarrow [S]\, I)$$

reducing to

$$I \wedge P \;\Rightarrow\; [S]\, I$$

As can be seen, this proof obligation has exactly the same shape as that obtained above for the pre-condition: in both cases, the predicate P (be it a pre-condition or a guard) will end up as an *hypothesis* of the proof of $[S]\,I$ (together, of course, with the other hypotheses corresponding to the invariant I).

Abstract System Refinement

What is interesting to notice now is what happen in the case of the refinement of the guarded substitution, **select** P **then** S **end**, by another guarded substitution, say **select** Q **then** T **end**, under an extra gluing invariant J. This yields:

$$I \wedge J \;\Rightarrow\; (Q \Rightarrow [T]\neg(P \Rightarrow [S]\neg J))$$

that is

$$I \wedge J \wedge Q \;\Rightarrow\; [T](P \wedge \neg[S]\neg J)$$

that is, by distributing the application of the substitution T over the conjunction

$$I \wedge J \wedge Q \;\Rightarrow\; [T]P \;\wedge\; [T]\neg[S]\neg J$$

reducing to the following (since the variables involved in T are necessarily distinct from those involved in P, thus the predicate $[T]\,P$ is equivalent to P):

$$I \wedge J \wedge Q \;\Rightarrow\; P$$

$$I \wedge J \wedge Q \;\Rightarrow\; [T]\neg[S]\neg J$$

The first of the above statements shows that the concrete guard Q is necessarily *stronger* than the abstract guard P: this contrasts with the result we have obtained above for pre-conditions, which was exactly the opposite. In that case, since the concrete pre-condition must always, as we have seen, be weaker than its abstract counterpart, we have the possibility to completely remove it in a final implementation (so that the concept of pre-condition appears in the constructing process only, not in the final implementation).

The situation with the guard is not as comfortable: once an event has got a non-trivial guard, it must at least be kept in a refinement; as a consequence, it cannot be removed as was the case for the pre-condition; on the contrary, it can only be strengthened. This means that when an event is refined, it potentially occurs *less frequently* than its abstraction.

Time Refinement

In our development, we would like to make the constant assumption that we will have *no simultaneity* between the execution of any two events: a very comfortable situation indeed when it is envisaged to make some formal reasoning involving events that modify the data of our abstract system. It would be quite complicated to envisage that two (or more events) could happen simultaneously.

The idea is thus that each event we would like to observe and formalize is so small ("atomic") that its execution can be considered to take no time. In other words, we shall always suppose that the "grain" of atomicity of our events is *always* as small as we like, so that there is no direct interference between any two events modifiying the same piece of data. We can thus assume that such events will always do so in a certain order, *never* simultaneously. This is certainly very strong an assumption, so that we wonder whether it is possible to consider that it has any practical ground.

This objection is certainly valid if we consider that the time is a "medium" with a constant measurement unit. In this case, we are obliged to consider, *right from the beginning*, the smallest grain we need and thus define, again right from the beginning, an enormous number of distinct events. This is clearly an hopeless situation.

In order to solve the above problem, we have no choice but to consider that the time can be abstracted (and thus refined). The level of refinement of the time is directly connected to the size of the unit, with which we measure it. The idea being that only one event can take place within one unit of time. Thus when the unit is large, the corresponding time is very abstract, and when the unit is small, the corresponding time is very concrete.

The most abstract time is thus a "time" whose unit is as large as we want: for instance, we can imagine a certain time with only one unit of measurement between the Big Bang and 1997. In this case of course, only a single event can take place, the one by which the state of the universe in 1997 is produced at once. Notice that this interesting result is clearly obtained "in no time", thus making our above assumption about an ideal granularity being present even within this extremely strong abstraction.

Notice that we shall never deal with the time dimension directly, we shall rather use our events themselves to express the dimensionality of time. An immediate consequence of this fact is that the refinement of an abstract system not only refines its existing abstract events, but also introduces some *new events* that were not perciptible in the abstraction simply because our time glasses were not powerful enough. Time is *stretched* when moving from an abstraction to its refinement, in exactly the same way as space is stretched when moving from a normal vision to one that is performed through a microscope. The stretching of time reveals some "time details", some new events.

At this point, we might ask ourselves what is to be proved concerning such a new event when it is introduced in the refinement of an abstract system. The answer is very simple indeed, such a new event simply refines the substitution that does nothing, that is the substitution **skip**. Notice that **skip** has no guard so that it can be refined by a guarded event, whatever the guard. In the abstraction, although we could not "see" it, this event indeed existed, it was simply *hidden* under the folding of time, thus having no effect on the abstract state.

For instance, suppose that we have an abstract system, with an invariant I, now refined together with the introduction of a new event, say **select** Q **then** T **end**, under the extra gluing invariant J. The proof obligation for this new event is then the following (since, by definition, the predicate $[\mathsf{skip}]\neg J$ is equivalent to $\neg J$):

$$I \wedge J \;\Rightarrow\; (Q \;\Rightarrow\; [T]\neg\neg J)$$

that is

$$\boxed{I \wedge Q \;\Rightarrow\; (J \;\Rightarrow\; [T]\,J)}$$

We have just what we expect. The new event must preserve the gluing invariant J under the assumption of the abstract invariant I and that of its guard Q (the firing condition of the event).

More on Generalized Substitutions

We shall now end up this small survey of B by recalling the various forms of generalized substitutions we shall use in the sequel, together with the corresponding axioms. We have already seen the pre-condition, **pre**, the guard, **select**, and the substitution that does nothing, **skip**, together with the following axioms:

[pre P then S end] R	$\Leftrightarrow$	$P \wedge [S]\,R$
[select P then S end] R	$\Leftrightarrow$	$P \Rightarrow [S]\,R$
[skip] R	$\Leftrightarrow$	R

The next two forms of generalized substitutions we shall use in the sequel are the multiple substitution, introduced by the classical operator ":=" and the specification statement (invented by C.C.Morgan [7]), introduced by the operator ":". We have the following axioms (of course, generalizable to more variables):

$$[x := E]\,R(x) \quad\Leftrightarrow\quad R(E)$$

$$[x : P(x)]\,R(x) \quad\Leftrightarrow\quad \forall(x)\cdot(\,P(x) \Rightarrow R(x)\,)$$

The operational interpretation of the multiple substitution is, of course, the classical assignment of imperative programming languages (although rarely multiple). The specification statement $x : P(x)$ can be interpreted as assigning to the variable(s) x *any* value E that is such that the predicate $P(E)$ holds. This can be made clear by considering the refinement proof obligation of $x : P(x)$ by $x' := E$ (under the gluing invariant $x = x'$), which yields $P(E)$ as one would expect.

$$
\begin{array}{l}
\quad [x' := E]\,\neg\,[x : P(x)]\,\neg\,(x = x') \\
\Leftrightarrow \\
\quad [x' := E]\,\neg\,\forall x \cdot (\,P(x) \Rightarrow \neg\,(x = x')\,) \\
\Leftrightarrow \\
\quad [x' := E]\,\exists x \cdot (\,P(x) \wedge x = x'\,) \\
\Leftrightarrow \\
\quad [x' := E]\,P(x') \\
\Leftrightarrow \\
\quad P(E)
\end{array}
$$

Sometimes we would like to mention the "before value" of some variables in the predicate P of a specification statement. For this, the classical convention is to denote the before value of a variable x by subscripting it with a 0. The more general specification statement is thus $x : P(x_0, x)$. And the corresponding axiom is the following:

$$[x : P(x_0, x)]\, R(x) \;\Leftrightarrow\; \forall x' \cdot (\, P(x, x') \;\Rightarrow\; R(x')\,)$$

5 Method of Formal Construction

In the sections to come after this one, we are going to construct the Bounded Retransmission Protocol in a systematic fashion using B. In this section, we present our general plan for doing this.

Rather than presenting the formal texts of our B specification and refinements in exactly the way they are entered in Atelier-B [8] (the tool that supports the B Method), we shall take a few liberties with the outer syntax, in order to make our presentation easier to comment and to read. We feel free to do so, since the development that is presented here has been entered into Atelier B and entirely proved with it. It can be consulted through the web pages of Atelier B [11].

Each section is made of four sub-sections that are the following:

1. *Purpose*: This sub-section contains an informal technical description of what is to be done (that is, specified or refined) and why.

2. *Variables*: This sub-section contains a list of the variables together with their informal denotations.

3. *Invariant*: This sub-section contains a formal description of the various invariant clauses, together with an informal presentation of their meaning and purpose.

4. *Events*: This sub-section contains the formal description of the various events, togther with an informal presentation of their guards and actions.

Notice that we might find here and there a few additional sub-sections in case of specific needs. This is the case during the fifth refinement with the problem of *termination*, and during the seventh refinement with the problem of *deadlockfreeness*.

The presentation contains one specification followed by seven refinements. Clearly, this is not the only possibility. In fact, we investigated many variants before agreeing on this one. In order to put some rationale on the way we constructed the protocol, we tried to have a *general theme* for each refinement. Such themes have all to do with the *gradual distribution* of various aspects of the protocol that are, by definition, totally centralized in the specification. We have thus the distribution of the termination (first and second refinements), the distribution of the data (third, fourth, and fifth refinements), and the distribution of the control (sixth and seventh refinements).

The presentation does not contain any explicit proofs (just sometimes a few hints). Again, this is because the formal construction has been proved entirely with Atelier B. It resulted in a total of 170 elementary proofs, among which 137 (that is 80%) where proven automatically by the tool (no human interventions), while 33 needed a manual intervention to instruct the prover. We must say that we used the prover of Atelier B a lot during the formal development. It helped us directing our investigations for the discovery of the best strategy. In fact, we used it very much in the way an engineer uses a pocket calculator to figure out what the order of magnitude of the main dimensions of its planned construction is.

6 Formal Specification of the BRP

6.1 Purpose

The specification consists in expressing what the protocol needs in order to be initiated properly, and what it has achieved once it is completed. At this stage we have thus two events called ini_brp and brp. The former corresponds to the initialization of both sites so that they are able to start communicating in a proper way. The latter is supposed to "execute" the entire protocol in no time.

The protocol is initialized by an event able to be fired, of course, when the protocol is not active. This event has the effect of initializing the Sender's file (a non-empty file) and resetting the Receiver's file. It also makes the protocol active on both sites so that both sites are now willing to cooperate in order to execute the protocol.

The formal specification consists then in explaining what kind of *knowledge* each site may believe in *at the end of the protocol.* This corresponds essentially to the following requirements:

(a) The protocol is a file transfer protocol. Its goal is to transfer a certain non-empty sequential file, totally or partially, from one site to another.

(b) At the end of the protocol, each site knows that the protocol is indeed completed. In other words, it is not possible for one site to believe that the protocol is still active while the other believes that the protocol is completed. This assertion is, of course, only valid at this very early stage; in the sequel, we shall see that, because of the distribution of its execution, the protocol will, of course, not neccessarily be completed simultaneously on both sites.

(c) Each site may end up in any of the two situations: either it knows that the protocol has terminated successfully (that the Sender's file has been entirely copied on the Receiver's site), or it knows that the protocol has aborted before being successfully terminated. However, as we shall see below, this knowledge is not necessarily shared by both sites.

(d) When the Sender knows that the protocol has successfully terminated then the Receiver knows it also. Conversely, when the Receiver knows that the protocol has aborted then the Sender knows it also.

(e) However, it is possible for the Sender to believe that the protocol has aborted while the Receiver knows that it has terminated successfully. This is due to the fact that the last acknowledgment sent by the Receiver might be lost. Notice, however, that this *cause* cannot be formally written at this stage.

(f) When the Receiver believes that the protocol has successfully terminated, this is because the Sender's file has been entirely copied on the Receiver's site. Conversely, when the Receiver believes that the protocol has aborted, then the Sender's file has certainly *not been copied entirely* on the Receiver's site.

6.2 Variables

- $a0$ denotes the Sender's file

- $b0$ denotes the Receiver's file

- $ts0$ denotes the activation bit of the Sender. When $ts0$ is equal to UP, this means either that the protocol execution has not yet started or that it is in progress (from the point of view of the Sender). When $ts0$ is equal to $\overline{UP}$, this means that the protocol execution is completed (again, from the point of view of the Sender).

- $tr0$ denotes the activation bit of the Receiver. We follow for $tr0$ the same conventions as those adopted for $ts0$.

- $ss0$ denotes the final status of the protocol from the point of view of the Sender. It can take the value OK (meaning that the entire file has been sent and acknowledged) or the value $\overline{OK}$ (meaning that the protocol execution has aborted).

- $sr0$ denotes the final status of the protocol from the point of view of the Receiver. It can take the value OK (meaning that the entire file has been received) or the value $\overline{OK}$ (meaning that the protocol execution has aborted).

Note that the symbol UP, $\overline{UP}$, OK, and $\overline{OK}$ are just paintings for 0 and 1, whose rôle is just to enhance readibility. So, clearly, the variables $ts0$, ..., $sr0$ (as well as their further refinements) are bits, that is members of the set $\{0, 1\}$.

6.3 Invariant

The invariant states that the Sender's file $a0$ is represented by a *non-empty* sequence of data items belonging to a certain set S (formally, the set of such sequences is denoted by the construct $\mathsf{seq1}\,(S)$). Likewise, the Receiver's file $b0$ is represented by a sequence of elements belonging to S (formally, the set of such sequences is denoted by the construct $\mathsf{seq}\,(S)$). Moreover this sequence $b0$ is a prefix of $a0$. The following three clauses thus essentially correspond to the informal requirement **(a)** of section 6.1.

$$a0 \in \mathsf{seq1}\,(S)$$

$$b0 \in \mathsf{seq}\,(S)$$

$$b0 \subseteq a0$$

The variables $ts0$, $tr0$, $ss0$, and $sr0$ are represented by a series of bits. The following four clauses thus formalize part of requirements **(b)** and **(c)**.

$$\begin{array}{l} ts0 \in \{ UP, \overline{UP} \} \\ tr0 \in \{ UP, \overline{UP} \} \\ ss0 \in \{ OK, \overline{OK} \} \\ sr0 \in \{ OK, \overline{OK} \} \end{array}$$

Since, as we shall see, the protocol is executed "in no time", then we can assert, as an invariant clause, that the Receiver's file is empty as long as the protocol is active on both sites, namely:

$$ts0 = UP \ \wedge \ tr0 = UP \quad \Rightarrow \quad b0 = [\,]$$

Note that the symbol "[]" denotes the empty sequence.

The next clause of the invariant states that, provided the protocol has terminated successfully at the Sender's site, then it has also terminated successfully at the Receiver's site. Conversely, when the protocol has terminated abnormally at the Receiver's site then it also has terminated abnormally at the Sender's site. This is essentially requirement **(d)**.

$$\begin{array}{l} ts0 = \overline{UP} \ \wedge \ ss0 = OK \ \Rightarrow \ tr0 = \overline{UP} \ \wedge \ sr0 = OK \\ tr0 = \overline{UP} \ \wedge \ sr0 = \overline{OK} \ \Rightarrow \ ts0 = \overline{UP} \ \wedge \ ss0 = \overline{OK} \end{array}$$

Notice that these clauses are compatible with the following clause (which is *not* an invariant clause), corresponding to requirement **(e)**. We remind the reader that this requirement said that it was possible for the Sender to have terminated abnormally (that is, $ts0 = \overline{UP} \ \wedge \ ss0 = \overline{OK}$) while the Receiver has terminated successfully (that is, $tr0 = \overline{UP} \ \wedge \ sr0 = OK$).

$$ts0 = \overline{UP} \ \wedge \ ss0 = \overline{OK} \ \wedge \ tr0 = \overline{UP} \ \wedge \ sr0 = OK$$

Finally, it is stated that provided the protocol has terminated at the Receiver's site then the fact that this termination is successful is equivalent to the fact that the Sender's file has been entirely copied into the Receiver's site. This is essentially requirement **(f)**.

$$tr0 = \overline{UP} \;\Rightarrow\; (\, sr0 = OK \;\Leftrightarrow\; b0 = a0\,)$$

6.4 Events

Our first event concentrates the initialization of the protocol within a single event. It is, of course, to be fired only when the protocol is inactive on both sites, that is when the condition $ts0 = \overline{UP} \;\wedge\; tr0 = \overline{UP}$ holds. Its effect is to choose (non-deterministically) a Sender's file $a0$ to be copied from one site to the other. Notice that the choice of a specific Sender's file $a0$ is not part of our protocol: this is certainly made by an "upper layer", and thus not studied here. Another action of this event is to make the Receiver's file empty. Finally, the protocol is made active on both sites. Of course, this initialization happens "in no time".

$$\begin{array}{lll}
\textsf{ini_brp} & \widehat{=} & \textsf{select} \\
 & & \quad ts0 = \overline{UP} \;\;\wedge\;\; tr0 = \overline{UP} \\
 & & \textsf{then} \\
 & & \quad a0, b0, ts0, tr0 : \left(\begin{array}{l} a0 \in \mathsf{seq1}\,(S) \;\;\wedge \\ b0 = [\,] \;\;\wedge \\ ts0 = UP \;\;\wedge \\ tr0 = UP \end{array}\right) \\
 & & \textsf{end}
\end{array}$$

As can be seen, we have used in the action part of this event a "specification statement" of the form $x : P(x)$ where x is a list of variables and $P(x)$ is a predicate on these variables. We remind the reader that it can be read as follows: "modify the variables x in such a way that $P(x)$ holds".

Our next event, **brp** concentrates the entire protocol execution in a single event. It has a guard stating that the protocol is active on both sites.

The action consists in choosing non-deterministically the various state variables (except the Sender's file, which is left unchanged)) in such a way that (1) the Receiver's file $b0$ becomes a (possibly full) prefix of the Sender's file $a0$, (2) the two status bits $ss0$ and $sr0$ become compatible with the invariant, and (3) the two terminating bits $ts0$ and $tr0$ indicate the termination of the protocol on both sites.

$$
\begin{array}{lll}
\mathsf{brp} & \triangleq & \mathsf{select} \\
& & \quad ts0 = UP \quad \wedge \quad tr0 = UP \\
& & \mathsf{then} \\
& & \quad b0, ss0, sr0, ts0, tr0 : \left(\begin{array}{l}
b0 \subseteq a0 \quad \wedge \\
ss0 \in \{\, OK, \overline{OK} \,\} \quad \wedge \\
sr0 \in \{\, OK, \overline{OK} \,\} \quad \wedge \\
ss0 = OK \;\Rightarrow\; sr0 = OK \quad \wedge \\
sr0 = OK \;\Leftrightarrow\; b0 = a0 \quad \wedge \\
ts0 = \overline{UP} \quad \wedge \\
tr0 = \overline{UP}
\end{array} \right) \\
& & \mathsf{end}
\end{array}
$$

7 First Refinement

7.1 Purpose

Our first refinement consists in *distributing* the final phase of the protocol on both sites. The abstract event, brp, of the specification is now splitted into four more events called end_s, abt_s, end_r, abt_r. Those events suffixed by s are concerned with the Sender, whereas those suffixed by r are concerned with the Receiver. Those events prefixed by end are concerned with a successful termination, whereas those prefixed by abt are concerned with an aborted execution of the protocol.

In this refinement, we perform our first *time-stretching.* It concerns the "time" of the Sender as well as that of the Receiver. In fact, we can consider that the new events end_s, abt_s, end_r, abt_r were all "hidden" under the abstract event brp of the specification (this is the reason why they should all refine the abstraction skip). As will be seen, the concrete protocol terminates *earlier* than its abstraction.

7.2 Variables

The concrete variables are identical (up to renaming) to their abstract counterparts. But we introduce a new variable, which is the following

- $tg1$ denotes the "global" termination bit. When $tg1$ is equal to $\overline{UP}$, this means that the protocol execution is certainly completed on *both* sites, and when it is equal to UP, then nothing can be said. This variable will not be used in the proper events of the protocol. This is just a technicality that is due to the fact that the concrete protocol terminates earlier than the abstract one.

7.3 Gluing Invariant

Next is the invariant that "glues" together the preceding abstraction and our first refinement. We first declare our new global terminating bit $tg1$:

$$tg1 \in \{UP, \overline{UP}\}$$

As we shall see, this variable will be used in the guard of the event ini_brp under the form of the condition $tg1 = \overline{UP}$. As the concrete guard is necessary stronger than the abstract one, we need the following invariant clause:

$$tg1 = \overline{UP} \quad \Rightarrow \quad ts0 = \overline{UP} \quad \wedge \quad tr0 = \overline{UP}$$

Likewise, this variable will be used in the guard of the event brp under the form of the condition $tg1 = UP$ (together with the fact that the protocol has terminated on both sites as indicated by the condition $ts1 = \overline{UP} \wedge tr1 = \overline{UP}$). Again, as the concrete guard is stronger that the corresponding abstract one, we need the following invariant clause:

$$tg1 = UP \quad \Rightarrow \quad ts0 = UP \quad \wedge \quad tr0 = UP$$

The new variable $a1$ denotes the Sender's file, it is the same as the abstract file $a0$. The variable $b1$ is not necessarily identical to its abstraction $b0$ (in fact, as we shall see, $b1$ is "younger" than $b0$), but it is necessary *included* in the Sender's file.

$$a1 = a0$$

$$b1 \subseteq a1$$

The various bits are also redeclared because they are distinct from their abstract counterparts (again, they are younger).

$$ts1 \in \{UP, \overline{UP}\}$$

$$tr1 \in \{UP, \overline{UP}\}$$

$$ss1 \in \{OK, \overline{OK}\}$$

$$sr1 \in \{OK, \overline{OK}\}$$

The next clauses of our gluing invariant deal with the time-stretching effect. The first one says that the abstract protocol termination takes place *after* the concrete protocol termination. The second one says that after the abstract termination, then the abstract and concrete Receiver's file, as well as the abstract and concrete bits, are identical.

$$ts0 = \overline{UP} \;\wedge\; tr0 = \overline{UP} \;\Rightarrow\; ts1 = \overline{UP} \;\wedge\; tr1 = \overline{UP}$$

$$ts0 = \overline{UP} \;\wedge\; tr0 = \overline{UP} \;\Rightarrow\; b1, ss1, sr1 \;=\; b0, ss0, sr0$$

The last three clauses are copies in the concrete space of the last three invariant clauses of the abstraction.

$$ts1 = \overline{UP} \;\wedge\; ss1 = OK \;\Rightarrow\; tr1 = \overline{UP} \;\wedge\; sr1 = OK$$

$$tr1 = \overline{UP} \;\wedge\; sr1 = \overline{OK} \;\Rightarrow\; ts1 = \overline{UP} \;\wedge\; ss1 = \overline{OK}$$

$$tr1 = \overline{UP} \;\Rightarrow\; (sr1 = OK \;\Leftrightarrow\; b1 = a1)$$

7.4 Events

Our first event, **ini_brp** is very close to its abstraction (up to the renaming of the state variables). Its only difference is concerned with the guard, which now states that the concrete global activation bit of the protocol is, of course, inactive.

$$\begin{array}{lll}\mathsf{ini\_brp} & \widehat{=} & \mathsf{select} \\ & & \quad tg1 = \overline{UP} \\ & & \mathsf{then} \\ & & \quad a1, b1, tg1, ts1, tr1 : \left(\begin{array}{l} a1 \in \mathsf{seq1}(S) \;\wedge \\ b1 = [\,] \;\wedge \\ tg1 = UP \;\wedge \\ ts1 = UP \;\wedge \\ tr1 = UP \end{array}\right) \\ & & \mathsf{end}\end{array}$$

Our next event, **end_s**, corresponds to the successful termination of the protocol on the Sender's site. This event can only occur when the protocol is active at the Sender's site, that is when the condition $ts1 = UP$ holds, and when the Receiver has already terminated successfully, that is when the condition $tr1 = \overline{UP} \wedge sr1 = OK$ holds. Notice that this latter fact is known magically (for the moment) by the Sender, which will normally have no direct access to the termination and status bits of the Receiver. In the present abstraction, however, it is possible for an event "situated" on one site to "look over the shoulder" of the other site. In other words, the guard of our event is *not yet fully localized.*

$$\textsf{end\_s} \;\widehat{=}\; \begin{array}{l} \textsf{select} \\ \quad ts1 = UP \quad \wedge \quad tr1 = \overline{UP} \quad \wedge \quad sr1 = OK \\ \textsf{then} \\ \quad ts1,\ ss1 \ := \ \overline{UP},\ OK \\ \textsf{end} \end{array}$$

The second event, **abt_s**, corresponds to the abnormal termination at the Sender's site. It can occur *at any time* while the protocol is active at the Sender's site, that is when the condition $ts1 = UP$ holds. In fact, as we shall see below, this guard will be refined (strengthened) in the next refinement, where we shall take account of the interrupt of the timer.

$$\textsf{abt\_s} \;\widehat{=}\; \begin{array}{l} \textsf{select} \\ \quad ts1 = UP \\ \textsf{then} \\ \quad ts1,\ ss1 \ := \ \overline{UP},\ \overline{OK} \\ \textsf{end} \end{array}$$

Our next event, **end_r**, corresponds to the successful termination of the protocol at the Receiver's site. It can only occur, of course, when the protocol is active at the Receiver's site. This event is still very primitive in that its effect is to copy the entire file *at once* into the Receiver's site.

$$\textsf{end\_r} \;\widehat{=}\; \begin{array}{l} \textsf{select} \\ \quad tr1 = UP \\ \textsf{then} \\ \quad b1,\ tr1,\ sr1 \ := \ a1,\ \overline{UP},\ OK \\ \textsf{end} \end{array}$$

The next event, **abt_r**, corresponds to the abnormal termination of the protocol at the Receiver's site. It can only occur when the protocol is active at the Receiver's site, that is when the condition $tr1 = UP$ holds, and when the protocol has already aborted on the Sender's site, that is when the condition $ts1 = \overline{UP} \wedge ss1 = \overline{OK}$ holds. Again this latter fact corresponds to a certain knowledge known magically by the Receiver. Like the preceding one, this event is still very primitive in that its effect is to copy a certain (strict) part of the file *at once* into the Receiver's site.

$$
\begin{array}{lll}
\mathsf{abt\_r} & \widehat{=} & \mathsf{select} \\
 & & \quad tr1 = UP \quad \wedge \quad ts1 = \overline{UP} \quad \wedge \quad ss1 = \overline{OK} \\
 & & \mathsf{then} \\
 & & \quad b1, tr1, sr1 \ : \ (b1 \subset a1 \quad \wedge \quad tr1 = \overline{UP} \quad \wedge \quad sr1 = \overline{OK}) \\
 & & \mathsf{end}
\end{array}
$$

Our last event, brp, is a refinement of the unique event of our abstraction. It can only occur when both sites have admitted the termination of the protocol, that is when the condition $ts1 = \overline{UP} \wedge tr1 = \overline{UP}$ holds and when the global termination bit is not set, that is when the condition $tg1 = UP$ holds. The event just sets the global termination bit to $\overline{UP}$. This shows very clearly the time-stretching effect of this refinement. Here the more concrete protocol has terminated *earlier* than its abstraction. This is because we now have a *distributed termination* between both partners.

$$
\begin{array}{lll}
\mathsf{brp} & \widehat{=} & \mathsf{select} \\
 & & \quad tg1 = UP \quad \wedge \quad ts1 = \overline{UP} \quad \wedge \quad tr1 = \overline{UP} \\
 & & \mathsf{then} \\
 & & \quad tg1 \ := \ \overline{UP} \\
 & & \mathsf{end}
\end{array}
$$

From now on, we shall not present further refinements of this event as they will all be identical to this one (up to the renaming of the state variables).

8 Second Refinement

8.1 Purpose

In this refinement, our intention is to make precise in which circumstances the protocol may abort either on the Sender's site or on the Receiver's site. For this, we introduce a re-try counter that lies within the range $0\,..\,(MAX+1)$. This will allow for at most MAX re-transmissions for each data item (this does not include the first, normal, transmission). For further details on how the counter is used, see below the comments preceding each event of this refinement.

We introduce the new event, out_s, corresponding to the occurence of the interrupt of the Sender's timer, and the new event rcv_s whose rôle, *for the moment*, is only to reset the re-try counter. As will be seen below, this event will be the one corresponding to the reception by the Sender of the acknowledgment issued by the Receiver.

8.2 Variables

Besides the renaming of the abstract variables, the only new variable is the following:

- $c2$ denotes the re-try counter.

8.3 Gluing Invariant

The files as well as the various bits are identical to their abstractions.

$$a2, b2, ts2, tr2, ss2, sr2, tg2 \;=\; a1, b1, ts1, tr1, ss1, sr1, tg1$$

We now declare the re-try counter and state its main property, which tells us that when time is ready for the concrete Receiver to abort, that is when the counter $c2$ is equal to $MAX+1$, then this fact is *equivalent* to the fact that the abstract Sender has aborted.

$$c2 \in 0\,..\,(MAX+1)$$

$$c2 = MAX+1 \;\Leftrightarrow\; ts1 = \overline{UP} \;\wedge\; ss1 = \overline{OK}$$

Going from the first predicate $c2 = MAX + 1$ to the second one is due, as we shall see, to the refinement of abt_r (whose guard is stronger than that of its abstraction). While going from $ts1 = \overline{UP} \ \wedge \ ss1 = \overline{OK}$ to the first one is due to the concrete cause of the Receiver's abortion, namely the passage of time in such a way that $c2$ becomes equal to $MAX + 1$.

8.4 Events

Our first event, ini_brp, is almost the same as its abstraction. It only initializes the re-try counter $c2$ to 0.

$$
\begin{array}{lll}
\textsf{ini\_brp} & \widehat{=} & \textsf{select} \\
 & & \quad tg2 = \overline{UP} \\
 & & \textsf{then} \\
 & & \quad a2, b2, tg2, ts2, tr2, c2 : \begin{pmatrix} a2 \in \textsf{seq1}(S) \quad \wedge \\ b2 = [\,] \quad \wedge \\ tg2 = UP \quad \wedge \\ ts2 = UP \quad \wedge \\ tr2 = UP \quad \wedge \\ c2 = 0 \end{pmatrix} \\
 & & \textsf{end}
\end{array}
$$

The next event, rcv_s, is a new one. It can be fired when the protocol is active at the Sender' site, that is when the condition $ts2 = UP$ holds. Its effect (for the moment) is to reset the re-try counter $c2$.

$$
\begin{array}{lll}
\textsf{rcv\_s} & \widehat{=} & \textsf{select} \\
 & & \quad ts2 = UP \\
 & & \textsf{then} \\
 & & \quad c2 := 0 \\
 & & \textsf{end}
\end{array}
$$

Up to the renaming of the variables, the next event is almost the same as its abstraction. As in the preceding event, the re-try counter is reset.

$$
\begin{array}{lll}
\textsf{end\_s} & \triangleq & \textsf{select} \\
& & \quad ts2 = UP \quad \wedge \quad tr2 = \overline{UP} \quad \wedge \quad sr2 = OK \\
& & \textsf{then} \\
& & \quad ts2,\ ss2,\ c2 \ := \ \overline{UP},\ OK,\ 0 \\
& & \textsf{end}
\end{array}
$$

The next event, **out_s**, is a new one. It formalizes the occurence of the interrupt of the Sender's timer. It can be enabled when the protocol is active at the Sender's site, that is when the condition $ts2 = UP$, and when the re-try counter is smaller than the predefined constants MAX. As can be seen, we do not formalize the continuous passage of time in the timer, we rather measure the time with a discrete unit that corresponds to the Sender's timer delay. At this stage, the effect of this event is just to increment the re-try counter.

$$
\begin{array}{lll}
\textsf{out\_s} & \triangleq & \textsf{select} \\
& & \quad ts2 = UP \quad \wedge \quad c2 < MAX \\
& & \textsf{then} \\
& & \quad c2 \ := \ c2 + 1 \\
& & \textsf{end}
\end{array}
$$

The next event, **abt_s**, can be enabled when the protocol is active at the Sender's site, that is when the condition $ts2 = UP$ holds. This condition was the only one that was required in the abstraction. We now require that the re-try counter has reached the value MAX. This guard is thus more precise (stronger) than the one we had in the abstraction. The effect of the event is to abort the protocol as in the abstraction, but also to increment the re-try counter $c2$ one more time, so that it takes the value $MAX + 1$. As can be seen below in the guard of the event **abt_r**, we shall use this value $MAX + 1$ to enable that event. This "synchronization" of both timers by means of the counter $c2$ is an artefact used to simulate the passage of the "universal time" that is, of course, the same for both the Sender and the Receiver.

```
abt_s  ≙  select
              ts2 = UP  ∧  c2 = MAX
          then
              ts2, ss2, c2 := ‾UP‾, ‾OK‾, c2 + 1
          end
```

Up to the renaming of the variables, the next event is the same as its abstraction.

```
end_r  ≙  select
              tr2 = UP
          then
              b2, tr2, sr2 := a2, ‾UP‾, OK
          end
```

The next event, abt_r, corresponds to the abortion of the protocol at the Receiver's site. In fact this event is caused by the occurrence of the interrupt of the Receiver's timer. It is to be fired when the protocol is active at the Receiver's site, that is when the condition $tr2 = UP$ holds. Moreover the passage of time is indicated by the extra guarding condition $c2 = MAX + 1$. This test, by which the Receiver's timer seems to have a "direct access" to the re-try counter of the Sender, is an artefact corresponding to the maximum "delay" that is required to fire the Receiver's timer. In the reality, the passage of the "universal time" ensures magically the synchronization of both timers. We notice that half of the third invariant clause stated above, that is

$$c2 = MAX + 1 \;\Rightarrow\; ts1 = \overline{UP} \;\wedge\; ss1 = \overline{OK}$$

is due to the fact that the present concrete guard must imply the corresponding abstract one (remember, a concrete guard is always *stronger* than its abstract counterpart).

$$\mathsf{abt\_r} \;\widehat{=}\; \begin{array}{l} \mathsf{select} \\ \quad tr2 = UP \quad \wedge \quad c2 = MAX + 1 \\ \mathsf{then} \\ \quad b2, tr2, sr2 \; : \; (b2 \subset a2 \quad \wedge \quad tr2 = \overline{UP} \quad \wedge \quad sr2 = \overline{OK}) \\ \mathsf{end} \end{array}$$

Notice finally that a 0 value for the constant MAX does not invalidate the behaviour of the protocol. It only does not give it the possibility to handle a faulty channel (abortion is immediate, no re-tries are possible).

9 Third Refinement

9.1 Purpose

The goal of this refinement is to split the abstract events **end_r** and **abt_r** into a third event called rcv_r. The idea is that the file is not copied any more at once from the Sender's site to the Receiver's site as it was done in the preceding abstraction by the events **end_r** and **abt_r**. In fact, in this refinement, the effect of the new event rcv_r is to copy a *single data item*. Notice, however, that this transportation is done at once. In other words, the incrementation of the Receiver's file is performed simultaneously with the decrementation of the Sender's file (we have no Data Channel yet).

In this refinement, some time-stretching effect still takes place, but only on the Receiver's site. This effect is particularly important, since, in the preceding abstraction, a *single* occurrence of the event **end_r** (or **abt_r**) was able copy at once all (or part) of the file. Now, it is necessary to have a *repetitive* series of occurrences of the event rcv_r to do the same thing. In the abstraction, this new event was hidden under the occurrence of the events **end_r** or **abt_r**.

9.2 Variables

The concrete variables are the same as in the preceding refinement (up to renaming).

9.3 Gluing Invariant

The re-try counter and the various bits are identical to their abstractions.

$$c3, ts3, tr3, ss3, sr3, tg3 \;=\; c2, ts2, tr2, ss2, sr2, tg2$$

The next clauses contain the declarations of the two concrete files $a3$ and $b3$. Notice that $a3$ does not denote any more the untouched file of the Sender, but rather the part of it that has not yet been copied on the Receiver's site. This is shown in the last clause below.

$$a3 \in \mathsf{seq}(S)$$

$$b3 \in \mathsf{seq}(S)$$

$$b3 \frown a3 = a2$$

Note that the operator "$\frown$" denotes sequence concatenation.

Finally, the last two clauses state (1) that, when the protocol is in progress at the Receiver's site, then the Receiver's file is a strict prefix of the Sender's file (this clause is necessary to prove that the concrete version of the event abt_r indeed refines its abstraction), and (2) that, when the protocol has terminated at the Receiver's site, then the Receiver's file is the same as its abstraction (in other words, that both abstract and concrete executions "reconciles" at the end).

$$tr3 = UP \;\Rightarrow\; b3 \neq a2$$

$$tr3 = \overline{UP} \;\Rightarrow\; b3 = b2$$

Notice that the clauses $b3 \frown a3 = a2$ and $tr3 = UP \Rightarrow b3 \neq a2$ allows one to derive the following extra clause:

$$tr3 = UP \;\Rightarrow\; a3 \neq [\,]$$

9.4 Events

Up to the renaming of the variables, our first event is the same as its abstraction.

$$\begin{array}{lll}
\mathsf{ini_brp} & \hat{=} & \mathsf{select} \\
& & \quad tg3 = \overline{UP} \\
& & \mathsf{then} \\
& & \quad a3, b3, tg3, ts3, tr3, c3 : \left(\begin{array}{l} a3 \in \mathsf{seq1}\,(S) \quad \wedge \\ b3 = [\,] \quad \wedge \\ tg3 = UP \quad \wedge \\ ts3 = UP \quad \wedge \\ tr3 = UP \quad \wedge \\ c3 = 0 \end{array} \right) \\
& & \mathsf{end}
\end{array}$$

Up to the renaming of the variables, the next event is the same as its abstraction.

$$\begin{array}{lll}
\mathsf{rcv_s} & \hat{=} & \mathsf{select} \\
& & \quad ts3 = UP \\
& & \mathsf{then} \\
& & \quad c3 := 0 \\
& & \mathsf{end}
\end{array}$$

Up to the renaming of the variables, the next event is the same as its abstraction.

$$
\begin{array}{lll}
\textsf{end\_s} & \widehat{=} & \textsf{select} \\
& & \quad ts3 = UP \quad \wedge \quad tr3 = \overline{UP} \quad \wedge \quad sr3 = OK \\
& & \textsf{then} \\
& & \quad ts3,\ ss3 \ := \ \overline{UP},\ OK \\
& & \textsf{end}
\end{array}
$$

Up to the renaming of the variables, the next event is the same as its abstraction.

$$
\begin{array}{lll}
\textsf{out\_s} & \widehat{=} & \textsf{select} \\
& & \quad ts3 = UP \quad \wedge \quad c3 < MAX \\
& & \textsf{then} \\
& & \quad c3 \ := \ c3 + 1 \\
& & \textsf{end}
\end{array}
$$

Up to the renaming of the variables, the next event is the same as its abstraction.

$$
\begin{array}{lll}
\textsf{abt\_s} & \widehat{=} & \textsf{select} \\
& & \quad ts3 = UP \quad \wedge \quad c3 = MAX \\
& & \textsf{then} \\
& & \quad ts3,\ ss3,\ c3 \ := \ \overline{UP},\ \overline{OK},\ c3 + 1 \\
& & \textsf{end}
\end{array}
$$

The next event, rcv_r, is a new one. It can only occur when the protocol is active at the Receiver's site, that is when the condition $tr3 = UP$ holds, and when the size of the not yet copied portion, $a3$, of the Sender's file is greater than one, that is when the condition $\textsf{tail}(a3) \neq []$ holds. This means that we have not yet reached the last data item of the Sender's file. The effect of the event is to copy the next data item into the Receiver's file. Notice how the invariant clause $b3 \frown a3 = a2$ acts as a "loop invariant" in the repetitive occurrence of this event. This is essentially the reason why this new event indeed refines skip.

$$\begin{array}{lll} \text{rcv\_r} & \widehat{=} & \textsf{select} \\ & & \quad tr3 = UP \quad \wedge \quad \textsf{tail}\,(a3) \neq [\,] \\ & & \textsf{then} \\ & & \quad b3,\ a3 \ := \ b3 \leftarrow \textsf{first}\,(a3),\ \textsf{tail}\,(a3) \\ & & \textsf{end} \end{array}$$

Note that the expression "first(s)" denotes the first element of a non-empty sequence s. Note also that the expression "tail(s)" denotes the sequence s (supposed to be non-empty) whose first element has been removed. Finally, the expression "$s \leftarrow x$" denotes the sequence s appended with the element x.

Note that the extra invariant clause that we have derived at the end of the preceding subsection, namely $tr3 = UP \Rightarrow a3 \neq [\,]$ allows us to use safely tail$(a3)$ and first$(a3)$, requiring that $a3$ is not empty.

The next event, end_r, is enabled when the protocol is active at the Receiver's site, that is when the condition $tr3 = UP$ holds, and when the size of the not yet copied portion, $a3$, of the Sender's file is exactly equal to one, that is when the condition tail$(a3) = [\,]$ holds. This means that we have now reached the last data item of our initial file and that the protocol will thus be able to terminate successfully at the Receiver's site. Notice that this event is not any more copying the entire file from one site to the other (as its abstraction did), it just transports the last item of the file.

$$\begin{array}{lll} \text{end\_r} & \widehat{=} & \textsf{select} \\ & & \quad tr3 = UP \quad \wedge \quad \textsf{tail}\,(a3) = [\,] \\ & & \textsf{then} \\ & & \quad b3,\ a3,\ tr3,\ sr3 \ := \ b3 \leftarrow \textsf{first}\,(a3),\ \textsf{tail}\,(a3),\ \overline{UP},\ OK \\ & & \textsf{end} \end{array}$$

Again the extra invariant clause that we have derived at the end of the preceding subsection, namely $tr3 = UP \Rightarrow a3 \neq [\,]$ allows us to use safely tail$(a3)$ and first$(a3)$, requiring that $a3$ is not empty.

This new version of the event abt_r does not transport part of the file any more as its abstraction was doing (again, the file is now copied piece by piece by the events rcv_r and end_r).

```
abt_r  ≙  select
              tr3 = UP   ∧   c3 = MAX + 1
          then
              tr3, sr3 := ¬UP, ¬OK
          end
```

10 Fourth Refinement

10.1 Purpose

The goal of this refinement is to distribute the transportation of the data items between the Sender and the Receiver. In the preceding abstraction, the events rcv_r and end_r were doing both the incrementation of the Receiver's file and the decrementation of the Sender's file simultaneously. In this refinement, the incrementation is still done by rcv_r and end_r, while the decrementation is now done by the events rcv_s and end_s. The alternative control of execution between the Sender and the Receiver is ensured by a *centralized* alternating bit.

In this refinement, the time-stretching only occur at the Sender's site.

10.2 Variables

Besides the renamings of the abstract variables, the only new variable is the following:

- $r4$ denotes a bit allowing us to control the alternative firing of the events rcv_r and rcv_s. When $r4$ is equal to 0, this means that the control is on the Receiver's site, whereas 1 means that the control is on the Sender's site.

10.3 Gluing Invariant

The Receiver's file, the re-try counter, and the various bits are identical to their respective abstractions.

$$b4, c4, ts4, tr4, ss4, sr4, tg4 = b3, c3, ts3, tr3, ss3, sr3, tg3$$

We now declare the variable $a4$ and the alternating bit $r4$.

$$a4 \in \mathsf{seq}\,(S)$$

$$r4 \in \{0, 1\}$$

Next is the relationship between $r4$, $a4$, and $a3$. When $r4$ is equal to 0 (the control being then, by convention, at the Receiver's site) then both variables $a4$ and $a3$ are identical. In other words, the concrete and the abstract files are both "in the same phase". When $r4$ is equal to 1 (the control being then, by convention, on the Sender's site) then the concrete variable $a4$ has one more element than the abstraction $a3$. The concrete variable $a4$ is younger than its abstraction (this is again an effect of the streching of time). This is because the last data item of $a4$ has not yet been removed from it (this will be done either by rcv_s or end_s).

$$r4 = 0 \;\Rightarrow\; a4 = a3$$

$$r4 = 1 \;\Rightarrow\; a4 \neq []$$

$$r4 = 1 \;\Rightarrow\; \mathsf{tail}\,(a4) = a3$$

10.4 Events

Our first event is almost the same as its abstraction. We only initialize the bit $r4$ to 1 (control at the Sender's site).

$$
\begin{array}{lll}
\mathsf{ini\_brp} & \widehat{=} & \mathsf{select} \\
& & \quad tg4 = \overline{UP} \\
& & \mathsf{then} \\
& & \quad \begin{array}{l} a4, b4, tg4, ts4, \\ tr4, c4, r4 \end{array} \; : \; \left(\begin{array}{l} a4 \in \mathsf{seq1}\,(S) \quad \wedge \\ b4 = [\,] \quad \wedge \\ tg4 = UP \quad \wedge \\ ts4 = UP \quad \wedge \\ tr4 = UP \quad \wedge \\ c4 = 0 \quad \wedge \\ r4 = 1 \end{array} \right) \\
& & \mathsf{end}
\end{array}
$$

The event rcv_s is to be enabled when the protocol is active on the Sender's site, that is when the condition $ts4 = UP$ holds, when the alternating bit $r4$ is equal to 1, and when the size of the not yet copied part of the Sender's file is greater than one, that is when the condition $\mathsf{tail}\,(a4) \neq [\,]$ holds. The effect of the event is to modify the alternating bit and to decrement the Sender's file.

$$
\begin{array}{lll}
\mathsf{rcv\_s} & \widehat{=} & \mathsf{select} \\
& & \quad ts4 = UP \quad \wedge \quad r4 = 1 \quad \wedge \quad \mathsf{tail}\,(a4) \neq [\,] \\
& & \mathsf{then} \\
& & \quad r4,\ a4,\ c4 \ := \ 0,\ \mathsf{tail}\,(a4),\ 0 \\
& & \mathsf{end}
\end{array}
$$

Note that, in this event, the invariant clause $r4 = 1 \Rightarrow a4 \neq [\,]$ allows us to use safely the expression $\mathsf{tail}\,(a4)$ requiring that $a4$ is not empty.

The next event, end_s, is to be enabled under the same first two conditions as the event rcv_s, and when the size of the Sender's file is exactly equal to 1. Its effect is to decrement (for the last time) the Sender's file and to terminate the protocol successfully at the Sender's site.

$$
\begin{array}{lll}
\mathsf{end\_s} & \triangleq & \mathsf{select} \\
& & \quad ts4 = UP \quad \wedge \quad r4 = 1 \quad \wedge \quad \mathsf{tail}\,(a4) = [\,] \\
& & \mathsf{then} \\
& & \quad a4,\ r4,\ ts4,\ ss4,\ c4\ :=\ \mathsf{tail}\,(a4),\ 0,\ \overline{UP},\ OK,\ 0 \\
& & \mathsf{end}
\end{array}
$$

Again, in this event, the invariant clause $r4 = 1 \Rightarrow a4 \neq [\,]$ allows us to use safely the expression $\mathsf{tail}\,(a4)$ requiring that $a4$ is not empty.

Up to the renaming of the variables, the next event is the same as its abstraction.

$$
\begin{array}{lll}
\mathsf{out\_s} & \triangleq & \mathsf{select} \\
& & \quad ts4 = UP \quad \wedge \quad c4 < MAX \\
& & \mathsf{then} \\
& & \quad c4\ :=\ c4 + 1 \\
& & \mathsf{end}
\end{array}
$$

Up to the renaming of the variables, the next event is the same as its abstraction.

$$
\begin{array}{lll}
\mathsf{abt\_s} & \triangleq & \mathsf{select} \\
& & \quad ts4 = UP \quad \wedge \quad c4 = MAX \\
& & \mathsf{then} \\
& & \quad ts4,\ ss4\ :=\ \overline{UP},\ \overline{OK} \\
& & \mathsf{end}
\end{array}
$$

The next event, rcv_r does the same thing as its abstraction except that it does not decrement the Sender's file (this is now done by the event rcv_s). Notice that the guard of this event is clearly not yet localized because of the reference to $\mathsf{tail}\,(a4)$. The action of this event is not localized either because of the reference to $\mathsf{first}\,(a4)$. Also notice the reference to the still global alternating bit $r4$ in the guard.

$$\text{rcv\_r} \;\widehat{=}\; \begin{array}{l} \textsf{select} \\ \quad tr4 = UP \quad \wedge \quad r4 = 0 \quad \wedge \quad \textsf{tail}\,(a4) \neq [\,] \\ \textsf{then} \\ \quad r4,\; b4 \;:=\; 1,\; b4 \leftarrow \textsf{first}\,(a4) \\ \textsf{end} \end{array}$$

Notice the guards $tr4 = UP$ and $r4 = 0$, the invariant clauses $r4 = 0 \Rightarrow a4 = a3$ and $tr4 = tr3$, and, finally, the invariant clause of the preceding abstraction $tr3 = UP \Rightarrow a3 \neq [\,]$, all these conditions together imply that the condition $a4 \neq [\,]$ holds. This justifies that, in this event (and in the next one), we have a safe usage of expressions such as $\textsf{tail}\,(a4)$ and $\textsf{first}\,(a4)$ requiring, precisely, that $a4$ is not empty.

The next event, **end_r**, does the same thing as its abstraction except that it does not decrement the Sender's file (this is now done by the event **end_s**). As for the preceding one, this event is clearly not yet localized.

$$\text{end\_r} \;\widehat{=}\; \begin{array}{l} \textsf{select} \\ \quad tr4 = UP \quad \wedge \quad r4 = 0 \quad \wedge \quad \textsf{tail}\,(a4) = [\,] \\ \textsf{then} \\ \quad r4,\; b4,\; tr4,\; sr4 \;:=\; 1,\; b4 \leftarrow \textsf{first}\,(a4),\; \overline{UP},\; OK \\ \textsf{end} \end{array}$$

Up to the renaming of the variables, the next event is the same as its abstraction.

$$\text{abt\_r} \;\widehat{=}\; \begin{array}{l} \textsf{select} \\ \quad tr4 = UP \quad \wedge \quad c4 = MAX + 1 \\ \textsf{then} \\ \quad tr4,\; sr4 \;:=\; \overline{UP},\; \overline{OK} \\ \textsf{end} \end{array}$$

11 Fifth Refinement

11.1 Purpose

The goal of this refinement is first to partially localize the events rcv_r and end_r. We remind the reader that, in the preceding abstraction, the actions of these events were able to directly modify the Receiver's file with the data item of the Sender's file. In order to solve this problem, we now introduce the Data Channel under the form of a variable denoting the message containing the sent item. We also introduce two more events called snd_s and snd_s1, whose rôle is to effectively send the data item through the Data Channel. We also noticed in the preceding abstraction, that the guards of the events rcv_r and end_r were not localized: a test needed to be performed in order to know whether or not the data item was the last one. In order to solve this problem, we send together with the data item a boolean information telling us whether or not the sent item is the last one.

A second category of problem is also tackled by this refinement. It is concerned with the possible re-transmission of a data item. This is to be performed when the corresponding acknowledgment is missing. For this, we introduce one more event, rcv_r1 that formalizes the reception of a re-transmission.

11.2 Variables

Besides the renamings of the abstract variables, the new variables are the following:

- $f5$ denotes the data item that is sent from the Sender to the Receiver.
- $g5$ denotes a bit that tells us whether the transmitted item is the last one in the Sender's file. When $g5$ is equal to $LAST$, this means that the transmitted item is indeed the last one.
- $o5$ denotes the centralized control counter that allows us to distribute the control in turn to the events snd_s (and snd_s1) when it is equal to 0, to rcv_r (and rcv_r1 and end_r) when it is equal to 1 and, finally, to rcv_s (and end_s) when it is equal to 2.

11.3 Gluing Invariant

In this refinement the files, the re-try counter, and the various bits are identical to their respective abstractions.

$$a5, b5, c5, r5, ts5, tr5, ss5, sr5, tg5 \;=\; a4, b4, c4, r4, ts4, tr4, ss4, sr4, tg4$$

We have the following declaration of the new variables:

$$f5 \in S$$
$$g5 \in \{LAST, \overline{LAST}\}$$
$$o5 \in \{0, 1, 2\}$$

The next three clauses relate the new variables $f5$ and $g5$ to the Sender's file $a5$. This is only valid when the control counter is equal to 1, that is when the control is to be given to the Receiver.

$$o5 = 1 \;\Rightarrow\; a5 \neq [\,]$$
$$o5 = 1 \;\Rightarrow\; (g5 = LAST \;\Leftrightarrow\; \mathsf{tail}\,(a5) = [\,])$$
$$o5 = 1 \;\Rightarrow\; f5 = \mathsf{first}\,(a5)$$

The next clause states that when the control is on the Sender site that is when the condition $ts5 = UP \wedge o5 \doteq 0$ holds then the not yet copied part of the Sender's file is not empty.

$$ts5 = UP \;\wedge\; o5 = 0 \;\Rightarrow\; a5 \neq [\,]$$

Finally, the last clause relates the centralized control counter $o5$ to the alternating bit $r5$. It says that when the centralized control counter is equal to 2 (where the control is to be given the acknowledgment part of the Sender), then indeed the abstract alternative bit $r5$ has the same meaning.

$$o5 = 2 \;\Rightarrow\; r5 = 1$$

11.4 Events

The first event is almost the same as its abstraction (there is no need to initialize neither $f5$ nor $g5$). We only initialize $o5$ to 0.

$$
\begin{array}{lll}
\mathsf{ini\_brp} & \hat{=} & \mathsf{select} \\
& & \quad tg5 = \overline{UP} \\
& & \mathsf{then} \\
& & \quad \begin{array}{l} a5, b5, tg5, ts5, \\ tr5, c5, r5, o5 \end{array} : \left(\begin{array}{l} a5 \in \mathsf{seq1}(S) \quad \wedge \\ b5 = [\,] \quad \wedge \\ tg5 = UP \quad \wedge \\ ts5 = UP \quad \wedge \\ tr5 = UP \quad \wedge \\ c5 = 0 \quad \wedge \\ r5 = 1 \quad \wedge \\ o5 = 0 \end{array} \right) \\
& & \mathsf{end}
\end{array}
$$

The next event, **snd_s**, is a new event. It is to be enabled when the protocol is active on the Sender's site, that is when the condition $ts5 = UP$ holds, when the centralized control counter $o5$ is equal to 0, and when the sender's file has more than one element in it. The effect of this event is to modify the centralized control counter $o5$ and to put the next data item on the Data Channel; finally, it sets the bit $g5$ indicating that the data item $f5$ is not the last one in the Sender's file.

$$\begin{array}{lll} \mathsf{snd\_s} & \widehat{=} & \mathsf{select} \\ & & \quad ts5 = UP \quad \wedge \quad o5 = 0 \quad \wedge \quad \mathsf{tail}\,(a5) \neq [\,] \\ & & \mathsf{then} \\ & & \quad o5,\ f5,\ g5\ :=\ 1,\ \mathsf{first}\,(a5),\ \overline{LAST} \\ & & \mathsf{end} \end{array}$$

The guard $ts5 = UP \wedge o5 = 0$ together with the invariant clause $ts5 = UP \wedge o5 = 0 \Rightarrow a5 \neq [\,]$ imply that the condition $a5 \neq [\,]$ holds, thus justifying the usage in this event (and in the next one) of expressions such as $\mathsf{tail}\,(a5)$ and $\mathsf{first}\,(a5)$.

The next event, **snd_s1**, is also a new one. It is very close to the preceding one. It is to be fired when the sent item is indeed the last one in the Sender's file

$$\begin{array}{lll} \mathsf{snd\_s1} & \widehat{=} & \mathsf{select} \\ & & \quad ts5 = UP \quad \wedge \quad o5 = 0 \quad \wedge \quad \mathsf{tail}\,(a5) = [\,] \\ & & \mathsf{then} \\ & & \quad o5,\ f5,\ g5\ :=\ 1,\ \mathsf{first}\,(a5),\ LAST \\ & & \mathsf{end} \end{array}$$

Up to the renaming of the variables, the next event is almost the same as its abstraction. The only difference is concerned with the test of the new centralized control counter $o5$ and its modification.

$$\begin{array}{lll} \mathsf{rcv\_s} & \widehat{=} & \mathsf{select} \\ & & \quad ts5 = UP \quad \wedge \quad o5 = 2 \quad \wedge \quad \mathsf{tail}\,(a5) \neq [\,] \\ & & \mathsf{then} \\ & & \quad o5,\ r5,\ a5,\ c5\ :=\ 0,\ 0,\ \mathsf{tail}\,(a5),\ 0 \\ & & \mathsf{end} \end{array}$$

The guard $o5 = 2$, the invariant clauses $o5 = 2 \Rightarrow r5 = 1$, $r5 = r4$, and $a5 = a4$, and, finally, the invariant clause of the preceding abstraction $r4 = 1 \Rightarrow a4 \neq [\,]$, all these conditions together imply that the condition $a5 \neq [\,]$ holds. This allows us to use safely the expression $\mathsf{tail}\,(a5)$ in this event (and in the next one).

Up to the renaming of the variables, the next event is almost the same as its abstraction. The only difference is concerned with the test of the new centralized control counter $o5$ and its modification in the action part of the event.

$$
\begin{array}{lll}
\textsf{end\_s} & \widehat{=} & \textsf{select} \\
& & \quad ts5 = UP \quad \wedge \quad o5 = 2 \quad \wedge \quad \textsf{tail}\,(a5) = [\,] \\
& & \textsf{then} \\
& & \quad o5,\ a5,\ r5,\ ts5,\ ss5,\ c5 \ := \ 0,\ \textsf{tail}\,(a5),\ 0,\ \overline{UP},\ OK,\ 0 \\
& & \textsf{end}
\end{array}
$$

Up to the renaming of the variables, the next event is almost the same as its abstraction. The only difference is concerned with the test of the new centralized control counter $o5$ and its modification in the action part of the event. Notice that the guard $o5 \neq 0$ corresponds to the fact that the control has left the Sender. This justifies the possibility for the Sender's timer to be able to be waken up.

$$
\begin{array}{lll}
\textsf{out\_s} & \widehat{=} & \textsf{select} \\
& & \quad ts5 = UP \quad \wedge \quad o5 \neq 0 \quad \wedge \quad c5 < MAX \\
& & \textsf{then} \\
& & \quad o5,\ c5 \ := \ 0,\ c5 + 1 \\
& & \textsf{end}
\end{array}
$$

Up to the renaming of the variables, the next event is almost the same as its abstraction. The only difference is concerned with the test of the new centralized control counter $o5$ and its modification in the action part of the event.

$$
\begin{array}{lll}
\textsf{abt\_s} & \widehat{=} & \textsf{select} \\
& & \quad ts5 = UP \quad \wedge \quad o5 \neq 0 \quad \wedge \quad c5 = MAX \\
& & \textsf{then} \\
& & \quad o5,\ c5,\ ts5,\ ss5 \ := \ 0,\ c5 + 1,\ \overline{UP},\ \overline{OK} \\
& & \textsf{end}
\end{array}
$$

The next event, rcv_r, corresponds to a partial localization of the event rcv_r (the localization is only partial in that the global data $o5$ and $r5$ are still in use).

The fact that the guarding condition $r5 = 0$ holds indicates that the coming item $f5$ is indeed an item that is received for the first time by the Receiver (certainly not an item that has already been received).

$$
\begin{array}{lll}
\mathsf{rcv\_r} & \triangleq & \mathsf{select} \\
 & & \quad tr5 = UP \quad \wedge \quad o5 = 1 \quad \wedge \quad r5 = 0 \quad \wedge \quad g5 = \overline{LAST} \\
 & & \mathsf{then} \\
 & & \quad o5,\ r5,\ b5\ :=\ 2,\ 1,\ b5 \leftarrow f5 \\
 & & \mathsf{end}
\end{array}
$$

The next event, **end_r**, corresponds to a partial localization of the event **end_r**.

$$
\begin{array}{lll}
\mathsf{end\_r} & \triangleq & \mathsf{select} \\
 & & \quad tr5 = UP \quad \wedge \quad o5 = 1 \quad \wedge \quad r5 = 0 \quad \wedge \quad g5 = LAST \\
 & & \mathsf{then} \\
 & & \quad o5,\ r5,\ b5,\ tr5,\ sr5\ :=\ 2,\ 1,\ b5 \leftarrow f5,\ \overline{UP},\ OK \\
 & & \mathsf{end}
\end{array}
$$

The next event, rcv_r1, is a new one. It is a special case of the event rcv_r corresponding to a re-transmission. This is indicated by the guarding condition $r5 = 1$. Nothing is to be done except modifying the centralized control counter $o5$. Notice that the guard of this event does not mention $g5$. This is because it is necessarily equal to $\overline{LAST}$ (indicating that the data item is *not* the last one) because the last item has certainly been catched by the event end_r (fired when the condition $r5 = 0$ holds), whose effect is to set $tr5$ to $\overline{UP}$. As a consequence, a loss of the last acknowledgment has the effect of eventually aborting the Sender. This is because any re-transmission sent by the Server will not be acknowledged by the Receiver, since it considers that it has already terminated the protocol successfully. We can prove (and in fact we did so with Atelier B) that the following holds:

$$
tr5 = UP \ \wedge \ o5 = 1 \ \wedge \ r5 = 1 \ \Rightarrow \ g5 = \overline{LAST}
$$

Here is then the event rcv_r1:

$$
\begin{array}{lll}
\mathsf{rcv\_r1} & \widehat{=} & \mathsf{select} \\
& & \quad tr5 = UP \quad \wedge \quad o5 = 1 \quad \wedge \quad r5 = 1 \\
& & \mathsf{then} \\
& & \quad o5 \ := \ 2 \\
& & \mathsf{end}
\end{array}
$$

Up to the renaming of the variables, the next event is the same as its abstraction.

$$
\begin{array}{lll}
\mathsf{abt\_r} & \widehat{=} & \mathsf{select} \\
& & \quad tr5 = UP \quad \wedge \quad c5 = MAX + 1 \\
& & \mathsf{then} \\
& & \quad tr5,\ sr5 \ := \ \overline{UP},\ \overline{OK} \\
& & \mathsf{end}
\end{array}
$$

11.5 Termination of the Protocol

During this refinement, we have reached a stage where further refinements will not any more "refine the time". In other words, we shall not introduce any extra events belonging to the protocol itself. For this reason, we are now in a position to prove that the protocol indeed terminates.

The technique for proving termination is very much the same as that used to prove that a loop terminates. In the case of a loop, the idea is to exhibit a *variant expression* that is supposed to "decrease" at each execution of the body of the loop. Most of the time, the variant is a natural number expression. The most general variant is an expression that is "decreased" by a *well-founded relation.* It can be shown that the relation "<" on natural numbers is well founded. More generally, the lexicographical ordering on sequences of natural numbers is also such a well-founded relation.

In our case, we are going to show that a certain sequence of natural number expressions decreases lexicographically in each event action. The problem, of course, is to construct the sequence in question.

In order to do so, we define a matrix, where each row is labelled with an event, and each column is labelled with a state variable appearing in some guard. Each element of the matrix is labelled with a "$\nearrow$", a "$\searrow$", a "=", or a "?" symbol. A "$\nearrow$" means that the corresponding event is increasing the corresponding variable, a "$\searrow$" means that the event decreases the variable, a "=" means that the event does not change the variable, and finally a "?" means that we do not know what the event does on this variable. Here is the first matrix, which we can construct. Notice that we have not written in this matrix a column labelled with $r5$. This is because this variable behaves at least like that labelled with $o5$ as far as the decreasing or the increasing is concerned (sometimes, $r5$ remains untouched, however, while $o5$ changes).

	$tr5$	$ts5$	size ($a5$)	$g5$	$o5$	$c5$
end_s	=	$\nearrow$	$\searrow$	=	$\searrow$	?
end_r	$\nearrow$	=	=	=	$\nearrow$	=
abt_r	$\nearrow$	=	=	=	=	=
abt_s	=	$\nearrow$	=	=	$\searrow$	$\nearrow$
rcv_r	=	=	=	=	$\nearrow$	=
rcv_s	=	=	$\searrow$	=	$\searrow$	?
snd_s	=	=	=	?	$\nearrow$	=
snd_s1	=	=	=	?	$\nearrow$	=
out_s	=	=	=	=	$\searrow$	$\nearrow$
rcv_r1	=	=	=	=	$\nearrow$	=

As can be seen, some columns in this matrix are filled up with the "$\nearrow$" sign only, together with some "=" symbols: this is the case for columns 1 and 2. Another column is filled up with the "$\searrow$" symbol only, together with some "=" symbols: this is column 3. For each of these columns, this means that any event acts on the corresponding variables in a *regular fashion*. By means of the type of the variable in question (defining its domain of variation), we can thus exhibit an expression that is always non-increasing within each event. In our case, this corresponds to the following natural number expressions:

$$(1 - tr5) + (1 - ts5) + \mathsf{size}\,(a5)$$

As a result, we can now remove the rows corresponding to the "$\nearrow$" or "$\searrow$" symbols in the choosen columns, as well as the choosen columns themselves. This yields the following reduced matrix:

	$g5$	$o5$	$c5$
rcv_r	$=$	$\nearrow$	$=$
snd_s	?	$\nearrow$	$=$
snd_s1	?	$\nearrow$	$=$
out_s	$=$	$\searrow$	$\nearrow$
rcv_r1	$=$	$\nearrow$	$=$

The last column can be choosen now, together with the following natural number expression:

$$MAX + 1 - c5$$

This results in the following reduced matrix:

	$g5$	$o5$
rcv_r	$=$	$\nearrow$
snd_s	?	$\nearrow$
snd_s1	?	$\nearrow$
rcv_r1	$=$	$\nearrow$

We can clearly choose the column labelled with $o5$, which has all its row labelled with a"$\nearrow$" symbol. This means that each of the remaining events *strictly decreases* the following natural number expression: $2 - o5$. Putting all this together, we obtain the following sequence, made of three natural number expressions, which is certainly decreasing lexicographically on each event:

$$[\,(1 - tr5) + (1 - ts5) + \mathsf{size}\,(a5),\ MAX + 1 - c5,\ 2 - o5\,]$$

Again, notice that the fact that the last column of the last matrix is *entirely* filled up with the "$\nearrow$" symbol (or the "$\searrow$" symbol) is *absolutely indispensable* in order to ensure what is needed to prove the termination of the protocol, that is a *strictly* decreasing lexicographical ordering.

In order to formally prove that the protocol is terminating, we can use the following *abstraction*:

$$\textsf{decrease} \;\hat{=}\; tr51, ts51, a51, c51, o51 : \left(\begin{array}{l} tr51 \in \{0,1\} \quad \wedge \\ ts51 \in \{0,1\} \quad \wedge \\ a51 \in \textsf{seq}\,(S) \quad \wedge \\ c51 \in 0\,..\,MAX+1 \quad \wedge \\ o51 \in \{0,1,2\} \quad \wedge \\ s \prec t \end{array}\right)$$

where

$$s \;=\; [\,(1-tr51)+(1-ts51)+\textsf{size}\,(a51),\;\; MAX+1-c51,\;\; 2-o51\,]$$

$$t \;=\; [\,(1-tr51_0)+(1-ts51_0)+\textsf{size}\,(a51_0),\;\; MAX+1-c51_0,\;\; 2-o51_0\,]$$

Notice that the symbol "$\prec$" denotes the strictly decreasing lexicographical ordering. We remind the reader that a state variable subscripted with a 0 denotes the "before value" of this variable.

To be completely formal, we would then have to prove that each event of the protocol (as defined in this refinement) indeed refines the previous "event", together with the following gluing invariant:

$$a51, c51, o51, ts51, tr51 \;=\; a5, c5, o5, ts5, tr5$$

As refinement is transitive, we can then be sure that further refinements of the protocol events will still refine the "event" **decrease**. Termination will thus be insured automatically and there is no need to later re-prove it. This is the reason why termination can be proved *early*, that is as soon as all events of a protocol constructed in this way have effectively been introduced. This is not the same for deadlockfreeness, however. This is because refinement makes the concrete guards stronger than their abstractions. As a consequence, deadlock is always a risk when one refines a series of events. In other words, deadlockfreeness has always to be proved during the *last* refinement.

12 Sixth Refinement

12.1 Purpose

The goal of this refinement is to distribute the alternating bit $r5$, which was used to distribute the control between the Sender and the Receiver. For this we use an alternating bit on each site. Moreover, the Sender's alternating bit is sent through the Data Channel to the Receiver (by means of a message) for comparison with the Receiver's own alternating bit.

12.2 Variables

Besides the renamings of the abstract variables (except $r5$), we have three new variables:

- $p6$ is the alternating bit of the Sender.
- $q6$ is the alternating bit of the Receiver.
- $h6$ is the message transporting $p6$ through the Data Channel.

Notice that the centralized variable $r5$ has disappeared. The only centralized variable that remains is the centralized control counter $o6$, which will be removed during the next refinement.

12.3 Gluing Invariant

The abstract variables except $r5$ are all unchanged.

$$a6, b6, c6, o6, ts6, tr6, ss6, sr6, tg6 \;=\; a5, b5, c5, o5, ts5, tr5, ss5, sr5, tg5$$

We have the following declaration of the new variables:

$$p6 \in \{0,1\}$$

$$q6 \in \{0,1\}$$

$$h6 \in \{0,1\}$$

Next are the invariant clauses stating the properties of the new bits.

$$p6 = q6 \Leftrightarrow r5 = 0$$

$$o6 \neq 0 \Rightarrow h6 = p6$$

12.4 Events

The first event is almost the same as its abstraction. We only initialize both alternating bits $p6$ and $q6$ to 0. There is no need for initializing the bit $h6$.

$$
\begin{array}{lll}
\textsf{ini\_brp} & \triangleq & \textsf{select} \\
 & & \quad tg6 = \overline{UP} \\
 & & \textsf{then} \\
 & & \quad \begin{array}{l} a6, b6, tg6, ts6, tr6, \\ c6, o6, p6, q6 \end{array} : \left(\begin{array}{l} a6 \in \textsf{seq1}\,(S) \;\; \wedge \\ b6 = [\,] \;\; \wedge \\ tg6 = UP \;\; \wedge \\ ts6 = UP \;\; \wedge \\ tr6 = UP \;\; \wedge \\ c6 = 0 \;\; \wedge \\ o6 = 0 \;\; \wedge \\ p6 = 0 \;\; \wedge \\ q6 = 0 \end{array} \right) \\
 & & \textsf{end}
\end{array}
$$

Up to the renaming of the variables, the next event is almost the same as its abstraction. The only difference is concerned with the copy of the alternating bit

$p6$ of the Sender into the corresponding Data Channel message $h6$.

$$
\begin{array}{lll}
\mathsf{snd\_s} & \triangleq & \mathsf{select} \\
& & \quad ts6 = UP \quad \wedge \quad o6 = 0 \quad \wedge \quad \mathsf{tail}\,(a6) \neq [\,] \\
& & \mathsf{then} \\
& & \quad o6,\ f6,\ g6,\ h6\ :=\ 1,\ \mathsf{first}\,(a6),\ \overline{LAST},\ p6 \\
& & \mathsf{end}
\end{array}
$$

Up to the renaming of the variables, the next event is almost the same as its abstraction. The only difference is concerned with the copy of the alternating bit $p6$ of the Sender into the corresponding Data Channel message $h6$.

$$
\begin{array}{lll}
\mathsf{snd\_s1} & \triangleq & \mathsf{select} \\
& & \quad ts6 = UP \quad \wedge \quad o6 = 0 \quad \wedge \quad \mathsf{tail}\,(a6) = [\,] \\
& & \mathsf{then} \\
& & \quad o6,\ f6,\ g6,\ h6\ :=\ 1,\ \mathsf{first}\,(a6),\ LAST,\ p6 \\
& & \mathsf{end}
\end{array}
$$

Up to the renaming of the variables, the next event is almost the same as its abstraction. The only difference is concerned with the modification of the Sender alternating bit $p6$ instead of the global alternating bit $r5$ in the abstraction.

$$
\begin{array}{lll}
\mathsf{rcv\_s} & \triangleq & \mathsf{select} \\
& & \quad ts6 = UP \quad \wedge \quad o6 = 2 \quad \wedge \quad \mathsf{tail}\,(a6) \neq [\,] \\
& & \mathsf{then} \\
& & \quad o6,\ p6,\ a6,\ c6\ :=\ 0,\ 1 - p6,\ \mathsf{tail}\,(a6),\ 0 \\
& & \mathsf{end}
\end{array}
$$

Up to the renaming of the variables, the next event is almost the same as its abstraction. The only difference is concerned with the modification of the Sender alternating bit $p6$ instead of the global alternating bit $r5$ in the abstraction.

$$
\begin{array}{lll}
\textsf{end\_s} & \widehat{=} & \textsf{select} \\
& & \quad ts6 = UP \;\;\wedge\;\; o6 = 2 \;\;\wedge\;\; \textsf{tail}\,(a6) = [\,] \\
& & \textsf{then} \\
& & \quad o6,\; a6,\; p6,\; ts6,\; ss6,\; c6 \;:= \\
& & \qquad\qquad 0,\; \textsf{tail}\,(a6),\; 1 - p6,\; \overline{UP},\; OK,\; 0 \\
& & \textsf{end}
\end{array}
$$

Up to the renaming of the variables, the next event is the same as its abstraction.

$$
\begin{array}{lll}
\textsf{out\_s} & \widehat{=} & \textsf{select} \\
& & \quad ts6 = UP \;\;\wedge\;\; o6 \neq 0 \;\;\wedge\;\; c6 < MAX \\
& & \textsf{then} \\
& & \quad o6,\; c6 \;:=\; 0,\; c6 + 1 \\
& & \textsf{end}
\end{array}
$$

Up to the renaming of the variables, the next event is the same as its abstraction.

$$
\begin{array}{lll}
\textsf{abt\_s} & \widehat{=} & \textsf{select} \\
& & \quad ts6 = UP \;\;\wedge\;\; o6 \neq 0 \;\;\wedge\;\; c6 = MAX \\
& & \textsf{then} \\
& & \quad o6,\; c6,\; ts6,\; ss6 \;:=\; 0,\; c6 + 1,\; \overline{UP},\; \overline{OK} \\
& & \textsf{end}
\end{array}
$$

Up to the renaming of the variables, the next event is almost the same as its abstraction. The only difference is concerned with the test of the global alternating bit $r5$, which is now replaced by the comparison between the coming bit $h6$ and the Receiver's bit $q6$. We also have, in the action part, the modification of the Receiver alternating $q6$ instead of the global alternating bit $r5$ in the abstraction.

$$
\begin{array}{lll}
\textsf{rcv\_r} & \triangleq & \textsf{select} \\
& & \quad tr6 = UP \quad \wedge \quad o6 = 1 \quad \wedge \quad h6 = q6 \quad \wedge \quad g6 = \overline{LAST} \\
& & \textsf{then} \\
& & \quad o6,\ q6,\ b6\ :=\ 2,\ 1 - q6,\ b6 \leftarrow f6 \\
& & \textsf{end}
\end{array}
$$

Up to the renaming of the variables, the next event is almost the same as its abstraction. The only difference is concerned with the test of the global alternating bit $r5$, which is now replaced by the comparison between the coming bit $h6$ and the Receiver's bit $q6$. We also have, in the action part, the modification of the Receiver alternating $q6$ instead of the global laternating bit $r5$ in the abstraction.

$$
\begin{array}{lll}
\textsf{end\_r} & \triangleq & \textsf{select} \\
& & \quad tr6 = UP \quad \wedge \quad o6 = 1 \quad \wedge \quad h6 = q6 \quad \wedge \quad g6 = LAST \\
& & \textsf{then} \\
& & \quad o6,\ q6,\ b6,\ tr6,\ sr6\ :=\ 2,\ 1 - q6,\ b6 \leftarrow f6,\ \overline{UP},\ OK \\
& & \textsf{end}
\end{array}
$$

Up to the renaming of the variables, the next event is almost the same as its abstraction. The only difference is concerned with the test of the global alternating bit $r5$, which is now replaced by the comparison between the coming bit $h6$ and the Receiver's bit $q6$.

$$
\begin{array}{lll}
\textsf{rcv\_r1} & \triangleq & \textsf{select} \\
& & \quad tr6 = UP \quad \wedge \quad o6 = 1 \quad \wedge \quad h6 \neq q6 \\
& & \textsf{then} \\
& & \quad o6\ :=\ 2 \\
& & \textsf{end}
\end{array}
$$

Up to the renaming of the variables, the next event is the same as its abstraction.

$$
\begin{array}{lll}
\textsf{abt\_r} & \triangleq & \textsf{select} \\
& & \quad tr6 = UP \quad \wedge \quad c6 = MAX + 1 \\
& & \textsf{then} \\
& & \quad tr6,\ sr6 := \overline{UP},\ \overline{OK} \\
& & \textsf{end}
\end{array}
$$

13 Seventh Refinement

13.1 Purpose

The purpose of this last refinement is to distribute our remaining centralized variable, namely the centralized control counter $o6$. This will be done by means of three more bits. The first two of these bits correspond to the activation of the Data Channel and to the activation of the Acknowledgment Channel respectively (as can be seen, the Acknowledgment Channel is only represented by its activation bit, no further data is needed). The third bit corresponds to an interrupt that is internal to the Sender's site and that is able to wake up the events snd_s and snd_s1.

We also introduce two extraneous events called loose_d and loose_a. The former corresponds to the loss of information within the Data Channel, and the latter to the loss of information within the Acknowledgment Channel. Such events *are not* proper events of the protocol. They formalize the faulty behaviour of the two channels, under the form of "attacking deamons" that can be fired at any time.

13.2 Variables

Besides the renamings of the abstract variables (except $o6$), we have three new variables:

- $i7$ is the activation bit of the Data Channel. When it is equal to 1 then the channel is active, otherwise it is inactive.

- $j7$ is the activation bit of the Acknowledgment Channel. When it is equal to 1 then the channel is active, otherwise it is inactive.

- $k7$ is the internal interrupt of the Sender. When $k7$ is equal to 1, the control can be given to the sending events of the Sender.

Notice that our last centralized variable $o6$ has disappeared. The control is now entirely distributed between the Sender and the Receiver.

13.3 Gluing Invariant

The abstract variables, except $o6$, are all unchanged.

$$a7, b7, c7, ts7, tr7, tg7 = a6, b6, c6, ts6, tr6, tg6$$

$$ss7, sr7, p7, q7, h7 = ss6, sr6, p6, q6, h6$$

We have the following declarations for the new variables:

$$i7 \in \{0,1\}$$

$$j7 \in \{0,1\}$$

$$k7 \in \{0,1\}$$

Next are the invariant clauses stating the properties of the new bits.

$$ts7 = UP \Rightarrow (k7 = 1 \Leftrightarrow o6 = 0)$$

$$tr7 = UP \wedge i7 = 1 \Rightarrow o6 = 1$$

$$ts7 = UP \wedge j7 = 1 \Rightarrow o6 = 2$$

13.4 Events

This event is almost the same as its abstraction. We only initialize the control bits $i7$, $j7$ and $k7$ to 0, 0, and 1 respectively

$$
\begin{array}{lll}
\textsf{ini\_brp} & \triangleq & \textsf{select} \\
& & \quad tg7 = \overline{UP} \\
& & \textsf{then} \\
& & \quad \begin{array}{c} a7, b7, tg7, ts7, tr7, \\ c7, p7, q7, i7, j7, k7 \end{array} : \left(\begin{array}{l} a7 \in \textsf{seq1}\,(S) \quad \wedge \\ b7 = [\,] \quad \wedge \\ tg7 = UP \quad \wedge \\ ts7 = UP \quad \wedge \\ tr7 = UP \quad \wedge \\ c7 = 0 \quad \wedge \\ p7 = 0 \quad \wedge \\ q7 = 0 \quad \wedge \\ i7 = 0 \quad \wedge \\ j7 = 0 \quad \wedge \\ k7 = 1 \end{array} \right) \\
& & \textsf{end}
\end{array}
$$

Up to the renaming of the variables, the next event is almost the same as its abstraction. The only difference is concerned with the test of the centralized control counter $o6$, which is now replaced by that of the internal interrupt activation bit $k7$. In the action part, the bit $k7$ is reset and the activation bit $i7$ of the Data Channel is set.

$$
\begin{array}{lll}
\textsf{snd\_s} & \triangleq & \textsf{select} \\
& & \quad ts7 = UP \quad \wedge \quad k7 = 1 \quad \wedge \quad \textsf{tail}\,(a7) \neq [\,] \\
& & \textsf{then} \\
& & \quad f7,\ g7,\ h7,\ k7,\ i7 \ := \ \textsf{first}\,(a7),\ \overline{LAST},\ p7,\ 0,\ 1 \\
& & \textsf{end}
\end{array}
$$

Up to the renaming of the variables, the next event is almost the same as its abstraction. The only difference is concerned with the test of the centralized control counter $o6$, which is now replaced by that of the internal interrupt activation bit

$k7$. In the action part, the bit $k7$ is reset and the activation bit $i7$ of the Data Channel is set.

$$
\begin{array}{lll}
\mathsf{snd\_s1} & \triangleq & \mathsf{select} \\
& & \quad ts7 = UP \quad \wedge \quad k7 = 1 \quad \wedge \quad \mathsf{tail}\,(a7) = [\,] \\
& & \mathsf{then} \\
& & \quad f7,\ g7,\ h7,\ k7,\ i7 \ := \ \mathsf{first}\,(a7),\ LAST,\ p7,\ 0,\ 1 \\
& & \mathsf{end}
\end{array}
$$

Up to the renaming of the variables, the next event is almost the same as its abstraction. The only difference is concerned with the test of the centralized control counter $o6$, which is now replaced by that of the activation bit $j7$ of the Acknowledgment Channel. In the action part, the internal interrupt activation bit $k7$ is set, and the activation bit $j7$ is reset.

$$
\begin{array}{lll}
\mathsf{rcv\_s} & \triangleq & \mathsf{select} \\
& & \quad ts7 = UP \quad \wedge \quad j7 = 1 \quad \wedge \quad \mathsf{tail}\,(a7) \neq [\,] \\
& & \mathsf{then} \\
& & \quad p7,\ a7,\ k7,\ j7,\ c7 \ := \ 1 - p7,\ \mathsf{tail}\,(a7),\ 1,\ 0,\ 0 \\
& & \mathsf{end}
\end{array}
$$

Up to the renaming of the variables, the next event is almost the same as its abstraction. The only difference is concerned with the test of the centralized control counter $o6$, which is now replaced by that of the activation bit $j7$ of the Acknowledgment Channel. In the action part, the bit $k7$ is set, and the activation bit $j7$ is reset.

$$
\begin{array}{lll}
\mathsf{end\_s} & \triangleq & \mathsf{select} \\
& & \quad ts7 = UP \quad \wedge \quad j7 = 1 \quad \wedge \quad \mathsf{tail}\,(a7) = [\,] \\
& & \mathsf{then} \\
& & \quad p7,\ a7,\ k7,\ j7,\ c7,\ ts7,\ ss7 \ := \\
& & \qquad\qquad\qquad 1 - p7,\ \mathsf{tail}\,(a7),\ 1,\ 0,\ 0,\ \overline{UP},\ OK \\
& & \mathsf{end}
\end{array}
$$

Up to the renaming of the variables, the next event is almost the same as its abstraction. The only difference is concerned with the guard, which is now made more precise.

It requires that the internal interrupt activation bit $k7$ is reset, together with the activation bits $i7$ and $j7$ of both channels. This formalizes that some data has been lost. We thus insist on the *cause* of the timer interrupt, namely the loss of a message either on the Data Channel or on the Acknowledgment Channel (such losses are realized by the deamons loose_d and loose_a).

However, as one can see, the condition $i7 = 0$ is put as an alternative only, the other one being $tr7 = \overline{UP}$. This is because, besides the loss of a message in the Data Channel, there exists *another cause* for the Sender's timer to reach its delay: this is when the Receiver has successfully terminated, but when the last acknowledgment has been lost. Then the Sender re-transmitt the last data item (hence the condition $i7 = 1$ holds), but never gets any answer, since the Receiver is not active any more: so in the absence of any loss, this bit would remain equal to 1 for ever. In fact, we discovered the necessity for this special guard ($tr7 = \overline{UP} \;\vee\; i7 = 0$) rather than the simplest one $i7 = 0$ when we tried to prove deadlockfreeness. Without that guard, a deadlock is possible.

```
out_s  ≙  select
              ts7 = UP  ∧  k7 = 0  ∧  j7 = 0  ∧
              (tr7 = ¬UP  ∨  i7 = 0)  ∧  c7 < MAX
          then
              k7, c7 := 1, c7 + 1
          end
```

Up to the renaming of the variables, the next event is almost the same as its abstraction. The same discussion as the one presented for the previous event is also valid for this event.

```
abt_s  ≙  select
              ts7 = UP  ∧  k7 = 0  ∧  j7 = 0  ∧
              (tr7 = ¬UP  ∨  i7 = 0)  ∧  c7 = MAX
          then
              k7, c7, ts7, ss7 := 1, c7 + 1, ¬UP, ¬OK
          end
```

Up to the renaming of the variables, the next event is almost the same as its abstraction. The only difference is concerned with the test of the centralized control counter $o6$, which is now replaced by that of the activation bit $i7$ of the Data Channel. In the action part, the activation bits $i7$ of the Data Channel is reset and the activation bit of the Acknowledgment Channel is set.

$$
\begin{array}{lll}
\textsf{rcv\_r} & \triangleq & \textsf{select} \\
 & & \quad tr7 = UP \;\; \wedge \;\; i7 = 1 \;\; \wedge \;\; h7 = q7 \;\; \wedge \;\; g7 = \overline{LAST} \\
 & & \textsf{then} \\
 & & \quad q7,\ b7,\ i7,\ j7 \ := \ 1 - q7,\ b7 \leftarrow f7,\ 0,\ 1 \\
 & & \textsf{end}
\end{array}
$$

Up to the renaming of the variables, the next event is almost the same as its abstraction. The only difference is concerned with the test of the centralized control counter $o6$, which is now replaced by that of the activation bit $i7$ of the Dta Channel. In the action part, the activation bits $i7$ of the Data Channel is reset and the activation bit of the Acknowledgment Channel is set.

$$
\begin{array}{lll}
\textsf{end\_r} & \triangleq & \textsf{select} \\
 & & \quad tr7 = UP \;\; \wedge \;\; i7 = 1 \;\; \wedge \;\; h7 = q7 \;\; \wedge \;\; g7 = LAST \\
 & & \textsf{then} \\
 & & \quad q7,\ b7,\ tr7,\ sr7,\ i7,\ j7 \ := \\
 & & \qquad\qquad 1 - q7,\ b7 \leftarrow f7,\ \overline{UP},\ OK,\ 0,\ 1 \\
 & & \textsf{end}
\end{array}
$$

Up to the renaming of the variables, the next event is almost the same as its abstraction. The only difference is concerned with the test of the centralized control counter $o6$, which is now replaced by that of the activation bit $i7$. In the action part, the activation bits $i7$ of the Data Channel is reset and the activation bit of the Acknowledgment Channel is set.

$$
\begin{array}{lll}
\textsf{rcv\_r1} & \triangleq & \textsf{select} \\
 & & \quad tr7 = UP \;\; \wedge \;\; i7 = 1 \;\; \wedge \;\; h7 \neq q7 \\
 & & \textsf{then} \\
 & & \quad i7,\ j7 \ := \ 0,\ 1 \\
 & & \textsf{end}
\end{array}
$$

Up to the renaming of the variables, the next event is the same as its abstraction.

$$
\begin{array}{lll}
\mathsf{abt\_r} & \widehat{=} & \textsf{select} \\
 & & \quad tr7 = UP \quad \wedge \quad c7 = MAX + 1 \\
 & & \textsf{then} \\
 & & \quad tr7,\ sr7 \ := \ \overline{UP},\ \overline{OK} \\
 & & \textsf{end}
\end{array}
$$

The next event, loose_d, is a new one. It simply resets the activation of the Data channel, thus formalizing that the data (if any) that is stored in it has disappeared suddendly.

$$\mathsf{loose\_d} \ \widehat{=} \ i7 := 0$$

The next event, loose_a, is a new one. It simply resets the activation of the Acknowledgment channel, thus formalizing that the acknowledgment (if any) that is stored in it has disappeared suddendly.

$$\mathsf{loose\_a} \ \widehat{=} \ j7 := 0$$

13.5 Deadlockfreeness

Since we have now reached our last refinement, we have indeed the possibility to prove that our system of events is deadlockfree. This is only justified of course provided the protocol is active on one site or the other, that is when the following condition holds:

$$ts7 = UP \quad \vee \quad tr7 = UP$$

The absence of deadlock is expressed by taking the disjunction of the guards involved in the protocol. After some simplifications, we obtain the following to prove:

$$
\begin{array}{l}
(\,ts7 = UP \quad \wedge \quad k7 = 1\,) \quad \vee \\[2ex]
(\,ts7 = UP \quad \wedge \quad j7 = 1\,) \quad \vee \\[2ex]
(\,ts7 = UP \quad \wedge \quad k7 = 0 \quad \wedge \quad j7 = 0 \quad \wedge \\
\qquad\qquad\qquad (tr7 = \overline{UP} \quad \vee \quad i7 = 0) \quad \wedge \quad c7 \neq MAX + 1\,) \quad \vee \\[2ex]
(\,tr7 = UP \quad \wedge \quad i7 = 1\,) \quad \vee \\[2ex]
(\,tr7 = UP \quad \wedge \quad c7 = MAX + 1
\end{array}
$$

Of course, the proof can take advantage of some of the invariant clauses. It turns out that the only two that are needed are the following (they come, up to the renaming of the variables, from the first and second refinements):

$$ts7 = \overline{UP} \quad \wedge \quad ss7 = OK \quad \Rightarrow \quad tr7 = \overline{UP}$$

$$c7 = MAX + 1 \quad \Leftrightarrow \quad ts7 = \overline{UP} \quad \wedge \quad ss7 = \overline{OK}$$

Taking account of the typing of the bits, we have thus the following to prove:

$$
\begin{array}{l}
\;(\,ts7 = UP \quad \vee \quad tr7 = UP\,) \quad \wedge \\
\;(\,ts7 \neq UP \quad \wedge \quad ss7 = OK \quad \Rightarrow \quad tr7 \neq UP\,) \quad \wedge \\
\;(\,c7 = MAX + 1 \quad \Leftrightarrow \quad ts7 \neq UP \quad \wedge \quad ss7 \neq OK\,) \\
\Rightarrow \\
\;(\,ts7 = UP \quad \wedge \quad k7 = 1\,) \quad \vee \\
\;(\,ts7 = UP \quad \wedge \quad j7 = 1\,) \quad \vee \\
\;(\,ts7 = UP \quad \wedge \quad k7 \neq 1 \quad \wedge \quad j7 \neq 1 \quad \wedge \\
\qquad\qquad\qquad (tr7 \neq UP \quad \vee \quad i7 \neq 1) \quad \wedge \quad c7 \neq MAX + 1\,) \quad \vee \\
\;(\,tr7 = UP \quad \wedge \quad i7 = 1\,) \quad \vee \\
\;(\,tr7 = UP \quad \wedge \quad c7 = MAX + 1\,)
\end{array}
$$

This statement is easily discharged by a simple propositional calculus prover.

14 Comparing the final construction with the pseudocode

As promised, we are now going to compare what we have obtained with the pseudocode that we wrote in section 3 as a refinement of the informal description given in section 2.

Before doing this comparision, however, we shall explain how we can put together a number of events having "almost" the same guards. We shall present this in rather a formal way to begin with. Suppose that we have two events of the following forms:

$$\begin{array}{lll} \mathsf{evt\_1} & \widehat{=} & \mathsf{select}\ P \land Q\ \mathsf{then}\ S\ \mathsf{end} \\ \mathsf{evt\_2} & \widehat{=} & \mathsf{select}\ P \land \neg Q\ \mathsf{then}\ T\ \mathsf{end} \end{array}$$

In fact any two events in our description of the protocol constitute an *alternative* in the execution. We can thus combine them equivalently by means of the choice operator, $[\!]$. This yields the following:

$$\begin{array}{lll} \mathsf{evt\_12} & \widehat{=} & \mathsf{select}\ P \land Q\ \mathsf{then}\ S\ \mathsf{end} \\ & & [\!] \\ & & \mathsf{select}\ P \land \neg Q\ \mathsf{then}\ T\ \mathsf{end} \end{array}$$

We remind the reader of the axiom defining the choice operator [5]:

$$[S \mathbin{[\!]} T]R \Leftrightarrow [S]R \land [T]R$$

According to this axiom and to the axiom of the guard, as presented in section 4, and after a few calculations, we obtain the following:

$$
\begin{array}{lll}
\textsf{evt\_12} & \widehat{=} & \textsf{select} \\
 & & \quad P \\
 & & \textsf{then} \\
 & & \quad \textsf{select } Q \textsf{ then } S \textsf{ end} \\
 & & \quad [\!] \\
 & & \quad \textsf{select } \neg Q \textsf{ then } T \textsf{ end} \\
 & & \textsf{end}
\end{array}
$$

The inner body of the previous event can be further simplified thanks to the following obvious definition:

$$
\textsf{if } Q \textsf{ then } S \textsf{ else } T \textsf{ end} \quad \widehat{=} \quad
\begin{array}{l}
\textsf{select } Q \textsf{ then } S \textsf{ end} \\
[\!] \\
\textsf{select } \neg Q \textsf{ then } T \textsf{ end}
\end{array}
$$

This yields

$$
\begin{array}{lll}
\textsf{evt\_12} & \widehat{=} & \textsf{select} \\
 & & \quad P \\
 & & \textsf{then} \\
 & & \quad \textsf{if } Q \textsf{ then } S \textsf{ else } T \textsf{ end} \\
 & & \textsf{end}
\end{array}
$$

As a means of further simplification, we shall adopt the following factorization

$$
\begin{array}{l}
\textsf{if } Q \textsf{ then} \\
\quad x, y := E, F \\
\textsf{else} \\
\quad x, y := E, G \\
\textsf{end}
\end{array}
\quad = \quad
\begin{array}{l}
x := E \;||\\
\textsf{if } Q \textsf{ then} \\
\quad y := F \\
\textsf{else} \\
\quad y := G \\
\textsf{end}
\end{array}
$$

or the following one

```
if Q then                    x := E ||
  x, y := E, F               if Q then
else                   =       y := F
  x := E                     end
end
```

We are now ready to combine some of our events. By combining **snd_s** and **snd_s1** and renaming the combination **snd_s** again, we obtain:

```
snd_s  ≙  select
              ts7 = UP  ∧  k7 = 1
          then
              f7, h7, k7, i7 := first(a7), p7, 0, 1  ||
              if  tail(a7) ≠ []  then
                  g7 := ¬LAST
              else
                  g7 := LAST
              end
          end
```

This is to be compared to the following pseudo-code:

```
snd_s  ≙  select
              snd_s is waken up
          then
              Acquire data from Sender's file;
              Store acquired data on Data Channel;
              Store Sender's bit on Data Channel;
              Start Sender's timer;
              Activate Data Channel
          end
```

As can be seen, and as expected, the formal code is more precise than its pseudo-code counterpart. We have taken account of the difference of treatments between the last data item and the other ones. Also the mechanism by which this event is waken up (the bit $k7$) is explicit. Note that $k7$ also acts as an initializer for the Sender's timer.

By combining rcv_s and end_s and renaming the combination rcv_s again, we obtain:

```
rcv_s  ≙  select
              ts7 = UP   ∧   j7 = 1
          then
              p7, a7, k7, j7, c7 := 1 − p7, tail(a7), 1, 0, 0  ||
              if tail(a7) = [] then
                  ts7, ss7 := ¬UP, OK
              end
          end
```

This is to be compared to the following pseudo-code:

```
rcv_s  ≙  select
              Acknowledgment Channel interrupt occurs
          then
              Remove data from Sender's file;
              Reset re-try counter;
              Modify Sender's bit;
              Wake up event snd_s;
              Reset Acknowledgment Channel interrupt
          end
```

Here again we make, in the formal code, a clear distinction between the last data item and the others. In fact, we make explicit the successful termination of the protocol.

By combining out_s and abt_s and renaming the combination out_s, we obtain:

```
out_s  ≙  select
              ts7 = UP  ∧  k7 = 0  ∧  j7 = 0  ∧
              (tr7 = ‾UP‾  ∨  i7 = 0)
          then
              k7, c7 := 1, c7 + 1  ||
              if c7 = MAX then
                  ts7, ss7 := ‾UP‾, ‾OK‾
              end
          end
```

This is to be compared with the following pseudo-code:

```
out_s  ≙  select
              Sender's timer interrupt occurs
          then
              if re-try counter is too big then
                  Abort protocol on Sender's site
              else
                  Increment re-try counter;
                  Wake up event snd_s
              end
          end
```

The formal code is very close to the pseudo-code. Notice that, in the formal code, we "wake" up snd_s and increment the counter even when the protocol is aborted.

By combining rcv_r, end_r and rcv_r1 and renaming the combination rcv_r again, we obtain:

```
rcv_r  ≙  select
              tr7 = UP  ∧  i7 = 1  ∧  h7 = q7  ∧  g7 = 1
          then
              i7, j7 := 0, 1  ||
              if h7 = q7 then
                  q7, b7 := 1 − q7, b7 ← f7  ||
                  if g7 = LAST then
                      tr7, sr7 := ¬UP, OK
                  end
              end
          end
```

This is to be compared with the following pseudo-code:

```
rcv_r  ≙  select
              Data Channel interrupt occurs
          then
              Acquire Sender's bit from Data Channel;
              if Sender's bit = Receiver's bit then
                  Acquire data from Data Channel;
                  Store data on Receiver's file;
                  Modify Receiver's bit;
                  if data is not the last one then
                      Start Receiver's timer
                  end
              end;
              Reset Data Channel interrupt;
              Activate Acknowledgment Channel
          end
```

Again here the formal code is quite close to the pseudo-code. In particular both of them have two conditionals. In the formal code the successful termination of the protocol is made explicit, whereas, in the pseudo-code the initialization of the Receiver's timer is made explicit (this is certainly a weakness of the formal code).

Finally, the next event is to be compared to the very close pseudo-code that follows.

$$\text{abt\_r} \ \hat{=} \ \textsf{select} \ tr7 = UP \ \wedge \ c7 = MAX + 1 \ \textsf{then} \ tr7,\ sr7 := \overline{UP},\ \overline{OK} \ \textsf{end}$$

abt_r ≙ **select**
Receiver's timer interrupt occurs
then
Abort protocol on Receiver's site
end

REFERENCES

(1) L. Helmink et al. *Proof-Checking a Data Link Protocol.* CWI Report CS-R9420 (1994)

(2) CCITT Fascicle VIII.3. *CCITT Recommendation X25. Interface between DTE and DCE for Terminals Operating in the Packet Mode on Public Data Networks.* (1988)

(3) K. Havelund and N. Shankar. *Experiments in Theorem Proving and Model Checking for Protocol Verification.* Formal Method Europe (1995)

(4) G. Tel. *Introduction to Distributed Algorithms.* Cambridge University Press (1994)

(5) J.-R. Abrial. *The B-Book: Assigning Programs to Meanings.* Cambridge University Press (1996)

(6) E.W. Dijkstra. *A Discipline of Programming.* Prentice Hall (1976)

(7) C.C. Morgan. *The Specification Statement.* ACM Transaction on Programming Languages and Systems, 10(3) (1988)

(8) Steria-Digilog. *Atelier B Version 3.2* (1997)

(9) R.J.R. Back and R. Kurki-Suonio. *Decentralization of Process Nets with Centralised Control.* 2nd ACM SIGACT-SIGOPS Symp. on Principles of Distributed Computing (1983)

(10) J.-R. Abrial. *Extending B Without Changing it (for Developing Distributed Systems).* First B Conference. Nantes (1996)

(11) Steria-Digilog. *Atelier B Web Pages.* http://www.atelierb.societe.com (1997)

Formal Derivation of a Protocol in CSP

Burghard von Karger

March 7, 1997

Abstract

We study a distributed version of Misra's phase synchronization problem in which a number of parallel processes have to execute their phases in lockstep. Unlike Misra we do not postulate a global communication mechanism. Instead we assume that the processes are distributed over a tree-shaped network and that all communications must be local.

The emphasis of this lecture is on derivation: The program is developed hand in hand with its proof. We use Hoare's CSP both as specification language and as implementation language.

1 Introduction

The first rigorous attempts at software correctness were concerned with verifying that some —already written— program meets its specification. Because the analysis was applied to the finished code, this became known as 'post-mortem verification'. A major drawback is that post-mortem verification usually requires reconstructing the thought process that led to the program in the first place. Also, when the proposed program is not correct (or just cannot be shown to be) quite wasteful backtracking is caused. For this reason, later efforts have concentrated on developing programs hand in hand with their proofs. *Formal derivation*, as it is called, has turned out to be a much more economic way of constructing provably correct programs. It also has great educational value, teaching and encouraging the use of succesive refinement and the writing of programs that are clear and easily seen to be correct.

The pioneers of program derivation and verification worked on sequential programs. Nowadays, we are even more concerned with the correctness of *distributed* systems, whose behaviour is often so non-deterministic that failures cannot be reproduced, and eliminating them by testing is hopeless. Since distributed algorithms are even harder to analyze than sequential ones, we would expect formal derivation to be more important than ever. In practice we seem to observe the converse. There is an ever-growing multitude of proof systems and model checking tools, but hardly ever are we shown how to *arrive* at correct distributed programs.

It is true that model checking can nowadays perform verification tasks that are well beyond the scope of pen-and-paper reasoning —and much more reliably, too— and this may be a reason why derivation of distributed algorithms is so unfashionable. But let us not forget that programs must be arrived at in some way. Guessing is not a good way, and derivation is its only alternative.

In contrast to post-mortem verification, which one tries to mechanize because it is tedious and boring, program derivation is a creative activity performed by human beings. Therefore, it should be based on a reasoning method that works well for humans, rather than for computers. Among these, equational (or inequational) rewriting stands out: it is the way we have been manipulating terms, ever since we learned algebra and calculus at school. Ideally, the specification is massaged by gradual transformations until it becomes an executable program, and each step is justified by an algebraic identity of the underlying model of computation. Equational reasoning is a compositional approach to software construction, thanks to the simple fact that equals may be substituted for equals in any context. Once we have shown that a program equals a certain specification (or another program), this knowledge may possibly be reused within several other designs.

A prerequisite for equational reasoning is that specifications and programs can be seen as elements of a common domain, and even that programs can be combined with specification operators and vice versa. Such a mixed calculus exists for CSP (see [5, 1]) which is why we shall use it as our vehicle for program derivation.

2 Problem statement

The following problem statement is adapted from [4]. Given a number of programs (called *phases*) $S_0, \ldots S_k$ we are asked to construct a set of $k+1$ parallel processes $P_0, \ldots, P_k$ such that each P_i repeatedly executes phase S_i, subject to the following conditions:

Synchronization No phase may enter its $(n+1)$st execution before all phases have completed their nth execution.

Absence of Blocking No process will be permanently blocked from executing its $(n+1)$th phase if all processes have completed their nth phase.

Geometry Each phase is to run on its own processor. The processors are connected in a tree-shaped network and communication is restricted to local synchronization.

Misra's phase synchronization problem does not have the third requirement which is why the task considered in our paper is qualified as "distributed". Misra also lists various applications of phase synchronization, which we do not repeat here. In Misra's formulation it is not assumed that all phases of a given process are identical. Indeed this property is not required for our development either; we have made the assumption solely because it facilitates the presentation.

3 CSP, in brief

CSP, which is short for 'Communicating Sequential Processes', is a small language designed by C.A.R. Hoare for reasoning about parallel programs with synchronous —possibly multi-party— interactions [2]. We use a version that includes sequential composition as a primitive operation (which, in the original CSP, is treated as a combination of parallel composition, synchronization, and hiding). On the other hand, we shall recall only those operators that we need for solving the given problem.

Let $\mathcal{A}$ be a set, the set of all names of *actions* that a process may possibly engage in. Associated with every process P is its *alphabet* $\alpha.P \subseteq \mathcal{A}$ which determines the actions that require participation of P.

Two processes P and Q can be composed *sequentially* to form a new process $P; Q$, provided they have the same alphabet, and we then have

$$\alpha.(P;Q) = \alpha.P = \alpha.Q\,. \tag{1}$$

The basic synchronization primitive is a process that participates in one given action and then terminates. Such a process is completely determined by the name of this action a, say, and by its alphabet $A \subseteq \mathcal{A}$ (it is required that $a \in \mathcal{A}$), and so we could denote it $(A : \mathbf{a})$. In practice, its alphabet can usually be inferred from rule (1) and we will just write $\mathbf{a}$ (in bold font).

Processes may also be combined in *parallel* and in this case the alphabet of the result is the union of the individual alphabets. A parallel process $P_1 \parallel \ldots \parallel P_n$ can proceed with an action in its alphabet when, and only when, all P_i's that can participate (have the action in their alphabet) are willing to do so. For example,

$$(\mathbf{a}\ ;\ P_1) \parallel \ldots \parallel (\mathbf{a}\ ;\ P_n) \;=\; \mathbf{a}\ ;\ (P_1 \parallel \ldots \parallel P_n)\,.$$

but, assuming that $a \in \alpha.Q$ and $b \notin \alpha.P$

$$(\mathbf{a}\ ;\ P) \parallel (\mathbf{b}\ ;\ Q) \;=\; \mathbf{b}\ ;\ ((\mathbf{a}\ ;\ P) \parallel Q)\,,$$

because the action a, which requires participation of both components, is initially blocked.

The *hiding* operator is used for abstracting from synchronization actions that are for internal use only and should not be participated in by the environment. For example,

$$((\mathbf{a}\,;\mathbf{b}) \parallel (\mathbf{b}\,;\mathbf{c})) \setminus b$$

$$=$$

$$(\mathbf{a}\,;\mathbf{b}\,;\mathbf{c}) \setminus b$$

$$=$$

$$\mathbf{a}\,;\mathbf{c} \quad ,$$

where it is understood that $\alpha.(\mathbf{a}\,;\mathbf{b}) = \{a, b\}$ etc. Note that hiding subtracts from the alphabet of a process:

$$\alpha.(P \setminus a) = (\alpha.P) \setminus \{a\} \, . \tag{2}$$

Since the order in which actions are hidden is irrelevant, we admit hiding a finite set of actions as a single operation.

The algebraic identities between processes that we use in this paper will be obvious to anyone who has worked with CSP, and we hope that they will be acceptable to other readers as well. Their full justification requires the construction of a denotational model in which they can be formally proved. Moreover, the denotational semantics ought to be justified by proving its congruence with a straightforward operational one. Such work has been done elsewhere [5, 6, 7, 1]. This paper is not about the theory of process algebra but about its application to program derivation.

4 Specification

The process which we wish to construct must begin with running each phase just once. The most general (least deterministic, least implementation-biased) way of doing so is by running them in parallel

$$S_0 \parallel \ldots \parallel S_k \, .$$

After having achieved this, the remainder of the process must again satisfy the original requirements. Thus the most general specification solves the equation

$$X \;=\; (S_0 \parallel \ldots \parallel S_k)\,;\, X \, . \tag{3}$$

Assuming that at least one of the phases involves an action —otherwise our programming assignment is meaningless and trivial— the recursive equation (3) is of the so-called *guarded* type (each recursive call is preceded by at least one action). The nice thing about guarded recursions is that they have unique solutions. In general, the unique solution of a guarded recursion $X = S; X$ is denoted by S^{∞} — an unending repetition of S's. Thus the first constraint —synchronization— will be met by any process semantically equal to (or better than)

$$(S_0 \parallel \ldots \parallel S_k)^{\infty} \, . \tag{4}$$

It is obvious that (4) also satisfies the absence-of-blocking constraint. Note that (4) is a perfectly valid CSP process and if there were no further constraints, our task would be complete already. The remaining requirement —geometry— is not *functional* in the sense that it does not constrain the behaviour of the process we have to construct. Instead it prescribes a certain syntactical shape — it must be so that, say, an appropriate compiler could convert it in a straightforward manner into code that runs on the given network of processors.

5 Solution strategy

To solve a difficult programming task it is usually a good idea to start by looking at a simplified version. Often a general solution can be found by extending one that works for a special case. When a recursive structure is involved, such as the shape of the given network, the most common simplification is a restriction of the recursion depth.

The nature of our problem provides further incentive for a recursive approach. For the specification can be seen as describing a process that runs on a single node —the root, say— and we must distribute its workload down the tree. What we propose is to start by sharing the work between the root and its successors. Once we have found how to achieve this, we will hopefully be able to repeat the game recursively until we reach the leaves.

Before we can start, we have to deal with two small technicalities. One is to give a mathematical description of the target network; the other concerns the introduction of fresh action names.

6 Tree structure

Let the processors be numbered from 0 to k, with 0 at the root. We write $i \rightarrow j$ to indicate that processor j is a child of processor i. Because the network is a tree, each node j has exactly one i with $i \rightarrow j$, a fact that we will use silently in the sequel. Also, $i \rightarrow^* j$ signifies that processor j is a node in the subtree starting at processor i. Let

$$T_i \;=\; \| \{S_j \mid i \rightarrow^* j\} \qquad (\text{for } 0 \leq i \leq k)\,, \tag{5}$$

the parallel composition of all phases to be placed into the subtree starting at node i. Since parallel composition is associative and

$$\{j \mid i \rightarrow^* j\} = \{i\} \cup \{k \mid \exists j : i \rightarrow j \rightarrow^* k\}$$

we have

$$T_i \;=\; \| \, (\{S_i\} \cup \{T_j \mid i \rightarrow j\}) \qquad (\text{for } 0 \leq i \leq k)\,. \tag{6}$$

7 Alphabet extension

In order to orchestrate the phases over the network we need to insert auxiliary actions that serve the sole purpose of synchronization. To avert any danger of interference with the real work, the auxiliary actions must be chosen to be different from any that occur within the phases. A notion of an action occurring in a process could be defined inductively on syntactic structure, but all that we shall need is the following identity

$$\| \{X_i; \mathbf{a}; Y_i \mid i \in I\} \;=\; \| \{X_i \mid i \in I\} \;;\; \mathbf{a} \;;\; \| \{Y_i \mid i \in I\}\,, \qquad (7)$$

provided that a does not occur in any of the X_i's.

Note that the typing rule (1) requires that a be in the alphabet of all X_i's and Y_i's. In the typical situation, where the X_i's and Y_i's are given and a is introduced to synchronize them, this is not the case. Therefore we need an operation that extends the alphabet of a process with a fresh action. We require, for every process P and every action $a \notin \alpha.P$ the existence of a process P^{+a} satisfying

$$\alpha.(P^{+a}) \;=\; \alpha.P \cup \{a\} \qquad (8)$$

$$(P^{+a}) \setminus a \;=\; P \qquad (9)$$

$$a \text{ does not occur in } P^{+a}. \qquad (10)$$

In fact, the newly introduced operator is uniquely determined by (8–10), and it distributes over both sequential and parallel composition:

$$(P;Q)^{+a} \;=\; P^{+a}\,;\,Q^{+a} \quad \text{and} \quad (P \parallel Q)^{+a} \;=\; P^{+a} \parallel Q^{+a}\,, \qquad (11)$$

provided $a \notin \alpha.P \cup \alpha.Q$. For convenience, we shall write P^{+ab} instead of P^{+a+b}.

8 Load distribution — a first step

In the first step we plan to distribute the entire workload among the processor at the root node and its immediate successors, as suggested by (6) with $i = 0$. In other words, we shall charge node 0, the root, with executing phase S_0 whereas each node i with $0 \to i$ gets the parallel composition T_i of all the phases belonging to its subtree (see (5)). Each node will execute its task repeatedly, and a single synchronization at the end of each turnaround is sufficient to keep them in lockstep. The task to be performed by the entire tree —the original specification— is

$$Q_0 \;=_{\text{def}}\; (T_0)^\infty\,. \qquad (12)$$

The process that we plan to place at the root is an infinite repetition of phase S_0 plus one synchronization per cycle

$$P_0 \;=_{\text{def}}\; (S_0^{+a_0}; \mathbf{a}_0)^\infty\,, \qquad (13)$$

where a_0 is a fresh action name. The task allocated to each of the subtrees is

$$Q_i =_{\text{def}} (T_i^{+a_0}; \mathbf{a}_0)^\infty \qquad (\text{for } 0 \to i)\,. \tag{14}$$

It is now time to verify that our initial design is correct in the sense that the parallel combination of P_0 with the Q_i's implements Q_0.

Fact 1 $Q_0 \;=\; \| (\{P_0\} \cup \{Q_i \mid 0 \to i\}) \setminus a_0$.

Proof. Since $Q_0 = (T_0)^\infty$, which is the unique solution of the equation $X = T_0; X$, it is sufficient to show that the right hand side of Fact 1 also solves the same equation. For convenience in writing down the calculation, let us assume that $\{i \mid 0 \to i\} = \{1, \ldots, \ell\}$ so that we may employ infix notation.

$$\begin{aligned}
& (P_0 \,\|\, Q_1 \,\|\, \cdots \,\|\, Q_\ell) \setminus a_0 \\
= \quad & \{ \text{ Expand, using (13) and (14) } \} \\
& ((S_0^{+a_0}; \mathbf{a}_0; P_0) \,\|\, (T_1^{+a_0}; \mathbf{a}_0; Q_1) \,\|\, \cdots \,\|\, (T_\ell^{+a_0}; \mathbf{a}_0; Q_\ell)) \setminus a_0 \\
= \quad & \{ \text{ Apply (7), which is allowed by (10) } \} \\
& ((S_0^{+a_0} \,\|\, T_1^{+a_0} \,\|\, \cdots \,\|\, T_\ell^{+a_0}) \;;\; \mathbf{a}_0 \;;\; (P_0 \,\|\, Q_1 \,\|\, \cdots \,\|\, Q_\ell)) \setminus a_0 \\
= \quad & \{ \text{ Distributivity of } {}^{+a_0} \text{ over parallel (11) } \} \\
& ((S_0 \,\|\, T_1 \,\|\, \cdots \,\|\, T_\ell)^{+a_0} \;;\; \mathbf{a}_0 \;;\; (P_0 \,\|\, Q_1 \,\|\, \cdots \,\|\, Q_\ell)) \setminus a_0 \\
= \quad & \{ \text{ (5) with } i = 0 \} \\
& (T_0^{+a_0} \;;\; \mathbf{a}_0 \;;\; (P_0 \,\|\, Q_1 \,\|\, \cdots \,\|\, Q_\ell)) \setminus a_0 \\
= \quad & \{ \text{ Hiding } a_0 \text{ distributes over ; and is inverse to } {}^{+a_0} \text{ (9) } \} \\
& T_0 \;;\; (P_0 \,\|\, Q_1 \,\|\, \cdots \,\|\, Q_\ell) \setminus a_0 \;.
\end{aligned}$$

This completes the proof of Fact 1.

9 Further down the line

Our first step towards load distribution has left us with the task of implementing the Q_i's for all i with $0 \to i$ (see (14)). Some of these may be leaves. If i is a leaf then $S_i = T_i$ (by (5)) and Q_i need not be decomposed any further. For the sake of tidiness, we are going to define a process called P_i for each node i, so let

$$P_i =_{\text{def}} Q_i \;=\; (S_i^{+a_0}; \mathbf{a}_0)^\infty \qquad (\text{for } 0 \to i \wedge i \text{ is a leaf})\,. \tag{15}$$

Now let's assume that $0 \to i$ and i is an inner node. Then the work that we have assigned to processor i —namely the process Q_i— needs to be distributed further down the line. We had hoped that at this point we would be able simply to repeat the first step. However, we are not in the same position as before. The original problem was to implement the infinite repetition Q_0 each cycle of which is an parallel composition

$$Q_0 \;=\; (T_0)^\infty \;=\; (S_0 \,\|\, \ldots \,\|\, S_k)^\infty$$

whereas now we have to implement a repetition in which each cycle is composed sequentially from two processes

$$Q_i \;=\; (T_i^{+a_0}; \mathbf{a}_0)^\infty \,.$$

The new task is a little harder. Not only do we have to ensure that each of the phases that T_i is composed from terminates before $\mathbf{a}_0$ starts (first synchronization), but also that none of them starts again before $\mathbf{a}_0$ is done (second synchronization). The same name a_i can be used for both synchronizations. So let a_i be a fresh action and define

$$P_i \;=_{\text{def}}\; (S_i^{+a_0 a_i} \;;\; \mathbf{a}_i \;;\; \mathbf{a}_0 \;;\; \mathbf{a}_i)^\infty \qquad (\text{for } 0 \to i \;\wedge\; i \text{ not a leaf}) \;. \tag{16}$$

We recall that, by convention, the alphabet of a process $\mathbf{a}_j$ can be omitted when it can be inferred from a sequential predecessor or successor. This is why the alphabet extension operators need only be applied to S_i. Every child j of node i must assume the task of executing T_j and synchronize twice after each cycle.

$$Q_j \;=_{\text{def}}\; (T_j^{+a_i} \;;\; \mathbf{a}_i \;;\; \mathbf{a}_i)^\infty \qquad (\text{for } 0 \to i \to j)\,. \tag{17}$$

Like in the first round we claim that Q_i can be implemented as a parallel composition of P_i and the Q_j's with $i \to j$:

Fact 2 Assume that $0 \to i$ and i is not a leaf. Then

$$Q_i \;=\; \|\,(\{P_i\} \cup \{Q_j \mid i \to j\}) \;\setminus\; a_i \,.$$

The proof of Fact 2 is entirely analogous to the one of Fact 1 and we omit it. Of course, we *do* write down all the proofs when deriving an algorithm. The proof of Fact 2 is omitted not because it is trivial, but because it contributes nothing to the didactical objective of this paper.

We are now left with implementing the Q_j's of (17). This new task has indeed the same shape as the previous one. The only difference between (14) and (17) is that, in the latter, the main task is followed by two actions rather than one. Since these two actions will reside on the same processor, no additional synchronization is required. We therefore conjecture that the specification (17) can serve for arbitrary nodes (except the root) rather than only for children of the root (recall that Q_j describes the task inherited by the subtree starting at j). Thus we generalize definition (17) to

$$Q_j \;=_{\text{def}}\; (T_j^{+a_i} \;;\; \mathbf{a}_i \;;\; \mathbf{a}_i)^\infty \qquad (\text{for } i \to j \;\wedge\; i \neq 0)\,, \tag{18}$$

For any such j that is a leaf we have $T_j = S_j$ (see (5)), so that Q_j can be taken as the process for node j:

$$P_j =_{\text{def}} Q_j = (S_j^{+a_i} ; \mathbf{a}_i ; \mathbf{a}_i)^\infty \qquad (\text{for } i \to j \ \wedge \ i \neq 0 \ \wedge \ j \text{ is a leaf}) . \quad (19)$$

Finally assume that j is an inner node with $i \to j$ and $i \neq 0$. We are aiming for Fact 3 below. Since Q_j now has one more $\mathbf{a}_i$ than before (compare (14) and (17)), we must also put one more $\mathbf{a}_i$ into P_j than before (cf. (16))

$$P_j =_{\text{def}} (S_j^{+a_i a_j} ; \mathbf{a}_j ; \mathbf{a}_i ; \mathbf{a}_i ; \mathbf{a}_j)^\infty \qquad (\text{for } i \to j \ \wedge \ i \neq 0 \ \wedge \ j \text{ not a leaf}) . \quad (20)$$

Fact 3 Assume that $i \to j$ where j is not a leaf and $i \neq 0$. Then

$$Q_j = \| (\{P_j\} \cup \{Q_{j'} \mid j \to j'\}) \setminus a_j .$$

The proof is, of course, the same as for Fact 1 and 2. Facts 1–3 add up to the following

Fact 4 For every node i that is not a leaf we have

$$Q_i = \| (\{P_i\} \cup \{Q_j \mid i \to j\}) \setminus a_i . \quad (21)$$

Recall that Q_0 was the specification. In conjunction with the definitions (13, 15, 16, 19, 20) of the P_i's, Fact 4 allows us to unfold the specification recursively into the final program.

In fact, the recursive description (21) can be unnested, using the identity

$$(X \setminus a) \| Y = (X \| Y) \setminus a \qquad \text{provided } a \notin \alpha.Y . \quad (22)$$

This yields the final program in closed form:

The final program Let $A =_{\text{def}} \{a_i \mid 0 \leq i \leq k \ \wedge \ i \text{ is not a leaf}\}$, which is the set of all synchronization actions we introduced during development. Then

$$Q_0 = (P_0 \| \ldots \| P_k) \setminus A , \quad (23)$$

where

$$\begin{array}{lll}
P_0 & = & (S_0^{+a_0}; \mathbf{a}_0)^\infty \\
P_j & = & (S_j^{+a_0}; \mathbf{a}_0)^\infty \qquad (\text{for } 0 \to j \ \wedge \ j \text{ is a leaf}) \\
P_j & = & (S_j^{+a_0 a_j} ; \mathbf{a}_j ; \mathbf{a}_0 ; \mathbf{a}_j)^\infty \qquad (\text{for } 0 \to j \ \wedge \ j \text{ not a leaf}) \\
P_j & = & (S_j^{+a_i} ; \mathbf{a}_i ; \mathbf{a}_i)^\infty \qquad (\text{for } i \to j \ \wedge \ i \neq 0 \ \wedge \ j \text{ is a leaf}) \\
P_j & = & (S_j^{+a_i a_j} ; \mathbf{a}_j ; \mathbf{a}_i ; \mathbf{a}_i ; \mathbf{a}_j)^\infty \qquad (\text{for } i \to j \ \wedge \ i \neq 0 \ \wedge \ j \text{ not a leaf}) .
\end{array}$$

The following picture shows the result for a tree with six nodes.

$$P_4 = (S_4^{+a_3} \,;\, \mathbf{a}_3 \,;\, \mathbf{a}_3)^\infty \qquad\qquad P_5 = (S_5^{+a_3} \,;\, \mathbf{a}_3 \,;\, \mathbf{a}_3)^\infty$$

$$P_3 = (S_3^{+a_2 a_3} \,;\, \mathbf{a}_3 \,;\, \mathbf{a}_2 \,;\, \mathbf{a}_2 \,;\, \mathbf{a}_3)^\infty$$

$$P_1 = (S_1^{+a_0} \,;\, \mathbf{a}_0)^\infty \qquad\qquad P_2 = (S_2^{+a_0 a_2} \,;\, \mathbf{a}_2 \,;\, \mathbf{a}_0 \,;\, \mathbf{a}_2)^\infty$$

$$P_0 = (S_0^{+a_0} \,;\, \mathbf{a}_0)^\infty$$

10 Some final remarks

Many specifications contain operations that cannot be implemented directly, at least not efficiently. For example the safe operation of a chemical plant may be described by specifying what an explosion is and then stating that it must *not* happen. Negation is not implementable — otherwise the task would be easy —, so it must be eliminated in the design of a controlling program. The elimination of a non-implementable operator is hardest when it occurs at the outermost level. It is therefore useful for an algebraic theory to provide rules for distributing such operators inwards, thereby decomposing a large task into smaller ones. In the phase synchronization problem we distributed the infinite repetition operator inwards until it could be implemented by local loops. A second look at our calculation reveals that the entire derivation boils down to the repeated application of a single distribution rule and it is well worthwhile to record this rule for future reuse. For the sake of simplicity we write the rule for just two processors with two synchronization points; it should be obvious how to generalize it to an arbitrary number of processors and synchronization points.

$$((P_1 \parallel Q_1);(P_2 \parallel Q_2))^{\infty} \;=\; ((P_1^{+a};a;P_2^{+a};a)^{\infty} \parallel (Q_1^{+a};a;Q_2^{+a};a)^{\infty}) \setminus a\,,$$

where it is assumed that $a \notin \alpha.P_1 = \alpha.P_2$ and $a \notin \alpha.Q_1 = \alpha.Q_2$.

Good correctness proofs for concurrent programs are still rare, and sometimes one is tempted to suspect a one-to-one correspondence between examples and theories. To help reversing the trend we have re-used an existing process algebra. But even so, there is still something to learnt from each example which should be reflected in the theory. The lesson from the present case study is the importance of sequential composition, even for process algebras. Without it, it would have been impossible to express the specification as a solution to the equation

$$X \;=\; (S_0 \parallel \ldots \parallel S_k)\,;\,X\,,$$

and the problem would have been quite intractable. The necessity of sequential composition for effective reasoning about protocols has also been recognized by W. Janssen and J. Zwiers in [3]. They introduce a third composition operator, called *layering* which is a hybrid of sequential and parallel composition. Layering is appropriate for reasoning about phases that are logically, but not necessarily physically, successive.

The reader is invited to compare our derivation to the phase synchronization problem with the one given by Wim Feijen elsewhere in this book which uses shared variables as its communication primitives. One disadvantage of the CSP solution is its higher level of abstraction, because more implementation work is required to actually run it. We might even be accused of having 'reduced the problem to itself' because the remaining work —simultaneous synchronization of a node with all its children— is itself a phase synchronization problem, and something like Misra's algorithm must be used to actually implement it in terms of shared memory. On the other hand, it is possible to argue that this conversion, which happens on a local scale, should be performed by a compiler and that it is preferable to design the global algorithm without unnecessary implementation bias.

By concentrating on the global structure we obtained a very straightforward derivation; unlike Feijen's derivation we did not require the introduction and subsequent elimination of counters. Also our specification (4) is much more straightforward. We stress this point because the history of software engineering shows that the most dangerous and costly design errors occur during specification. By their very nature, specifications cannot be proved correct and so the best protection against mishap is utter clarity and simplicity.

A nice thing about CSP is that we do not have to provide an extra proof for the absence of blocking. There was no blocking in the specification, the final program is equal to it, and nothing more needs saying. Of course, this only works because the semantics of CSP does take blocking into account. This is very different from most other verification methods that separate the concerns of (partial) correctness and deadlock-freeness and where the latter proof obligation often is the harder.

11 Credits

This paper was written during a stay in Eindhoven which was made possible by a Philips stipendium. My warmest thanks go to Wim Feijen from whom I learned the problem, and who (together with Daniela Buhăceanu and Nettie van Gasteren) showed me how to solve it in the Owicki-Gries formalism. I am also indebted to the Eindhoven Tuesday Afternoon Club for reading and commenting on an earlier draft. Tony Hoare made helpful comments on my lecture; it was his idea to view the derivation as distributing the infinity operator over parallel composition.

References

[1] R. Berghammer and B. von Karger. Towards a design calculus for CSP. *Science of Computer Programming*, 25, 1996. to appear.

[2] C. A. R. Hoare. *Communicating Sequential Processes.* Series in Computer Sciences. Prentice-Hall, 1985.

[3] W. Jansen and J. Zwiers. Protocol design by layered decomposition. In *FTRTFT '92*, LNCS 571, 1992.

[4] J. Misra. Phase synchronization. Notes on UNITY: 12-90, available from URL `http://www.cs.utexas.edu/users/psp/notesunity.html`.

[5] E. Olderog and C. A. R. Hoare. Specification oriented semantics for communicating processes. *Acta Inf.*, 23:9–66, 1986.

[6] A. W. Roscoe and C. A. R. Hoare. The laws of `occam` programming. *Theoretical Computer Science,*, 60:177–229, December 1988.

[7] B. von Karger. Plotkin, Hoare and Smyth order: On observational models for CSP. In E.-R. Olderog, editor, *Programming Concepts, Methods and Calculi*, volume A-56 of *IFIP Transactions*, pages 383–402. North-Holland, 1994.

Part 2

Languages and Models for Program Design

The first two parts of this book showed the necessity and the value of formal notation, manipulation rules and methods for proof and program design. Now we look at a number of languages for program design and at their models. As program design starts from a first description of the intended properties of a software system and extends to its implementation, we are interested in languages and models for this whole spectrum. Different languages are intended for different phases of program design. And different models of one phase are intended for different purposes. It is therefore particularly interesting to formally relate such models.

David Gries, Friedrich L. Bauer, C.A.R. Hoare

Manfred Broy

C.A.R. Hoare

Genesis of Algorithmic Languages

Friedrich L. Bauer

The history of Algorithmic Languages shows two worlds: On the one side mathematical usage, on the other side peculiarities stemming from the emerging mechanization of hand calculation and the invention of programming, the method of describing once and for all what to do.

These two worlds have opened the two main streams of functional ('function-oriented') and procedural ('process-oriented') programming, both being subject to further considerable proliferation. There are important additional aspects like concurrency (multiple-control-state vs. single-control-state languages) and recursion. More recently, the particular flavor of object-oriented programming arose. In the functional world, there is also a still widely unexplored niche of descriptive style, with logic programming as a precursor.

Algorithmic languages have ornamented Computer Science with many scientific accomplishments. The article shows how this came about.

Part I brings the historical elements of the control structure of algorithmic languages, mainly machine instructions, plugboard connections, and calculation sheets. Part II discusses transitions between rigid programs, calculation forms, and forests of formulae and the introduction of loops and iteration; it ends with recursion and universality. Part III deals with object ('data') structures. Finally, Part IV provides a taxonomy of historically worthwhile programming languages based on the previously developed concepts and using the following dichotomies

descriptive	vs.	*executable*
nondeterministic, liberal	vs.	*deterministic, totalitarian*
functional, applicative	vs.	*procedural, imperative*
concurrent, parallel	vs.	*sequential, serial*
lucid, transparent	vs.	*opaque, obscure*

Contents

Part III : Object structures

7 Untyped versus typed objects

7.1 The Gödelization trend in logic
7.2 Many-sortedness in computer science
7.3 Universal algebra

8 From hypercomplex numbers and data types to algebraic specifications and Herbrand models

8.1 Structured objects
8.2 Storage structures
8.3 Programming languages dealing with structures
8.4 Programming with pointers
8.5 Functional programming
8.6 Algebraic specification and the Herbrand model
8.7 Object-oriented programming

Part IV : A taxonomy of algorithmic languages

9 Dichotomies

9.1 Executable versus descriptive
9.2 Deterministic versus liberal
9.3 Procedural versus functional
9.4 Sequential versus concurrent
9.5 Opaque versus lucid

10 Some more outstanding topics

10.1 Early low-level languages
10.2 Two-dimensional notation
10.3 Curry and McCarthy
10.4 The way evolution goes

Part I : Historical elements

1 The origins of the words program, plan, algorithm

1.1 Program. The word *program* (British people often write *programme*) denotes today in common usage 'a list of planned activities, a plan for further action, a written display of goal-oriented principles, stipulated inferences, and previewed run-offs'.

The word is of Greek-Latin descent and came up in 19th century with the prevailing meaning 'public announcement, proclamation, agenda'. While the two latter meanings, dominating in 'party program' and 'theater program', are rather specific, the meaning of *program(me)* in today's computer science is closer to the etymology of *pro-gráphein* 'prescribe'.

Greek *graphma, gramma* means 'precept, prescription, specification, instruction, rule, order, law'. *Prescribe* (German *vorschreiben*, French *prescrire*, Russian *predpisivat*) stands for 'to state what must happen or be done in certain conditions'; *precept* (German *Vorschrift*, French *précepte*, Russian *predpisanie*) stands for 'a guiding rule on which behaviour, a way of thought or action is based, governing action or procedure'. *Action* as well as *procedure* means 'something done, effected or accomplished'.

Where did the term *program(me)* in its computer science meaning turn up first? One would not expect it earlier than in Babbage's work, but in fact in his paper of Dec. 26, 1837, where the Analytical Engine, the 'mill', is outlined, the word is not used; nor does it show up in the work of Ada Countess Lovelace, who speaks on the other hand of 'weaving algebraic patterns'. Babbage speaks of 'small pieces of formulae previously made by the first cards', meaning in today's terms a short program punched by the machine itself.

Likewise, Percy E. Ludgate (1909) and Leonardo Torres y Quevedo (1914) have no terminological need for 'programs'. Torres, however, uses the term *automaton* and says 'the automaton acts exactly as an intelligent being who follows certain rules'; his point is that it chooses the action and follows one or another path.

Alan Turing, too, did not speak of programs for his *Gedankenmaschine*. Thus, even as late as the Harvard Conference in January 1947, a 'Symposium on Large-Scale Digital Calculating Machinery', words like 'tape-reading' and 'card-feeding' are used in a description of Aiken's Mark I, without mentioning a name for what is written on these tapes or cards, and 'a tape has been coded' can be found there. In a description of the ENIAC, we find 'master programmer' and 'The program circuits of the accumulator are stimulated by program pulses carried over program lines, which are grouped together in program trunks ...' . Closer to present understanding is 'The divider-square rooter operates as a programming device ...' and 'The programming of the ENIAC is a fairly difficult problem'. The description of the Bell Lab's Relay Computing System

(Stibitz) speaks of 'a routine order represented in perforated code on the routine tape'. George R. Stibitz uses 'program' in the sense of today's 'task', like 'the program of calculating π to a thousand decimal places'. John W. Mauchly speaks of 'preparation of problems' and of 'subroutines' instead of 'subprograms'. In England, around 1950, Alick E. Glennie and Christopher Strachey spoke of AUTOCODE routines. However, Alan M. Turing mentions: 'Thus, we eliminate additional apparatus simply by putting in more programming'.

There is not yet an indication of the term 'programming languages'. But around 1950 the term 'program' comes into wide use: 'The logic of programming electronic digital computers' (A. W. Burks 1950), 'Programme organization and initial orders for the EDSAC' (D. J. Wheeler 1950), 'A program composition technique' (H. B. Curry 1950), 'la conception du programme' (Corrado Böhm 1952), and finally 'Logical or non-mathematical programmes' (C. S. Strachey 1952).

1.2 Computation plan. In the definition above of 'program', the term 'plan' is used. *Plan* means (Merriam-Webster) 'a method devised for making or doing something or achieving an end' and thus is rather parallel to *program*. In fact, the late German inventor Konrad Zuse (22.6.1910–18.12.1995) uses *plan* (German *Plan*) systematically in his German Patent Application Z23 139 IX/42m from April 11, 1936.

Heinz Rutishauser in Zürich took over from Zuse the term *Rechenplan*. The use of *Rechenplan* in Germany for this purpose was not entirely new; however more frequently the terms *Rechenschema*, *Rechenschablone* or *Rechenformular* and *Formblatt* would be used, in particular in connection with printed forms (*Vordruck*). Zuse used *Formular* ('calculation sheet') elsewhere for what we may call now 'computation plan' (see Sect. 2.1 and Fig. 1).

There is an interesting parallel: 'Planning and Coding of Problems for an Electronic Computing Instrument' was the title of the famous and highly influential Goldstine-von Neumann report of 1947. The 'flow diagram' of Goldstine and von Neumann was called *Programmablaufplan* in German; there is also in German the word *Datenflußplan* ('data flow diagram').

1.3 Algorithm. The word *algorithm* comes even later into general use. Its origin points to Al-Choresmi, the Uzbekian scholar who lived around 800 at the court of the caliphs and whose books, translated into Latin, were highly influential for the development of mathematics in Europe since 1200. The expression 'algorithm of multiplication' (with denary or binary numbers) was still used by Leibniz, and the 'algorithm of extracting the square root' had to be learned at secondary schools. Over the centuries, an algorithm for forming the greatest common divisor of two natural numbers, described already by Euclid, kept the word alive. Algorithms were usually stated in verbal form or even by showing only an illustrative example. Only with Markov algorithms (1953), did the word come again into wide use in its very general meaning; Householder and Faadeeva, around 1950, did not yet use it in their monographs on numerical analysis.

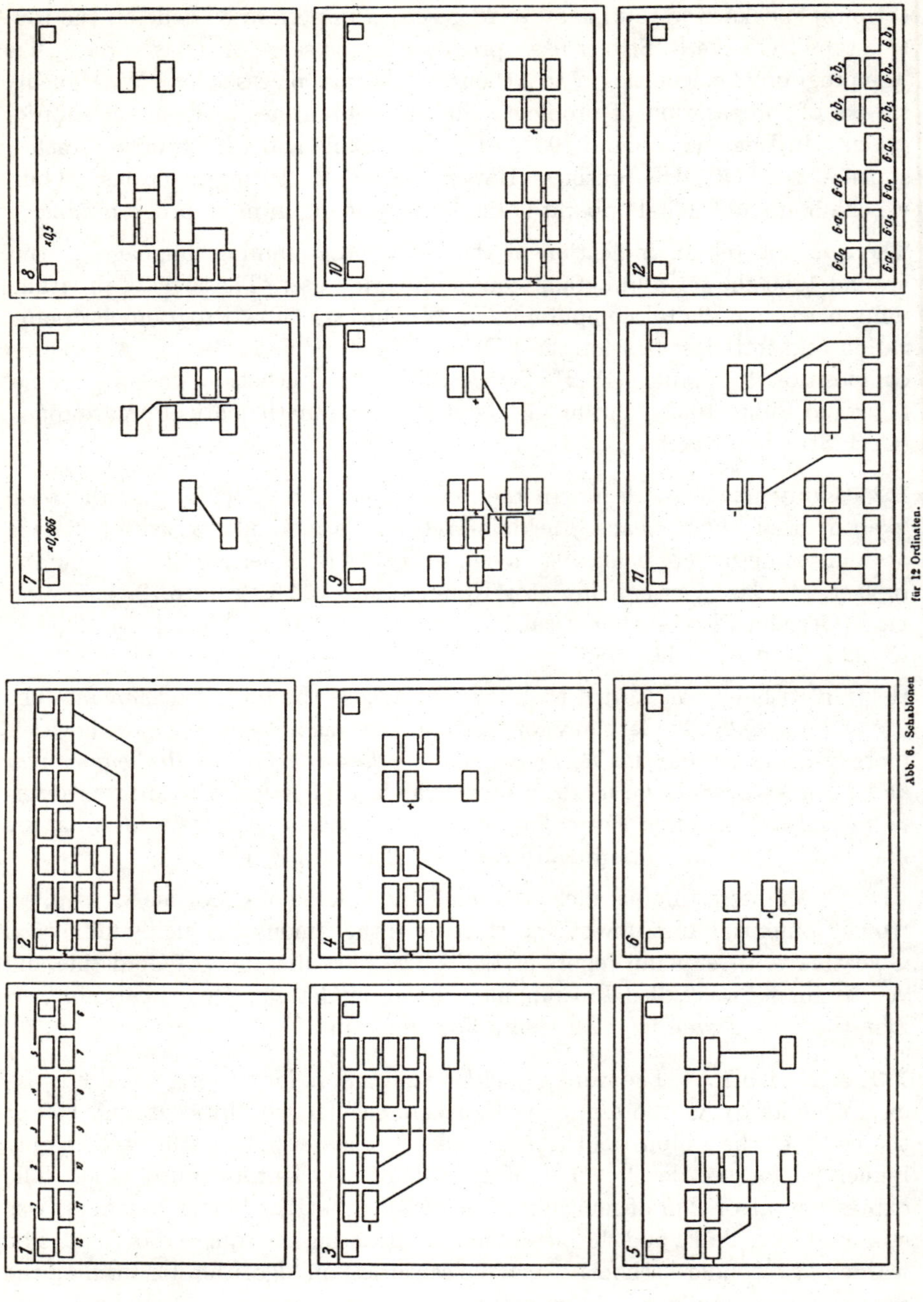

Fig. 1 Stencils for numerical Fourier transform

While *program* and *computation plan* rather synonymously concentrate more on the record, *algorithm* is more directed toward the essential content and indicates some level of abstraction. Thus, two rather different looking programs can describe the same algorithm. Mathematicians therefore frequently prefer the term. With these reservations in mind, we are going to use the words *program* and *computation plan* interchangingly, while *algorithm* means a class of programs and/or plans that differ only in notation.

2 Early elements of programs and computation plans

As long as *programs* or *computation plans* were described verbally or very informally in other ways, there was no need to use the word 'language'.

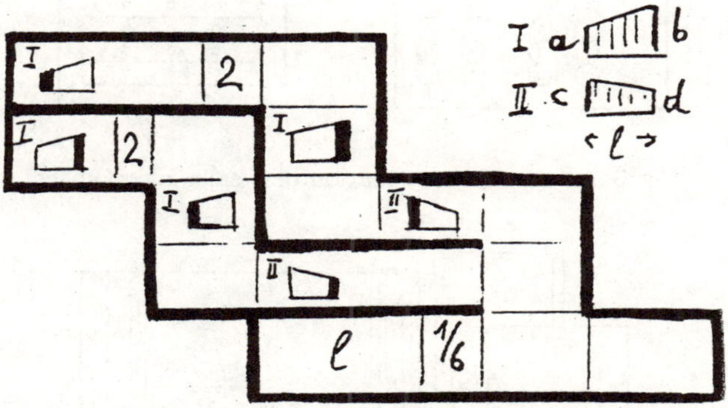

Fig. 2 A calculation sheet (*Formular*) Zuse worked with in the early 1930s:
$_0\int^1 M^I \cdot M^{II} \cdot ds = \frac{l}{6} \cdot ((2 \cdot a + b) \cdot c + (a + 2 \cdot b) \cdot d)$
[from Zuse 1984]

2.1 Programming by calculation sheets. Computation plans in the form of *calculation sheets* had been in use in geodetic, nautical, and astronomical institutions as well as in engineering bureaus for a long time. A two-line sheet for the calculation of a Newton correction using the Horner scheme and a sheet for calculating Fourier transforms were common knowledge before 1900.[1] Engineers used special sheets, e.g., for statics calculation. These sheets had data fields for recording the intermediate results of some calculation, the individual steps of which were to be performed using a 4-species calculator; connecting lines indicated the sequencing of the operations to be performed. An example is given in Fig. 1, taken from a collection of 12 stencils for a Fourier transform with 12 ordinates, constructed — along the lines of Carl Runge — around 1930

[1] In 1900, Jakob Lüroth writes: "Kehrt dieselbe Rechnung oft wieder, so empfiehlt es sich, sich einen Vordruck machen zu lassen, der neben den Formeln auch noch die Bezeichnung der einzelnen Zahlen enthält, sodass man, wenn man die Rechnung später wieder zur Hand nimmt, leicht ersehen kann was jede Zahl bedeutet."

in Alwin Walther's Darmstadt Institute. In some areas of scientific computation, such as geodesy and orbit astronomy, rather complicated calculations were organized in this way; with the particular motive to minimize the number of recordings, which was supported by the use of multiple-carriage-machines like the THALES-GEO, the BRUNSVIGA DUPLA (Hamann 1925), and the BRUNSVIGA TRINKS-TRIPLEX R (Trinks 1926).

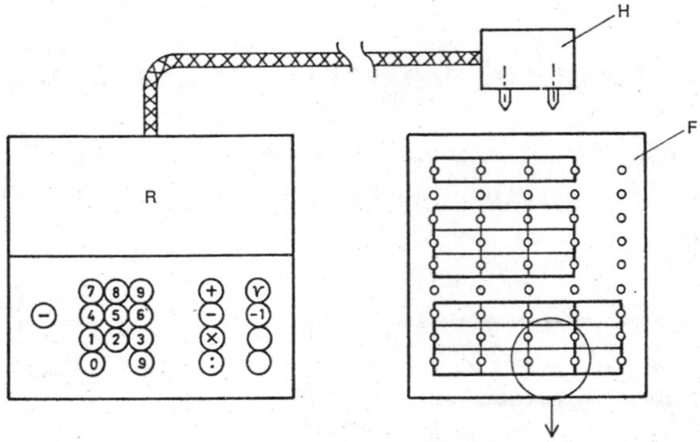

Fig. 3 Zuse's mechanization of a calculation sheet

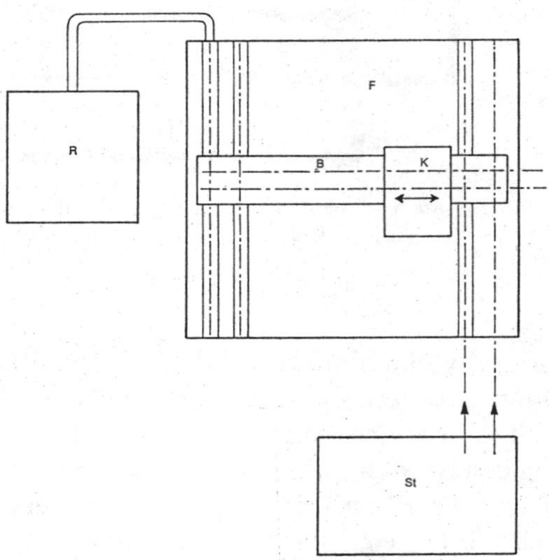

Fig. 4 Zuse's automation of the mechanized calculation sheet

In 1934, Konrad Zuse, then a student, started thinking about a program-controlled computer under the pressure of the tiring numerical work he had to do in civil engineering statics. The example he gives is a calculation of two rectangular moment areas (Fig. 2). He thought of mechanizing the work on the calculation sheet by manually connecting a 4-species calculator R with the appropriate fields of the sheet (Fig. 3), these fields being equipped with suitable gadgets for writing in and reading out numbers. These gadgets would amount to what is today called a one-time assignment store. Later, he even considered a movable bridge with a crane-crab (*Laufkatze*) K whose movement is controlled by a special unit St (Fig. 4), which follows a path between the data fields on the calculation sheet. Calculation sheets supplemented with connecting lines between the data fields, labeled to prescribe uniquely the operation, could well be considered to be written in a language, actually more in a two-dimensional picture language. An example is given by the compact calculation sheet for the expression (c) $\sqrt{a^2+b^2}$ (Fig. 5, left) and the corresponding tree-like *calculation form* (Fig. 5, right), which is topologically equivalent and obtained from a formula by parsing it, that is to say by forming its *Kantorovič tree*.

Note that Zuse in 1936 had abandoned calculation sheets in favor of machine instructions, a development step understandable from the pressure of available technology. The more abstract calculation forms, however, are a first attempt to arrive at an algorithmic language.

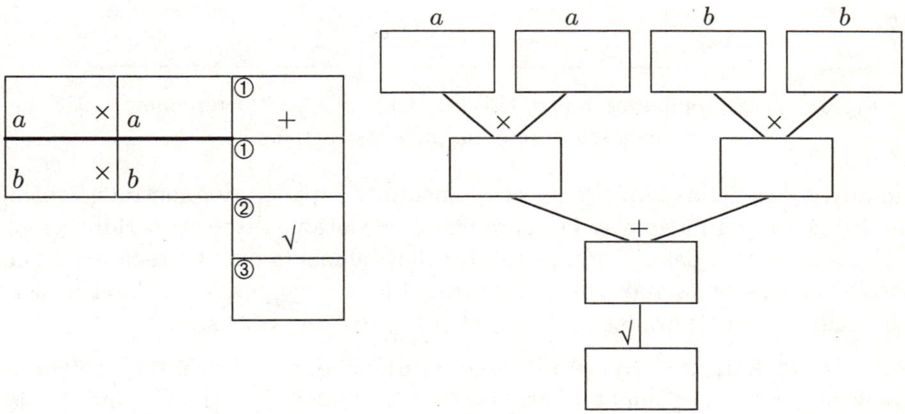

Fig. 5 Calculation sheet (*Formular*) and tree-like calculation form for $\sqrt{a^2+b^2}$

2.2 Programming by machine instructions. Babbage, in discussing, e.g., how to calculate Bernoulli numbers, uses as building elements machine instructions such as '$V_4 \times V_4 = V_{10}$' and implements them with operation signs like × appearing on an 'operation card' and subscript numbers like (4, 4, 10) appearing on a separate 'variable card'. Sequencing such punched cards would hardly have been called a language. But already in 1837, Babbage speaks of 'formulae placed on cards'.

When Zuse arrived in 1936 at a calculating machine that worked rather like the Babbage machine by sequencing computation steps with numbers fetched from and brought to a store, not too surprisingly he used quite independently a three-address notation similar to the one Babbage used. This can be seen from Zuse's patent application Z23 139 IX/42m from April 11, 1936 (Fig. 6). His *computation plan* (*Rechenplan*) is strictly sequential and has no branching.

Operation No.	Adress of cell for operands		Type of basic arithmetic operation	Adress of cell containing result
	1	2		
1.)	1	5	Mult.	0
2.)	0	9	„	10
3.)	2	6	„	0
4.)	0	7	„	11
5.)	3	4	„	0
6.)	0	8	„	12
7.)	1	6	„	0
8.)	0	8	„	13
9.)	2	4	„	0
10.)	0	9	„	14
11.)	3	5	„	0
12.)	0	7	„	15
13.)	10	11	Add.	0
14.)	0	12	„	0
15.)	0	13	Subt.	0
16.)	0	14	„	0
17.)	0	15	„	0 = Result

Fig. 6 Zuse's computation plan (*Rechenplan*) for a 3×3 determinant (1936) — a sequence of machine instructions

Howard Aiken, in his 1937 'Prospectus', mentions 'a paper tape control in which mathematical formulae may be represented by suitable disposed perforations', but does not give details. He also states that 'numbers may be removed from the calculating units and temporarily stored in storage positions. Twelve such positions should be available'. See Sect. 5.1 under *re-usable store.*

2.3 Programming by plugboard connections. In 1914, Torres y Quevedo used a 'diagram' to show how his 'automaton' acts (Fig. 7); indeed his program is an electric circuit. Typically, Torres thought of sequences of actions the automaton performs along the paths between branchings. That programs can already split into trees of actions or that actions can perform collaterally until they are collected into one path is beyond the descriptional tools he was accustomed to using.

Programming (*planning*) by circuitry was later the daily business of Hollerith type tabulating machines. Again, the connection plans for the plugboard of a tabulating machine (Fig. 8) did not justify the word 'language'.

3 Formula, program, and calculation-form languages

After suitable normalization, sequences of machine instructions as well as calculation sheets may be considered as languages ('machine languages'). Typically, mathematical formulae form a language.

A. W. Burks ('An intermediate program language as an aid in program synthesis', 1951) was among the first to use the term 'program(ming) language'.

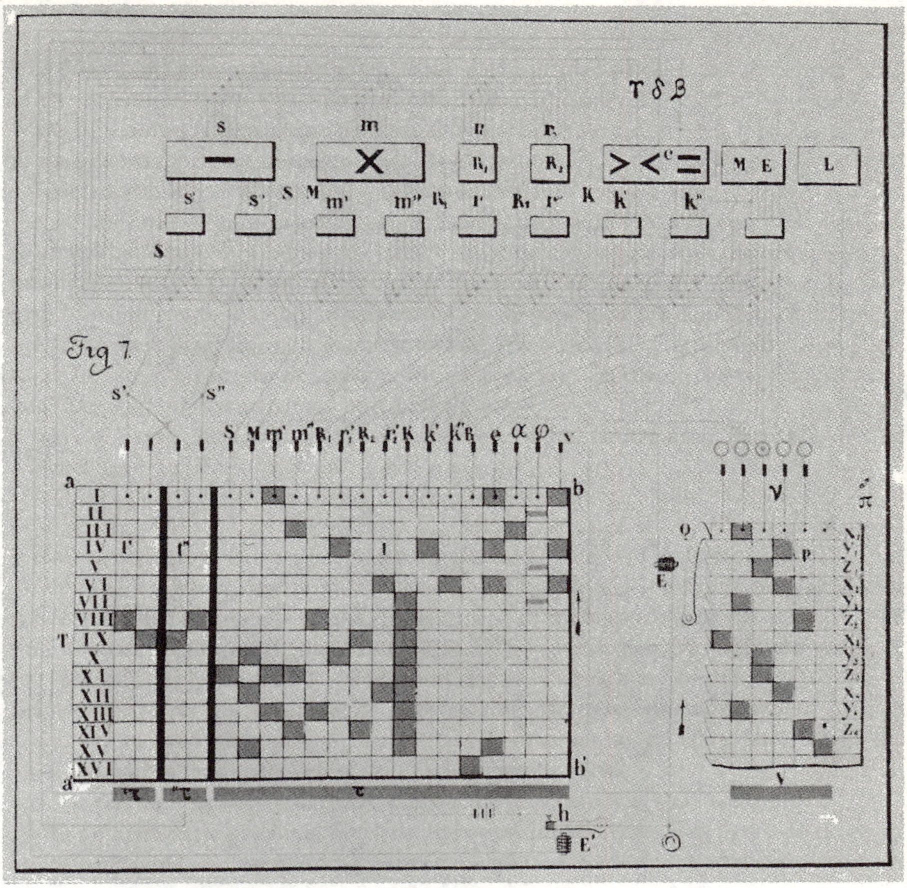

Fig. 7 Torres y Quevedos program circuitry

3.1 Forests of formulae. Donald Knuth said 'Mathematicians would traditionally present the control mechanism of algorithms informally, and the computation steps involved would be expressed by means of equations [with appropriate formulae on the right hand side]'. In fact, the (one-dimensional, parenthesized) *formula notation* commonly used in mathematics implicitly contains the control. Compared to sequences of *machine instructions* in three-address notation, it is quite abstract in so far as it omits unnecessary details. In the example of Fig. 6, Zuse encoded the 9 coefficients $a_{i,k}$, $1 \leq i \leq 3, 1 \leq k \leq 3$, of the matrix by their 9 'addresses' $3 \cdot (i-1)+k$. Moreover, for 16 intermediate results with the addresses 10 through 25 and for the final result with the address 26, the computation steps are explicitly given. Compared with this, the Cramer formula that describes the computation

$$a_{1\,1} \cdot a_{2\,2} \cdot a_{3\,3} + a_{1\,2} \cdot a_{2\,3} \cdot a_{3\,1} + a_{1\,3} \cdot a_{2\,1} \cdot a_{3\,2}$$
$$-a_{1\,1} \cdot a_{2\,3} \cdot a_{3\,2} - a_{1\,2} \cdot a_{2\,1} \cdot a_{3\,3} - a_{1\,3} \cdot a_{2\,2} \cdot a_{3\,1}$$

is more compact, and we would rather use it to describe the algorithm of the Cramer rule.

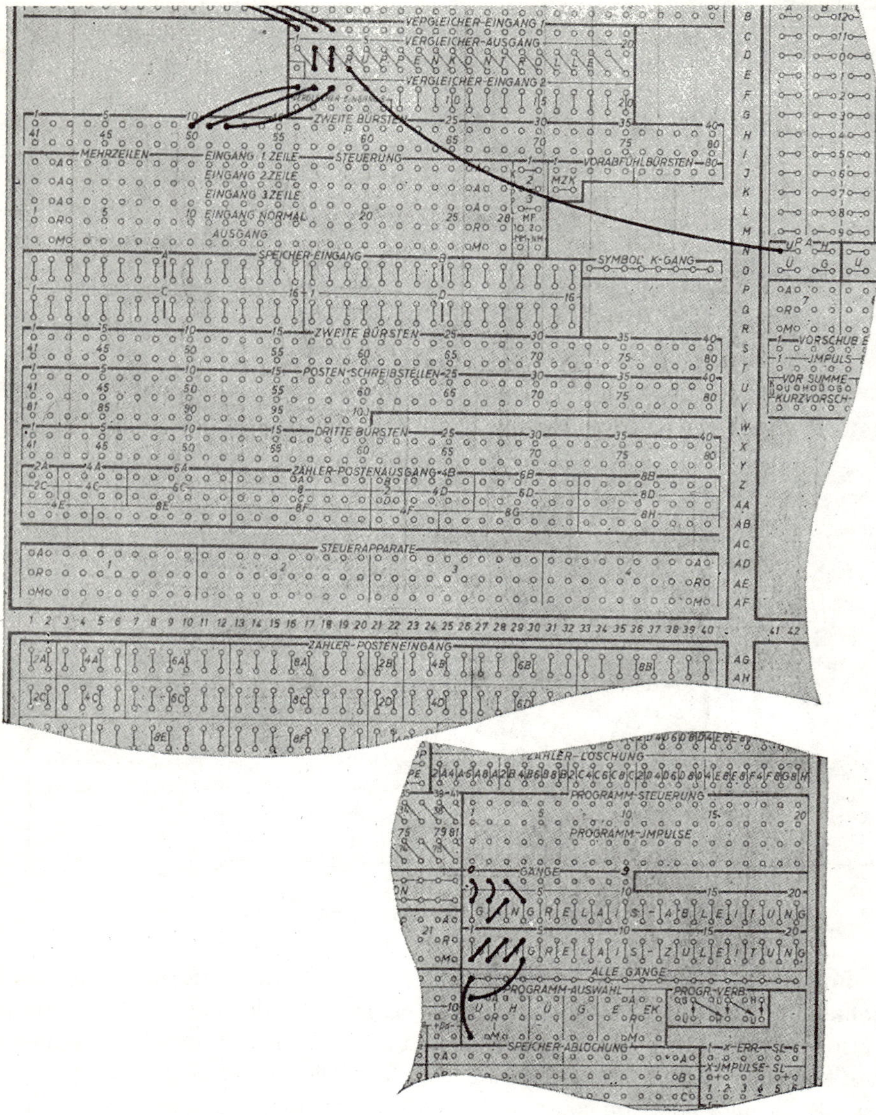

Fig. 8 Connection plan for the plugboard of a tabulating machine

3.2 Sequential programs. In 1944, Konrad Zuse, in a first step toward his *Plankalkül*, introduced a slightly more abstract form of machine instructions: sequences of equations with atomic formulae (W. J. Eckert 1948: 'commands', John Mauchly 1947: 'instructions'; Howard Aiken 1946: 'coding'). This we characterize with the term *sequential program.* Two examples, corresponding to the Euclidean length and to the solution of a quadratic equation, are given in Fig. 9. All the *computation steps* there are shown explicitly.

$$
\text{(c)}\quad
\begin{aligned}
a &\Rrightarrow V_1\\
b &\Rrightarrow V_2\\
V_1 \times V_1 &\Rrightarrow V_3\\
V_2 \times V_2 &\Rrightarrow V_4\\
V_3 + V_4 &\Rrightarrow V_5\\
\sqrt{V_5} &\Rrightarrow V_6\\
V_6 &\Rrightarrow c
\end{aligned}
\qquad\qquad
\text{(q)}\quad
\begin{aligned}
a &\Rrightarrow V_1\\
b &\Rrightarrow V_2\\
V_1/2 &\Rrightarrow V_3\\
V_3 \times V_3 &\Rrightarrow V_4\\
V_4 - V_2 &\Rrightarrow V_5\\
\sqrt{V_5} &\Rrightarrow V_6\\
V_3 \times (-1) &\Rrightarrow V_7\\
V_7 + V_6 &\Rrightarrow V_8\\
V_7 - V_6 &\Rrightarrow V_9\\
V_8 &\Rrightarrow x_1\\
V_9 &\Rrightarrow x_2
\end{aligned}
$$

Fig. 9 Two examples of Zuse's programs (1944) for
(c) $c = \sqrt{a^2 + b^2}$ (cf. Fig. 5) and (q) $x_{1,2} = -\frac{a}{2} \pm \sqrt{(\frac{a}{2})^2 - b}$

Typically, Zuse wrote the atomic formula on the left-hand side, as is suggested by the dynamic character of the *computation step.* To make this evident, we use here the 'yields-sign' (German *Ergibt-Zeichen*) $\Rrightarrow$ that Zuse introduced later with the *Plankalkül.*[2] We find in such a sequential program (Knuth: 'straight-line program', Zuse: 'lineares Programm') *initializing computation steps* ('read instructions') at the very beginning, followed by a sequence of *arithmetic computation steps,* with *finishing computation steps* ('display instructions, print instructions') at the very end. Zuse wanted the machine to interpret $a \Rrightarrow V_1, b \Rrightarrow V_2$ as a request for two numbers to be typed in and $V_8 \Rrightarrow x_1$, $V_9 \Rrightarrow x_2$ as a request for two numbers to be displayed by the machine. The sundry V_i, denoting the intermediate results, he calls boldly 'Variable' (*variables*); they carry the addresses of his 1936 notation as subscripts. For brevity, we will call this a *rigid program* (Zuse: *starrer Rechenplan*). With this expression, Zuse wanted to indicate that addresses cannot be changed, only the content of the storage positions they designate can be changed.

Note that some of the computation steps could be rearranged: On the left-hand side, the third and the fourth computation step could be swapped; on the

[2] John Backus, having used the orthodox order in FORTRAN, adamantly refused to introduce the Zuse *Ergibt-Zeichen* into ALGOL.

right-hand side, the sixth and the seventh computation step could be swapped. Furthermore, the display instruction $V_8 \Rrightarrow x_1$ can be moved up one step; in both programs, the read instruction $b \Rrightarrow V_2$ can be delayed one step. In all these cases, the result is the same, even in the presence of rounding. In the formula, this freedom in the order of execution is inherent. Sequential programs are deterministic descendants of formulae.

The very first programming languages, before FORTRAN (1954–1956), were based too heavily on sequential atomization to be honestly called algorithmic languages. But they paved the way.

Obviously, the sequential (rigid) programs considered so far are examples of *single-control-state programs* in the following sense: At any moment of calculation, it is possible to point to the computation step that is just busy; apart from a transition interval, at most one computation step and always some computation step is busy.

In machines, the transition between states is done by the execution of machine instructions; fittingly one speaks in this context also of *imperative* programming languages. The *machine language* of early machines was almost exclusively an imperative one. 'Sequential' was the magic word at the start of 'Large-Scale Digital Calculating Machinery' (H. Aiken: *Automatic Sequence Controlled Calculator*, W. J. Eckert: *Selective Sequence Electronic Calculator*), sequential atomization was the goal. This turned out to be a dead-end road.

3.3 Multiple-control-state programs. It just happens that human performance, restricted as it is with respect to mental concentration, makes it advisable to "do just one thing at any time", so at any moment the human calculator is occupied with just one task. If asked "where are you working", he (or she) will always be in one and only one state. This worthwhile caution with respect to human operators was fatuously applied also to mechanization and automatization — quite unnecessarily, since machines do not think and therefore can be much more reliable than humans.

There is an obvious generalization of these *single-control-state* programs to *multiple-control-state* programs, which describe what happens if a group of human operators works jointly and under division of labor on one task. Each person will have his (or her) *individual* program, but will also need an overall *organization* program. Such work was done, e.g., in 1938 in New York in a large project for the calculation of tables, using hitherto unemployed persons; and during World War II, Alwin Walther in Darmstadt organized a group of women from the Aids Corps to do a network flow of calculations. No documentation of the organizational details has survived.

In 1934, Konrad Zuse stood on the threshold of multiple-control-state programs. The calculation sheet of Fig. 2 for (z) $\frac{l}{6} \cdot ((2 \cdot a + b) \cdot c + (a + 2 \cdot b) \cdot d)$ allows two persons to work jointly. They both have access to the input data a, b, c, d and they calculate within two channels; only at the end must one person perform the addition, the calculation of $l/6$, and the final multiplication — they may both perform this as a check. Had he seen it as feasible to construct such a machine, Zuse could have written this as in Fig. 10 using a split into two 'concurrent', but still independent sequences of calculation, which are 'collected' later. Such independent sequences of calculation are called *processes*. The notation used here is from the German Standards Organization, DIN 66001.

$$\begin{aligned} a &\Rightarrow V_1 \\ b &\Rightarrow V_2 \\ c &\Rightarrow V_3 \\ d &\Rightarrow V_4 \end{aligned}$$

$$\text{(z)} \qquad \begin{array}{r} V_1 \times 2 \Rightarrow V_{15} \\ V_{15} + V_2 \Rightarrow V_{16} \\ V_{16} \times V_3 \Rightarrow V_{17} \end{array} \qquad\qquad \begin{array}{r} V_2 \times 2 \Rightarrow V_{25} \\ V_{25} + V_1 \Rightarrow V_{26} \\ V_{26} \times V_4 \Rightarrow V_{27} \end{array}$$

$$\begin{aligned} V_{17} + V_{27} &\Rightarrow V_5 \\ l &\Rightarrow V_6 \\ V_6/6 &\Rightarrow V_7 \\ V_5 \times V_7 &\Rightarrow V_8 \\ V_8 &\Rightarrow res \end{aligned}$$

Fig. 10 A multiple-control-state program for Zuse's calculation sheet (cf. Fig. 2). A process is split into two processes, which are merged later into one process.

While a single human normally finds it difficult to concentrate simultaneously on a number of processes (even Caesar is reported as having managed only a few), machines, as stated above, should not suffer like humans from 'one-track minds'. And since a mere group of persons was used successfully in such circumstances, even more so should a machine have no difficulties. Konrad Zuse did not live in the more fortunate surroundings of Bell Laboratories or the Harvard Physics Department, where he could have arrived at a multi-process (relay) machine.

By the way, Howard Aiken was also close to a multi-process machine. In his 1937 'Prospectus', he speaks of four arithmetic units and four control panels, 'each capable of controlling the entire machine or any of its parts. Thus, for complicated problems the entire sources could be thrown together, for simpler problems fewer resources are required and several problems could be in progress at the same time'. Whether this idea included splitting and merging of processes, remains open.

Many simple and everyday calculations can be shortened (in time) by multi-process calculation, as Fig. 11 shows for the task of evaluating a polynomial in three concurrent, interwoven processes. Nevertheless, multi-process programs were frowned upon by engineers and programmers in the early days. John Mauchly wrote (1947): 'However, the machine should be kept serial so far as the operator is concerned. That is, no two instructions which the operator gives the machine are to be carried out at the same time. Any particular instruction which the operator gives, however, may involve the simultaneous operation of numerous parts'. Hesitation to look into the problems of multi-process calculation led the authors of the article 'Programmierung' in Steinbuch's *Taschenbuch der Nachrichtenverarbeitung*, 2nd edition 1967, to omit the flow diagram symbols of DIN 66001 for 'Aufspaltung bei Parallelbetrieb, Sammlung bei Parallelbetrieb, Synchronisationsschnitt'.

Note that the multiple-control-state programs considered in this section are rigid programs, too.

3.4 Simply looped programs. Sequential programs so far had the property that any equation was busy only once (straight-line programs). The same holds for multiple-control-state programs. If, however, the whole rigid program is to be repeated, with read instructions for new data each time, we have a *simply looped program*, a rigid program with a single loop — the term coming from the way a paper tape is made endless by gluing together its beginning and its end. In Sect. 5, loop structures in general are studied more closely.

3.5 Calculation forms. Quite complicated calculations can be represented by not necessarily tree-like *calculation forms*. This is shown in Fig. 11 for the concurrent evaluation of a polynomial. Note that again the same freedom in the order of execution exists as in the formula. The formula on the one side and the compact calculation sheet (above) and its topologically equivalent calculation form (below) on the other side are on the same level of abstraction; both of them describe more properly the algorithm than a machine-oriented sequence of disguised n-address instructions.

3.6 Data flow diagrams. A calculation form is closely related to a data flow diagram: the data flow diagram describes a stream of data to be read in and put out by repeated actions, corresponding to a simply looped program. Data flow diagrams can be executed in a systolic way, *provided every atomic formula has its own processing unit.* Early punch card machine complexes frequently had aspects of such systolic nature, common today in miniature hardware.

3.7 Algorithmic language. Algorithmic languages are (or should be) programming languages under some abstractions from notational details. This gives formulae and calculation forms an advantage. Moreover, formula notation looks slims compared to calculation forms. Thus, so far, formula notation seems to be the best candidate for proper formulation of algorithms. This attitude is certainly found among logicians and pure mathematicians. In the world of functions, propositions, and predicates, there is no need and also no room for showing states explicitly: *genuine functional programming is state-free.* Whereas Torres y Quevedo used the plan shown in Fig. 7 to indicate his program, a mathematician is quite satisfied to write $\alpha = a \times x \times (y - z)^2$ instead; and

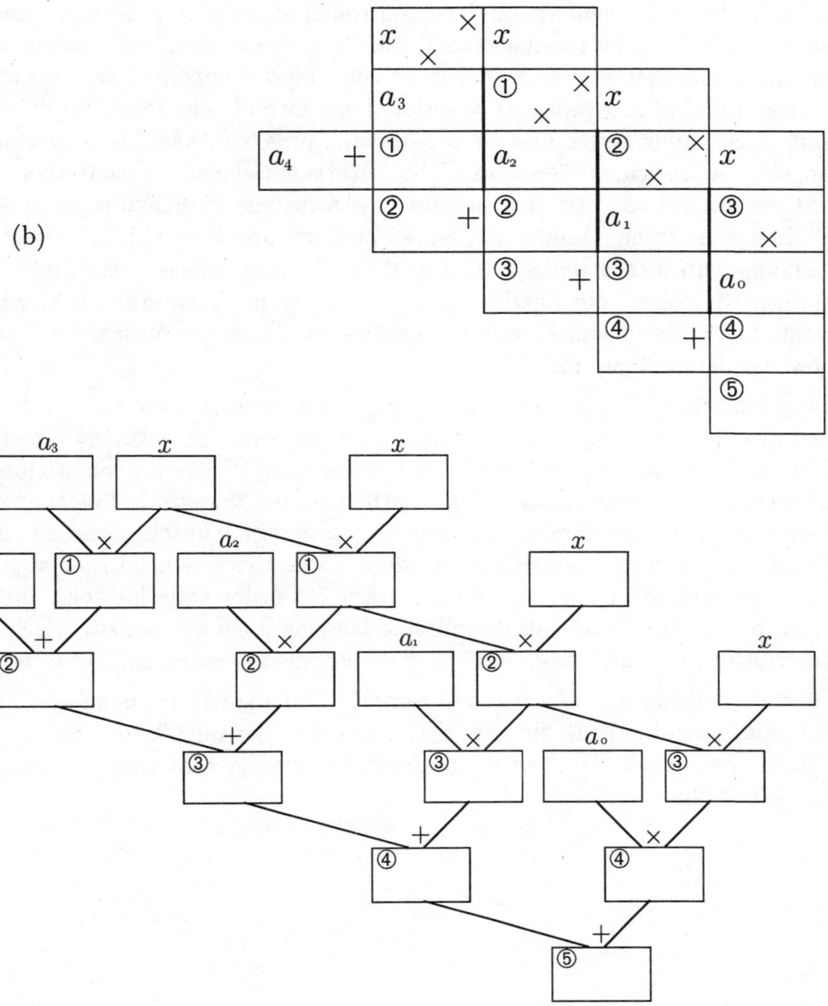

Fig. 11 Calculation sheet (*Formular*) and non-tree-like calculation form for the concurrent evaluation of a polynomial $a_4 + a_3 \times x + a_2 \times x^2 + a_1 \times x^3 + a_0 \times x^4$. For example, in step ②, $a_4 + (a_3 \times x)$, $a_2 \times (x^2)$, and $x \times (x^2)$ are formed; the + and × signs indicate the operation upon neighboring fields.
[From Bauer, Goos 1982]

for Zuse's *Rechenplan* (Fig. 9, (q)), the mathematician finds it at least helpful, possibly even preferable, to be told that it means

(q) compute the roots x_1 and x_2 of $x^2 + a \times x + b = 0$.

$\{x : x^2 + a \times x + b = 0\}$ is the natural logical description of this problem.

Why, then, did Zuse bother to write the 'complicated' program with explicit states, and how could he even be proud to have done so? It is because the machine he had invented could be controlled step by step by his imperative program, but not by the mathematician's functional program. He was wrong, in that the mathematician Rutishauser, only a few years later, was able to write a *programming program* that translated the formula the mathematician likes and understands better into a state-oriented program that Zuse's machine Z4, working at Zürich, understood. Thus, Rutishauser showed that there exists the meaningful concept of a *functional* algorithmic language capable of controlling a machine. Bauer and Samelson even obtained a 1957 patent for a *formula-controlled machine* that directly obeys programs of this simple, purely applicative, form. Zuse had given up too early his naive approach, when he wanted to build machines with the available technology. With the *Plankalkül*, however, he made up for the loss.

That a mathematician prefers the formulae notation is of course to a large extent also due to the fact that the mathematician is used to this; any time human perception is involved, habits play a decisive role. Thus, mathematicians may be misled by their prejudice. If the mathematician were to declare that he also considers the description (q) above to be a program, caution would be advised: clever mathematical systems today may break this down into an algorithm, but they will not do it with an arbitrary 5th-order equation with unknown coefficients. The history of algorithmic languages cannot yet say much about descriptive programs, since some of their aspects are still a subject of research.

Formula notation is so far a good example of an algorithmic language, but calculation forms have the didactic advantage that the calculation can be carried out on the spot.[3] We shall see in Sects. 4.2 and 4.4 that they are even more expressive than formulae.

---

[3] Compare the *Formularmaschine* (Bauer-Goos, Informatik I, 4th edn., Sect. 2.3.3).

Part II : Loops, iteration and recursion

4 Transitions between rigid programs, formulae, and calculation forms

4.1 Formulae from rigid programs. Clearly, every sequential (single-control-state, rigid) program can be transformed into a set of display instructions with formulae on their left-hand side by eliminating all the intermediate results; the formulae containing only input data (*arguments*) occurring on the left side of read instructions. However, in case of Fig. 9, this complete elimination process will result in the two display instructions

$$\text{(q)}\quad \begin{array}{l} (a:2)\times(-1)+\sqrt{(a/2)\times(a/2)-b} \Rightarrow x_1 \\ (a:2)\times(-1)-\sqrt{(a/2)\times(a/2)-b} \Rightarrow x_2 \end{array}$$

which contain some duplication of computational effort. Therefore, even in the purest functional notation, it is useful to keep some intermediate results, in our case

$$\text{(q)}\quad \begin{array}{r} \sqrt{(a/2)\times(a/2)-b} \Rightarrow V_6 \\ (a:2)\times(-1) \Rightarrow V_7 \\ V_7+V_6 \Rightarrow x_1 \\ V_7-V_6 \Rightarrow x_2 \end{array} .$$

or even better

$$\text{(q)}\quad \begin{array}{r} a/2 \Rightarrow V_3 \\ \sqrt{V_3\times V_3-b} \Rightarrow V_6 \\ V_3\times(-1) \Rightarrow V_7 \\ V_7+V_6 \Rightarrow x_1 \\ V_7-V_6 \Rightarrow x_2 \end{array} .$$

The same holds for multiple-control-state programs like that of Fig. 10: Complete elimination of all intermediate results produces

$$\text{(z)}\quad ((2\times a+b)\times c+(a+2\times b)\times d)\times(l/6) \Rightarrow res\ .$$

4.2 Calculation forms from formulae. Furthermore, every such formula may be made into a calculation form by parsing it, that is to say by building up its Kantorovič tree; a set of display instructions is thus reduced to a forest of Kantorovič trees rooted with the result names. For the formula $\sqrt{a^2+b^2}$ (calculation of the Euclidean length), Fig. 5 has shown how the calculation form is obtained by decorating the Kantorovič tree with boxes.

For the pair (q) of formulae for x_1 and x_2 in Sect. 4.1 , or rather for the set of five computation steps there, using the common intermediate results V_3, V_6, and V_7 , by building the five corresponding Kantorovič trees and joining them, the calculation form of Fig. 12 is obtained. It is no longer a tree.

(q)

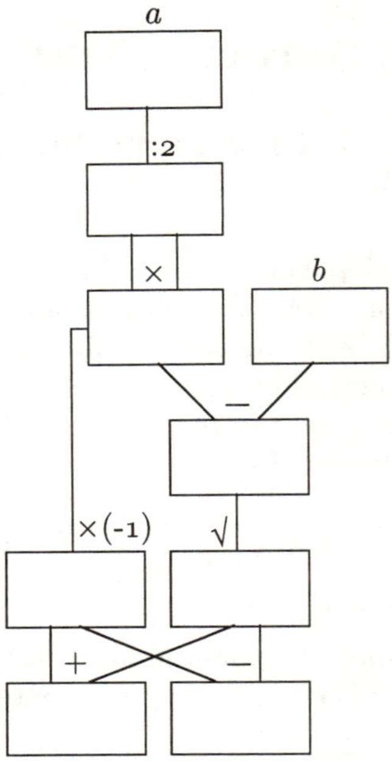

Fig. 12 Calculation form (*Formular*) for $x_{1,2} = -\frac{a}{2} \pm \sqrt{(\frac{a}{2})^2 - b}$

(c)

$$a \Rightarrow V_1 \qquad b \Rightarrow V_2$$
$$V_1 \times V_1 \Rightarrow V_3 \qquad V_2 \times V_2 \Rightarrow V_4$$

$$V_3 + V_4 \Rightarrow V_5$$
$$\sqrt{V_5} \Rightarrow V_6$$
$$V_6 \Rightarrow c$$

Fig. 13 Multiple-control-state program for $c = \sqrt{a^2 + b^2}$

4.3 Rigid programs from calculation forms. Finally, for a given calculation form, there are usually several rigid programs corresponding to it, in particular several sequential ones resulting from sequentialization of multiple-control-state programs. For the calculation form in Fig. 5, Fig. 13 shows such a program, obtained in an obvious way *by numbering the boxes in a tree-consistent order.* The single-control-state program (c) of Fig. 9 and some others obtained from it by swapping certain computation steps are all descendants.

Likewise, Fig. 14 shows a multiple-control-state program for the solution of a quadratic equation, obtained from the calculation form of Fig. 12. In this special case, there is a split at the beginning and a merge at the end, while every intermediate merge is followed immediately by a split — this has been called

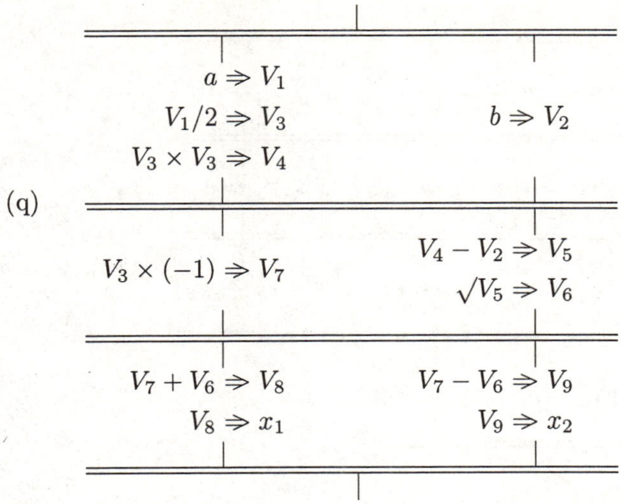

Fig. 14 Multiple-control-state program for $x_{1,2} = -\frac{a}{2} \pm \sqrt{(\frac{a}{2})^2 - b}$

a *synchronization cut* (German 'Synchronisationsschnitt'). Moreover, some of the atomic computation steps can now be condensed; Fig. 15 shows the result.

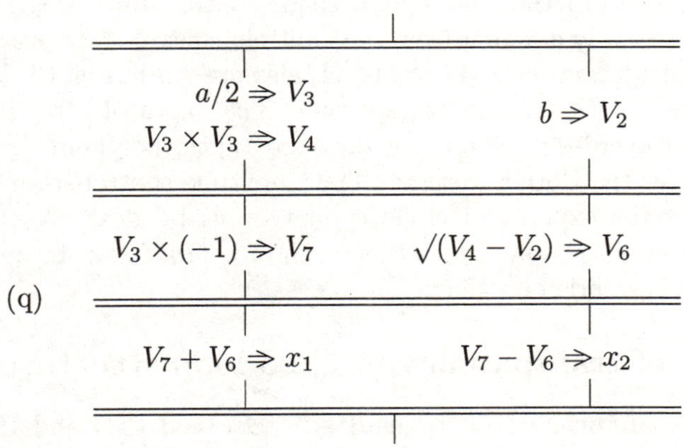

Fig. 15 Condensed two-process program for $x_{1,2} = -\frac{a}{2} \pm \sqrt{(\frac{a}{2})^2 - b}$

4.4 Richness of the class of calculation forms. Not every calculation form can be made into a forest of single formulas without losing information: The class of calculation forms is richer than the class of formulae. This can be seen from the example of Fig. 11 . In fact, $a_4+a_3\times x+a_2\times x^2+a_1\times x^3+a_0\times x^4$ is a formula in the sense used so far provided a^b is meant to be a binary atomic function.

But this is not what textbooks on numerical analysis tell us. They want us to pass over to $a_4+(a_3+(a_2+(a_1+a_0\times x)\times x)\times x)\times x$, using the distributive law

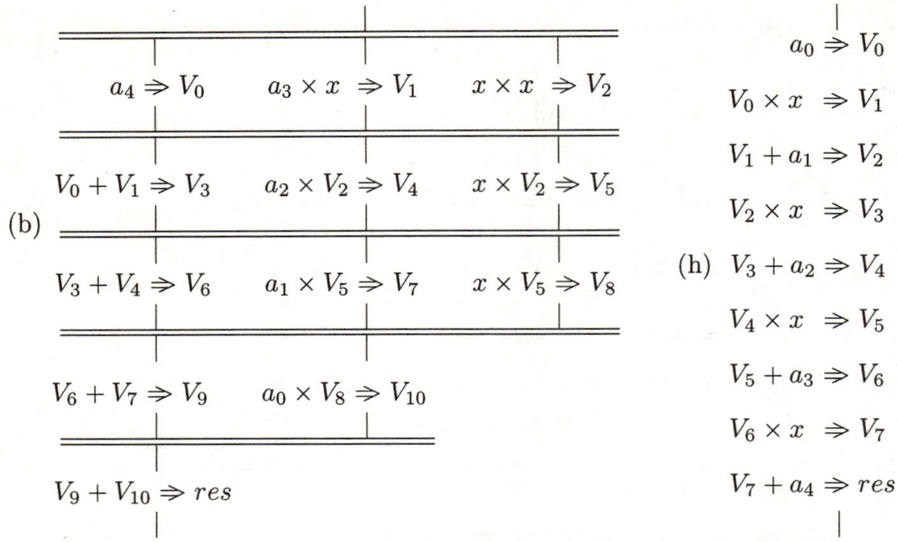

Fig. 16 Program for (b) concurrent and (h) Horner evaluation of a polynomial $a_4 + a_3 \times x + a_2 \times x^2 + a_1 \times x^3 + a_0 \times x^4$.

— the Horner form of the polynomial, from which we may obtain the classical Horner calculation form that leads to a single-control-state program (h) in Fig. 16. This is certainly different from the multiple-control-state program that belongs to the calculation sheet (b) in Fig. 11, also present in Fig. 16. It is again of the special form with synchronization cuts. The depth of computation for this scheme is of the order n, where n is the degree of the polynomial, while it is of the order $2n$ for the Horner method. Thus, multiple-control-state programs can save time at the expense of circuitry, a very useful deal. A brute-force attack by circuitry may even reduce the depth of computation to order $\log n$ (communication by Christoph Herrmann, Passau).

5 The loss of transparency in fixed loop structures

5.1 The stack of intermediate results. Between 1934 and 1936, Zuse realized that storing numbers on the mechanized calculation sheet is complicated and expensive and thus considered re-usable storage. Re-usable storage was known in particular in desk-calculating machines: the accumulator had this property, and from multiple-carriage machines to machine complexes of the sort Comrie used in the 1920s, multiple storage was found. Zuse realized this in 1936, but he failed to utilize it systematically. In fact, it turns out that he could have replaced the programs of Fig. 9 by the ones in Fig. 17. In his *Plankalkül* draft of 1945, he wrote explicitly that $Z + 1 \Rightarrow Z$ was short for $Z_i + 1 = Z_{i+1}$. Burks, in 1950, used the somewhat unusual form $i \rightarrow i+1$ for incrementing the variable i .

By the way, Aiken — unlike Zuse — was aware in 1937 of the work of Babbage, who had already considered re-using storage. This is demonstrated in the

$$
\text{(c)}\quad
\begin{aligned}
a &\Rightarrow V_1\\
b &\Rightarrow V_2\\
V_1 \times V_1 &\Rightarrow V_1\\
V_2 \times V_2 &\Rightarrow V_2\\
V_1 + V_2 &\Rightarrow V_1\\
\sqrt{V_1} &\Rightarrow V_1\\
V_1 &\Rightarrow c
\end{aligned}
\qquad\qquad
\text{(q)}\quad
\begin{aligned}
a &\Rightarrow V_1\\
b &\Rightarrow V_2\\
V_1/2 &\Rightarrow V_1\\
V_1 \times V_1 &\Rightarrow V_3\\
V_3 - V_2 &\Rightarrow V_2\\
\sqrt{V_2} &\Rightarrow V_2\\
V_1 \times (-1) &\Rightarrow V_1\\
V_1 + V_2 &\Rightarrow V_3\\
V_1 - V_2 &\Rightarrow V_1\\
V_3 &\Rightarrow x_1\\
V_1 &\Rightarrow x_2
\end{aligned}
$$

Fig. 17 Zuse programs (cf. Fig. 9) with *Working-Variables*

example given by Lady Lovelace (Fig. 18): She said, 'variables cease to be *Result-Variables*, and become mere *Working-Variables*'. She used superscripts to distinguish the states the variable can be in; the stratified record could as well have been written with collective assignments. From Fig. 17, one obtains in this way Fig. 19, using bracketed superscripts as Rutishauser did.

First Series of Operations	Second Series of Operations	Third Series, which contains only one (final) Operation
${}^1V_{10} \times {}^1V_0 = {}^1V_{20}$	${}^1V_{11} \times {}^1V_0 = {}^1V_{31}$	${}^1V_{21} + {}^1V_{31} = {}^2V_{21}$, and
${}^1V_{10} \times {}^1V_1 = {}^1V_{21}$	${}^1V_{11} \times {}^1V_1 = {}^1V_{32}$	V_{31} becomes = o
${}^1V_{10} \times {}^1V_2 = {}^1V_{22}$	${}^1V_{11} \times {}^1V_2 = {}^1V_{33}$	
${}^1V_{10} \times {}^1V_3 = {}^1V_{23}$	${}^1V_{11} \times {}^1V_3 = {}^1V_{34}$	

Fourth Series

$$
\begin{cases}
{}^1V_{32} \div 2 + {}^1V_{22} = {}^2V_{22} = BA_2 + \tfrac{1}{2}B_1A_1 & \\
{}^1V_{32} \div 2 + {}^1V_{20} = {}^2V_{20} = BA + \tfrac{1}{2}B_1A_1 & = C_0
\end{cases}
$$

$$
\begin{cases}
{}^1V_{33} \div 2 + {}^1V_{23} = {}^2V_{23} = BA_3 + \tfrac{1}{2}B_1A_2 & = C_3{}^*\\
{}^1V_{33} \div 2 + {}^2V_{21} = {}^3V_{21} = BA_1 + B_1A + \tfrac{1}{2}B_1A_2 & = C_1
\end{cases}
$$

$$
\begin{cases}
{}^1V_{34} \div 2 + {}^0V_{24} = {}^1V_{24} = \tfrac{1}{2}B_1A_3 & = C_4\\
{}^1V_{34} \div 2 + {}^2V_{22} = {}^3V_{22} = BA_2 + \tfrac{1}{2}B_1A_1 + \tfrac{1}{2}B_1A_3 & = C_2
\end{cases}
$$

Fifth and Final Series of Operations

$$
\begin{aligned}
{}^2V_{20} \times {}^0V_{40} &= {}^1V_{40}\\
{}^3V_{21} \times {}^0V_{41} &= {}^1V_{41}\\
{}^3V_{22} \times {}^0V_{42} &= {}^1V_{42}\\
{}^2V_{23} \times {}^0V_{43} &= {}^1V_{43}\\
{}^1V_{24} \times {}^0V_{44} &= {}^1V_{44}
\end{aligned}
$$

Fig. 18 Programs with *Working-Variables* considered by Ada Lovelace (B.V. Bowden)

$$\text{(c)}\quad \begin{aligned} a &\Rightarrow V_1^{(1)} \\ b &\Rightarrow V_2^{(1)} \\ V_1^{(1)} \times V_1^{(1)} &\Rightarrow V_1^{(2)} \\ V_2^{(1)} \times V_2^{(1)} &\Rightarrow V_2^{(2)} \\ V_1^{(2)} + V_2^{(2)} &\Rightarrow V_1^{(3)} \\ \sqrt{V_1^{(3)}} &\Rightarrow V_1^{(4)} \\ V_1^{(4)} &\Rightarrow c \end{aligned}$$

$$\text{(q)}\quad \begin{aligned} a &\Rightarrow V_1^{(1)} \\ b &\Rightarrow V_2^{(1)} \\ V_1^{(1)}/2 &\Rightarrow V_1^{(2)} \\ V_1^{(2)} \times V_1^{(2)} &\Rightarrow V_3^{(1)} \\ V_3^{(1)} - V_2^{(1)} &\Rightarrow V_2^{(2)} \\ \sqrt{V_2^{(2)}} &\Rightarrow V_2^{(3)} \\ V_1^{(2)} \times (-1) &\Rightarrow V_1^{(3)} \\ V_1^{(3)} + V_2^{(3)} &\Rightarrow V_3^{(2)} \\ V_1^{(3)} - V_2^{(3)} &\Rightarrow V_1^{(4)} \\ V_3^{(2)} &\Rightarrow x_1 \\ V_1^{(4)} &\Rightarrow x_2 \end{aligned}$$

Fig. 19 Zuse programs in the Ada Lovelace notation with storage state superscripts

Finally, Samelson and Bauer showed in 1955 that with their stack technique a program with the minimal number of re-usable stores for intermediate results can be deduced in a simple and elegant way from the formula [see Appendix].

Thus, introduction of *Working-Variables* (the term *variable* is now really appropriate, and Aiken's 'temporarily stored', Sect. 2.1 applies) saves storage, usually a tremendous saving since the stack depth of a formula goes with the order of the square root of its length. On the other hand, re-use of store means *loss of transparency*: Most of the intermediate results are no longer available at the end of the computation, and the newer results cast a shadow on the older ones. Since, superficially, only the end results matter, and for simply looped programs these are all displayed or printed, this loss was thought to be tolerable. But it is important to deal with some of the intermediate results in case of loops (Sect. 5.6) or recursion not only for checking the exactness of the arithmetic unit, but much more for proving the correctness of the algorithm against the specification.

5.2 The Bauer-Samelson machine. The stack of intermediate results (*Zahlkeller*) in the machine Bauer and Samelson patented in 1957 was a piece of hardware. But we wanted the machine also to be directly formula-controlled, and we introduced for this purpose a similar hardware stack for pending arithmetic instructions (*Befehlskeller*). The infix formula language we used was part of the patent. In the end, such a machine was not built, and we kept the functional input language but rewrote the hardware design into a translating program for a von-Neumann-type machine. The flexibility we gained by this triggered our development toward ALGOL 58 as shown in the Lugano Draft of Fall 1957. Thus, ALGOL could have become a genuine functional, state-free language. It did not, since it was corrupted — a typical committee effect — by the influence of machine-language programming, petrified in FORTRAN.

5.3 A Pandora's box. Quite interestingly, in his 1936 *Gedankenmaschine* Turing used a re-usable store. Naively, since he wanted to work with a minimal machine, this was only consequent; and the fact that any real calculation with this machine would drive practical people crazy was considered an advantage. Still, for a genuine logician, the loss of transparency was hard to bear.

For stored-program machines, a re-usable store meant that not only the data but also the program store could be changed. This advantage in case a rigid program was used over and over with a few changing numbers, e.g., with sequences of real numbers or with vectors, was immediately seen by physicists like Mauchly. That it opened a Pandora's box, however, since such a 'von Neumann program' could also change itself, was not seen so immediately. The good thing in the box was the fact that this, *in conjunction with backward jumps*, made the Eckert–Mauchly–von Neumann machine of 1944/1945 technically universal in Church's sense. The situation was recognized more accidentally by von Neumann and also by Mauchly and made public in 1946 and 1947 in vague form. We shall come back in Sect. 6 to the story of achieving universality.

5.4 Fixed-loop structures. A widely used class of non-universal languages is historically connected with single-control-state looped programs — to be more precise, with programs with a fixed loop structure comprising a finite number of loops. The origin of this dates back centuries, with the construction of clocks and musical instruments having a rotating barrel usually equipped with pegs that triggered action — a *single-loop machine*. It was soon generalized to *multiple-loop machines*, for example clocks with a second loop for the quarter hours, and by the middle of the 20th century, multiple-loop machines flourished, examples being automatic multiplication and division in mechanical desk calculators.

In the line of Babbage–Ludgate–Zuse machines up to 1941, fixed loop structure was coupled with rigid programs (Zuse's *starre Rechenpläne* also implied a fixed loop structure) — even Zuse's Z3 of 1941 had only one loop; however, the tape, spliced to a circle, could be exchanged manually. Zuse planned to equip his Z4 in 1944 with a few program tape readers, allowing a few loops. Rutishauser, in working with the Z4, was frequently forced to exchange program tapes manually in order to simulate missing loops.

5.5 Zuse's non-rigid programs. While Zuse had foreseen machine instructions for (conditionally) swapping the program tape reader, the more liberal Eckert–Mauchly–von Neumann machine architecture (still with single-control-state programs) allowed a system of loops to be set up with the help of a cleverly designed crutch, the conditional jump instruction. Since a von Neumann program, in contrast to the stubborn Aiken doctrine, could change itself, not only the path through the loop structure but also the loop structure itself, although always present and finite, could change during computation.

In his 'First Draft of a Report on the EDVAC'[4] from June 30, 1945, John von Neumann 'makes no mention of the consequences, from the point of view of

---

[4] 'Electronic Discrete Variable Automatic Computer'

either logical design or of programming, of the decision to store instructions in the internal memory'. So says Brian Randell, who continues: 'in the report it states that stored instructions and data are to be distinguished, and provisions are made for modifying only the address field of an instruction'. This was exactly what Mauchly had wanted in order to deal with sequences and vectors.

In his 1944 draft of a thesis, Zuse had arrived at the same conclusion: there is a need for what he called *free computation plans* (*Freie Rechenpläne*) or synonymously *non-rigid computation plans* (*Unstarre Rechenpläne*).[5] He exemplifies this by conditional execution and by variable, computable subscripts (*variable Indizes*), which latter amount to address modification and invoke universality. In the *Plankalkül*, written in the summer of 1945, this is outlined in some detail.

A full-fledged *Plankalkül* would have been universal, too. Had he been able to follow this line, Zuse would have arrived at a universal machine just as Eckert, Mauchly, von Neumann, and Goldstine did: by surprise. In 1945, programming the von Neumann machine is not more progressed than Zuse's non-rigid programs. The post-war situation, however, had the effect that in 1945 nobody took notice of Zuse, and all the fame went to Goldstine and von Neumann, who indeed prepared a big step forward to universality after 1947.

This brings U. S. developments into focus. Aiken at Harvard as well as Stibitz at Bell Labs did not contribute to programming languages or techniques. Aiken, in his Mark I from 1939–1944, still did not have conditional branching, although he could switch control tape readers. Seeber, who had left Aiken around 1945 to develop the 'Selective Sequence Electronic Calculator' for IBM, knew better: he 'found it useful to be able to perform arithmetic on, and then execute, stored instructions' (Phelps 1974). This parallels the development at the Moore School; originally Mauchly was also satisfied to be able with the help of a special address modification instruction to work with variable, computable subscripts — just as Zuse had in mind.

5.6 Iteration. Variable, computable subscripts and iteration are two sides of the same coin: A non-rigid program with *Working-Variables* allows iteration, i.e., a program that *works on some of its own results* in a loop. This frequently means tremendous saving in the program record, at the cost of time.[6]

An example is given (see Fig. 16 (h)) by the Horner evaluation of the polynomial $a_4 + a_3 \times x + a_2 \times x^2 + a_1 \times x^3 + a_0 \times x^4$. Condensation leads to the iteration

$$(a_i + V_i) \times x \Rightarrow V_{i+1} \ ,$$

---

[5] Zuse also speaks of *quasistarre Rechenpläne*, program schemes that could be adapted structurally to particular needs or situations *before the calculation started* — at that time, Zuse envisaged a special device for program preparation (*Planfertigungsgerät*). This, however, would have given no more than primitive recursion.

[6] For fast electronic computers with limited storage this is an advantage. On the other hand, Rutishauser frequently unwound loops in order to save time on the slow relay computer Z4, since program storage on film tape was cheap.

which is started with $0 \Rightarrow V_0$ and ended with $a_4 + V_4 \Rightarrow res$. Rutishauser, working with the Z4, introduced, as already mentioned, the storage state superscripts into an *equation*

$$(a_i + V^{(i)}) \times x = V^{(i+1)} ,$$

and interpreted this as 'silent superscripts' of a variable V_0 ,

$$(a_i + V_0^{(i)}) \times x = V_0^{(i+1)} ,$$

passing over to the *assignment* with a *Working-Variable* V_0

$$(a_i + V_0) \times x \Rightarrow V_0 .$$

Moreover, instead of closing the loop with a conditional jump instruction, in around 1950 he introduced explicit iteration (FOR statement), obtaining

$$\begin{array}{ll} & 0 \Rightarrow V_0 \\ \text{for } i = 0(1)3 & (a_i + V_0) \times x \Rightarrow V_0 \\ & a_4 + V_0 \Rightarrow res \quad . \end{array}$$

In the same way, the multiple-control-state program (b) of Fig. 16 can be changed into an iterative one; although this style was not in use before 1960.

The FOR statement, however, was too special and was later, under the influence of Dijkstra, expelled and replaced by the DO WHILE statement. Such a repetition element can be found already in Zuse's *Plankalkül*.

6 Recursion and universality

Quite innocently at first, logicians began in the 1930s to study programs. The reason was that after Kurt Gödel's revolutionary result in 1931 about incompleteness (*formale Unentscheidbarkeit*) of some systems, e.g., arithmetic, *computability* also became an issue. In 1936, two people (Fig. 20) independently conceived a convincing definition of what 'computable' should mean: Alan Mathison Turing (1912–1954) and Emil Leon Post (1897–1954). Strikingly, they came out with almost identical proposals: they both tried to imitate in very abstract form what a human being would do, reduced for simplicity to the most elementary actions. It was clear that such a proposal would have the character of a machine, a hypothetical machine, a *Gedankenmaschine*, not good for practical work but well suited to theoretical studies about it. The programming languages of the Turing machine (Fig. 21) and the Post machine are single-control-state languages. An attentive observer recognizes, however, that these machines, in contrast to the Babbage–Ludgate–Zuse line, store programs together with numbers and thus allow programs to be changed by themselves, an idea Zuse did not have and Aiken found so horrible that he fought it until the end of his life. Logicians showed that the Turing machine and the Post machine are equivalent. Why Post is rarely mentioned remains unclear; both Turing and Post were somewhat strange characters.

6.1 The Church Thesis. In the same year 1936, a very different approach was made by Alonzo Church. As a logician he found it natural to start with expressions and thus formulated a system (the 'Lambda Calculus') that could be

230 A. M. Turing [Nov. 12,

ON COMPUTABLE NUMBERS, WITH AN APPLICATION TO THE ENTSCHEIDUNGSPROBLEM

By A. M. Turing.

[Received 28 May, 1936.—Read 12 November, 1936.]

The "computable" numbers may be described briefly as the real numbers whose expressions as a decimal are calculable by finite means. Although the subject of this paper is ostensibly the computable *numbers*, it is almost equally easy to define and investigate computable functions of an integral variable or a real or computable variable, computable predicates, and so forth. The fundamental problems involved are, however, the same in each case, and I have chosen the computable numbers for explicit treatment as involving the least cumbrous technique. I hope shortly to give an account of the relations of the computable numbers, functions, and so forth to one another. This will include a development of the theory of functions of a real variable expressed in terms of computable numbers. According to my definition, a number is computable if its decimal can be written down by a machine.

In §§ 9, 10 I give some arguments with the intention of showing that the computable numbers include all numbers which could naturally be regarded as computable. In particular, I show that certain large classes of numbers are computable. They include, for instance, the real parts of all algebraic numbers, the real parts of the zeros of the Bessel functions, the numbers π, e, etc. The computable numbers do not, however, include all definable numbers, and an example is given of a definable number which is not computable.

Although the class of computable numbers is so great, and in many ways similar to the class of real numbers, it is nevertheless enumerable. In § 8 I examine certain arguments which would seem to prove the contrary. By the correct application of one of these arguments, conclusions are reached which are superficially similar to those of Gödel†. These results

† Gödel, "Über formal unentscheidbare Sätze der Principia Mathematica und verwandter Systeme, I", *Monatshefte Math. Phys.*, 38 (1931), 173–198.

The Journal of Symbolic Logic
Volume 1, Number 3, September 1936

FINITE COMBINATORY PROCESSES—FORMULATION 1

EMIL L. POST

The present formulation should prove significant in the development of symbolic logic along the lines of Gödel's theorem on the incompleteness of symbolic logics[1] and Church's results concerning absolutely unsolvable problems.[2]

We have in mind a *general problem* consisting of a class of *specific problems*. A solution of the general problem will then be one which furnishes an answer to each specific problem.

In the following formulation of such a solution two concepts are involved: that of a *symbol space* in which the work leading from problem to answer is to be carried out,[3] and a fixed unalterable *set of directions* which will both direct operations in the symbol space and determine the order in which those directions are to be applied.

In the present formulation the symbol space is to consist of a two way infinite sequence of spaces or boxes, i.e., ordinally similar to the series of integers $\cdots, -3, -2, -1, 0, 1, 2, 3, \cdots$. The problem solver or worker is to move and work in this symbol space, being capable of being in, and operating in but one box at a time. And apart from the presence of the worker, a box is to admit of but two possible conditions, i.e., being empty or unmarked, and having a single mark in it, say a vertical stroke.

One box is to be singled out and called the starting point. We now further assume that a specific problem is to be given in symbolic form by a finite number of boxes being marked with a stroke. Likewise the answer is to be given in symbolic form by such a configuration of marked boxes. To be specific, the answer is to be the configuration of marked boxes left at the conclusion of the solving process.

The worker is assumed to be capable of performing the following primitive acts:[4]

(a) *Marking the box he is in (assumed empty)*,
(b) *Erasing the mark in the box he is in (assumed marked)*,
(c) *Moving to the box on his right*,
(d) *Moving to the box on his left*,
(e) *Determining whether the box he is in, is or is not marked.*

The set of directions which, be it noted, is the same for all specific problems and thus corresponds to the general problem, is to be of the following form. It is to be headed:

Start at the starting point and follow direction 1.

Received October 7, 1936. The reader should compare an article by A. M. Turing, *On computable numbers*, shortly forthcoming in the *Proceedings of the London Mathematical Society*. The present article, however, although bearing a later date, was written entirely independently of Turing's. *Editor*.

[1] Kurt Gödel, *Über formal unentscheidbare Sätze der Principia Mathematica und verwandter Systeme I*, *Monatshefte für Mathematik und Physik*, vol. 38 (1931), pp. 173–198.

[2] Alonzo Church, *An unsolvable problem of elementary number theory*, *American Journal of Mathematics*, vol. 58 (1936), pp. 345–363.

[3] Symbol space, and time.

[4] As well as otherwise following the directions described below.

103

Fig. 20 Title pages of the papers of Turing and Post in 1936

called a functional machine, although he did not see it this way. As a functional language of application and abstraction, it included, well hidden, a feature that was somewhat controversial at that time and even later: recursion. After Ackermann had shown in 1928 that the great Hilbert was misled in his intuitive power when he thought that so-called primitive recursion was the *ultima ratio* and in this sense universal, general recursion became a dreaded instrument, in particular for those who did not understand it. Church then posed the thesis that there can be only one equivalence class of universal machines or algorithmic languages, and all fruitful definitions should stand the test of being shown equivalent to the Lambda Calculus. Church's Thesis, generally accepted today, unifies Turing–Post machines, Markov algorithms, and many more, including in particular stored-program machines with address modification (Sect. 5.3).

As far as we can see, recursion in programming (restricted to primitive recursion) was first envisaged by E. K. Blum (ADES II, 1956) on the recommendation of the logician Calvin C. Elgot.

6.2 A need for universality? A question that caused much trouble is the following: Do programming languages that are to be used in practice need to be universal? My experienced and shrewd friend Heinz Rutishauser thought for a long time that the answer was *No*, and I think he never convinced himself completely that the answer should be *Yes*. Samelson and I said *Yes, but ...*, indicating that full generality would have its price ('Do not let the simple but frequent case pay the price for the complicated but rare case'). We were afraid that full generality might hinder the acceptance of a programming language intended for practical use, like ALGOL. In the 1970s, however, further progress in hardware speed and costs made it clear that it would no longer matter to pay the price: the price could and should be paid. Today, most people agree with this.

Even hand-held programmable calculators are today universal in this sense. Simple calculators for the four first rules of arithmetic, of course, are not even repetitive — they are nevertheless useful in quick one-time calculations. The right instrument in the right place — that is the conclusion. But general recursion is still discriminated in some quarters: Wirth dissuades the reader from using recursion ('Wo Rekursion zu vermeiden ist') in *Algorithmen und Datenstrukturen*, 1975, and still in *Programming in Modula-2*, 1985. On the other hand, jump instructions are also frequently condemned in general ('Go to statement considered harmful', Dijkstra 1968), while only backward jumps could be said to be harmful.

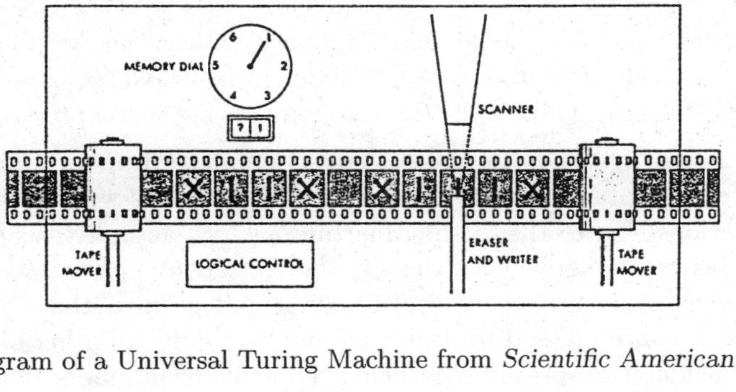

Fig. 21 Diagram of a Universal Turing Machine from *Scientific American*, April 1955

Part III : Object structures

7 Untyped versus typed objects

Early mechanical calculators dealt with integers — sometimes even with natural numbers only — in common decimal notation. Although already Napier and later Leibniz had advocated the building of mechanical calculators for binary numbers, only in the twentieth century were practical attempts made — by R. Valtat in 1931, by E.W. Phillips in 1936, and by Konrad Zuse roughly at the same time. Valtat and Phillips failed to achieve practical results, but Zuse succeeded in the mechanical Z1 (1938) and in its relay successor Z3 (1941). Electromechanical relays and in particular vacuum tubes led to a clear preference for binary circuitry for calculation, storage, and control, but genuine binary calculation had to fight against binary coded decimal calculation, used by Presper Eckert and John Mauchly in the ENIAC (one-out-of-ten code), by the stubborn Aiken (2-4-2-1 code), by Stibitz (biquintal code, Stibitz code), by Lake and Durfee in the Pluggable Sequence Relay Calculator (Hollerith code), by Wallace Eckert in the SSEC (Hollerith code), and by Sheldon and Tatum in the CPC (Hollerith code). Under the influence of von Neumann and Goldstine, Eckert and Mauchly used genuine binary calculation for the first time in the EDVAC — ten years after Zuse. From then on, binary calculation was the winner in the hardware.

7.1 The Gödelization trend in logic. In mathematical logic, however, natural numbers were used in an abstract sense that needed no coding, and, if so, the coding would rather be by unary numbers, following the Peano definition. Coding was considered trivial at this level, although it was quite substantial on the next level: natural numbers were frequently considered to be the basis for coding any other objects. In particular, Gödel, in his famous 1931 paper on the *Entscheidungsproblem*, introduced a correspondence (Gödelization) between well-formed expressions of an arbitrary formal system and natural numbers. Theoretically, formal systems could then be treated fully within number theory, although this frequently blurred the results. But practical considerations were irrelevant for Gödel; in fact Gödelization is an utterly impractical instrument, since the natural numbers are not expressly structured.

Turing was slightly closer to reality. His attempt to formalize what came to be called 'computable numbers' among the real numbers written in decimal or binary fractions lead to the 'Turing machine', which actually was described to work on binary numbers. Nevertheless, the *Entscheidungsproblem* application of Turing's machine used Gödelization again. It seemed that mathematical logicians were quite pleased to deal with numbers only, all other objects under possible consideration presumably being reducible to numbers.

7.2 Many-sortedness in computer science. Computer scientists, however, developed a more differentiated view. A classical calculator was a finite instrument and could handle only numbers up to a fixed number of, usually decimal, places. The first automated computers almost completely followed this line. What was later called 'fixed point calculation' was considered sufficient in all the early attempts mentioned above, up to and including the EDVAC, the prototypical 'von Neumann machine'; Zuse, who introduced 'floating point calculation' from the very beginning, was far ahead of them again, but the number of floating point numbers he could handle even in the Z4 also had a fixed upper limit.

Double-precision and triple-precision calculation, which came up in the EDVAC heritage by Wilkes, Wheeler, and Gill, did not change the situation either. It was again like it was with universality: the 'von Neumann machine', quite accidentally, turned out to be able to handle integers of any size, by chaining together numbers using the address field as a pointer to the successor and thus modeling the potential infinity of the natural numbers. But this is rather cumbersome, and even today, most high-level algorithmic languages restrict themselves to a finite set of natural numbers and leave unlimited precision calculation, which requires non-rigid object types, to special mathematical systems, like MAPLE or MATHEMATICA.

Thus, with numbers — usually integers and even limited in size as fixed-point numbers or floating-point numbers — many-sorted programming arose late in the 1950s. But this was not the end. Standard high-level algorithmic languages soon introduced other sorts of objects for convenience — one did not want to worry about how they were represented internally. Consequently, apart from fixed-point numbers and floating-point numbers, truth values were considered a sort (which, for reasons of clarity, should no longer be mapped into the number pair $\{0, 1\}$); alphabetic characters and other marks were given recognition as a sort (and for the same reasons should no longer be mapped into the number interval $\{0, \ldots, 255\}$ when restricted to the ISO 8-bit code).

7.3 Universal algebra. Under Bourbaki, about 1950, the 'algèbre universelle' originated from axiomatic algebra, and introduced general concepts like homomorphisms under the aspects of category theory. It worked at first mainly with homogeneous systems. About twenty years later (1968, 1974), Birkhoff and Lipson extended the classical one-sortedness of universal algebra to many-sortedness, introducing heterogeneous algebras, i.e., algebras of heterogeneous signature.

8 From hypercomplex numbers and data types to algebraic specifications and Herbrand models

Algebra is the science
of operations with arbitrary objects
(Jean Dieudonné)

So far, we have been concerned with program structures. Let us now look at how object (data) structures and functionals made their way into present-day programming, culminating in algebraic specifications and in so-called object-oriented programming.

8.1 Structured objects. Descartes introduced numerical coordinates ('cartesian coordinates'), pairs and triples of real numbers to deal with geometry, but no algebraic abstraction was intended — which is shown by the prevailing use even today of the term *analytical geometry*. With Gauss, Hamilton, and Cayley, certain concrete structures came into algebra: pairs for complex numbers, quadruples for quaternions, octuples for octonions. The aim of these hypercomplex numbers was to introduce constructively 'higher systems of numbers', thereby keeping as much as possible of the laws of the usual realm of numbers ('principle of permanency') and giving up only some of them: linear order, commutativity, and even associativity. E. Study with his 'dual numbers' even allowed nilpotency. All these structures were rigid tuples, were direct products. These developments culminated in Grassmann algebras and Clifford algebras, important in the theory of Lie groups and differential topology.

When axiomatic algebra arrived under Grassmann and Peano, under Kummer, Dedekind, and Frobenius, many new realms were laid out besides the realm of numbers, but the models that were investigated were predominantly structurally simple matrices, i.e., tuples of tuples; and 'representation theory', introduced by Frobenius, meant by implication matrix representations. This was so even under Emmy Noether, Artin, Hasse, and van der Waerden.

Such was the situation in mathematics before the period 1950–1965, when the computer stormed into prominence. Structured objects of traditional mathematics were still mostly handled using the rather uniform machinery of matrices. And where physics needed structured objects to describe quantum effects and elementary particles mathematically, again matrices were the first choice (matrix mechanics by Heisenberg, Dirac matrices by Dirac).

The revolutionary idea, however, was to introduce structured objects as objects of some specified sort. This was already done (without success) by Zuse in his *Plankalkül*[7]. Then, around 1970, non-rigid object types other than McCarthy's lists (see Sect. 10.3) were defined *recursively* by C.A.R. Hoare and others. Recursion marked another breakthrough, and in the early 1970s, recursive definitions of classes of abstract structured objects ('abstract data types') paved the way to take theoretical aspects of many-sortedness seriously.

Inevitably, Hoare's building principles of 'abstract data types' — finite enumeration, forming of pairs (if done repeatedly: forming of tuples), and forming of arrays and sets — were later unmasked to be only ritual cases of term-generated models for algebras with specified signatures of operations, defined by characteristic predicates. However, Zuse's basic idea of the *Plankalkül*, that every sort of objects essential in computer science should be built ultimately as a structure over the binary elements $\{\mathbf{O}, \mathbf{L}\}$ and thus that computing should be completely reducible to structures of binary elements, is too narrow a view.[8]

---

[7] "Unter Struktur einer Angabe wird ihr komponentenmässiger Aufbau ohne Hinblick auf die Bedeutung der einzelnen ... Komponenten verstanden." (Zuse 1945, p. 3).

[8] Zuse did not even take full advantage of his binary principle: dyadic lists in the sense of McCarthy are missing in the *Plankalkül*.

8.2 Storage structures. One of the most significant characteristics of computers is that they have memories for storing some (more than one) or even many numbers. Konrad Zuse, in his very first attempts of 1934–1935 in Germany, started from engineers' calculation sheets (see Sect. 2.3), which are usually structured according to the computation plan and thus have a rather widely varying structure. But technological limitations forced him into using for programs and data the simplest storage structure one could build: a tuple of storage elements; and as Babbage had already done, he numbered them by 'addresses', thus giving the store a simple linear structure. Independently, Eckert and Mauchly took the same path since the addressable, re-usable store could be easily implemented with existing hardware, e.g., with mercury delay lines. Zuse, like others, had an electromechanical 'selection pyramid' to choose the position of a storage element, in his case a relay. But rotating magnetic drums and spooled magnetic tapes made the linear selection mode (the *Zugriff*) even more natural. Non-numerical applications of the computer developed quickly, and whenever structured objects came up in a computation task, a reduction of their natural structure to a tuple structure, i.e., an embedding into an addressable store which is in mathematics a linear representation, had to take place.

Most computer programmers, in particular those with modest education, took this for granted. They could see and knew at least superficially how their computer worked and accepted this subconsciously in their programming style — what else could they do short of building their own machines?

More complex data storage structures were first tried in practice in connection with attempts to speed up computation by parallelization. The ILLIAC IV was an example. But only very independent thinkers like Zuse went further. His *Plankalkül* of 1945 was a theoretical study which in a situation more agreeable than the postwar misery of Germany could have led to a highly advanced machine. Zuse considered applications that went beyond the tuple-matrix scheme, like what he called pair lists. Mathematically speaking (although Zuse did not master this jargon), these pair lists used relational algebra as a basis for the representation of objects and operations in his hypothetical machine. Nobody so far has revived the idea, although it would be easy to build a relational algebra chip in modern semiconductor technology.

8.3 Programming languages dealing with structures. While computer hardware in the 1960s was still dominated by the addressable store, complicated structures had to be dealt with by software — not a bad solution, in fact. Again, it is fascinating to see how deeply experienced programmers were corrupted by the machine. As said above, the von Neumann machine typically had an addressable, re-usable store, and instead of introducing axiomatically a variable for a complete structured object, another pioneer, Heinz Zemanek in Vienna and his VDM group in Vienna around Peter Lucas, Hans Bekič, and Cliff Jones started in 1968 constructively with structured stores, that is, stores that were given a certain structure that automatically caused objects to be structured in the same way. The first examples studied were stacks and queues, where the linear machine storage was interpreted to form a sequence of storage cells with particular, restricted access properties. Some hardware storage devices, like magnetic tapes, immediately implemented such a storage structure.

The next logical step was abstraction. Work with 'abstract data types' as mentioned in Sect. 8.1 was studied by Hoare in 1970, 1972, Liskov and Zilles in 1974, Guttag in 1975, and Horning in 1977. Guttag introduced the concept of 'sufficient completeness' for the characterization of the defined type, the 'type of interest'. Liskov's CLU was the first language with direct support for data abstraction.

8.4 Programming with pointers. Free definition of 'Data Types' was introduced in the late 1960s, e.g., with PASCAL by Wirth but also with ALGOL 68, in a particular way: with a semantics based on the pointer ('reference') concept invented by Tony Hoare in 1965. This meant that selective updating did get appropriate attention regarding efficient execution. In some applications, working with 'data types' in the form of aggregates of storage cells still dominates the programming field, although it is today as obsolete as assembly code. But use of references saves storage by avoiding duplication.[9]

To deal with more general structures, the address mechanism of existing computers was interpreted to give the references, and a reference (or as many references as needed) was glued to every proper part of a structured object. These references determined the structural relation. Tony Hoare called such a conglomerate of a data-storing element and elements storing the references a record. A complete set of records interconnected by references (a strongly connected component) implemented the structured object. Superficially, this has its merits: selective updating is as easy as possible. Unfortunately, it is equally easy to produce wrong programs this way.

8.5 Functional programming. Looking back on ALGOL 68 today (30 years after the 1966 Kootwijk meeting when ALGOL X came into the hands of Aad van Wijngaarden) is like returning to the dark Middle Ages, full of mysteries like conjuring rites of coercion and casting or superstition about unions. Yet, ALGOL 68 was better than its present reputation. Some cleaning-up has occurred in the meantime. The programming language C looks in fact like a procedural derivative of ALGOL 68, still ugly and inconsistent.

But also some decisive steps have been made since. One of the most important ones is the liberation from the von Neumann type of machine, which should no longer guide programming. This emancipation was started by John Backus ('Can Programming be Liberated from the von Neumann Style? A Functional Style and its Algebra of Programs', 1978). It was also the essence of the work in the mid-1970s of S.N. Zilles, J.V. Guttag, and the ADJ group (J.W. Thatcher, E.G. Wagner, and J.B. Wright whose guru was J.A. Goguen) that brought a genuine functional philosophy into the use of structured objects.

---

[9] Behind the use or non-use of references lies a distinction between sets (preferred by orthodox logicians) and bags (which should be more natural for computer scientists).

The functional style Backus advocates does not allow a re-usable store. Whatever is stored may be used, once or as many times as needed, but it cannot be overwritten: at the same address or name the same object is always found. This seemed intolerable in the early days when storage was very expensive and its size was limited by physical constraints. The situation is quantitatively different today, thanks to progress in semiconductor technology. Still, genuine functional programming, as opposed to procedural programming which allows overwriting (the use of *working* variables in the sense of Lady Lovelace), has a long way to go against the prejudice of the established programming world.

With functional programming, structured objects can be introduced, as we shall see below, in a truly algebraic way. This is to say that all objects we deal with in computer science must be *con-structed* first — otherwise they would not exist for the computer. Twenty years ago, I called this the *Principle of Generation.* I consider it one of the fundamental concepts of computing science.

The way the objects are constructed in functional programming obviously determines the structure they have, once and for all. Thus, the Herbrand universe of an axiomatic system is the subject of computing science for this system, and Jacques Herbrand (who died in 1930, unfortunately at very young age) is the Patron Saint of functional programming.

8.6 Algebraic specification and the Herbrand model. Algebraic specification is purely axiomatic and thus on the highest possible level of abstraction. The axiom system used by a particular algebraic specification allows many interpretations or models; the power of the abstraction is tremendous. Checking the real problem against this axiomatic system is all that is needed.

On the other hand, algebraic specification promotes one model, the Herbrand model, as the norm. If the Herbrand model is empty or trivial, as is the case for many mathematical axiom systems in common use, then the algebraic specification is also empty or trivial. This is as it should be. To give an example: the group of rotations of the Euclidean plane, although a beloved subject of geometry, if axiomatized as a group in the common way, has no non-trivial Herbrand model. This axiomatic theory is therefore void from the point of view of computing science. This is fortunate, since no machine will do calculations with non-computable real numbers, and among the rotation angles almost all are non-computable. The militant constructionist is fully satisfied that no hocus-pocus will be attributed to the machine.

The tolerant constructionist, however, will attribute a kind of existence to the group of all rotations of the Euclidean plane in that this fiction has a wealth of subgroups that are enumerable or even finite. Any one of these subgroups can be constructively specified, for example the group of all rotations by multiples of an angle $2\pi/n$, where n is a natural number in binary representation. Adding one constant x of order n : $x^n = id$ to the specification of the group produces a non-trivial Herbrand model, a (finite) cyclic subgroup of the rotation group. All its explicit rotations can be calculated by a (possibly non-terminating) algorithm from the binary number n.

We can say this:

> *Computing science is that part of mathematics that deals with axiomatic systems having a non-trivial Herbrand model. It studies intensively the Herbrand model, or in cases where the specification is not categorical, all homomorphic images of the Herbrand model. 'Studying it intensively' means that software is built and abstract machines are set up for running this software. Thus, computing science goes to the bottom of the algebraic specification of the system in question, looking beyond its diversity.*

As an aside, it should be mentioned that functionals play an important role among the objects of algebraic specification, and the process of currying is important in the program development process. Manfred Broy has shown how the style of modeling distributed interactive processes can be reintroduced into functional programming, and this time in a safe way, by using systematically higher-order functionals.

8.7 Object-oriented programming. From an area where one would not expect it, from simulation, a new philosophy arose in 1967 with SIMULA: Individuals and the operations defined on them should be considered to be the entities called classes in SIMULA. Today, such an entity is somewhat confusingly called an *object*. Object-oriented programming is at present considered to be the highest level of sophistication. Although it combines in principle with a functional style, too — in this case, the classes are nothing but the signatures — it still is frequently grafted on a state-oriented style working with pointers.

In any case, programming in problem-adapted complex object structures not only makes life easier, it also helps save storage. Furthermore, introducing corresponding problem-adapted complex operations helps to save time. Thus, the development has an economic advantage. SIMULA (1967, Nygaard and Dahl) was followed by SMALLTALK (1971, Alan Kay) and typed EIFFEL (1985, Meyer).

Step by step, most algorithmic languages, in particular those that are not purely functional, developed a richer reservoir of sorts and even the capability to define new sorts without using the crutch of pointers. Although many aspects of what is called today object-oriented programming are particularly well suited to pointer-free, purely functional, programming languages, we see today object-oriented extensions to classical imperative programming languages: OBERON (based on PASCAL), MODULA-3 (based on MODULA), C^{++} and OBJECTIVE-C (based on C), followed recently by JAVA. But more and more object-oriented functional languages are coming into the field, and there is not enough room here to discuss them. See the paper by John C. Mitchell in this volume.

Part IV : A taxonomy of algorithmic languages

9 Dichotomies

Algorithmic languages can be classified according to some very deep-lying distinctions. Some properties (of the many that deserve to be studied[10]) have been mentioned already; they are now discussed in more detail:

descriptive	vs.	*executable*
implicit conditions		*explicit expressions*
nondeterministic, liberal	vs.	*deterministic, totalitarian*
ambiguous result		*unique result*
functional, applicative:	vs.	*procedural, imperative:*
formulae without explicit states		*commands with explicit states*
concurrent, parallel:	vs.	*sequential, serial:*
multiple-control-state		*single-control-state*
lucid, transparent:	vs.	*opaque, obscure:*
one-time store		*re-usable store*

9.1 Executable versus descriptive. All the programming systems and algorithmic languages considered so far are *executable*: programs indicate — in whatever way — what operations are to be performed. In contrast, *descriptive* languages can be transformed only under suitable conditions into executable ones. Whether these conditions hold, is, in general, undecidable. Therefore, the class of useful descriptive (pre-algorithmic) languages is quite restricted. The description $x^2 + a \times x + b = 0$ of Zuse's *Rechenplan* (Fig. 9), like many other equational ones, is certainly pre-algorithmic. In particular, many problems involving function inversion are basically equationally descriptive.

Some of the effective forms of the *descriptive programming style* go under the names *predicative programming* or *logical programming*. Executing such a program means giving a proof of an *existence formula*, e.g., over the integers

$$\exists x.\ x^2 + a \times x + b = 0 \quad ,$$

although more conservative logic uses for this purpose the Bernays η-operator,

$$\eta x.\ x^2 + a \times x + b = 0 \quad .$$

10 A number of interesting other details of the history of programming languages cannot be discussed here in the given framework. Also, some technical aspects, e.g., encapsulation, remain undiscussed here. Further diversification is based on aspects like morphology (typed versus untyped), modularity (information hiding versus inheritance), scope (lexical versus dynamic), definedness (total versus partial with exception handling), strictness (eager versus lazy evaluation), program and data structuring (controlled use versus free use of labels and of pointers).

Descriptive programming was used by Zuse in the *Plankalkül*[11]. It is particularly appropriate in the specification process, which should always precede a programming process; this was also studied at some length in the Munich CIP project. Other specification languages are CLEAR (Burstall, Goguen 1977) and LOTUS (1983), an extension of CCS.

Logical programming is necessarily, even in first-order logic, subject to Gödel type limitations of decidability and therefore needs syntactical restrictions in order to function safely. Particularly simple, although in some cases inconvenient, is the restriction to Horn clauses, which is safe but nevertheless preserves universality; it has found application and some admirers in the PROLOG family of languages and in other inference systems.

Related are special algorithmic languages for mathematical proofs and in connection with this for program specification and development; among others LCF ('Logic for Computable Functions') and SML by Milner. There are many derivatives, e.g., NUPRL (Constable 1986), DEVA (Sintzoff 1990), and LEGO (Burstall 1990).

Other forms of descriptive programming that show up in a variety of mathematics-oriented applications do not yet have established algorithmic languages in the proper sense.

9.2 Deterministic versus liberal. Likewise, all the programming systems and algorithmic languages considered so far are *deterministic* in that operations have uniquely defined results. In contrast, there are *nondeterministic* languages, with operations that can in the same situation give different results that are equally acceptable for some specific purpose. Since many applications have to deal with such liberal situations, in particular in distributed systems, it is an interesting class of algorithmic languages in a wider sense. While non-deterministic automatons are a well established part of theoretical computer science, the general public seemed to be reluctant for quite some time to use non-determinism in algorithmic languages. An early exception, apart from random functions, is the guarded conditional introduced by Dijkstra in 1975, an example being

$$\textbf{if } x \geq -1 \textbf{ then } x \;[]\; x \leq 1 \textbf{ then } -x \textbf{ fi} \quad .$$

Non-determinism is also involved in using the Bernays η-operator. The deterministic counterpart is meant with the unique existence quantifier

$$\exists_1 x.\ a \times x + b = 0 \quad ;$$

conservative logic uses for this purpose the ι-operator,

$$\iota x.\ a \times x + b = 0 \quad .$$

[11] Zuse uses $\acute{x}\,(x^2 + a \times x + b = 0)$ to denote exhaustive determination. In his 1949 publication, the operator μ showing up in Fig. 22 under ⑤ is not the minimum determination operator, but simply means 'the next element of the sequence that fulfills the condition'. Zuse changed the meaning μ had in the book of Hilbert–Bernays, which he consulted in 1937.

Most important, liberal programming with non-determinism allows underspecification, which is a particularly useful technique in software engineering.

9.3 Procedural versus functional. Most of the programming systems and algorithmic languages considered so far are *procedural*, using computation steps operating on *working* variables. Zuse's original calculation sheets would have been an exception had they been developed into practical programming instruments. The idea of using algebraic formulae as pieces within an otherwise imperative programming language came up with Heinz Rutishauser in 1951, followed by Laning and Zierler (1953); this work fertilized other work that become highly important: IT by Perlis (since 1956), AT3 (later MATH-MATIC) by Katz (since 1956), and in particular FORTRAN by Backus (since 1954). In competition, Laning, Hankins, and Werner developed MAC (1958).

By and by, imperative programming languages were blended with functional features of application and somehow abstraction. This led via FORTRAN and ALGOL 58 to ALGOL 60 and then to the Algol family of ALGOL 68 (1968), PASCAL (1969) and ELAN (1974); PASCAL branched again into ADA and MODULA-2; COMAL was an upgrading of BASIC and PASCAL.

BASIC (1965) by Kurtz and Kemeny and C (1969-1973) by Ritchie and Thompson are low-level relapses of the ALGOL idea, the latter originating from untyped BCPL (1967) by Richards, which in turn was based on Strachey's CPL.

COBOL, a 'business oriented' sister of FORTRAN from 1959, is a relic that, blended with FORTRAN, lead to PL/I (1964). Several languages specialized on particular sorts, e.g., on Boolean values (LOGALGOL, 1961), on text strings (SNOBOL, 1963), and on matrices (APL, 1962).

The usual restriction of applicative or functional programs to an equivalent of one-time-assignment storage does not inflict recursiveness; however, it means a 'waste' of storage that until recently seemed intolerable in many practical cases.

9.4 Sequential versus concurrent. In the beginning, under the fascination with existing machines, imperative programming was strictly single-control-state programming. Rutishauser opens this line in 1951, surpassed by Böhm (1952), who for the first time described the translator even in its own language and moreover remarked that his language was universal. Laning and Zierler (1953), Backus (1954), Perlis (1956), and Katz (1956) were all on the sequential trip, and this went into ALGOL 58, ALGOL 60 and even PASCAL.

There are also special imperative, rightfully sequential programming languages for real-time applications, like PEARL and BASEX.

Under the work of Dijkstra and Hoare, multi-control-state programming (*parallel programming, concurrent programming*) captivated the hearts of many, but not all programmers, and entered some programming languages, starting with ALGOL 68 and leading to OCCAM, based on more theoretical calculi like CSP (Hoare), CCS (Milner), and COSY (Lauer). In a corner lives the subculture of Petri nets, propagated by A.W. Holt, H. Saint, R. Shapiro and S. Warshall (1968), and J. Dennis (1970).

A unification of functional and concurrent programming elements is a goal that *Robin Milner* has recently approached with his concept of the Π-*calculus* of agents, paralleling the Lambda Calculus of terms.

9.5 Opaque versus lucid. In procedural programming, a *working* variable may change its value even without mentioning it. In purely functional applicative programming, this may not happen — there is 'referential transparency', there are no 'side-effects'.

Opaqueness will not happen in imperative languages if only one-time-assignments ('single assignments') occur. However, if a statement like

$$a \times a \Rightarrow a$$

occurs in the imperative world, it can be made harmless in the functional world by introducing, as was done in the data flow idea (Jack Dennis, Gilles Kahn) of LUCID, appropriate superscripted variables that reflect the storage state history:

$$a^{(i)} \times a^{(i)} = a^{(i')} \quad \text{where } i' \text{ denotes the state successor of } i \ .$$

For single-control-state programs, integers will suffice for the subscript family; this was already advocated by Rutishauser in the early 1950s. For multiple-control-state-oriented programs, the state history is no longer linear but is a general digraph; the subscript family should reflect this. There is a deep connection between these state digraph systems and the Kripke structures of the modal logics that correspond to the branching time situations of program execution — connections that show up in the work of *Fred Kröger* and *Zohar Manna*.

Another difficulty arises with structured objects in an imperative programming language: a *working* variable for a structure is not the same as the corresponding structure of program variables. Efficiency arguments suggest frequently updating single components instead of overwriting the whole *working* variable.

Pointer realizations that are used in this context lead to tricky situations; the field has been studied carefully by *Bernhard Möller*. Since storage is getting cheaper and cheaper, the storage-consuming functional programming style, which avoids these risks, is increasingly advantageous.

10 Some more outstanding topics

We have dealt with practical programming languages that developed from early seeds in the '50s , grew in the '60s, and matured in the '70s and '80s. Twenty of the early specific programming languages originating between 1945 and 1956 — contributions by Zuse; Goldstine and von Neumann; Curry; Mauchly; Burks; Rutishauser; Böhm; Glennie; Hopper; Laning and Zierler; Backus; Brooker; Shura-Bura, Kamynin, and Ljubimski; Ershov; Grems and Porter; Elsworth; Blum; Perlis; Katz; Bauer and Samelson — were thoroughly discussed by D.E. Knuth and L.T. Pardo at the History of Computing Conference in 1976. Altogether, Jean E. Sammet has listed over 700 programming languages that were developed, proposed, and described up to 1966. Meanwhile, nobody will count them any more. The landscape of programming languages is as crowded as the sky is with satellites.

10.1 Early low-level languages. Among the twenty early ones, most are imperative and very close to machine languages — so-called 'low level'. This list starts in 1946 with the informal flow diagram language introduced by Goldstine and von Neumann and is carried on by assemblers designed by Mauchly (SHORT CODE, 1949), Burks (INTERMEDIATE PL,1950), Glennie (AUTOCODE, 1952), Hopper (A-2, 1953), Brooker (MARK I AUTOCODE, 1954), Grems and Porter (BACAIC, 1955), and Elsworth (KOMPILER 2, 1955) — each for a different machine. This in particular makes them somewhat arbitrary. A rather ambitious assembler was the PACT system of 1956 by Melahn et al.

Machine independence, however, was an important goal from the beginning; programming languages that did not depend too much on particular machines were called 'high level'. Freedom from servitude to machines and their manufacturers has been the key motivation in the development of programming languages.

10.2 Two-dimensional notation. At the beginning of Knuth's list is the *Plankalkül* of Konrad Zuse. It shows many advanced features, as demonstrated by an example in Zuse's short publication in 1949 (Fig. 22) dealing with the problem of determining the well-formedness of an arithmetic expression, a problem that Kleene tackled and solved in 1934 and that Zuse may have adopted from the Berlin logician Kurt Schröter. The algorithm performs a repeated test, indicated by $W[...]$, that runs over the length of the expression to be investigated, and uses the *next*-operator μ , indicated by ⑤. In Zuse's two-dimensional notation, the lines V, K, S may show respectively a subscript, a component number ('index'), and the sort of structure. In PASCAL-like notation, we have, e.g.,

⑤ $\mathit{next}\ x \,|(x \in V_0{:}m{\times}\sigma \ \wedge\ x \neq V_0[0]{:}\sigma) \ =:\ Z_1{:}\sigma$;
⑥ $Sq(Z_0{:}\sigma,\, Z_1{:}\sigma)\ \wedge\!=:\ R_0{:}\mathit{Boolean}$.

Zuse's early *Plankalkül* shared the fate of the two Russian contributions, the *programming programs* of Shura-Bura, Kamynin, and Ljubimski and of Ershov. They have not had the impact they deserved, possibly due to world-political circumstances. Interestingly, they show certain parallels in notation; *Plankalkül* and the Liapunov notation used by the Russians have a two-dimensional outlook. For the input devices available in the '50s, this was not very attractive and was one of the reasons that Rutishauser, Samelson, and myself did not follow Zuse notationally.[12]

[12] We did so, however, in our language philosophy as far as we thought appropriate for a compiler builder, but strangely Zuse did not appreciate this. ALGOL 68 went already in some respects beyond public willingness to accept it; this aversion would have been aggravated had we allowed some of the *Plankalkül* aspects that appeared as a nightmare to the compiler builder and to the user, who wanted run-time-efficient programs.

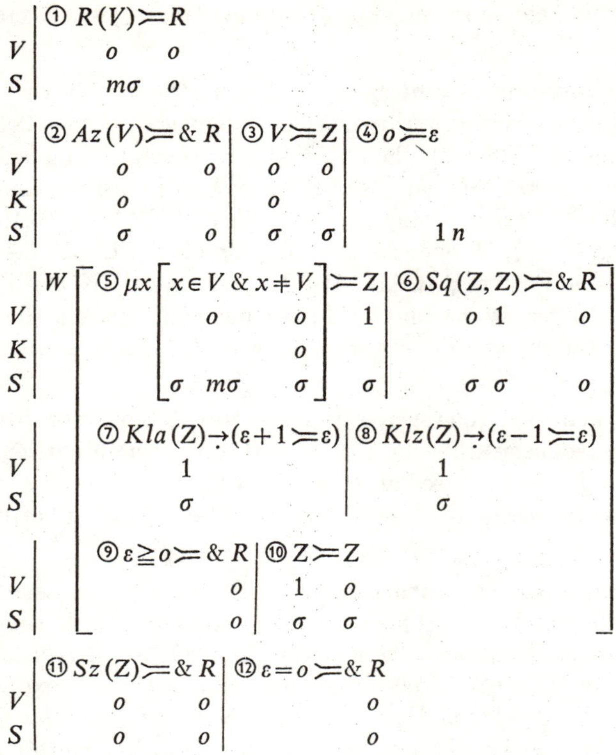

Fig. 22 Original Zuse *Plankalkül* example from 1949

More recently, progress in computer technology has brought to us window systems, picture drawing, and image processing. Two-dimensional notation would no longer find obstacles. Yet, neither picture languages nor even graph languages have so far been used for programming.

10.3 Curry and McCarthy. Pure functional algorithmic languages arrived rather late. An exception stems from the logician H. B. Curry. His rudimentary programming language of 1948 — it was not even given a name — was deeply connected with the idea of 'combinators' (Schönfinkel 1924, Curry 1930), and is much more functional and less imperative than all those discussed above. His functions can have several arguments and several results, so common notation cannot be used. Apart from this strangeness, function composition (*program composition*) is his main instrument, and he does not refrain from recursion. Function composition is a particularly pure form of functional programming. Curry's language, however, was not put to the test of practical use.

Functional algorithmic languages based on combinators are rare. They avoid even parameters in applications and know nothing but function composition. Curry's attempts might have led into this, had Curry continued them. John Backus, with FP, has for a while advocated variable-free programming. More recently, however, Curry's combinators have found new interest.

In most cases, functional programming is explicitly *applicative*. This is true for LISP, designed in 1958 by John McCarthy. It has only one sort of individual: lists. Lists are formed by something, namely lists again or atomic symbols. Its individuals are therefore recursively defined, and general recursion also gives all the power to the operations.

The unique peculiarity of LISP is that its only data structure can also express the structure of the language itself, which, like most programming languages, has naturally a tree character. Thus, a LISP program can take another LISP program as data — Rutishauser's idea of a programming program and Böhms idea of describing a language in its own terms is made explicit, although neither Rutishauser nor Böhm happened to be sympathetic to LISP. Actually, a LISP program can 'calculate itself', a LISP interpreter can be written in LISP. A special function *eval* is used for this purpose; it is implemented by a LISP interpreter which is to be hand-coded in machine language or C (or cross-compiled on a machine for which a LISP compiler already exists). The power of this function *eval* was recognized by Steve R. Russell in 1959. This 'bootstrapping' idea of the Count Münchhausen was taken over from mathematical logic (an 'unstratified language' is a language that is its own metalanguage) and brought into the programming world by Saul Gorn before 1958. A general abstract theory of such 'self-referential languages' still does not seem to exist.

Pure LISP, or its successor SCHEME, has no *working* variables, i.e., it does not countenance about re-usable stores, storage elements that can be overwritten: It is absolutely state-free. Declarations are made in Lambda notation, so a bridge to the ancient ideas of Church is established. Jean Sammet has correctly spoken of it as 'one of the most important ideas in programming'. Credit for this should also go to H. Gelernter and N. Gerberich. Unfortunately, McCarthy's work was too late to be fed into the Zurich ALGOL Conference in May 1958. And in the Paris ALGOL Conference in January 1960, McCarthy was not strong enough to push his concept into ALGOL 60 against a majority of state-oriented people, led by Perlis and van Wijngaarden.

LISP was for a long time the uncontested champion in its class. More recently, LOGO was created, which has advantages when used for didactic purposes. LISP, an untyped language, is often used in AI applications. In introductory courses at universities, typed variants like ML (Milner 1979), MIRANDA (Turner 1986), and HASKELL (Hudak, Wagner 1990) are often preferred; GOFER (Mark P. Jones 1991) is a recent contribution, which demonstrates the usefulness of polymorphism.

LISP has also found object-oriented extensions, like COMMON LISP and CLOS. The lack of state brings the idea of object-orientation much better into effect.

Today, functional algorithmic languages are gaining importance in computer science education, GOFER being an example. Nevertheless, imperative programming, although several times said to be moribund, still dominates in practical use and in numbers. However, this means nothing about quality.

10.4 The way evolution goes. Sometimes, people speak of two worlds — the world of imperative programming and the world of functional programming, and even an imaginary world of descriptive programming could be postulated. But these worlds are not isolated. The Munich project CIP studied in great detail the transitions between them.

If we select in the taxonomy of Sect. 9 the right-hand-side properties, we have with

opaque, sequential, procedural, deterministic, executable

a characterization of algorithmic languages that embraces machine languages, assembler languages, and a few higher, but not very high-level languages like BASIC. They are all acceptable in their own right, but it may be said that they are dinosaurs. On the other hand, a

lucid, concurrent, functional, liberal

programming language can be a fine piece of software; and if it also allowed descriptive elements, it could be a marvel. It should be clear that this is the way the evolutionary development goes, regardless of the present situation on the market. Equipped now with a wealth of compilation techniques, furnished with most elegant development environments, and backed by powerful and still inexpensive hardware, it is time to approach

the ultimate
lucid, concurrent, functional, liberal, descriptive
algorithmic language of the next century.

Postscript

In November 1955, the joint work of F. L. Bauer and K. Samelson on the use of stacks in formula translation was first presented (by K. Samelson) at the Colloquium on Problems of Computing Techniques, Dresden. Since the publication (VEB Deutscher Verlag der Wissenschaften Berlin 1957) is not easily accessible, a scanned and reprinted version can be found in the appendix following the bibliography.

References

Aiken, H.H. (1937) Proposed automatic calculating machine. In: *B. Randell (1973), pp. 191–197*.

Aiken, H.H., Hopper, G.M. (1946) *The automatic sequence controlled calculator. In: B. Randell (1973), pp. 199–218*.

Aiken, H.H. *(ed.), Proc. of a Symposion on Large Scale Digital Calculating Machinery, 7-10 Jan. 1947. Cambridge (Mass.): Harvard Univ. Press 1948*.

Babbage, C. *(1836) Note, found by M. Trask in Babbage's notebook (9 July 1836). In: B. Randell (1973), pp. 349, 353*.

Babbage, C. *(1837) On the Mathematical Powers of the Calculating Engine (26 Dec. 1837). In:* B. Randell *(1973), pp. 17–52*.

Bauer, F.L., Samelson, K. *(1957)* Verfahren zur automatischen Verarbeitung von kodierten Daten und Rechenmaschine zur Ausübung des Verfahrens. German Patent Application Nr. *B44 122/IX/42m* from March *30, 1957*. German Patent Nr. *1094019*, US Patent No. *3047228*.

Bauer, F.L., Bottenbruch, K., Graeff, P., Läuchli, P., Paul, M., Penzlin, F., Rutishauser, H. and Samelson, K. *(1958a)* Formelübersetzungsprojekt Zürich-München-Darmstadt - Projektstufe *1:* Festlegung der Formelsprache - Interner Bericht No. 2c. Unpublished memorandum, pp. *21 + 23, 1958*.

Bauer, F.L., Bottenbruch, Rutishauser, H., Samelson, K., under cooperation of Graeff, P., Läuchli, P. and Paul, M. *(1958b)* Proposal for a Universal Language for the Description of Computing Processes. In: J.W. Carr (ed.), Computer Programming and Artificial Intelligence, University of Michigan Summer School 1958, pp. 355-373.

Bauer, F.L., Wössner, H. *(1972)* The Plankalkül of Konrad Zuse. Commun. ACM Vol. *15 (1972), pp. 678–685*.

Bauer, F.L., Wössner, H. *(1982)* Algorithmic Language and Program Development. Berlin: Springer *1982*.

Bauer, F.L., Goos, G. *(1982)* Informatik I *(3. ed.)*. Berlin: Springer *1982*, p. *91*.

Bauer, F.L., Goos, G. *(1991)* Informatik I *(4. ed.)*. Berlin: Springer *1991*, p. *131*.

Blum, E.K. *(1956)* Automatic Digital Encoding System II. Proc. ACM Nat. Conf. *6 (1956) 4 pp*.

Böhm, Corrado *(1952)* Calculatrices digitales: Du déchiffrage de formules logico-mathématiques par la machine même dans la conception du programme. Ph.D. dissertation, ETH Zurich *1952*.

Böhm, Corrado *(1954)* Calculatrices digitales: Du déchiffrage de formules logico-mathématiques par la machine même dans la conception du programme. Annali di Mathematica Pure ed Applicata *37 (1954)* pp. *175-217*.

Bowden, B. V. (1953) Faster than Thought. London: Pitman 1953.

Broy, Manfred (1986) An assessment of programming styles: Assignment-oriented languages versus functional and applicative languages. Fundamenta Informaticae IX (1986) pp. 169–204.

Burks, Arthur W. (1950) The logic of programming electronic digital computers. Indust. Math. Vol. 1 (1950) pp. 36–52.

Burks, Arthur W. (1951) An intermediate program language as an aid in program synthesis. ERA Report, U of Michigan, Ann Arbor (1951) ii+15 pp.

Dahl, O.-J., Dijkstra, E.W., Hoare, C.A.R. (1972) Structured Programming. London: Academic Press 1972.

Goguen, J. A., Thatcher, J. W., Wagner, E. G. and Wright, J. B. (1975) Abstract Data Types as Initial Algebras and Correctness of Data Representations, in: Proceedings of Conference on Computer Graphics, Pattern Recognition and Data Structure, Mai 1975.

Guttag, John (1975) The Specification and Application to Programming of Abstract Data Types, Technical Report CSRG-59, Computer Systems Research Group, University of Toronto, Canada, 1975.

Hilbert, David and Bernays, Paul (1934) Grundlagen der Mathematik Bd. 1. Berlin: Springer 1934.

Hoare, C.A.R. (1965) Record Handling. Algol Bull. 21 (1965) pp. 39–69. Extended version in: Genuys, F. (ed.): Programming Languages. London: Academic Press 1968, pp. 291–347.

Hoare, C.A.R. (1970) Notes on Data Structuring. International Summer School on Data Structures and Computer Systems, Marktoberdorf 1970. Extended version in: Dahl, Dijkstra, Hoare 1972, pp. 83–174.

Hoare, C.A.R. (1972) Proof of Correctness of Data Representations. Acta Informatica Vol. 1 (1972) pp. 271–278.

Hoare, C.A.R. (1973) Recursive Data Structures. Stanford University Report CS-73-400 (1973). Extended version in: International J. Computer Information Sci. Vol. 4 (1975) pp. 105–132.

Knuth, Donald E. and Trabb Pardo, Luis (1977) Early development of programming languages. In: Encyclopedia of Computer Science and Technology Vol. 7. New York: Dekker 1977, pp. 419-493.

Liskov, Barbara H. and Zilles, S. N. (1974) Programming with abstract data types. Proc. ACM Conference on Very High-Level Languages. SIGPLAN Notices Vol. 9:4 (1974) pp. 50–59.

Liskov, Barbara H. and Zilles, S. N. (1975) Specification Techniques for Data Abstraction. IEEE Trans. Software Eng. Vol. 1 (1975) pp. 7–19.

Lucas, Peter et al. (1968) Method and notation for the formal definition of programming languages. Techn. Report 25.085, IBM Lab. Vienna, Dec. 1968.

Ludgate, Percy E. (1909) On a Proposed Analytical Machine. In: B. Randell (1973), pp. 73-88.

Lüroth, Jakob (1900) Vorlesungen über numerisches Rechnen. Leipzig: Teubner 1900.

Mauchly, John W. (1947) Preparation of Problems for EDSAC-type Machines. In: Proc. of a Symposion on Large Scale Digital Calculating Machinery, 7-10 Jan. 1947. Cambridge (Mass.): Harvard Univ. Press 1948.

Post, Emil Leon (1936) Finite Combinatory Processes – Formulation 1. J. Symbolic Logic Vol. 1 (1936) pp. 103-105.

Randell, Brian, ed. (1973) The Origins of Digital Computers. Berlin: Springer 1973.

Samelson, K. (1955) Probleme der Programmierungstechnik. Intern. Kolloqu. über aktuelle Probleme der Rechentechnik – Dresden 1955. Berlin 1957.

Schröter, Kurt (1943) Axiomatisierung der Frege'schen Aussagenkalküle. Forschungen zur Logik und zur Grundlegung der exakten Wissenschaften, N.F. Vol. 8 (1943); also J. Symb. Logic Vol. 9 (1944) p. 69.

Torres y Quevedo, Leonardo (1914) Essais sur l' Automatique. Sa définition. Étendue théorique de ses applications. Revue de l' Académie des Sciences de Madrid, 1914.

Turing, Alan M. (1936) On Computable Numbers, with an Application to the Entscheidungsproblem. Proc. London Math. Soc. (2), Vol. 42 (1936) pp. 230-267.

Turing, Alan M. (1937) On Computable Numbers, with an Application to the Entscheidungsproblem. A Correction. Proc. London Math. Soc. (2), Vol. 43 (1937) pp. 544–546.

Zemanek, Heinz (1968) Abstrakte Objekte. Elektron. Rechenanlagen Vol. 10 (1968) pp. 208–216.

Zilles, S.N. (1974) Algebraic Specification of Data Types. MIT Lab. for CS, Progress Report XI (1974) pp. 52–58.

Zuse, Konrad (1945) Der Plankalkül. In: Berichte der Gesellschaft für Mathematik und Datenverarbeitung, Vol. 63, part 3, 1972 (manuscript prepared in 1945).

Zuse, Konrad (1948-1949) Über den allgemeinen Plankalkül als Mittel zur Formulierung schematisch-kombinativer Aufgaben. Arch. Math. Vol. 1 (1949), pp. 441-449.

Zuse, Konrad (1952) Der Programmator. Z. Angew. Math. Mech. Vol. 32 (1952) p. 246.

Zuse, Konrad (1959) Über den Plankalkül. Elektron. Rechenanlagen Vol. 1 (1959) pp. 68-71.

Zuse, Konrad (1984) Der Computer – Mein Lebenswerk. Berlin: Springer 1984, 3. Aufl. 1993.

Zuse, Konrad (1993) The Computer – My Life. Berlin: Springer 1993.

Intern. Kolloquium ü. Probl. d. Rechentechnik, Dresden 1955

PROBLEME DER PROGRAMMIERUNGSTECHNIK

Von Dr. K. SAMELSON, München [1]

Die Aufgabe der programmgesteuerten Rechenmaschine ist die aller Maschinen, dem Menschen Arbeit abzunehmen. Aus dem Namen Rechenmaschine geht hervor, daß es sich um Rechenarbeit handelt. Der Zusatz „programmgesteuert" besagt, daß neben der eigentlich rechnenden Maschine eine zweite Maschine (das Steuerwerk) steht, die die Rechenmaschine mit Zahlen und Rechenvorschriften versorgt und die Ergebnisse der einzelnen Rechenoperationen zu weiterer Verwendung entgegennimmt. Als Arbeit für den Benutzer verbleibt nur noch die Anfertigung eines detaillierten Arbeitsplanes für das Steuerwerk, die man als Programmfertigung bezeichnet.

Die Prinzipien der Anfertigung von Rechenprogrammen finden sich zum Beispiel in der grundlegenden Abhandlung von GOLDSTINE – v. NEUMANN: 'Planning and Coding for an electronic computing instrument' [1].

Es handelt sich dabei im wesentlichen um den Entwurf von Flußdiagrammen und deren Übersetzung in Maschinenbefehle.

Der Vorgang der Programmfertigung mit diesen Hilfsmitteln ist dann etwa folgender:

Der Mathematiker liefert eine Darstellung des Problems in Form eines algorithmischen Plans. Der Programmierer (der natürlich mit dem Mathematiker identisch sein kann) verfertigt danach ein Flußdiagramm und aus diesem an Hand der Befehlsliste der Rechenmaschine das explizite Programm-Manuskript. Der Codierer stellt danach das Eingabe-Programm in verschlüsselter Form her, und der Eingabemechanismus setzt dieses zum internen Maschinenprogramm um.

Als die ersten großen Maschinen zufriedenstellend zu arbeiten begannen, zeigte sich sehr bald, daß mit diesen Methoden das Problem der Programmfertigung in keiner Weise gelöst war. Der Zeitaufwand bei der Programmfertigung stand in einem solchen Mißverhältnis zu den mit den Maschinen erzielbaren Rechenzeiten, daß es einen großen Stab von Programmierern erforderte, um nur eine Maschine voll auszulasten, während auf der anderen Seite die Programmierung eine recht mühselig zu erlernende Kunst war.

Natürlich hatte man von vornherein erkannt, daß eine Programmierung auf Vorrat möglich und notwendig ist. Aus dieser Erkenntnis entstand der Begriff des „Bibliotheksprogramms", eines Programms, das einen bestimmten abgeschlossenen Algorithmus darstellt und jederzeit (und an beliebiger Stelle im Speicher) verwendbar ist. Damit reduziert sich die Programmierungsarbeit für den betreffenden Algorithmus auf die Lösung der Aufgabe, dem Bibliotheksprogramm im Augenblick des Aufrufs die Ausgangsdaten zur Verfügung zu stellen, die es bearbeiten soll. Diese

---

1 Arbeitsgruppe für elektronische Rechenanlagen der Technischen Hochschule, München

Aufgabe, auf die später noch einmal zurückzukommen sein wird, ist nicht so trivial, wie sie im ersten Moment erscheinen mag, muß aber auch nur ein für allemal gelöst werden.

Die Tatsache allerdings, daß die Möglichkeiten des „Auf-Vorrat-Programmierens" noch sehr viel weiter reichen, hat wohl als erster RUTISHAUSER in Zürich erkannt, als er in seiner Abhandlung, über automatische Planfertigung [2] darauf hinwies, daß die Maschine selbst den Schlüssel zur Lösung des Problems der Programmfertigung darstellt. Aus dem RUTISHAUSERschen Gedanken des organisatorischen Programms, des sogenannten Superplans, sind die Methoden und Hilfsmittel entstanden, die man heute unter dem Namen „automatische Programmierung" (automatic coding) zusammenfaßt.

Diese Hilfsmittel sind neben den bereits erwähnten Bibliotheksprogrammen, die jetzt als geschlossene Unterprogramme bezeichnet werden sollen, die sogenannten „offenen Unterprogramme", die wohl von WILKES [3] erstmalig angegebene „algebraische Adresse" (floating address) und schließlich der mit diesen vorbereiteten Hilfsmitteln arbeitende programmerzeugende Superplan, in der angelsächsischen Literatur gewöhnlich als „compiler" bezeichnet.

Mit diesen Hilfsmitteln stellt sieh nun der Vorgang der Programmfertigung folgendermaßen dar:

Der Mathematiker fertigt einen Rechenplan an und übergibt ihn dem Programmierer. Dieser stellt danach ein Programm her, das zumindest in großen Teilen nicht aus Befehlen in Maschinencode besteht, für das Steuerwerk der Maschine also nicht verständlich ist und daher als Pseudoprogramm bezeichnet wird (seine Befehle entsprechend als Pseudobefehle). Dieses Pseudoprogramm verschlüsselt der Codierer und übergibt es der Maschine, aber nicht dem Steuerwerk, sondern einem Vorentschlüsseler, dem programmerzeugenden Superplan, der nun aus dem Pseudoprogramm mit den ihm zur Verfügung stehenden Hilfsmitteln, in erster Linie eben der algebraischen Adresse, den offenen Unterprogrammen und den geschlossenen Bibliotheksprogrammen, das eigentliche Maschinenprogramm erzeugt und nun entweder zur sofortigen Durchrechnung oder zum vorherigen Abdruck bringt.

Wir wollen bei der Diskussion mit den Hilfsmitteln beginnen, und zwar zunächst mit der algebraischen Adresse, die die höchst lästige explizite Angabe von Speicheradressen unnötig macht. Hier gibt es verschiedene Systeme; ich möchte mich aber auf die Darstellung eines einzigen beschränken, das nach einem Vorschlag von SCHECHER für PERM vorgesehen ist, aber sicher auch schon bei anderen Anlagen verwendet wird. Im wesentlichen handelt es sich dabei darum, daß man die Rechengrößen (Zahlen) und diejenigen Befehle, die durch Sprünge aufgerufen werden (z.B. die Anfangsbefehle von Unterprogrammen) durch Symbole kennzeichnet, die von der Maschine mitgelesen werden. Sie bewirken, daß die Maschine bzw. der Superplan in einer eigens dafür reservierten Gruppe von Speicherzellen, einem „Telephonbuch", das Symbol und die tatsächliche Speicheradresse der gekennzeichneten Größe, also Namen und Rufnummer, festhält.

Die Befehle, die sich auf die so gekennzeichneten Größen beziehen, lauten nun auf Namen, enthalten also als Adressenteil das entsprechende Symbol, und der Superplan muß dafür sorgen, daß der Name rechtzeitig durch die Rufnummer ersetzt wird.

Er muß also das Telephonbuch durchsuchen, bis er den angegebenen Namen findet, und dann die Rufnummer ablesen und zur Verwendung weitergeben. Damit ist der Superplan „algebraische Adresse“ mit dem „Telephonbuch“ nichts anderes als eine programmiert simulierte Koinzidenzschaltung von derselben Art wie die verdrahtete Speichersucheinrichtung.

Als nächsten Punkt wollen wir nun die offenen Unterprogramme besprechen. Hier handelt es sich um im allgemeinen kurze Folgen von Maschinenbefehlen, die so häufig vorkommen, daß es sich lohnt, sie auf Vorrat zu programmieren, und die kurz genug sind, um a) sie zusammen mit dem Programmerzeuger permanent zu speichern und b) sie jedesmal, wenn sie in einem Programm gebraucht werden, an der betreffenden Stelle als Ganzes einzufügen, so daß Aufruf- und Rückkehrsprung unnötig sind.

Jedes solche Programm wird im Pseudoprogramm durch einen eigenen Pseudobefehl bezeichnet, der dann von dem Programmerzeuger durch das betreffende Unterprogramm ersetzt wird, unter Umständen mit Einbeziehung von algebraischen Adressen. Beispiele sind etwa die Quadratwurzel, die elementaren Transzendenten einer Variablen oder (besonders beliebt) das Rechnen mit komplexen Zahlen, nebenbei ein Fall dafür, daß auch Befehle in Maschinencode, nämlich die Speziesoperationen, als Pseudobefehle auftreten kännen.

Solche offenen Unterprogramme eignen sich natürlich besonders gut zur festen Verdrahtung. Tatsächlich ist auch ein spezielles, besonders einfaches organisatorisches Programm dieser Art bei fast allen modernen Maschinen verdrahtet vorhanden, nämlich die Abzählung der Komponenten einer indizierten Variablen (Vektor, Matrix usw.). Sie wird bewirkt durch Addition eines dem Koinzidenzregister vorgeschalteten sogenannten Indexregisters auf die im Befehl stehende Adresse, und zwar im Moment der Ausführung des Befehls. Die genannte Adresse selbst ist Bezugsadresse für den ganzen Zahlenblock, der durch die indizierte Variable repräsentiert wird, und bleibt im Speicher unverändert. Es gibt noch mehr Beispiele dieser Art. Man könnte daher sogar die offenen Unterprogramme etwas überspitzt charakterisieren als diejenigen Maschinenoperationen, deren Verdrahtung man vergessen hat.

Daran anschließend kann man gleich die geschlossenen Bibliotheksprogramme als diejenigen Maschinenoperationen bezeichnen, deren Verdrahtung man nicht mehr geschafft hat, weil sie zu komplex sind. Hier handelt es sich stets um abgeschlossene numerische (oder auch evtl. organisatorische) Prozesse. Vom Prozeß her gesehen unterscheiden sie sich von den offenen Unterprogrammen dadurch, daß sie so lang sind, daß man a) sie nicht mehr mit dem Programmerzeuger zusammen permanent speichern und b) sie bei Gebrauch einmal im Arbeitsspeicher der Maschine unterbringen will, selbst wenn sie in einem Rechenprogramm mehrere Male hintereinander gebraucht werden. Sie müssen infolgedessen durch einen Sprung aufgerufen werden, der gleichzeitig alle Parameter des Prozesses mitbringt, und werden durch einen Rückkehrsprung wieder verlassen, was sie rein äußerlich von den offenen Unterprogrammen unterscheidet. Zusammenfassend kann man also die Unterprogramme als durch Programme simulierte Maschinenoperationen (Maschinenbefehle) betrachten.

Als letzter Punkt bleibt noch der programmerzeugende Superplan selbst. Seine Aufgabe ist, wie schon gesagt, das Pseudoprogramm zu entschlüsseln und mit den ihm zur Verfügung stehenden Hilfsmitteln in ein Maschinenprogramm zu übersetzen. Hier unterscheidet man nun im allgemeinen zwei verschiedene Typen, das compilierende, umrechnende Programm, und das interpretierende, ausdeutende Programm. Das erstere entschlüsselt das Pseudoprogramm, schreibt das Resultat, das Maschinenprogramm, einfach nieder und hat damit seine Aufgabe erledigt. Es ist also einfach ein simulierter Codierer bzw. Code-Umsetzer. Das zweite, also das interpretierende Programm, entschlüsselt im Extremfall jeden Pseudobefehl für sich, läßt ihn sofort ausführen und liest dann erst den nächsten Pseudobefehl. Damit ist es nicht nur ein Code-Umsetzer, sondern ein simuliertes Steuerwerk.

Im allgemeinen gibt man heute den compilierenden Programmen den Vorzug. Der Grund dafür ist leicht ersichtlich: Das compilierende Programm stellt das Rechenprogramm vorweg ein, es hat also nur den statischen Anfangszustand des Programms aufzubauen; jeder Pseudobefehl wird einmal entschlüsselt und dann vergessen. Das interpretierende Programm dagegen muß die gesamte Dynamik des Rechenprogramms nachbilden, insbesondere jede Schleife in der verlangten Vielfachheit durchfahren; jeder Pseudobefehl wird so viele Male, entschlüsselt, wie er effektiv ausgeführt werden muß. Das führt natürlich im allgemeinen zu einen ganz ungeheuren Mehraufwand an Zeit und ist daher im höchsten Grade unrationell.

Im Grunde ist es aber falsch, die beiden Techniken (im Unterschied zu den entsprechenden Superplänen) in Gegensatz zueinander zu stellen. Tatsächlich treten sie stets beide nebeneinander auf und ergänzen sich, und die Frage ist nur, an welcher Stelle der Übergang von der einen zur anderen vollzogen wird. Für den Compiler ist es seiner statischen Natur wegen nicht möglich, in die Schleifen des Programms einzugreifen. So ist schon die Einbringung der Parameter in mehrfach zu verwendende geschlossene Bibliotheksprogramme für ihn nicht mehr zu bewältigen und erst recht nicht die Umadressierung von indizierten Variablen. Hier ist die interpretative Technik durchaus brauchbar, allerdings als Erweiterung der Steuerung selbst.

Denn das Steuerwerk ist ja selbst nichts anderes als ein verdrahtetes interpretatives Programm, und wenn jeder Befehl inspiziert, unter bestimmten Umständen abgeändert und dann auf jeden Fall ausgeführt werden soll, dann legt man diesen Vorgang am vernünftigsten dahin, wo sowieso inspiziert wird, nämlich eben in das Steuerwerk. Damit erreicht man, daß die lnterpretierung nur dort vorgenommen wird, wo sie gebraucht wird, während alle nicht änderungsbedürftigen Befehle unbesehen ausgeführt werden können.

Für die Interpretation müssen natürlich gewisse Kriterien angegeben und im Steuerwerk überprüft werden. Jedoch wird es häufig nicht nötig sein, diese Kriterien dem einzelnen Befehl anzuheften.

Man kann sie statt dessen nach Art eines Wahlschalters einmal tasten und damit die normale Codeinterpretation ändern. So wird zum Beispiel bei der Münchener PERM (nach einem Vorschlag von R. PILOTY) der Rechenmodus (festes oder gleitendes Komma) eingestellt und tritt im einzelnen Rechenbefehl gar nicht in Erscheinung. Das gleiche könnte man natürlich auch für komplexes Rechnen oder Rechnen mit doppelter Genauigkeit vorsehen, wenn man die entsprechenden Speziesprogramme verdrahten will.

Zusammenfassend kann man also sagen, daß die vollständige Interpretierung von Pseudoprogrammen mit Hilfe von Programmen aus Zeitgründen unrationell ist, daß aber die Interpretation mit Hilfe von Erweiterungen des Steuerwerkes durchaus sinnvoll sein kann. Darüber hinaus ist es häufig möglich, daß man auch bei der Pseudocodeentschlüsselung durch Programme eine gemischte Technik verwenden muß in dem Sinne, daß das Pseudoprogramm abschnittsweise entschlüsselt und sofort durchgerechnet wird.

Dies ist zum Beispiel sicher dann der Fall, wenn das Pseudoprogramm ein geschlossenes Bibliotheksprogramm mehrmals aufruft und dieses vom Compiler mit Parametern versorgt werden soll, da dann die erste Rechnung schon durchgeführt sein muß, bevor die neuen Parameter eingesetzt werden können.

Ähnlich ist die Situation bei den Hilfsspeicherzellen, wie sie fast jedes geschlossene Programm neben den Zellen für Ausgangsdaten und für Ergebnisse braucht. Was in jedem Moment noch an für Zwischenrechnungen freien Speicherzellen zur Verfügung steht, hängt von der ganzen Vorgeschichte des Programms bis zum betrachteten Moment ab und läßt sich kaum vernünftig vom Compiler am Anfang festlegen. Es bleibt hier wohl nichts anderes übrig, als jeweils beim Übergang in ein neues abgeschlossenes Teilprogramm festzustellen, von wo ab der Speicher für Rechnungen frei ist, und diese Angabe dem Teilprogramm etwa in der Weise mit auf den Weg zu geben, daß man sie unter einem festen Variablennamen im „Telephonbuch" eintragen läßt, so daß also die zu der Variablen „Anfang freier Speicher" gehörige Rufnummer = Speicheradresse sich während der Rechnung laufend ändert.

Damit wollen wir die programmerzeugenden Superpläne und gleichzeitig das Gebiet der im wesentlichen gelösten Probleme der Programmierungstechnik verlassen. Wir kommen also zum nächsten Punkt, zum Pseudoprogramm, und damit zu *dem* Problem der Programmierungstechnik. Der Grund für diese kühn klingende Behauptung ist folgender: Mit der Verwendung von Pseudoprogrammen hat man sich vom speziellen Maschinencode völlig gelöst und ist in der Wahl der zu verwendenden Zeichen frei. Trotzdem ist es im allgemeinen wohl so, daß bei jeder Maschine, die über programmerzeugende Superpläne verfügt, nur der Maschinencode durch einen eigenen, ebenso individuellen Pseudocode ersetzt ist. Damit hat man wohl das Zeitproblem gelöst, nicht aber das der Verständigung zwischen den verschiedenen um je eine Maschine gruppierten Rechenzentren. Und dieses Problem bedarf unbedingt einer Lösung, die nur in der Einführung eines universellen Pseudocodes bestehen kann.

Es ist wohl kaum noch notwendig, darauf hinzuweisen, daß ein solcher Pseudocode im wesentlichen bereits existiert. Vielleicht darf daran erinnert werden, daß zu Beginn die Programmierung mit Flußdiagrammen und Befehlslisten als Kunst bezeichnet wurde. Nun, sie war es schon einmal vor 500 Jahren, für die Rechenmeister des ausgehenden Mittelalters und der Renaissance. Wir sollten uns darüber klar werden, daß die Programmierung als die formale Seite des Maschinenrechnens die Entwicklung der formalen Seite der Mathematik an sich völlig wiederholt, und daraus die nötigen Konsequenzen ziehen. Sie bestehen darin, daß man sich von vornherein bemüht, die Programmierung soweit wie nur irgend möglich der in jahrhundertelanger Entwicklung entstandenen Symbolik der Mathematik anzugleichen.

Erst wenn dies geschehen ist, ist der Schritt vom Zahlen- zum Buchstabenrechnen, von der Arithmetik zur „Algebra“, der in dem Übergang von der Handrechenmaschine zum programmgesteuerten Automaten enthalten ist, wirklich vollzogen. Dann genügt auch die allgemeine mathematische Ausbildung weitgehend zur Benutzung von Rechenautomaten, und die Scheu vieler Mathematiker vor der Benutzung von Rechenautomaten wird zusammen mit dem lästigen Zwang zur Erlernung der Programmierung verschwinden.

Für bereits existierende Maschinen wird damit die Frage gestellt, ob sich der programmerzeugende Superplan so weit ausdehnen läßt, daß er in der üblichen mathematischen Symbolik geschriebene Formeln in Maschinenbefehle übersetzen kann. Daß dies prinzipiell möglich ist, steht fest. Es gibt bereits einige solche Systeme, die allerdings hinsichtlich des Zeitbedarfs sehr ungünstig sein sollen. Trotzdem wollen wir die Frage, wie ein solcher Superplan aussehen muß, kurz diskutieren.

Zunächst einmal muß ein solches Programm sämtliche Zeichen, die das Eingabegerät (z. B. Fernschreiber) benützt, insbesondere die Betriebszeichen *Bu, Zi, Zl, Wr* identifizieren und sämtliche Maschinenbefehle aufbauen können. Als wichtigste Fähigkeit kommt dann die, rationale Ausdrücke beliebiger Kompliziertheit abbauen zu können. Was dabei die Variablen anbetrifft, so werden sie als Buchstaben sofort identifiziert und als algebraische Adressen behandelt. Es handelt sich also nun noch um die Interpretation der Vor- bzw. Operationszeichen $+$, $-$, $\cdot$, $/$, der Klammern und der '='-Zeichen.

Die Übersetzung der Operationszeichen in die entsprechenden Rechenbefehle sowie des Gleichheitszeichens in den Speicherbefehl ist dabei völlig unproblematisch. Die Adresse wird in jedem Fall durch die auf das Zeichen folgende Variable festgelegt. Die einzige Schwierigkeit bringen die gegebenenfalls erforderlichen Zwischenspeicherungen, die von den Klammerzeichen abhängen.

Hier zeigt sich nun sofort ein Nachteil der üblichen algebraischen Symbolik als Pseudokode bei Rechenmaschinen. Gemeint ist die Vorschrift für die Klammersetzung, wie sie sich aus der üblichen Formulierung des Distributivgesetzes herleitet, mit der darin enthaltenen Bevorzugung der Multiplikation.

Denn eine einfache Folge von Additions- und Multiplikationsbefehlen ohne Zwischenspeicherung stellt einen Ausdruck der Form $(((a \cdot b + c) \cdot d + e) \cdot fg + h) \cdot i + \ldots$ dar. (Dabei soll das Pluszeichen Addition und Subtraktion, das Malzeichen Multiplikation und Division repräsentieren.) Es wäre vom Standpunkt der Maschine aus vernünftig, solche Ausdrücke klammerfrei zu schreiben. Dies würde darauf hinauslaufen, die Operationszeichen 'Op' ($+$, $-$, $\cdot$ und $/$) bzw. die Kombinationen aus Operationszeichen und darauf folgenden Variablen 'Op a' als völlig gleichberechtigte Rechenoperatoren zu behandeln, die stets bis zum letzten vorangegangenen „Klammer-auf“-Zeichen bzw. bis zum Anfang hin wirken. Bei einer solchen Vorschrift muß immer dann, wenn auf ein Operationszeichen ein Ausdruck folgt, der vor Ausführung der Operation berechnet werden muß, auf das Operationszeichen ein Klammer-auf-Zeichen folgen, Damit besteht zwischen den Klammern und der Notwendigkeit der Zwischenspeicherung ein eindeutiger Zusammenhang. Die Zeichenfolge 'Op (' bedeutet stets: vor Ausführung der Operation 'Op' ist die Klammer zu berechnen. Der durch 'Op' bestimmte Rechenbefehl ist zu ersetzen durch eine

Zwischenspeicherung des momentanen Linksoperanden, also des Rechenergebnisses im AC; und ist erst nachzuholen, wenn das entsprechende Klammer-zu-Zeichen anzeigt, daß der Rechtsoperator fertig aufgebaut ist. An dieser Stelle ist der zurückgestellte Rechenbefehl mit dem Inhalt der zuletzt besetzten Zwischenspeicherzelle als Operand einzufügen; die Zelle selbst ist anschließend wieder frei verwendbar.

Diese einfache Zuordnung von Klammern und Zwischenspeicherungen geht bei der herkömmlichen Schreibweise wegen der darin enthaltenen Bevorzugung der Multiplikation und Division verloren. Denn hier müssen in Summen von Produkten (bzw. Quotienten) zunächst die Produkte völlig aufgebaut sein, bevor die Summation vorgenommen werden kann, ohne daß dies durch Klammerzeichen angezeigt wird.

Daher geht in die Vorschrift zur Zwischenspeicherung außer den Klammerzeichen auch noch die Reihenfolge der Operationszeichen ein.

Der einfachste Fall sind die klammerfreien Ausdrücke. Hier muß der Superplan bei Auftreten von '+a' prüfen, ob das nächste Operationszeichen ein '$\cdot$' oder '/' ist. Ist dies der Fall, so ist die Addition (wie oben) durch eine Zwischenspeicherung von AC, etwa nach H_i, und Herauslesen von a nach AC zu ersetzen, und der Additionsbefehl ist erst einzuschieben, wenn das nächste Pluszeichen auftritt.

Genau dasselbe hat man bei Ausdrücken in Klammern zu beachten. Die Addition ist nachzuholen, wenn sämtliche unmittelbar anschließenden Multiplikationen (Divisionen) ausgeführt sind. Dies bedeutet, daß zwar auch hier die Sequenz 'Op (' eine Zwischenspeicherung von AC nach H_i, und ein Zurückstellen der Rechenoperation Op (mit H_i als Operand) verlangt. Hinsichtlich der Auswertung des nachfolgenden Zeichens ')' besteht aber jetzt eine Alternative bezüglich Op. War 'Op' ein '$\cdot$' -Zeichen, so darf jetzt die Operation $H_i \cdot AC$ sofort eingesetzt werden, anschließend ist das nächste Zeichen zu lesen und zu interpretieren, wobei jetzt H_i als Hilfsspeicherzelle wieder verfügbar ist. War 'Op' aber ein '+' -Zeichen, so ist die Operation $H_i + AC$ zurückzustellen, bis der zweite Summand fertig aufgebaut ist. Folgt also auf das ')'-Zeichen ein '$\cdot$(', so ist eine weitere Zwischenspeicherung von AC nach H_{i+1} vorzunehmen und der Befehl $H_{i+1} \cdot AC$ zurückzustellen bis zum nächsten ')'. Erst wenn auf ein ')' entweder ein '+' oder das Gleichheitszeichen folgt, darf auch der zurückgestellte Additionsbefehl $H_i + AC$ eingefügt werden.

Eine kleine Schwierigkeit besteht noch bei den Operationen '$-$' und ':', weil hier zuerst der Minuend bzw. Dividend und anschließend der Subtrahend bzw. Divisor aufgebaut werden. Da bei Ausführung der Operation jeweils der erste Operand in AC, der letztere in MD stehen muß, verlangt dies entweder eine Zwischenspeicherung des letzteren Operanden und nachfolgendes Herauslesen beider Operanden in richtiger Reihenfolge oder besser eine Vertauschung der Befehlsfolgen, in denen die beiden Operanden aufgebaut werden, durch den Superplan, was relativ einfach zu bewerkstelligen ist. Das Gleichheitszeichen bzw. die Folge '$= r$' bedeutet Abschluß der Rechnung und ist mit 'speichere AC nach Zelle r' zu übersetzen.

Neben einem derartigen Superplan zum Aufbau rationaler Funktionen braucht man für ein algebraisches Code-System noch Superpläne zur Interpretation von algebraischen Adressen und Codesymbolen für offene und evtl. geschlossene Unterprogramme. An solchen Unterprogrammen sind als wichtigste zunächst die Indexabzählungen zu nennen (von RUTISHAUSER mit dem Symbol „für $i = \alpha(\beta)\gamma$" belegt), weiter z. B. Alternativen, die geläufigen transzendenten Funktionen, die Standardverfahren der numerischen Analysis, Druck-(*editing*) Programme usw. Alle diese Unterprogramme sollte man entweder, soweit vorhanden, durch die üblichen Symbole (Funktionszeichen usw.) kennzeichnen oder durch Worte der Umgangssprache, die man intern leicht auf eine bestimmte Buchstabenzahl begrenzen kann. Die Liste dieser Unterprogramme darf selbstverständlich nicht unveränderlich sein, sondern muß nach Bedarf erweitert werden können.

Es ist möglich, daß ein Eingabesuperplan der hier skizzierten Art für schon existierende Maschinen zu schwerfällig wäre. Das kann allein der Versuch entscheiden. Die Technik des Rechenmaschinenbaues ist aber heute so weit fortgeschritten, daß bei zukünftigen Maschinen die Verwendung eines solchen Superplans in verdrahteter Form möglich erscheint. Denn die Magnetkerntechnik ermöglicht auf der Basis der WILKESschen Idee der Mikro-Programmierung [4] die Verdrahtung von (offenen) Unterprogrammen beträchtlicher Anzahl in größerer Länge, und dies ist eine der Hauptvoraussetzungen für eine so radikale Erweiterung des Steuerverkehrs, wie sie ein voll algebraischer Code verlangt, der ja eine beträchtliche Anzahl einzelner Zeichen identifizieren und darüber hinaus Gruppen von Zeichen als Worte (im Sinne der Umgangssprache) interpretieren können müßte.

Literaturverzeichnis

[1] H. H. GOLDSTINE und J. v. NEUMANN, Planning and Coding of Problems for an Electronic Computing Instrument. Institute for Advanced Study, Princeton N. J., 1947.

[2] H. RUTISHAUSER, Automatische Rechenplanfertigung bei programmgesteuerten Rechenmaschinen. Birkhäuser, Basel 1952.

[3] M. V. WILKES, The Use of a 'Floating Address' System for Orders in an Automatic Digital Computer. Proc. Cambr. Philos. Soc. 49 (1953), 84-89.

[4] M. V. WILKES and J. S. STRINGER, Micro Programming and the Design of the Control Circuits in an Electronic Digital Computer. Proc. Cambr. Philos. Soc. 49 (1953) 230-238.

Erschienen in:
Bericht über das Internationale Mathematiker-Kolloquium,
Dresden, 22. bis 27. November 1955 „Aktuelle Probleme der Rechentechnik"

1957 VEB Deutscher Verlag der Wissenschaften Berlin
Herausgeber: N. J. Lehmann, Dresden
Korrektur von Druckfehlern: flb 960514

Mathematical Methods in System and Software Engineering*

Manfred Broy
Institut für Informatik, Technische Universität München
80290 München, Germany

Abstract

Today, there is still a remarkable gap between the techniques and methods used in practice in software engineering and the formal techniques worked out and advocated by academics. Our goal is to close that gap and to bring together the pragmatic and mostly informal ideas in systems and software engineering used in practice and the mathematical techniques for the formal specification, refinement, and verification of software systems. In practice, software engineers are used to working with

- a development method that describes the development process in detail and
- description formalisms that describe the system under development; these descriptions are often annotated diagrams.

The development process is often supported by CASE (Computer Aided Software Engineering) tools.

We present a mathematical, scientific basis for system and software engineering. In its core, there is a mathematical model of a system and formal techniques to describe it. We outline representative examples of diagrammatic description techniques as they are used in software engineering approaches in practice and show how they are formally related to the system model. These description techniques include in particular

- data models,
- process models,
- structure and distribution models,
- state transition models,
- interface models.

We define a translation of the description techniques into predicate logic. This allows us to combine techniques of formal specification and verification with pragmatic system description techniques. We show how to develop systems with the help of these description techniques in refinement steps. Thus, we demonstrate how software engineering methods can be backed up by mathematics. We discuss the benefits of such a mathematical and scientific foundation. These benefits go far beyond the benefits of the formal methods for the specification and verification of software.

---

*) This work was partially sponsored by the Sonderforschungsbereich 342 "Werkzeuge und Methoden für die Nutzung paralleler Rechnerarchitekturen", the BMBF project ARCUS and the industrial research project SysLab sponsored by Siemens Nixdorf and by the DFG under the Leibniz program.

1. Introduction

It is widely accepted by now that the development of software is an engineering discipline. To underline this we speak of *software engineering* and also of *system engineering* in cases where the software is embedded into a system context. Typically, all engineering disciplines today are based on theoretical and, in particular, on mathematical foundations. These theoretical foundations are used as a basis for a deeper understanding of the engineering task, for a body of rules, procedures, and engineering processes. There can be no doubt that software engineering has and needs its own theoretical and mathematical foundation like any other engineering discipline.

Software engineering, in practice, deals with the development of large and often complex information processing systems. This includes techniques for the description of requirements and systems in the disparate development phases. The main goals of software engineering are good quality in the engineering process and its results, the timely delivery of the end product, and high productivity. In this paper we ignore most of the economic aspects and sketch what a mathematical basis of the technical aspects of system and software engineering may look.

1.1 Formal Methods

Starting with the pioneering work of Backus, Bauer, Samelson, McCarthy, Petri, Strachey, Scott, Dijkstra, Floyd, Hoare, deBakker, Reynolds, Milner, the VDM group, and many others, formal methods in software development have been extensively investigated over the last 25 years. Today formal methods are a well-established subject in academia.

After denotational semantics, axiomatic specifications, and numerous logical calculi became available, many of the theoretical problems for modeling software engineering concepts by mathematical and logical techniques were solved (see also [Abrial 92]). More work is needed, however, along the lines of [Jones 86] and [Guttag, Horning 93] to make the theoretical work better accessible to software engineers working in practice. The fact that theoretical foundations are sufficient already today to cover the foundations of pragmatic methods is demonstrated, for instance, by [Hußmann 95] which provides a complete formalization of the method SSADM by translating it into SPECTRUM. SSADM is a pragmatic requirement engineering method used as the UK government's standard method for carrying out systems analysis and design stages of information technology development projects. SPECTRUM is an axiomatic specification language [SPECTRUM 93].

How many basic problems still have to be studied and how much foundational work is still needed and stimulated by practical problems can be seen, for instance, by the concept of state charts (see [Beeck 94] for a long list of open problems) for describing embedded process control systems. Also Statecharts (see [Harel 87]) are now widely used in industry - partly due to the fact that they are supported by a commercial CASE tool.

1.2 From Theory to Practice

Thirty years ago, for many concepts used in computing practice, a theoretical foundation was completely missing. This is no longer true today. A solid body of foundational work is now available, giving principal answers to many of the demanding questions of theoretical foundations of software engineering. We have denotational semantics; axiomatic specification techniques for handling data structures, functions, and statements; assertion calculi, logical calculi for parallel and distributed systems; and various logical concepts for dealing with interactive systems.

The next pertinent step is to increase experimentation and the transfer of these theoretical results to engineering practice. Yet, this needs the good will and professionalism of both practical software engineers and theoreticians. This means that also the theoreticians are requested to undertake serious efforts to get a better understanding of the needs of software engineers working in practice and of their advanced engineering concepts.

Development needs methodological guidance. In addition to the specification notation and logical calculi and maybe a few small, hopefully illustrative case studies, more methodological guidance is required to bring theoretical ideas closer to practical use. It is not realistic that everyone in practice develops their own ideas on how to use the methods provided by theoreticians in an appropriate and systematic manner.

1.3 Elements of Software Engineering

The development of large software systems involves a large number of people with quite different skills. The goal is to construct a technical solution to a client's problem. Accordingly, software engineering is a discipline that involves organizational and management aspects as well as technical aspects. A decisive issue is cost management.

Software engineering, like many other engineering disciplines in their early phases, has developed a rich and often confusing wealth of proposals for overcoming the difficulties of organizing large development processes. Here economic aspects are a major issue. Among others, these includes

- team organisation and management,
- cost and schedule prediction, statistical models to evaluate the quality of software,
- development process organisation and integration (phases, milestones, documentation),
- integration with existing software components,
- tool support.

It is naive to assume that managers responsible for large projects will switch to new more formal development methods for which these questions are not convincingly answered.

Of course, it is a long job to provide a complete and mathematically well-founded software development method, but only when we start to work in this direction might there be a chance to transfer theoretical results more effectively into practice. When working towards such a method, we will discover many challenging theoretical problems. Examples are the mathematical capture of the concepts of software architectures, design patterns, and the formalization of existing pragmatic description methods.

We do not treat economic aspects in the following at all and concentrate solely on technical aspects. These comprise:

- development process organization (phases, milestones, documentation),
- formulation of strategic goals and constraints,
- modeling, description, specification,
- quality guarantee,
- integration with existing software components, reuse,
- documentation,
- tool support.

Although we will concentrate in the following mainly on the technical aspects, one should keep in mind that there is a close relationship between the technical and the management aspects. The best technical solution is worthless if it is too costly or if the product cannot be finished in time when working with it. Moreover, most of the management and planning cannot be carried out without a deep understanding of the technical tasks.

Whenever it is necessary to assess the role of technical aspects with respect to management aspects, we will refer to them. We find the closest connection between management and technical issues in the organization of the development process, which is part of what is called the *process model.* A central notion in software engineering is the development method[1]. A method in software engineering comprises description techniques in its syntax, semantics, and pragmatics; rules for working with these techniques; development techniques; and general principles.

1.4 Overall Organization of the Paper

A software system is represented by a set of descriptions in the form of textual or graphical documents, and, at the same time, it is a product that can be brought to life and then show a complex dynamic behavior. The goal of the software development process is a model of the application described by formalisms for which an efficient execution is available. Therefore, like no other engineering discipline, software engineering deals with models, description formalisms, and modeling techniques. In the following we want to carefully distinguish between

- *mathematical models*: the mathematical structures forming the semantic conceptual model associated with a system or a software description formalism,
- *description techniques*: the notations of the descriptions, given by the syntax, graphics, or tables used in the documentation of a software system,
- *modeling techniques*: the activities and methods mapping, representing, and relating real life aspects of applications by using software description techniques.

Of course, these three aspects are closely related. Nevertheless, we want to observe these distinctions carefully in the following, since they are of major relevance for the understanding and foundation of software engineering. In particular, it is, important to distinguish between the description formalisms and the modeling idea for systems.

This paper is an attempt to give a comprehensive mathematical foundation to software engineering formalisms and methods in the form of mathematical models

[1] We distinguish between method and methodology. Methodology is the systematic study of methods.

and relations between them. These particular mathematical models should be understood as an instance of what we are aiming at rather than the only way to provide a mathematical foundation. We want to demonstrate, however, that a comprehensive family of models and well-founded description techniques for software engineering can be provided.

The paper is organized as follows. In Sect.2 we give a survey of the most important notions we deal with. In Sect.3–7 we then deal with the notions of a data model, a process, a component, a state machine, and a distributed system and give mathematical denotations for them. In Sect.8 we deal with development methods, refinement, and development process models. The conclusion follows a short section on implementation issues and tools.

Throughout this paper we use as our running example a simple adaptation from [Broy, Lamport 93] (see also [Broy, Merz, Spies 96]). It deals with a simple memory and a component to access it. The storage is faulty. This means that we can read from or write to the storage but as a result of reading or writing we may get failures. We assume that those failures are indicated by failure messages which allow us to do some exception handling either by indicating to the environment that something went wrong or by retrying. We specify such an unreliable memory and components that organize access to the unreliable memory. We use the faulty store as a running example for the specification of the data models, the process models, the component models, and the distributed system, as well as the state machines.

2. Mathematical Models and System Description

In the following sections we define mathematical models and description techniques for systems and system aspects. In this section we discuss the role of formal methods and mathematics in systems engineering.

2.1 System Aspects and Views

A complex information processing system cannot be described in a monolithic way if understandability is an issue. It is better to structure its description into a number of complementing views. These include:

- data models,
- system component models,
- distributed system models,
- state transition models,
- process models.

For each of these aspects of a system, a mathematical model is defined in the following, consisting of a syntactic and a semantic part. We use these models as a basis for giving meaning to more pragmatic description formalisms.

2.2 Description Formalisms in Software Engineering

In software engineering it is necessary to describe large information processing systems and their properties. Of course, this cannot be done by only providing one description formalism or a single document. Therefore software engineering methods suggest several complementarry description formalisms and documents. Each of these documents describes particular views on the system. Putting these views together should lead to a consistent and comprehensive system model. We speak of *view integration.*

Integration of description formalisms and their mathematics is an indispensable prerequisite for more advanced software development methods. In the following, we discuss some of the more widely used description formalisms in software engineering and their mathematics.

Description formalisms are used to represent models like programming languages are used to represent algorithms. It is therefore helpful to distinguish between description formalisms and modeling techniques. A description formalism is a textual or graphical formalism, such as a diagram or a table, for describing systems, system parts, system views, and system aspects. A model is a mathematical structure to be used as an image of certain aspects of a real-life system. Description formalisms allow us to represent such models and views as well as their properties by concrete syntactic means. Mathematical semantic models are mathematical structures that are used to give a precise meaning to description techniques.

Practical software engineers tend to overemphasise the importance of description formalisms in terms of syntax while theoreticians often underestimate the significance of the notation they use. In the engineering process, notation, be it graphics, tables, or texts, serves as a vehicle for expressing ideas, for analysis, communication, and documentation. To provide optimal support for these purposes both the presentation of the descriptions and the mathematical models behind them have to be well chosen. Depending on the purpose of the modeling and the education and experience of the user, the effectiveness of the presentations may be quite different, hence it is generally wise to have several presentations for the same description. Examples are text, diagrams, and tables.

When analyzing and describing systems, we distinguish between static and dynamic aspects. Static aspects of an information processing system are those that do not change over the lifetime of a system. Dynamic aspects have to do with behavior. We can often use the same description formalisms for the static and the dynamic aspects.

Dynamic aspects of components can be described either axiomatically, or in a property-oriented way, or in operational terms. Operational descriptions tend to be less abstract. Therefore, non-operational descriptions are often better suited in the early phases of development. Both property-oriented descriptions as well as operational descriptions can be based rigorously on logic, of course. To improve readability, we can also give graphical or more illustrative formalisms for describing properties of systems. For such descriptions, a translation into logical formulas should be provided.

2.3 The Role of Description Formalisms in the Development Process

Description formalisms and their underlying semantic models form the basis for formulating and documenting the results of the development process. We may understand the development process as the working out of documents describing the requirements and design decisions for the system in more and more details, adding implementation aspects until a desired implementation level is reached.

In the development process the various description formalisms serve mainly the following purposes:

- means for the analysis and requirement capture for the developer,
- basis for the validation of development results,
- communication between the engineers and the application experts,
- documentation of development results and input for further development steps.

Of course, the usefulness of a description formalism always has to be evaluated with respect to these goals. Software engineering has provided many different description techniques for each of the various aspects of a system. A proper relationship between these description techniques and a foundation for them are among the goals of the mathematics of software engineering.

2.4 The Role of Logic

A universal and precise discipline for the description of all kinds of properties and aspects of a system is mathematical logic. A logical formalism provides a formal syntax, its mathematical semantics and a calculus in terms of a set of deduction rules. The later can be used to derive further properties from a set of given properties.

Properly defined description formalisms also have a formal syntax and a mathematical semantics. Often, however, deduction rules are not provided. Nevertheless, for most description formalisms we may define transformation rules for manipulating them. In contrast to logical formalisms, the description formalisms of software engineering are not especially designed for formal manipulation by deduction and transformation rules.

Description techniques used in practice often do not have a mathematical semantics or even a proper informal description of their meaning.

In any case, there is a close relationship between description formalisms and mathematical logic. Strictly speaking, a description formalism formulates a property of a system. So it can be understood as a predicate. Consequently, we may look for rules that allow us to translate description formalisms into logical formulas. This allows us to use pragmatic description techniques without having to give up the precision of mathematical logic and the power of formal reasoning.

3. Data Models and Their Specification

In this section we introduce a mathematical model for data structures on which specification techniques can be based. Then we introduce perceptual description techniques for data structures and relate them to the mathematical model.

3.1 Mathematical Data Models

Data models are needed to represent the information and data structures involved in an application domain. They are also used to provide computing structures for representing the internal data of information processing systems. In general, data models capture mainly the structure of the data and their relationship, but not their characteristic operations. These, of course, should be an integral part of every data model. Therefore, we always understand a data model as family of data sets named by sorts together with their relationships and the basic characteristic operations and functions.

Families of sets and operations over them are mathematically treated in algebra. From a mathematical point of view, a data model is a heterogeneous algebra. Such an algebra is given by a family of carrier sets and a family of functions. More technically, we assume a set S of sorts[2] (often also called types or modes) and a set F of constants including function symbols with a fixed functionality

$$\mathbf{fct} : F \rightarrow S$$

The function **fct** associates with every function symbol in F its domain sorts and its range sort. We assume that the set of sorts S contains besides basic sorts (such as Bool, Nat, etc.) also tuples of sorts as well as functional sorts and even polymorphic sorts (see [SPECTRUM 93] for details). Both sets S and F provide only names. The pair (S, F) together with function **fct** that assigns functionalities to the identifiers in F is often called the *signature* of the algebra. The signature is the static part of a data model and provides a syntactic view of a data model.

In every algebra A with signature (S, F) we associate with every sort $s \in S$ a carrier set s^A (a set of data elements) and with every function symbol $f \in F$ a constant or function f^A of the requested sort or functionality. An algebra gives meaning to a signature and can be seen as the semantics of a data model. It is typical for mathematical structures modeling information processing concepts that they include static (syntactic) parts such as name spaces (in the algebraic case the signature) and semantic parts (in the algebraic case carrier sets and functions).

Data models are used to capture various aspects of software systems that often go much beyond pure data and information structure aspects. In other words, they may be used also to represent dynamic aspects of system behaviors or system structures and not only the static structures of data sets and selector functions. That such aspects can also be captured by data models is not surprising. Following our definitions, data models can be seen as algebras, and algebras are a very general mathematical concept. All kinds of mathematical structures and mathematical models of systems can be understood as heterogeneous algebras. Nevertheless, we suggest using the general concept of algebras for data models only. In the following we give more specific mathematical structures for other system aspects.

3.2 Description of Data Models by Sort Declarations

In the data view, we fix the relevant sorts and their characteristic functions for the system we want to describe or implement. In addition, we can describe the system states. This can be done by sort declarations as we find them in programming

[2] We believe, like many other computer scientists and software engineers, that the use of data sorts (typing) is very helpful in modeling application and software structures.

languages, by axiomatic data structure specifications, and/or by E/R-diagrams, especially when our system processes mass data. Often data models are used to describe the state of systems and their components.

In our example we use a sort CalMem and fix its data attributes by giving their sorts. The sort CalMem is described by the following sort declaration:

Sort CalMem = put (i: Location, d: MemVals) | get (i: Location)

In a sort declaration we introduce a sort (in the example above the sort CalMem) and describe it by a sort expression. The sort expression is formed from enumeration sorts, records, and variants. An enumeration sort is given by an expression of the form

$\{a_1, ..., a_n\}$

and introduces the identifiers a_1, ..., a_n as constants, which are the constructors of the enumeration sort.

A record sort has the form

con (sel_1: M_1, ..., sel_n: M_n)

where M_1, ..., M_n are sorts. It introduces con as a constructor function and sel_1, ..., sel_n as selector functions. Here, we assume the convention that we use the sort identifiers as selectors, if no selectors are mentioned explicitly.

A variant sort has the form

$R_1 | ... | R_n$

where R_1, ..., R_n are record or enumeration sorts. The constructors for these sorts are used as constructors and discriminators for the variant sort.

3.3 Description of Data Models by Logic

Data models can be described by the logical properties of the functions involved. Then we speak of the axiomatic or of the algebraic specification of data structures. The techniques of algebraic specification are now wellunderstood. Therefore we just give a simple example and refer to [Wirsing 90] for an overview and to [SPECTRUM 93] for an instance of a particular algebraic specification language.

We give two simple examples. We start with the polymorphic specification of the algebra of finite sets.

SPEC SET =

{ **sort** Set α,

∅ :	Set α,
_∪{_}, _\{_}:	Set α, α → Set α,
_=∅ :	Set α → Bool,
_∈_ :	α, Set α → Bool,

Set α **generated_by** ∅, ∪,

∅ =∅,

¬(s∪{x} =∅),

$\neg(x \in \emptyset)$,
$x \in s \cup \{x\}$,
$x \in s \Rightarrow x \in s \cup \{y\}$,
$x \neq y \Rightarrow x \in s \cup \{y\} = x \in s$,

$\emptyset \backslash \{x\} = \emptyset$,
$(s \cup \{x\}) \backslash \{x\} = s \backslash \{x\}$,
$x \neq y \Rightarrow (s \cup \{y\}) \backslash \{x\} = (s \backslash \{x\}) \cup \{y\}$,

$(s \cup \{x\}) \cup \{y\} = (s \cup \{y\}) \cup \{x\}$,
$x \in s \Rightarrow s \cup \{x\} = s$ }

Such sorts as enumeration sorts, records, and variants used in sort declarations given above can be schematically translated into algebraic specifications.

As a second example we define a simple object model by algebraic techniques. The object model describes for given classes α and given attributes for each class the object model in a polymorphic style. The sort Store denotes the set of object stores and the set Obj α denotes the set of object identifiers for the class α. More precisely, α is a record sort of the attributes of the class.

OBJECT_MODEL = {

sort	Store, Obj α,	
emptyStore :	Store,	*empty store*
update :	Store, Obj α, α → Store,	*update of an object*
newObj :	Store, α → Obj α,	*creation of an object*
newObjstore :	Store, α → Store,	*allocation of storage for a new object*
valid :	Store, Obj α → Bool,	*test, if object id is declared for a store*
deref :	Store, Obj α → α,	*dereferencing*

Axioms:

For the empty store:

valid(emptyStore, r) = false,

For selective update:

valid(σ, v) ⇒ valid(update(σ, v, a), r) = valid(σ, r) ∧
deref(update(σ, v, a), r) = **if** r = v **then** a **else** deref(σ, r) **fi**

For object storage allocation:

valid(σ, newObj(σ, a)) = false,

valid(newObjstore(σ, a), r) = (valid(σ, r) ∨ r = newObj(σ, a)),

valid(newObjstore(σ, a), r) ⇒
deref(newObjstore(σ, a), r) = **if** r = newObj(σ, a) **then** a **else** deref(σ, r) **fi**
}

Here Obj α is a polymorphic sort. This means that for every sort M we may form the sort Obj M.

3.4 Data Models of Component States

Data models can be used to model the state of a component. A mathematical notion of a component will be given in the following section.

Given a signature (S, F) as a basis, we define a state space by a signature (S, F ∪ V). The symbols in V are called the *variables* or the *attributes* of a state. They as well as the symbols in F, have a sort.

A widely used technique to describe state spaces is that of entity/relationshipdiagrams. In an entity/relationship diagram we find notations and symbols as shown in Fig. 1.

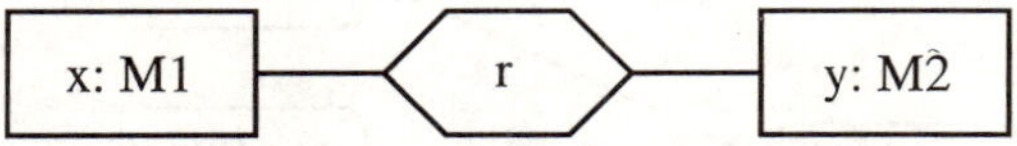

Fig. 1 Two entities x and y connected by one relationship

The diagram in Fig. 1 is used to express that x, r, y are state attributes. The shape of the graphical symbols indicates that x and y are entities with the sorts shown. These nodes are equivalent to the declarations

x: Set M1

y: Set M2

The shape of the r symbol indicates that r is a relationship. This corresponds to the declaration

r: Rel[M1, M2]

where Rel[M1, M2] is a polymorphic sort declared by:

sort Rel[α, β] = Set Pair(c1: α, c2: β)

Of course, we have to add invariants such that the relation r contains in every state only pairs (a, b) with elements a that are in x and elements b that are in y, respectively.

Formally, this is expressed by

$(a, b) \in r \Rightarrow a \in x \wedge b \in y$

We also allow inclusion of simple attributes in the state space that correspond to program variables.

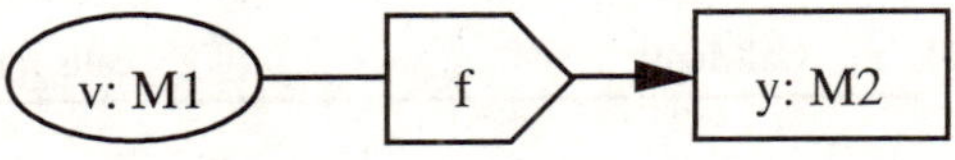

Fig. 2 Functional relationship

The diagram shown in Fig. 2 expresses that v is an attribute

v: M1

and f is a function attribute (a program variable)

f: M1 $\rightarrow$ M2

that maps v onto an element of the entity y: Set M2. If we replace the attribute by an entity, this expresses that every element in the set is mapped by f onto an element of y. Again we assume an invariant:

f(v) ∈ y

Data models can thus be used to describe the information structure of an application and the state space of components. In our example system, we use a component Store. The data view is described by an E/R-model. It is based on the sorts given in a table of sort declarations in Table 1. Figure 3 gives simple examples of a data structure diagram and an E/R-diagram.

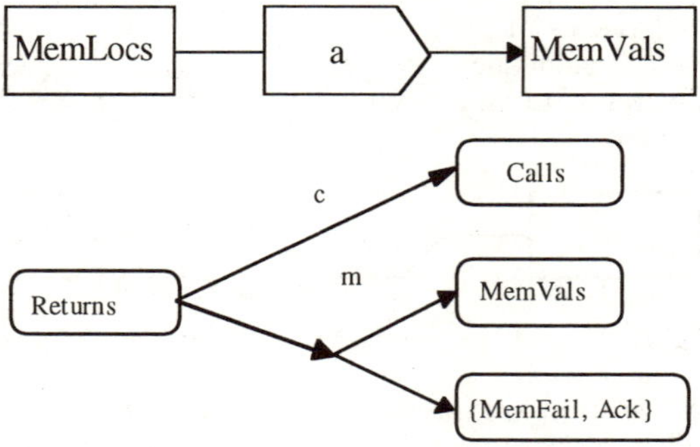

Fig. 3 Data Sort Diagram

With this data view of the state of the component Store the behavior of the component Store can be made more precise later. Each action of a component may change some of its attributes. We come back to this issue in Sect.5 on the description of behaviors.

Tab. 1 Data sorts and their declaration

MemLocs		memory locations
MemVals		memory values
PrIds		identifiers for processes
CalMem	= put (i: MemLocs, d: MemVals) | get (i: MemLocs)	memory calls
RetVals	= MemVals | {MemFail, Ack}	return values
RetMem	= ret (c: CalMem, rv: RetVals)	return messages of memory
Returns	= ren (c: Calls, m: RetVals)	return messages of the driver
Calls	= ca (pi: PrIds, mc: CalMem)	calls for the driver

Graphical representations and diagrams can be used as an illustrative, but nevertheless fully formal, description technique if a formal syntax and a mathematical semantics are provided. A useful form to do this is a translation into axiomatic specifications. This allows us to integrate formal axiomatic data description techniques and pragmatic graphical description techniques into one homogenous software engineering method. A careful translation of E/Rconcepts into axiomatic specifications is given by [Hettler 94]. Hettler distinguishes between E/Rtechniques for modeling static aspects (the static data model) and dynamic aspects of systems (such as the states of a data base). The sort MemLocs is described by a record sort listing its attributes. In the dynamic view, a program variable of sort Set MemLocs is associated with the entity MemLocs.

Axiomatic and algebraic specification techniques as well as E/Rdiagrams provide powerful and very general description formalisms. They allow us to describe data sorts and their characteristic operations in an abstract way not biased by the choice of a concrete representation. They follow the principle of information hiding of implementation information in a consequent manner.

4. System Components

In this section we introduce a mathematical concept of a system component and techniques for describing components. A component is considered in practice as a black box that is an encapsulation of related services according to a published specification. Our mathematical model of a component is consistent with this informal description.

4.1 Component Models:
Interface Models by Stream Processing Functions

We are interested in system models that allow us to represent systems in a modular way. We think of a system as being composed of a number of subsystems that we call *components*. Moreover, a system itself is a component again which can be part of a larger system. A component is a self-contained unit with a clear-cut interface. Through its interface it is connected with its environment. In this section we introduce a simple, very abstract mathematical notion of a system component.

For us, a (system) *component* is an information processing unit that communicates with its environment through a set of input and output channels. This communication takes place in a (discrete) time frame.

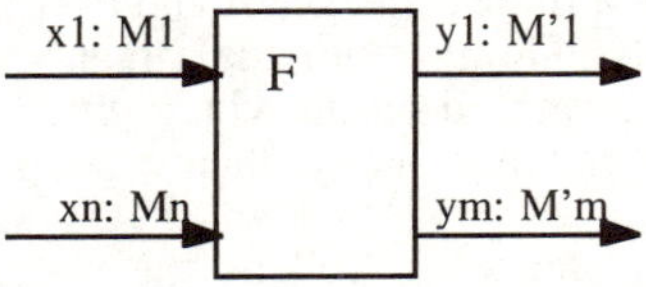

Fig. 4 Graphical representation of a component as a eata rlow node with input channels x1, ..., xn and output channels y1, ... , ym and their respective sorts M1, ... , Mn and M'1, ..., M'n

In software engineering, it is helpful to work with a *black box view* and a *glass box view* of a component. In a black box view we are only interested in the interface of a component with its environment. For this we have to describe the causal dependencies between the input and the output messages. In a glass box view we are interested in the internal structure of a component, which can either be given by its local state space together with a state transition relation or by its decomposition into subcomponents. We first give a model for the black box view.

Let I be the set of input channels and O be the set of output channels. With every channel in the set I ∪ O we associate a data sort indicating the sort of messages sent on that channel. Then by (I, O) the *syntactic interface* of a system component is given. For simplicity, we use just one set M of data sorts for messages on the channels in the sequel in order to keep the presentation simple. A graphical representation of a component with its syntactic interface is shown in Fig. 4.

Let M be a sort of messages and signals. By M^* we denote the set of finite sequences over the set of messages M, by M^∞ we denote the set of infinite sequences over the set of messages M. By ^ we denote the concatenation of sequences. The set M^∞ can be understood to be represented by the total mappings from the natural numbers $\mathbb{N}$ into M. Formally we define the set of timed streams by

$$M^{\aleph} =_{def} (M^*)^\infty$$

(where we write S^∞ for the function space $\mathbb{N}^+ \to S$ and $\mathbb{N}^+$ for $\mathbb{N} \setminus \{0\}$).For every set of channels C, every mapping $x: C \to M^{\aleph}$ provides a complete communication history. Note that $(C \to M^*)^\infty$ and $C \to (M^*)^\infty$ are isomorphic. Moreover, the set $M^{\aleph}$ is isomorphic to the set of streams over the set $M \cup \{\surd\}$ which contain an infinite number of time ticks (here $\surd$ denotes a time tick; we assume $\surd \notin M$).

We denote the set of valuations of the channels C by sequences

$$C \to M^*$$

by

$$\vec{C}^*$$

We denote the set of valuations of the channels in C by infinite timed streams

$$C \to M^{\aleph}$$

by

$$\vec{C}$$

We denote for every number $i \in \mathbb{N}$ and every stream $x \in M^{\aleph}$ by

$$x{\downarrow}i$$

the sequence of the first i sequences in the stream x. This represents the observed communication over the first i time intervals. By

$$\overline{x} \in M^* \cup M^\infty$$

we denote the finite or infinite stream that results from concatenating all the finite sequences in the stream x. $\overline{x}$ is a finite sequence if and only if only a finite number of sequences in x are nonempty. Going from the stream x to $\overline{x}$ provides a *time abstraction*. In the stream x we can find out in which time interval a certain message arrives, while in $\overline{x}$ we see only the messages in their order of communication without any indication of their timing.

We use both notations $x{\downarrow}$ and $\overline{x}$ as well as the concatenation introduced for streams x also for tuples and sets of timed streams by applying them pointwise.

Figure 4 describes the syntactic interface of a component with the input channels $x_1, \ldots, x_n$ of the sorts $M_1, \ldots, M_n$ and the input channels $y_1, \ldots, y_m$ of sorts $M'_1, \ldots, M'_m$. In the theoretical treatment we assume for simplicity always the same sort M.

We represent the behavior of a component with the set of input channels I and the set of output channels O by a set-valued function:

$$F: \vec{I} \to \wp(\vec{O})$$

This function yields the set of output histories F.x for each input history x. Given the input history x a component with the behavior F produces one of the output histories in F.x. We write F.x for F(x) to save brackets.

Only if a set-valued function on streams fulfils certain properties can we accept it as a representation of the behavior of a component. To give a precise definition of these requested properties we introduce a number of notions for set-valued functions on streams. A function

$$F: \vec{I} \rightarrow \wp(\vec{O})$$

is called

- *timed*, if for all i ∈ ℕ we have

  $$x\downarrow i = z\downarrow i \Rightarrow F(x)\downarrow i = F(z)\downarrow i$$

  Then the output in the time interval i only depends on the input received up to the i-th time interval. In the literature F is then called a *causal* function, too.

- *time guarded*, if for all i ∈ ℕ we have

  $$x\downarrow i = z\downarrow i \Rightarrow F(x)\downarrow i+1 = F(z)\downarrow i+1$$

  Time guardedness assumes in addition to timedness that reaction to input is delayed by at least one time unit.

- *realizable*, if there exists a time guarded function[3] f: $\vec{I} \rightarrow \vec{O}$, such that for all input histories x we have:

  $$f.x \in F.x$$

  By ⟦F⟧ we denote the set of time guarded functions f with f.x ∈ F.x for all x.

- *fully realizable* , if for all input histories x we have:

  $$F.x = \{f.x: f \in [\![F]\!]\}$$

We assume in the following that stream processing functions representing the behavior of components are always time guarded and fully realizable. For the set of functions

$$F: \vec{I} \rightarrow \wp(\vec{O})$$

that are time guarded and fully realizable we write

Ç[I, O]

Ç[I, O] denotes the set of all behaviors with the syntactic interface consisting of the input channels I and the output channels O. By Ç we denote the set of all behaviors with arbitrary syntactic interfaces.

4.2 The Logical Specification of System Components

We describe components by set-valued functions on communication histories. Set-valued functions are isomorphic to relations. A set-valued function or a relation on streams can be described by logical means with the help of logical formulas describing the relationship between the input and output streams.

Example: A simple store
We follow always the same scheme to describe a component. We describe the syntactic interface of the component Store by the data flow node given in Fig. 5. A store consists of a set of memory locations. For each memory location a value is

[3] Since functions can be seen as a special case of set-valued functions where the result sets contain exactly one element, the notion of time guardedness extends to functions.

stored. The values can be read and updated. Both reading and updating a value can either succeed or fail.

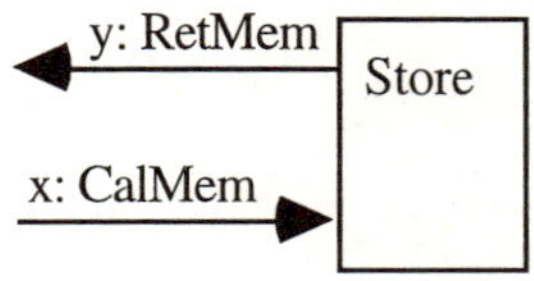

Fig. 5 Data flow node giving the syntactic interface of the store

Let MemLocs be the sort of memory locations and MemVals the sort of values that can be stored in the memory. The data model consists of the sorts of messages that are declared as in Sect.3.4 by Table 1. A straightforward formalization of the behavior reads as follows:

$$
\begin{aligned}
y \in \text{Store}.x \Leftrightarrow \quad & \#\bar{x} = \#\bar{y} \\
& \wedge \forall\, i \in \text{MemLocs},\ d \in \text{MemVals},\ k \in \mathbb{N}: \\
& \qquad \bar{x}.k = p \Rightarrow \bar{y}.k \in \{\text{ret}(p, \text{MemFail}), \text{ret}(p, \text{Ack})\} \\
& \quad \wedge \quad \bar{x}.k = g \Rightarrow \bar{y}.k \in \{\text{ret}(g, \text{MemFail}), \text{ret}(g, v)\} \\
& \textbf{where}\ \ p = \text{put}(i, d) \\
& \qquad\quad\ g = \text{get}(i) \\
& \qquad\quad\ v = \text{last}(\bar{y} : (k\text{-}1), i)
\end{aligned}
$$

Here v denotes the last successfully written value for the memory location i and

$\bar{x}.k$

denotes the k-th element in the stream $\bar{x}$. $\#\bar{x}$ denotes the number of elements in $\bar{x}$. $\bar{x}$:k denotes the prefix of the stream $\bar{x}$ of length k. The function last is specified as follows (let init_val be a given element of sort MemVal):

$$
\begin{aligned}
& \text{last}(\langle\rangle, i) = \text{init\_val}, \\
& \text{last}(z\hat{\ }\langle \text{ret}(\text{get}(j), w)\rangle, i) = \text{last}(z, i) \\
& \text{last}(z\hat{\ }\langle \text{ret}(\text{put}(j, d), w)\rangle, i) = \quad \textbf{if}\ i = j \wedge w = \text{Ack}\ \ \textbf{then}\ d \\
& \qquad\qquad\qquad\qquad\qquad\qquad\qquad\qquad\qquad\qquad \textbf{else}\ \text{last}(z, i) \\
& \qquad\qquad\qquad\qquad\qquad\qquad\quad\ \ \textbf{fi}
\end{aligned}
$$

last(z, i) returns the last value written for location i in the result stream z. The logical specification of components is a very purist description of the behavior of components. A more pragmatic specification of component Store is given in the following section by state transition diagrams.

5. State Transition Systems

In this section we introduce a mathematical concept of a state transition system and techniques for describing it. We suggest state transition diagrams (STDs) for describing of the behavior of components.

5.1 State Transition Models

State machines are a well-known kind of system model. They are used in both practice and theory. We work here with a specific version of state machines that corresponds well to our notion of a system component. By a state machine some internal detail of the system in terms of the states is provided. Therefore we say that state machines provide a *glass box view* of a system component. In the following section we provide another type of a glass box view of components described by the internal structure of a component.

A *communication pattern* for a set of channels C is given by a mapping $p \in C^*$ which assigns a finite sequence of messages to every channel in C. A mathematical state based system model uses state transitions. It includes in addition to the sets of the syntactic interface above a set

State

which denotes the set of states (the state space) of the component. We define a state transition machine by a state transition function

$$\Delta: \text{State} \times \vec{I}^* \to (\text{State} \times \vec{O}^* \to \text{Bool})$$

and a set

$$\text{State}_0 \subseteq \text{State}$$

of initial states.

A state transition machine is nondeterministic, in general. In each transition step it takes a state and a communication pattern of its input streams and produces a successor state and a communication pattern for its output streams. For this kind of state transition machine, we represent the set of possible updated states and outputs of a transition with the help of a predicate. Of course, sets of pairs of states and output patterns can be used here directly. A state machine models the behavior of an information processing unit with input channels from the set I and output channels from the set O in a time frame as follows. Given a family of finite sequences $x \in \vec{I}^*$ representing the sequence of input messages x(c) received in a time interval on the channel $c \in I$ of the component in state $\sigma \in$ State, every pair (σ', y) in the set $\delta(\sigma, x)$ represents a possible successor state and the sequence of output messages y(c) produced on channel $c \in O$ in the next time interval.

We associate a stream processing function with a state machine given by the transition function Δ using the following definition. More precisely, we associate a time guarded function F_σ with every state $\sigma \in$ State as defined by the following equation:

$$\begin{aligned} F_\sigma(x) = \{y \in \vec{O}: & \\ & (\exists\, i \in \vec{I}^*, o \in \vec{O}^*, \sigma' \in \text{State}, x' \in \vec{I}, y' \in \vec{O}: \\ & \qquad \overline{y} = o\hat{\ }\overline{y'} \wedge \overline{x} = i\hat{\ }\overline{x'} \wedge \Delta(\sigma, i).(\sigma', o) \wedge y' \in F_{\sigma'}(x')) \vee \\ & (\forall\, i \in \vec{I}^*: i \sqsubseteq \overline{x} \Rightarrow \forall\, o \in \vec{O}^*, \sigma' \in \text{State}: \neg\Delta(\sigma, i).(\sigma', o))\} \end{aligned}$$

If the state transition relation contains cycles then the definition of F_σ is recursive. In this case, we cannot be sure that by the equation above the behavior F_σ is uniquely specified. Therefore we define F_σ by the largest (in the sense of pointwise set inclusion) time guarded function that fulfils this equation.

The first (existentially quantified) part of the left-hand side of this defining equation handles the case where at least one of the input patterns applies. The second part treats the case where for the input stream x none of the input patterns applies. This case is mapped onto arbitrary output called *chaos*[4]. This definition is justified by the principle that, if nothing is specified for an input pattern, the system may react by arbitrary output. This definition, moreover, guarantees that the function F_σ is always fully realizable.

If we want to associate not a chaotic but a more specific behavior with input situations where no input pattern applies, we can work with default transitions (for instance time ticks) or simply drop the second clause in the definition. Working with chaos, however, has the advantage that adding input patterns for input for which no previous pattern applied is a correct refinement step in the development process.

Our model of the behavior of a component works with timed input and output streams. Since the input and output patterns as introduced above do not refer to the timing of the messages, in the definition we work with time abstractions of the input and output streams. Of course, we may also work with input patterns that refer to time. The explicit time concept in the behavior model of a component allows us also to deal with priorities in transitions.

We even permit transitions with empty input patterns called spontaneous transitions. If we work with our assumption that for a transition on every channel there is at least one time tick, then a spontaneous transition is only enabled if there are no messages but only time ticks on the input channels.

Note that we can give along these lines a precise treatment for sophisticated concepts like priorities and spontaneous transitions due to our carefully chosen semantic model that includes time. Without an explicit notion (at least on the semantic level) of time a proper semantic treatment of priorities or of spontaneous reactions is difficult or even impossible.

5.2 Description of State Transition Systems

We describe state transition systems by state transition diagrams. To demonstrate the use of state transition diagrams (STDs) for the description of the behavior of components we give a first simple example.

Example: A simple store
The state space of the state machine contains only one attribute declared by

$$a: \text{MemLocs} \rightarrow \text{Data}$$

Initially all memory cells have the initial value init_val which is a distinguished element of MemVals

$$\text{Initial: } a(i) = \text{init\_val}$$

In the case of the store we work only with one node in the state transition diagram. Then, we might do better not to use a state transition diagram at all, but work with a set of transitions very similar to TLA (see [Abadi, Lamport 90]). □

We do not give a formal translation of STDs to state transition systems. It is rather technical, but quite straightforward.

---

[4] If no input pattern applies, we assume that a specific behavior is not required.

5.3 Description of the Behavior of System Components

We have shown that state transition systems can be used to describe the behavior of system components. A state transition system can be described by a state transition diagram or equivalently by a state transition table. Such descriptions receive much more acceptance in practice than purely logical representations of state transition systems. However, without much overhead we may describe state transition systems by tables or diagrams and translate these into logic.

$$x{:}put(i, d) \,/\, y{:}ret(put(i, d), Ack)\ \{a'(i) = d \wedge \forall j, j \neq i{:}\ a'(j) = a(j)\}$$
$$x{:}put(i, d) \,/\, y{:}\ ret(put(i, d), MemFail)\ \{a' = a\}$$
$$x{:}get(i) \,/\, y{:}ret(get(i), a(i))\ \{a' = a\}$$
$$x{:}get(i) \,/\, y{:}ret(get(i), MemFail)\ \{a' = a\}$$

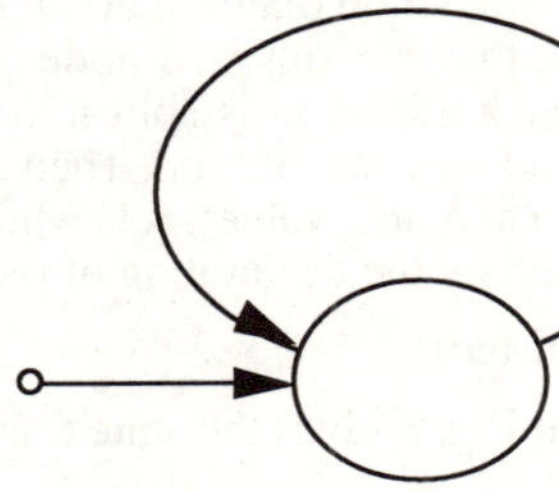

Fig. 6 State transition diagram

The transition rules can be gathered into a table. This is shown for our example of Fig. 6 by Table 2. This way we avoid long formulas in the state transition diagrams.

Tab. 2 Transition table

name	input	output	postcondition
Write	x:put(i, d)	y:ret(put(i, d), Ack)	$a'(i) = d \wedge \forall j, j \neq i : a'(j) = a(j)$
WriteFail	x:put(i, d)	y: ret(put(i, d), MemFail)	$a' = a$
Read	x:get(i)	y:ret(get(i), a(i))	$a' = a$
ReadFail	x:get(i)	y:ret(get(i), MemFail)	$a' = a$

We can use the table in connection with a state transition diagram as given by Fig. 7 which only refers to labels included in the table.

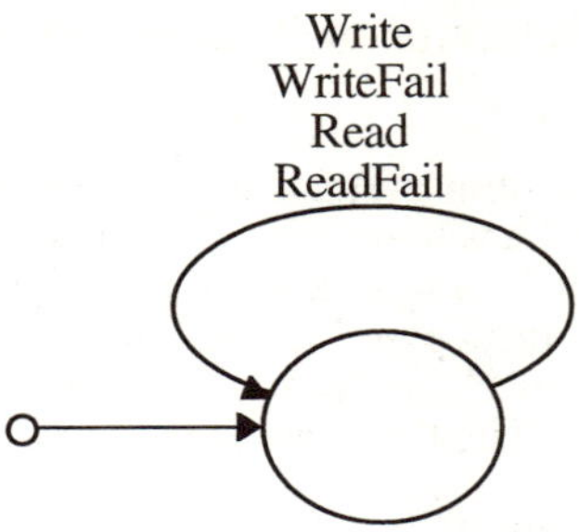

Fig. 7 State transition diagram with abbreviations for transitions

The state transition diagram of Fig. 7 contains only one node. So, as pointed out, it is an overkill to work with a state transition diagram at all. However, we may replace the state transition diagram by one that contains two nodes. We may also work with a state transition diagram where each transition is split into two, one that accepts the input and the next one which produces the output. Then we need a more detailed state space to store the location (and the value) for which reading or writing is required. So the state space is given by the declaration of the attributes

a: MemLocs → Data, c: CalMem

Table 3 lists all the transitions and Fig. 8 gives the state transition diagram.

Tab. 3 Transition table for the decoupled state transition diagram

name	condition	input	output	postcondition
Com		x:e	-	$c' = e \wedge a' = a$
Write	c = put(i, d)	-	y:ret(put(i, d), Ack)	$a'(i) = d \wedge$ $\forall j, j \neq i : a'(j) = a(j)$
WriteFail	c = put(i, d)	-	y: ret(put(i, d), MemFail)	$a' = a$
Read	c = get(i)	-	y:ret(get(i), a(i))	$a' = a$
ReadFail	c = get(i)	-	y:ret(get(i), MemFail)	$a' = a$

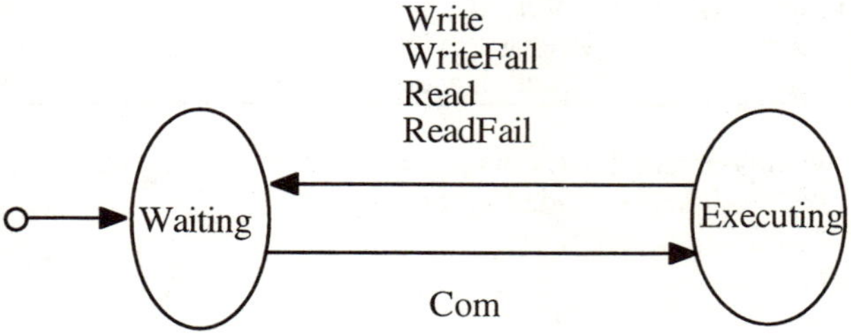

Fig. 8 State transition diagram for the store with transition rules separated into input and output

This refined version of the store has the same behavior as the store described by the diagram above. □

The example above shows a technique for detailing a state transition diagram by splitting its nodes and transition rules. If all nodes have only transition rules where either the input pattern or the output pattern is empty the state transition diagram is called *decoupled*. They correspond closely to the I/Omachines of [Lynch, Stark 89] where each state transition is labeled by exactly one input or output action.

So far, we have shown examples only for the description of components. In the next section we show how to put together components into a network such that they cooperate. Since state transition diagrams describe data flow components, for which a composition by data flow nets is wellunderstood, we can use this concept of composition immediately for state transition descriptions. We demonstrate this by a simple driver component that cooperates with the memory.

Example: Store driver
A driver is a component that controls access to the memory for many processes. It receives calls and forwards them to the memory. It may try again if a memory call fails. However, it may stop trying at any time, even if before it tried at all. We use our general specification scheme again.

The syntactic interface of the component Driver is described by a data flow node as it shown in Fig. 9.

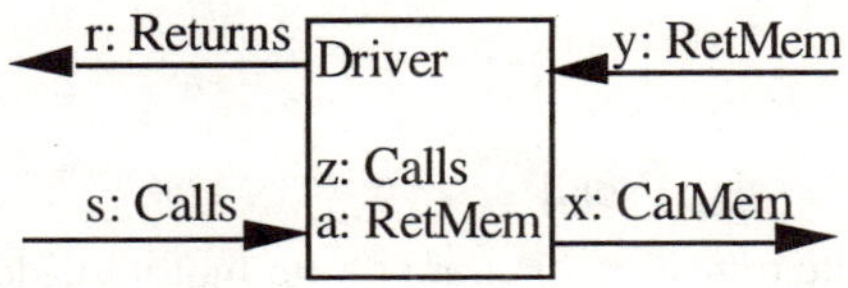

Fig. 9 Data flow node giving the syntactic interface and the state attributes of the driver

The state space of the Driver with its attributes is described as follows:

z: Calls, a: RetMem

The state transition diagram given in Fig. 10 specifies the behavior of the driver.

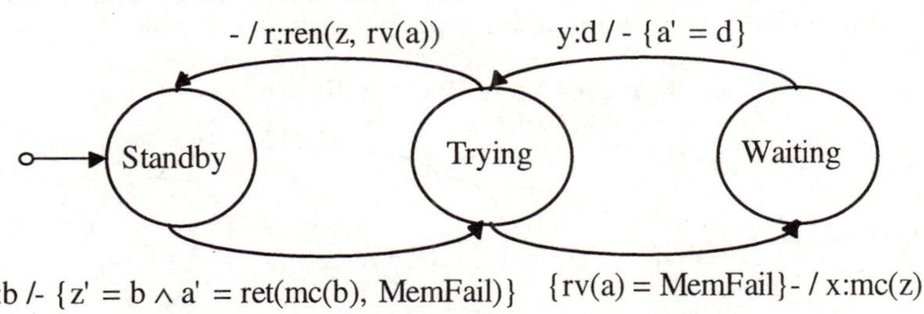

Fig. 10 State transition diagram for the driver

The driver is designed to cooperate with the store. The driver may try forever in the case of memory failures, if we do not assume any fairness assumptions. To get rid of this "unfair" behavior, we may add the liveness condition

$$\# \bar{r} = \# \bar{s}$$

to the component specification that expresses that every input leads to an output under the assumption

$$\# \bar{x} = \# \bar{y}$$

which expresses that all calls transmitted on x get eventually served. It is very helpful that the state transition description can be combined with logical equations to express liveness properties. □

The technique of state transition diagrams gets more flexible by working with sets of such parameterized diagrams.

Example: RPC
Our driver in the example above can handle only one request at a time. To be able to handle more than one request at a time (more precisely one request for each processor) we introduce a set PrIds of processors and define the state space as follows

$$v : PrIds \rightarrow state(z: Calls, a: RetMem)$$

We work with a multithreaded diagram (or more precisely with a family of single threaded diagrams). For all process identifiers $p \in PrIds$ we give one transition diagram.

We use the following straightforward abbreviation. For a record r of sort

$$cons(sel_1: M_1, ..., sel_n: M_n)$$

we use the convention that

with r: E

stands for

$$E[sel_1(r)/sel_1, ..., sel_n(r)/sel_n]$$

Figure 11 shows the state transition diagram of the multithreaded driver. Initially, all default outputs are memory failures:

Initial: rv(a(v(p))) = MemFail

Note that the syntactic interface of the multithreaded driver coincides with that of the single-threaded driver. □

In a multithreaded state transition diagram a set of nodes is active. Each transition moves the control of one of these nodes to another one. Chaotic default transitions are only allowed if for none of the active nodes a transition is enabled.

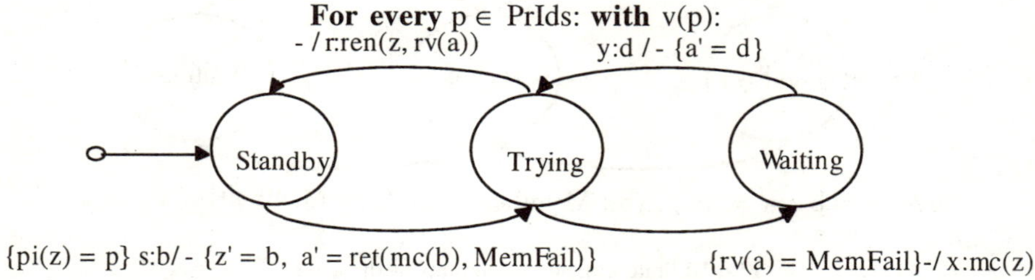

Fig. 11 State transition diagram for the multithreaded driver

Note: If we use a set of nonconnected state transition diagrams that work with disjoint sets of channels we can decompose the data flow node in a set of data flow nodes that work in parallel. Conversely, a data flow diagram where we have a transition diagram for each node is a multithreaded state transition system. □

More research is needed, however, to work out concepts that allow us to encode more powerful logical descriptions into tables and diagrams in a readable way (see [Parnas, Madley 91]).

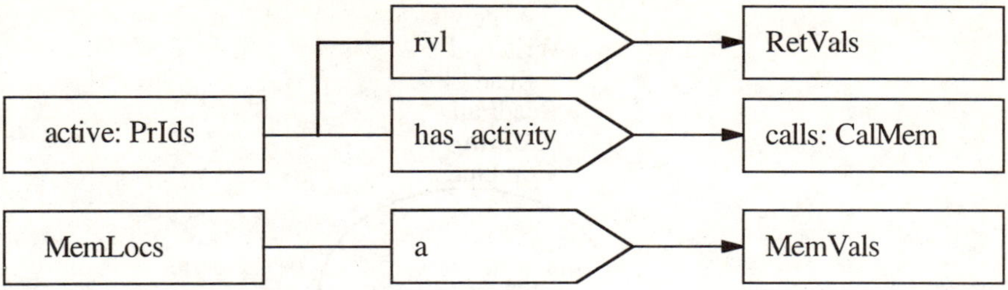

Fig. 12 Data model for the dtate of the multituser memory

For many applications it is appropriate to distinguish between regular and exceptional system activities. What is called regular and what is called exceptional has to be decided by application considerations. Often it is helpful to separate regular cases from exceptional cases when describing system behavior.
For the memory only one reader or writer can be active in the memory at a time. A memory where several readers/writers are active simultaneously is specified with the help of a more complex data model, shown in Fig. 12. Every active member is related to a call by the function has-activity.

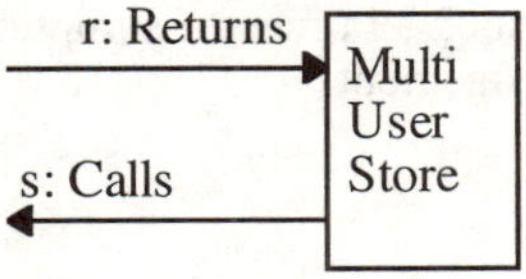

Fig. 13 Data Flow Node Giving the Syntactic Interface of the Store

Now we can describe the behavior of the multi-user memory by the state transition diagram. Figure 13 gives this transition diagram. The transition rules are gathered in Table 4.

Tab. 4 Transition table for Fig. 14

name	precondition	input	output	postcondition
Write	has_activity.p = put(i, d)	-	-	rvl(p) := Ack, a(i) := d
WriteFail	has_activity.p = put(i, d)	-	-	rvl(p) := MemFail
Read	has_activity.p = get(i)	-	-	rvl(p) := a(i)
ReadFail	has_activity.p = get(i)	-	-	rvl(p) := MemFail
ComIn	¬p ∈ active	r:ca(p, c)	-	has_activity(p) := c, rvl(p) := MemFail
ComOut	p ∈ active ∧ has_activity.p = c	-	s:ren(c, val(p))	¬p ∈ active'

We can use the table in connection with a state transition diagram as given by Fig. 6 which only refers to labels included in the table.

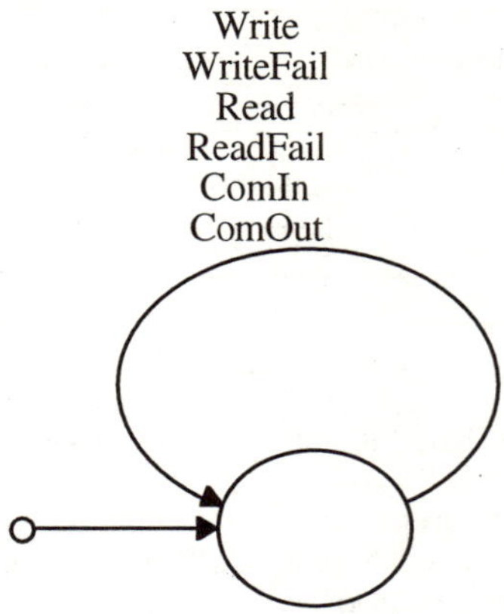

Fig. 14 State transition diagram for the multi-user store with abbreviations for transitions

Again we may add liveness conditions such as fairness requirements that make sure that every call eventually returns if we assume that the store is fair and eventually returns a positive acknowledgement if the driver tries long enough. Another possibility is to work without such a fairness assumption and let the driver retry until it receives a positive acknowledgement.

6. Distributed Systems and Their Description

In this section we introduce a mathematical notion of distributed systems and techniques for their description. Our notion is based on a data flow model of distributed systems.

6.1 Mathematical Models of Distributed Systems

An interactive distributed system consists of a family of interacting components often also called *agents* or *objects*. These components interact by exchanging messages on channels by which they are connected. A structural view onto a distributed system consists of a network of communicating components. Its nodes model components and its arcs communication lines (channels) on which streams of messages are sent. A glass box view of a system component can be represented by a distributed system or by a state machine. In the first case we speak of a *composed distributed system* and in the later case we speak of a *nondistributed system*. Figure 12 shows the data flow representation of a simple example of a distributed system.

We model distributed systems by data flow nets. Let N be a set of identifiers for components (each of which is represented by a data flow node) and O be a set of output channels. A distributed system (ν, O) with syntactic interface (I, O) is described by the mapping

$$\nu: N \rightarrow Ç$$

that associates with every node a component representing a behavior (an interface behavior). As a well-formedness condition we require that for all component identifiers $i, j \in N$ (with $i \neq j$) the sets of output channels of the components $\nu(i)$ and $\nu(j)$ are disjoint:

$$Out(\nu(i)) \cap Out(\nu(j)) = \emptyset$$

In other words there are no name clashes for the output channels, and thus each channel has a unique source. We denote the *set of nodes* of the net by

$$Nodes((\nu, O)) = N$$

We denote the *set of channels* of the net by Chan((ν, 0)), defined as follows:

$$Chan((\nu, O)) = O \cup \{In(\nu(i)): i \in N\} \cup \{Out(\nu(i)): i \in N\}$$

The set I of *input channels* of the net is defined as follows:

$$I = \{In(\nu(i)): i \in N\} \backslash \{Out(\nu(i)): i \in N\}$$

The channels in the set

$$\{Out(\nu(i)): i \in N\} \backslash O$$

are called *internal.* By these definitions, every channel is either internal, an output channel, or an input channel.

According to this definition a distributed system has like all other description techniques a static (syntactic) part and a dynamic (semantic) part. The set of component identifiers, the channels, and the way they connect components form the static part of a distributed system, while the function ν, which assigns behaviors to the components, represents the semantic.

In recent years the theoretical and practical interest in distributed *dynamic* systems has increased. Dynamic systems are also called systems with *mobile communication* or *mobile systems* for short. In a dynamic system the set of components and channels changes over its lifetime. Such systems can also be described by the type of system models introduced above. However, more refined models are needed where either the net is used as a state of the system that changes over time or possibly infinite nets are considered that comprise all components that may be created over the lifetime of the system. In the case of dynamic channel creation similar ideas can be used (see [Grosu 94], [Broy 95a], [Grosu et al. 95]).

6.2 Black Box Views onto Distributed Systems

Given a distributed system in the form of a data flow net with the set I of input channels we get an abstraction of it (which we call its black box view) by mapping the distributed system onto a component behavior F ∈ Ç[I, O].

Every data flow net defines a black box view given by a component behavior F via the following specification:

$$F(x) = \{y|_O: y|_I = x \wedge \forall\, i \in N: y|_{Out(\nu(i))} \in \nu(i)(y|_{In(\nu(i))})\}$$

For a function $g: D \rightarrow R$ and a set $T \subseteq D$ we denote by $g|_T: T \rightarrow R$ the restriction of the function g to the domain T. The formula above is based essentially on the idea that the output of the net is the restriction of a fixpoint[5] for all equations on the streams induced by the channel of the network.

[5] Given that we only consider delayed timed behaviors, it can be shown that in the deterministic case for each input there is always a unique fixpoint. In the nondeterministic case, there may be several fixpoints, of course, each of which corresponds to a computation of the net. Therefore a

6.3 The Description of Distributed Systems

A distributed system, as we defined it in the previous section, can be described graphically by a data flow network (called CD for communication diagram in GRAPES, see [Grapes 90], called block diagram in SDL, see [SDL 88]). A data flow network is a labeled directed graph. This way the structure of distributed systems can be represented.

Data flow nets are illustrative diagrams (see Fig. 12) that give a structural view of information processing systems. Their weak points are that they only describe the static, syntactic properties and do not indicate any semantic properties. They are only helpful as long as the set of components of a system is static. As soon as the sets of components get dynamic, it is impossible to provide a static structural view of systems by data flow nets. Nevertheless, even for object oriented techniques that include the dynamic creation and deletion of objects, data flow models can be useful. We can represent a complete class of objects in object oriented programming by a single data flow node. Each such node represents the set of objects of that class operating in parallel. It is not very helpful, of course, to represent the structure of a class itself by a data flow diagram, in general.

A data flow net graphically represents the components of a system and their communication interconnections by the channels. Thus a data flow net provides a static view on a set of components and their connections. For systems with a more dynamic structure as mentioned above, where the set of components as well as their interconnection by communication links do change over the lifetime of a system, a classical data flow net can only provide a snapshot. However, we can also use (even infinite) data flow models that incorporate all components (or classes of components) that may exist over the lifetime of the system. Then each component goes through three phases. In the beginning it is inactive. Then it gets active (in the slang of object orientation "it gets created"). After a while it may become inactive again.

The exact meaning of data flow diagrams is not always described precisely for the software engineering methods used in practice, although most of these methods advocate versions of data flow diagrams as part of their description techniques. Nevertheless, it is well known that we can formalize the meaning of data flow nets by stream processing functions [Broy 95b].

As mentioned above, the weak side of data flow nets is their limited ability to describe the behavior of systems. They indicate which components exist and may exchange messages, but this does not say much about the causal relationships between the exchanged messages.

Of course, we can annotate data flow diagrams with information about the sorts of messages exchanged on a data link. This provides a description of the syntactic interfaces of the components of a system.

The driver can be combined with the component Store. This composition is asynchronous as it is defined in asynchronous data flow. By combining the driver and the store, we obtain a component that again can be described by a state transition machine. Its state space consists of the product of all the state spaces of the subcomponents. In addition, we need buffers to store the messages sent on internal channels until they are received.

Such buffers can be avoided, of course, if we work with decoupled state transition diagrams defining state machines that do all their transitions with input from internal channels in two steps. In the first step the input is received and in the next step without input the output is produced. By such a refinement of state machines into decoupled state machines the buffers have to become part of the state space.

---

sophisticated logic of least fixpoints is not needed.

Example: Composing the driver and the store into a data flow net
The interface of the driver and the store fit together such that we may compose them as shown in Fig. 15. In cases where the channel identifiers do not match we have to rename them.

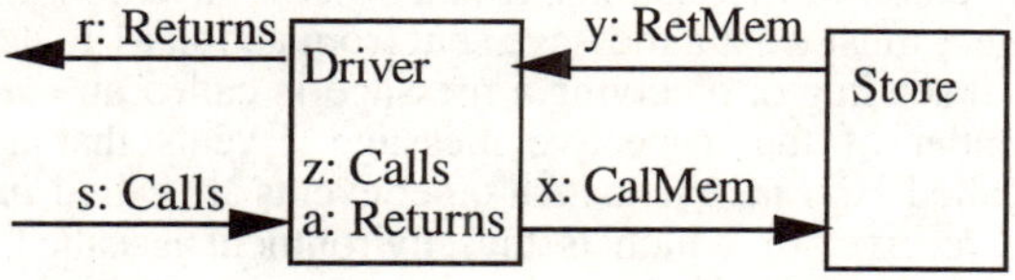

Fig. 15 Composition of the driver and the store

This composition can also be described by a textual syntax. As the diagram suggests, the channels x and y are hidden and not part of the interface of the black box view.

The driver and the second version of the memory component of Sect.5.3 are already in a form that allows us to work without buffers when combining their state spaces into a state space for the state transition description of the composed system. □

Of course, the buffer can only be completely avoided if the decoupling is broken down to the level of single steps and in every node every input can be accepted. But then no additional buffers are needed to store messages on input channels, provided the system runs fast enough to process all inputs immediately.

6.4 State Views of Distributed Systems

If all components of a distributed system are described by state machines we easily may obtain a state view of a distributed system. To keep our presentation simple we assume that for all components their sets of attributes are pairwise distinct, that their state transition description is decoupled, and that in every state every input can be immediately processed. Then we obtain a state view of a distributed system simply by joining the state views of all its components. The state transitions are triggered by the state transitions of the components and the exchange of messages. Independent machines can carry out transitions in parallel.

7. Processes

In this section we introduce the mathematical concept of a process and in addition techniques for describing processes. Processes are a concept describing the internal communication behavior of distributed systems.

7.1 Mathematical Models of Processes

The notion of a process is crucial in software engineering. This can be seen by the term information *processing*. The notion of a process is very general and used in many different ways in computer science. We understand a (discrete) process as a model of the flow of causally related activities in a distributed interactive system. It is represented by a mathematical model consisting of the actions and their causal relationships. A process is a family of actions that are in some causal relationship. In the case of interacting systems, an action is triggered by a number of messages that have been received. When it is carried out, it results in sending a number of messages that may cause further actions. If in a process an action a1 is directly causal for an action a2 there must exist a message sent from action a1 to action a2.

Each instance of sending or receiving a message is called an *event*. Each event is caused by the sender of the respective message. Events that are caused by the environment are called *external* events. All other events are called *internal* events. An internal event the receiver of which is the environment is called an *output* event. Special events are *time* events. They can be understood as messages that are sent by a timer.

Pioneering work on the notion of a process was done by Carl Adam Petri, who suggested so-called Petri nets as a fundamental technique for describing concurrent processes. Strictly speaking Petri nets are a kind of automaton that describe concurrent processes. An explicit description of processes is obtained by so-called *occurrence nets* [Reisig 86] which are Petri nets without cycles and conflicts.

In our model of a distributed system, a process is represented by an acyclic data flow net (ν_p, O_p) and a valuation function

$$\eta\colon \mathrm{Chan}((\nu_p, O_p)) \to M$$

that associates with each of the net exactly one message. Then an action is represented by a behavior relation that relates the messages in its input lines to the messages in its output lines.

As a result of our definition, a process is a special case of a data flow net. In contrast to general data flow diagrams, however, through each of its channels exactly one message is sent. So arcs correspond in process nets to communication events (message transmission) and nodes correspond to action events (receiving and sending of messages).

There are many variations of mathematical models for processes including trace sets and action structures [Broy 91]. We do not introduce a more explicit mathematical model of a process here but consider processes as a special form of distributed systems. The relationship of our notion of process to data flow nets is explained in detail in Sect.9.4.

7.2 The Description of Processes

An individual history of a system behavior (also called a run of a system) can be described by the set of events (exchanged messages) and their causal relationship. It is represented by a process. In software engineering, the concept of a process for illustrating individual cases of interaction is found for instance as a spin off of SDL [SDL 88] called *message sequence charts* [MSC 96] or in objectory [Booch 91] called *use cases*.

A sequence chart defines a trace of communication events for the system and for its components that take part in the interaction. Figure 16 provides an example of a sequence chart. It provides the same information as the process diagram given by Fig. 17.

Processes of interactive systems can be represented by acyclic data flow graphs where each node stands for exactly one action and each arc stands for one event of a message or a signal transfer from one action to another. So each action in the process is represented by exactly one node and each event is represented by exactly one arc.

Graphically, each event can be represented by an arc and by the two involved actions (components) (sender and receiver). Since causality between events is always realized through actions (components), an event e1 can only be directly causal (triggering) for an event e2 of another component if the sender of e2 is the receiver of e1.

This way we get component-based process views. Processes of our kind are obtained as specific subprocesses of the behaviors of structural system views. Of course, in a structural view several processes and tasks may be carried out in an interleaved manner.

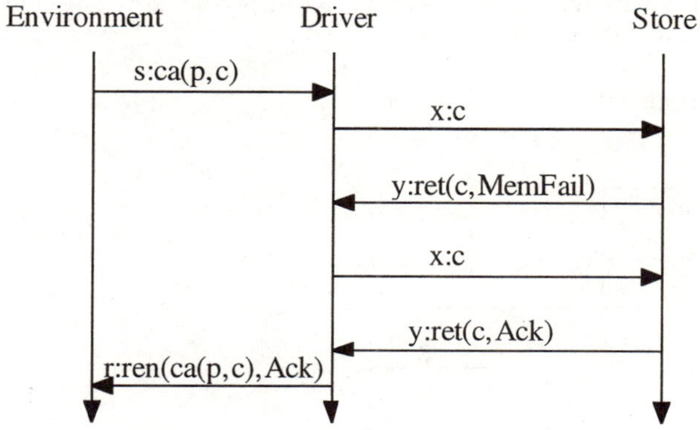

Fig. 16 Sequence chart of a case of interaction between the driver and the store for c = put(...)

In the processes we do not always explicitly include the data flow between the different activities of one component. However, this may be necessary if the process view is to be refined into a structural system view during system development. Similarly, access to a common data base may not be modelled explicitly in a first development step. However, such a modeling is necessary later to obtain a correct representation of the data dependencies between the actions involved. Figure 17 gives a process view of a standard process for our example application.

The behavior of a component connecting a number of input events with a number of output events occurring in a process view can be described again by a data flow network or by a state machine (called PD in [Grapes 90]). A formal definition of the interpretation of processes as runs of systems described by data flow graphs is given in Sect.8.4.

8. Modeling Techniques

Like programming languages for describing algorithms, description formalisms give a solid basis for representing models. There is as much freedom in how to use a description formalism as there is in using a programming language to provide an algorithm for a problem.

For the development of information processing systems we have to develop an adequate model of the information processing task. Since it is difficult or even impossible to describe a complex systems in full detail in one document, we structure the description by providing models and abstractions. An abstraction is a simplified description of the system. A useful abstraction allows us to concentrate on significant aspects of the system, reducing the overall complexity of the modeling task.

A *view* shows us the system under an individual perspective highlighting specific aspects. So a view is always an abstraction. However, since we generally work with a fixed collection of views, each of the views provides a uniform type of abstraction. Certain views are well accepted in software engineering. For instance, data views, which provide a view on the data aspects of an information processing system, are well received and can be found in terms of E/Rmodels in most development methods. Process modeling techniques are used in many methods, too. Recently, they have gained more attention. It is widely recognised now that process views can help significantly in understanding an application problem and properties of solutions, especially in the communication between the software engineer, the application expert, and the client.

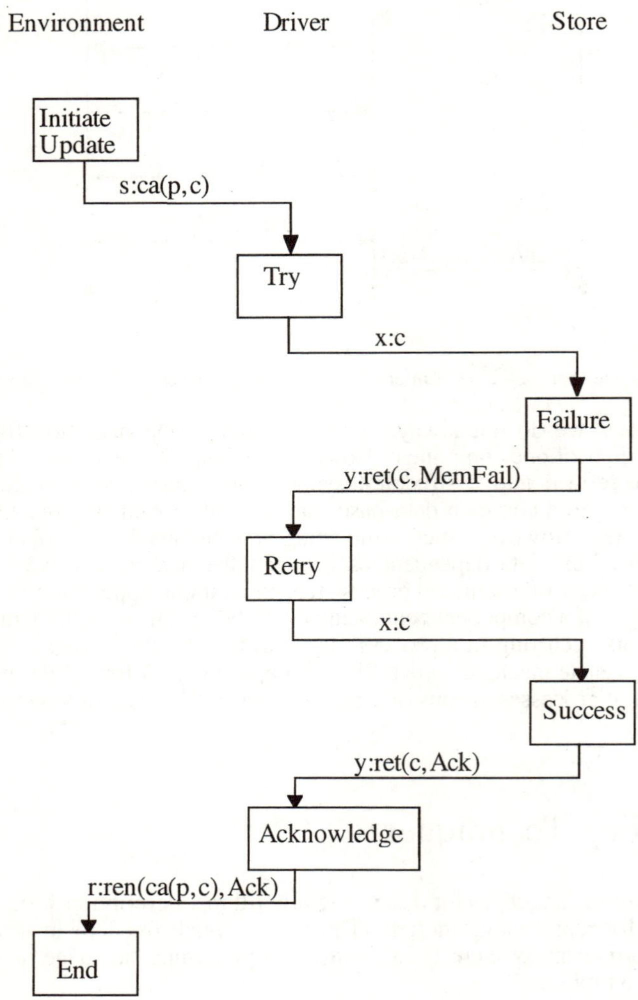

Fig. 17 Process description of a case of interaction for a call c = put(...)

Nevertheless, how to make systematic use of process modeling techniques in software and system development is still not fully understood. A process represents the history of the running of a system in terms of the actions performed by a system and their causal relationship. Since often the processes of a system are too large to show them in full detail the behavior of a system is illustrated by sample subprocesses (called use cases) handling individual cases. There is general agreement that besides data views and structural views this modeling of processes is essential for the development of interactive systems.

Process modeling techniques can be used in many application areas. In telecommunication, for instance, message sequence charts are advocated [MSC 96]. Object oriented techniques provide *use cases*, too. Process modeling is a helpful concept in most application areas of information processing. We believe, too, that a systematic use of process modeling techniques can be helpful, in principle, both in technical and in business applications.

In the early phases of system development often called system analysis or requirement engineering, we work out the following five complementary system views:

- data view (also called information model),
- process view (dynamic view),
- structural view (also called organizational or architectural model),
- behavior view (interface model, behavior history model, black box view),
- state view (state transition model).

Software engineering has provided many different description techniques for all these modeling aspects (see above). What is lacking in many methods is a well-worked out semantic and methodological integration of these views (see below). In particular, it is not always clear what the role of process views is in system and software engineering from a methodological point of view.

In the following, we sketch the role of the data view, the behavior view, and the process view in the course of development of a system. Structuring descriptions is one of the most important goals when we try to keep the description of large systems comprehensible. This is also supported by levels of abstraction. Ideas of how the levels of abstractions are connected by refinement notions are shown in Sect.9 on the development method.

8.1 Building the Data View and the State View

We suggest using data views mainly for modeling static aspects of the data relevant for a system. There are strictly objectoriented approaches that suggest mixing static and dynamic aspects of a system into one E/Rdiagram. They also suggest to modeling each business case by an object. This object includes all the attributes needed to represent the state of the individual business case and all the methods needed to change the business case attributes. This is suggestive as long as such an object is modeled as being passive. It becomes questionable as soon as such an object gets active and calls methods of other objects. This contradicts the principle of object orientation, which has the goal representing reality as "naturally" as possible.

We can also give a data view of the state of a distributed system (ν, O) by an E/Rdiagram that defines a number of components in terms of the set of components N and provides a state space for each of these components by a number of attributes with their sorts.

8.2 Process Modeling

In this section we discuss briefly how to use process modeling techniques in system development. We model actions in a process by a data flow node that formally represents a component. However, a component representing an action has an especially simple structure. On each of its input arcs it receives exactly one message and on each of its output arcs it issues exactly one message.

Note that we do not require a strict behavior where all input messages have to be available before all the output messages are generated. Some output messages may require only a subset of the input messages. In a complete modeling of an action all the information an action needs is provided by its input messages. Since an action is represented formally by a data flow component we can describe it by the same techniques that we use to describe components.

There is a rich body of research on the topic (Petri nets, process algebras like CCS and CSP, event structures, etc.) of how to describe systems by sets of processes. However, most of this work can only be used as a theoretical foundation and does not help much in concrete description tasks. So more work needs to be done to make these theoretical approaches practically useful.

8.3 Structural System Views

In a structural system view the structural decomposition of a system into a family of components is shown. A complete description of a structural system view by a data flow net where for each of its data flow nodes black box behaviors are given contains enough information to simulate system behaviors, provided the behavior of all components is described by means for which an operational interpretation is given. Structural system views are easily provided by data flow diagrams as shown in Fig. 12.

In a system development there may be several structural views, for instance, indicating the logical structure and the physical structure of a system. Building a structural view for a component for which we only have a black box view is a difficult task. This is why the importance of architectures is so much recognized, today. We have to decide upon the set of subcomponents, how they are connected and how they interact. For this process views may help. Finally, we have to give the black box views of the subcomponents.

8.4 Behavior Views of Components

A structural system view on a system provides a description of the overall system architecture. It also provides a description of the system behavior if for each component a behavior is fixed. We distinguish between a black box view and a glass box view of a component. In a black box view the behavior of a component of an interactive system can be represented either by a trace set or by a stream processing function. In a glass box view the behavior can either be represented by a state machine or by an interactive distributed system.

Often, it is advisable, first to take into consideration only the regular aspects of the behavior of components. We call this the behavior in nonexceptional cases. Only in a second step do we consider the exceptional cases. In other words, the behavior is refined to nonexceptional cases in a second step. We distinguish the following classes of exceptional cases:

- external input events that are expected, but do not occur (within certain time bounds),
- external input events that occur unexpectedly,
- external output events are expected that cannot be generated by the system because of lack of information.

The same distinctions can be made for the internal events of distributed systems. In any case, it is advisable for complex components to start with a simplified version and then to refine it to a more sophisticated one.

9. Development Method

A development method gives hints and suggestions as to how to use the description and modeling techniques in a systematic way to develop a computer-based information processing system. In software engineering, the development is carried out by going through several development phases. We speak of a *development process model* when these phases and the documents are fixed.

9.1 Development Process Model

In software engineering the discussion of the development process model plays a central role. This has clear reasons. The development process model governs the whole development process and is therefore decisive for the structure, the economies, and the quality of the development.

Often religious fights are fought in software engineering whether the waterfall model, the spiral model, the experimental or the evolutionary prototyping model, or object oriented development methods provide the best development process models. Not enough practical data are available and not enough theoretical work has been invested to analyze, evaluate, discuss, characterize, relate, and formally specify these different development models.

Mathematical techniques can help to relate the various views and documents in a development process. They help to understand whether a particular proceeding is appropriate and how the interaction between the various documents and views can be formalised and supported by methods and tools.

A more rigorous formalization of the development process is suggestive. Development processes are instances of processes as they are discussed in theoretical computer science. So we can apply all our notions from process theory here, too. In the discussions about development process models, it is often not recognised that we have to distinguish between the overall structure of the development documents and how they are related on one hand and the temporal order in which they are produced on the other. So we can talk about the statics (the logical dependencies between the documents) and the dynamics (the order in which the documents are produced) of the development method and the activities involved.

9.2 Refinement

In software engineering, systems are described by a number of complementary views and on different levels of abstraction. Of course, we need to integrate these views. Furthermore, we need clear mathematical relations between the different levels of abstraction. This is closely related to refinement notions. View integration is dealt with in Sect.9.4. In this section we treat refinement.

Much theoretical work has been devoted towards the refinement notion. After the more pragmatic and informal ideas of stepwise refinement developed by Dahl, Wirth, and many others in the 1970s, much formalization has taken place converging essentially into the following two concepts:

- *property refinement*: In property refinement a system is developed by adding further properties (requirements) and further system parts (enriching the signature). The basic mathematical notion of property refinement is logical implication (with respect to the logical properties) or set inclusion (with respect to the set of models). This allows us also to replace component specifications by logically equivalent component implementations.
- *representation refinement* (a special case is *data refinement*): In representation refinement we change the representation of a data model or of the states and messages of a system model. This can be done with the help of a function relating the two models.

For data models, representation refinement has been studied extensively over the last 25 years in the wake of stimulating papers by Hoare. For models of system components such studies have emerged only more recently.

9.2.1 Data Refinement

For data structure descriptions, property refinement is quite simple. Given a signature (S, F) and a set of algebras Alg, by property refinement we add further requirements to our specification. This way, the set of models Alg is reduced to a subset Alg´ ⊆ Alg.

More sophisticated is the idea of the refinement of the data representation. However, this is also wellunderstood now. To refine the representation of a sort Asort we need a representation sort Rsort and functions

α: Rsort → Asort abstraction function

ρ: Asort → Rsort representation function

that relate the elements of these two sorts. The function ρ provides a representation for each element in Asort. In general, we work with relations ρ or with sets of representation functions to be able to allow many representations for one abstract data element. For our purpose it is sufficient to work with one instance of a representation function.

The function α maps the elements in Rsort that are used as representations onto their abstractions in Asort. The functions α and ρ have to fit together to define a proper idea of abstraction and representation. This is expressed by the following equation:

$$\alpha.\rho.x = x$$

Moreover for all functions on the sort Asort we have to provide representations. For instance for a function

$$f: \text{Asort} \to \text{Asort}$$

we have to find a function

$$f': \text{Rsort} \to \text{Rsort}$$

such that for all elements x in Asort we obtain

$$f.x = \alpha.f'.\rho.x$$

This requirement can be generalized to functions with more complex functionalities.

9.2.2 State Transition System Refinement

Since state spaces are only special cases of data models, the refinement concepts for data models carry over to state spaces immediately. As long as the sorts of the input and output channels are not refined we can simply work with property refinement. If not only the state space is to be refined but also the sorts of the messages and the number of input and output channels we should use the concepts of component refinement as introduced in the following section.

9.2.3 Component Refinement

There are with three fundamental kinds of refinement of the interface view of system components:

- property refinement,
- interface refinement,
- glass box refinement.

By property refinement we do not change the syntactic interface of a component but add properties to its specification. This reduces the set of possible outputs.

In interface refinement we may change the syntactic interface, but insist on a well-specified relationship between the behavior of the refined component and the original one.

In glass box refinement we replace the black box view by a glass box view, which is provided either by a state machine or by a distributed system. The black box behavior associated with the glass box view is required to be a property refinement of the original black box view.

For our concept of a system model, *property refinement* is very simple. A component with interface view

$$\hat{f} : Ç[I, O]$$

is called a *refinement of a component*:

$f: Ç[I, O]$

if for all input streams $x \in \vec{I}$ we have:

$$\hat{f}(x) \subseteq f(x)$$

We write then

$$\hat{f} \subseteq f$$

Of course, this refinement notion can be carried over to state machines. For a systematic use of refinement concepts, *compositionality* (which is the mathematical version of *modularity*) of refinement is essential. Compositionality guarantees that, if we replace a component of a system by its refinement, we obtain a refinement of the overall system.

Property refinement does not change the syntactic interface of a component. An *interface refinement* changes the syntactic interface of a component, but provides a precise interpretation of the behavior of the refined component as behaviors of the given one. We introduce what is called an *upwards simulation* in the literature. We consider two components with different syntactic interfaces:

$$F \in Ç[I, O],$$
$$\hat{F} \in Ç[\hat{I}, \hat{O}].$$

If we want to consider the behavior $\hat{F}$ as an interface refinement of the behavior F we have to interpret all computations of $\hat{F}$ as behaviors of F. Therefore we introduce the concept of a *channel history refinement*. Let Z and $\hat{Z}$ be sets of channels. A pair of functions

$$A \in Ç[\hat{Z}, Z], \qquad R \in Ç[Z, \hat{Z}],$$

is called a *channel history refinement* if

$$R ; A = Id$$

where Id is the identity function and ";" is the usual functional composition (relational product). For an interface refinement we need two channel history refinements:

$$A_I \in Ç[\hat{I}, I], \qquad R_I \in Ç[I, \hat{I}],$$
$$A_O \in Ç[\hat{O}, O], \qquad R_O \in Ç[O, \hat{O}],$$

such that

$$R_I ; \hat{F} ; A_O \subseteq F$$

So for every input of the component F a computation of f is represented by a computation of $\hat{F}$.

We do not have to say much about glass box refinement. In the sections on state transition systems and distributed systems we have specified how to derive a black box view from a state transition description or a description of a distributed system. In glass box refinement we simply use this definition and reverse the direction of development. Starting from a black box description we work out a state transition description or a description of a distributed system whereby the black box view of these is a property refinement of the original black box description.

9.2.4 Refinement of Distributed Systems

A distributed system is refined by refining its components. For each of its component we may use a property refinement. It is shown in [Broy 92] that a refinement of individual conponents of a network is guaranteed to lead to a property refinement of the overall distributed system. We may refine each component either into a state transition machine or into a distributed system. Also this leads into the hierachical refinement of the distributed system.

Of course, we may also refine the communication histories for the internal channels of a distributed system by an interaction refinement. It is shown in [Broy 92] that this can be done in a modular way, too.

9.2.5 Process Refinement

Since, in our approach, processes are only a special case of distributed systems, the concepts of distributed system refinement can be used for process refinement as well. We can apply all three concepts of refinement of data flow nets, such as property refinement for actions, interface refinement and the glass box refinement for the refinement of processes. Glass box refinement allows us to replace an action by a process or to describe an action by a state machine.

A central notion of refinement is decomposition. We can use static or dynamic notions of decomposition of components, states, or processes.

9.3 The Modeling Universe

In the development of a system, we fix the five views mentioned above and work out corresponding models for representing them:

- information (data),
- interface,
- subsystem behavior by state change (state transitions),
- structure (organization, distribution),
- process (business cases, use cases, message sequence charts).

These views have to provide a consistent integrated view of the systems.

For all applications the details to be described by the corresponding modeling notions have to be agreed on and to be worked out during the development process. This means that the five complementary views are refined through several levels of abstaction.

9.4 View Integration: Processes and the Structural System View

Given a complete system description, say by a structural system view (a hierarchical data flow graph), and a state transition description for each atomic component (a data flow node) the overall behavior of a system is fixed. Then we can associate processes with such a system by unraveling the cycles of the data flow graph. This leads, in general, to infinite acyclic data flow graphs corresponding to a set of processes each representing a run of the system. In this unrolling of the data flow nodes the information flow between the actions of one component is made explicit.

A *process* is represented by an acyclic data flow net (ν_p, O_p) and a valuation function

$$\eta: \text{Chan}((\nu_p, O_p)) \to M$$

that associates with each of its arcs exactly one message.

In the view integration we have to relate the process view to the structural view of a distributed system. This means that we have to define when a process is a consistent view of a data flow net. To formalize this motive, we define what it means to call a process a *run* of a data flow net (ν, O) with input history $x \in \vec{I}$. We require that there is a function

$$\kappa: \text{Nodes}((\nu_p, O_p)) \to \text{Nodes}((\nu, O))$$

that associates with every action of the process a data flow component. Intuitively $\kappa(a)$ yields the component that models (carries out) the behavior of the action a.

Every arc in the process corresponds to communiction event. Each of these events occurs for one of the channels of the data flow node. To model this correspondence, we require that there exists a function

$$\chi: \text{Chan}((\nu_p, O_p)) \to \text{Chan}((\nu, O))$$

that associates with every arc in the process a channel in the data flow net such that for every action represented by a node $i \in \text{Nodes}((\nu_p, O_p))$ each channel in its set of input channels is mapped onto an input channel of the node onto which the node i is mapped. An analogous condition is required for the output channel:

$$\{\chi(c): c \in \text{In}(i)\} \subseteq \text{In}(\kappa(i))$$

$$\{\chi(c): c \in \text{Out}(i)\} \subseteq \text{Out}(\kappa(i))$$

Since several channels (representing communication events) in the process are mapped onto the channel in the data flow network we assume that for each channel $c \in \text{Chan}((\nu, O))$ a linear order on the sets of channels (arcs)

$$\{c' : \chi(c') = c\}$$

in the process is defined. This order is used to determine in which order the messages in the process are sent on the channels on the system. Of course, this order has to be consistent with the causality order in the process. This way we can associate a finite stream $\upsilon(c)$ of message s with each channel $c \in \text{Chan}((\nu, O))$ of the data flow net. It consists of the messages in the process associated with that channel.

The process (ν_p, O_p) is called a *run* of the system (ν, O) for input x, if there exists a valuation function

$$y: \text{Chan}((\nu, O)) \to M^{\varpi}$$

such that

$$y|_I = x \wedge \forall\, i \in N: y|_{\text{Out}(\nu(i))} = \nu(i)(y|_{\text{In}(\nu(i))})$$

and

$$\upsilon(c)|_{\text{In}(\nu(i))} \sqsubseteq \overline{y(c)}$$

In words, the process provides communication histories that form a prefix of a computation of the net.

In many cases, processes are not used to model initial segments of system computations. Then a process may consist of all messages of a particular subset M' of the set of messages M. Then we require

$$\upsilon(c)|_{In(v(i))} \sqsubseteq M' \copyright \overline{y(c)}$$

where M'©y denotes the family of streams obtained from y by filtering out all messages in the set M'. This allows us to model transactions by processes.

9.5 The Role of Process Views in the Development

In the development, process views are used as a means for documentation and analysis for the following purposes:

- In the early development phases, instances of processes and process views can help in the requirement capture. They help to illustrate representative use cases. They can be a valuable means of communication between an application expert and the system analysis expert.
- As soon as a complete structural system view is provided, process views can be used to illustrate the system behavior by simulations. One may also think of a model checking tool that checks whether only the required processes are possible behaviors of the structural system view.

We can associate a set of states with every process. Formally, the set of states can be represented by the set of prefixes that, in fact, represent all partial executions (snapshots) of a run of the system. Of course, such a representation of the states of a process is much too abstract and therefore not very helpful. However, by introducing appropriate names and representations for the states we can gain additional views on a process and relate them to the state transition view.

As we have seen, a component can be described by a state machine. The state machine can be systematically derived from the process view. For every component we can construct a local view over the process and represent this by a set of states. It can be understood as the projection of the state view of a process onto a component. Then we can describe the behavior of the component concerning the process by a state machine.

9.6 View Integration by Axiomatic Specifications

The different views developed in systems modeling finally have to be integrated into a consistent system description. View integration can be made on a purely mathematical level by joining together all the mathematical structures given by the various description methods.

Another possibility is to understand every description method as a logical statement about the system. Then all the logical statements provided by the description can be composed into one big axiomatic specification. Then consistency of a description coincides with logical consistency. Here axiomatic specification techniques are extremely helpful. They allow us to translate all views of a system into a set of axioms about a system. This technique is successfully applied in

[Hußmann 95]. There a widely used practical software development method is defined by translating it into axiomatic specifications.

10. Conclusions

Working out a pragmatic, practically usable method for the development of large software systems that is properly founded on a scientific, mathematical basis is one of the challenges for computer scientist and software engineers. Such a method would allow inclusion in a flexible way of formal methods for the specification, refinement, and verification of system parts. It would be the key for a deep tool support that goes beyond the pure preparation, storage, and retrieval of development documents.

We are at an exciting stage in software engineering and the integration of formal methods. A lot of the theoretical work required for the foundations has been done. What is needed is experimental integration and application.

In the SysLab project at the Technical University of Munich we are trying to follow this line and making a significant effort to establish a closer relationship between formal methods and pragmatic software engineering.

Acknowledgement

The thoughts presented above have benefited greatly from discussions within the SysLab team, the IFIP working group 2.3, software engineers from BMW, ESG, Siemens, Siemens Nixdorf, Digital, and many others. I thank Herbert Ehler and Ursula Hinkel for a careful reading and a number of comments.

References

[Abadi, Lamport 90]
M. Abadi, L. Lamport: Composing Specifications. Digital Systems Research Center, SRC Report 66, October 1990

[Abrial 92]
J.R. Abrial: On Constructing Large Software Systems. In: J. van Leeuwen (ed.): Algorithms, Software, Architecture, Information Processing 92, Vol. I, 103-119

[Beeck 95]
M. v. d. Beeck: A Comparison of State Charts Variants. In: H. Langmaack, W.-P. de Roever, J. Vytopil (eds): Formal Techniques in Real Time and Fault-Tolerant Systems. Lecture Notes in Computer Science 863. Berlin: Springer. 1994, 128-148

[Booch 91]
G. Booch: Object Oriented Design with Applications. Redwood City, CA: Benjamin-Cummings. 1991

[Broy 91]
M. Broy: Formalisation of distributed, concurrent, reactive systems. In: E.J. Neuhold, M. Paul (eds.): Formal Description of Programming Concepts. IFIP W.G. 2.2 advanced seminar, Rio de Janeiro 1989. Berlin: Springer. 1991, 319-361

[Broy 92]
M. Broy: Compositional Refinement of Interactive Systems. DIGITAL Systems Research Center, SRC 89, 1992

[Broy 95a]
M. Broy: Equations for Describing Dynamic Nets of Communicating Systems. In: E. Astesiano, G. Reggio, A. Tarlecki (eds): Recent Trends in Data Types Specification, 10th Workshop on Specification of Abstract Data Types joint with the 5th COMPASS Workshop, S.Margherita, Italy, May/June 1994 Lecture Notes in Computer Science 906. Berlin: Springer, 1995

[Broy 95b]
M. Broy: Advanced Component Interface Specification. In: Takayasu Ito, Akinori Yonezawa (Eds.). Theory and Practice of Parallel Programming, International Workshop TPPP'94, Sendai, Japan, November 7-9, 1994, Proceedings, Lecture Notes in Computer Science 907. Berlin: Springer, 1995

[Broy, Lamport 93]
M. Broy, L. Lamport: Specification Problem. http://www.research.digital.com/SRC/personal/Leslie_Lamport/dagstuhl/all.html

[Broy, Merz, Spies 96]
M. Broy, S. Merz, K. Spies (ed.): Formal Systems Specification. The RPC-Memory Specification Case Study. Lecture Notes in Computer Science 1169. Berlin: Springer, 1996

[Grapes 90]
GRAPES-Referenzmanual, DOMINO, Integrierte Verfahrenstechnik. Siemens AG, Bereich Daten-und Informationstechnik 1990

[Grosu 94]
R. Grosu: A formal foundation for concurrent object-oriented programming. Dissertation, Fakultät für Informatik, Technische Universität München, December 94

[Grosu et al. 95]
R. Grosu, K. Stølen, M. Broy: A Denotational Model for Mobile Data Flow Networks. To appear

[Guttag, Horning 93]
J.V. Guttag, J.J. Horning: A Larch Shared Language Handbook. Berlin: Springer, 1993

[Harel 87]
D. Harel: Statecharts: A Visual Formalism for Complex Systems. Science of Computer Programming 8, 1987, 231-274

[Hettler 94]
R. Hettler: Zur Übersetzung von E/R-Schemata nach SPECTRUM. Technischer Bericht TUM-I9409, TU München, 1994

[Hußmann 94]
H. Hußmann: Formal foundation of pragmatic software engineering methods. In: B. Wolfinger(ed.): Innovationen bei Rechen- und Kommunikationssystemen, Informatik aktuell, Berlin: Springer. 1994, 27-34

[Hußmann 95]
H. Hußmann: Formal Foundations for SSADM. Technische Universität München, Fakultät für Informatik, Habilitationsschrift 1995

[Jones 86]
C.B. Jones: Systematic Program Development Using VDM. Prentice-Hall 1986

[Lynch, Stark 89]
N. Lynch, E. Stark: A proof of the Kahn principle for input/output automata. Information and Computation 82, 1989, 81-92

[MSC 96]
Message Sequence Charts (MSC), Recommendation Z.120. Technical report, ITU-T, 1996

[Parnas, Madey 91]
D. L. Parnas, J. Madey: Functional Documentation for Computer Systems Engineering (Version 2). CRL Report 237. McMaster University, Hamilton, Ontario, Canada 1991

[Reisig 86]
W. Reisig: Petrinetze - Eine Einführung. Studienreihe Informatik; 2., überarbeitete Auflage. Berlin: Springer, 1986

[SDL 88]
Specification and Description Language (SDL), Recommendation Z.100. Technical report, CCITT, 1988

[SPECTRUM 93]
M. Broy, C. Facchi, R. Hettler, H. Hußmann, D. Nazareth, F. Regensburger, O. Slotosch, K. Stølen: The Requirement and Design Specification Language SPECTRUM. An Informal Introduction. Version 1.0. Part I/II Technische Universität München, Institut für Informatik, TUM-I9311 / TUM-I9312, May 1993

[Wirsing 90]
M. Wirsing: Algebraic Specification. Handbook of Theoretical Computer Science, Vol. B, Amsterdam: North-Holland 1990, 675-788

Unified Theories of Programming

C.A.R. Hoare
Oxford University Computing Laboratory, Wolfson Building, Parks Road, Oxford, UK

Abstract. Professional practice in a mature engineering discipline is based on relevant scientific theories, usually expressed in the language of mathematics. A mathematical theory of programming aims to provide a similar basis for specification, design and implementation of computer programs. The theory can be presented in a variety of styles, including

1. Denotational, relating a program to a specification of its observable properties and behaviour.
2. Algebraic, providing equations and inequations for comparison, transformation and optimisation of designs and programs.
3. Operational, describing individual steps of a possible mechanical implementation.

This paper presents simple theories of sequential non-deterministic programming in each of these three styles; by deriving each presentation from its predecessor in a cyclic fashion, mutual consistency is assured.

Keywords. Programming, programming methods, non-determinism, semantics, operational, algebraic, denotational

Chapter 1

Introduction

A scientific theory is formalised as a mathematical description of some selected class of processes in the physical world. Observable properties and behaviour of such a process can then be predicted from the theory by mathematical deduction or calculation. An engineer applies the theory in the reverse direction. A specification describes the observable properties and behaviour of some system that does not yet exist in the physical world; and the goal is to design and implement a product which can be predicted by the theory to meet the specification.

This paper proposes a mathematical treatment of computer programming in the simple non-deterministic programming language introduced by Dijkstra [1]. The theory is well suited for use by engineers, since it supports both stepwise development of designs from specifications and hierarchical decomposition of complex systems into simpler components which can be designed and implemented separately. Furthermore, it permits derivation of a complete set of algebraic laws to help in transformation of designs and optimisation of programs. Finally, an operational semantics is derived; this treats practical aspects of implementation and efficiency of execution.

The main goal of this monograph is to show simple ways in which the three presentations of the same language can be derived from each other by mathematical definition, calculation and proof. The denotational theory consists just of a number of separate mathematical definitions of the notations of the language as predicates describing observable aspects of program execution. These definitions can be individually formulated and understood in isolation from each other.

The individual algebraic laws can then be derived from the denotational definitions, without danger of inconsistency or other unexpected interactions. A normal form theorem gives insight into the degree of completeness of the laws, and permits additional laws to be proved without induction. An argument is given for the computability of programs expressed in the language.

An operational theory is equally easily derived from the algebraic. First an algebraic definition is given for the basic step (transition relation) of an abstract implementation; and then the individual transition rules can be proved separately and individually as algebraic theorems, again with reduced risk of complex or unexpected interactions. A concept of bisimulation then lifts the level of abstraction, to permit derivation of a minimal collection of algebraic laws. An alternative abstraction technique derives the observational presentation from the operational, by proving all its definitions as theorems.

This completes the circle: it means that any of the presentations can be accepted as a primary definition of the meaning of the language, and the others can be derived from it. Even better, different definitions can be safely and consistently used at different times and for different purposes. It is a characteristic of the most successful theories, in mathematics as well as in natural science, that they can be presented in several apparently independent ways, which are in a useful sense provably equivalent. For example, topology can be based on a certain kind of family of sets, or on neighbourhoods, or on a closure operator; and a theory of gravity can be based on Newtonian forces acting at a distance, or on field theory, or on Einsteinian geodesics. It is to be hoped that a multifaceted theory of programming can be developed to deliver similar benefits in the practice of software engineering.

The technical material presented in this monograph serves primarily as an illustration or proof of concept. If the concept is found attractive, there is plenty of scope for future work. The most obvious direction is to extend the programming language, for example by including local variables, procedures and parameters. Each extension must be presented in all three semantic styles, with a preferably derivational proof of consistency. An essential goal is to manage each newly introduced complexity by use of as much as possible of the existing theory and algebra: starting again from scratch is forbidden. Ideally, each new feature can be defined and introduced separately, in a way that permits them to be combined without further complexities of interaction. Mature mathematical theories like topology provide an excellent model of the kind of structure needed.

More ambitious extensions may also be of interest: perhaps timing, probability and even a range of paradigms of parallel programming – shared-store, dataflow, and communicating processes. The ultimate goal would be to cover all the interesting paradigms of computing, both declarative and procedural, both in hardware and in software. Again, the challenge is to reveal and profit from their commonality, and so to isolate and control their differences. The work involved is not likely to be easier than fundamental research in any other branch of science; it can be motivated only by the same intense curiosity to find the answer to fundamental questions. This must be supported by the same intense belief in the ultimate simplicity of basic law, and even the elegance of the reasoning on which it is based. The challenge is one that will require the cooperative endeavour of a whole generation of theoretical computing scientists.

The insights described here were obtained by a study of communication and concurrency in parallel processes, where the three semantic styles have been applied individually by independent schools of research to the same class of phenomena. The operational style was used first [2] to define the Calculus of Communicating Systems (CCS); the algebraic style took precedence in the definition [3] of the Algebra of Concurrent Processes (ACP), whereas the denotational style lies at the basis of the mathematical theory [4] of Communicating Sequential Processes (CSP). Many of the detailed differences between these three process theories originate from their different styles of presentation. This monograph attempts a synthesis by presenting the same programming language in all three styles. The choice of a simple sequential language may defuse some of the controversy that has accompanied research into process algebras.

Not a single idea in this paper is original. The concept of denotational semantics is due to Strachey and Scott [5], and the particular choice of ordering of non-deterministic programs is due to Smyth [6]. The embedding of programs as predicates is due to Hehner [7]. The language is essentially the same as that of Dijkstra [1]. The denotational theory is taken from Tarski's calculus of relations [8]. The treatment of recursion in specifications is given by Tarski's fixed point theorem [8] and for programs by Plotkin [9]. The algebraic treatment of the language has already been fully covered in [10]. Even the idea of consistent and complementary definitions of programming languages goes back at least to [11].

The intention of this monograph is to develop these familiar ideas in a smooth progression. In fact, the smoothness is the result of many laborious iterations, mercifully concealed from the reader. Even more, it is due to witting or unwitting inspiration, advice and assistance from many scientists, including E.W. Dijkstra, A.J.R.G. Milner, J.A. Bergstra, A.W. Roscoe, He Jifeng, G.D. Plotkin, J.W. de Bakker, J. Baeten, M. Hennessy, W.P. de Roever, P. Gardiner, S.A. Schneider, C.C. Morgan, A. Sampaio, B. von Karger, G. Lowe, F. Vaandrager, D.S. Scott, Joachim Parrow.

Acknowledgements

The research which is reported in this monograph was largely inspired and partially supported by the European Community under ESPRIT basic research actions CONCUR and PROCOS.

Chapter 2

Denotational Semantics

When a physical system is described by a mathematical formula, the free variables of the formula are understood to denote results of possible measurements of selected parameters of the system. For example, in the description of a mechanical assembly, it may be understood that x denotes the projection of a particular joint along the x-axis, $\dot{x}$ stands for the rate of change of x, and t denotes the time at which the measurement is taken. A particular observation can be described by giving measured values to each of these variables, for example:

$$x = 14\text{mm} \ \wedge \ \dot{x} = 7\text{mm/s} \ \wedge \ t = 1.5\text{sec}.$$

The objective of science is not to construct a list of actual observations of a particular system, but rather to describe all possible observations of all possible systems of a certain class. The required generality is embodied in mathematical equations or inequations, which will be true whenever their free variables are given values obtained by particular measurements of any particular system of that class. For example, the differential equation

$$\dot{x} = 0.5 \times x, \qquad\qquad \text{for } t \leq 3$$

describes the first three seconds of movement of a point whose velocity varies in proportion to its distance along the x axis. The equation is clearly satisfied by the observation given previously, because

$$7 = 0.5 \times 14 \quad \text{and} \quad 1.5 \leq 3.$$

In applying this insight to computer programming, we shall confine attention to programs in a high level language, which operate on a fixed collection of distinct global variables

$$x, y, \ldots z.$$

The values of these variables are observed either before the program starts or after it has terminated. To name the final values of the variables (observed after the program terminates), we place a dash on the names of the variables

$$x', y', \ldots, z'.$$

But to name the initial values of the variables (observed before the program starts), we use the variable names themselves, without decoration. So an observation of a particular run of a program might be described as a conjunction

$$x = 4 \quad \wedge \quad x' = 5 \quad \wedge \quad y' = y = 7.$$

This is just one of the possible observations of a program that adds one to the variable x, and leaves unchanged the values of y and all the other variables; or in familiar symbols, the program consisting of the single assignment

$$x := x + 1.$$

A general formula describing all possible observations of every execution of the above program is

$$x' = x + 1 \quad \wedge \quad y' = y \quad \wedge \ldots \wedge \quad z' = z.$$

Such a formula will henceforth be abbreviated by the programming notation which it exactly describes; for example, the meaning of an assignment is actually explained by the *definition*

$$x := x + 1 \quad =_{df} \quad x' = x + 1 \wedge y' = y \wedge \ldots \wedge z' = z.$$

Similarly, a program which makes no change to anything is written as $\mathbb{I}$ (pronounced "skip") and defined

$$\mathbb{I} \ =_{df} \ x' = x \wedge y' = y \wedge \ldots \wedge z' = z.$$

In words, an observation of the final state of $\mathbb{I}$ is the same as that of its initial state.

These definitions play the role of specifications, giving an observable criterion of correctness for any implementation of the language. Like all definitions, they cannot actually be proved false or even refuted by experiment: however, as we shall see later, they may fail to define the language that we want to implement and use.

Of course, high level programs are more usually (and more usefully) regarded as instructions to a computer, "*given* certain values of $x, y, \ldots, z$, *to find* values of $x', y', \ldots, z'$ that will make the predicate true". But for the purpose of our mathematical theory, there is no need to distinguish between descriptive and imperative uses of the same predicate.

2.1 Correctness and implication

In engineering practice, a project usually begins with a specification, perhaps embodied in a formal or informal contract between a customer and an implementor. A specification too is a predicate, describing the desired (or at least permitted) properties of a product that does not yet exist. For example, the predicate

$$x' > x \quad \wedge \quad y' = y$$

specifies that the value of x is to be increased, and the value of y is to remain the same. No restriction is placed on changes to any other variable. There are many programs that satisfy this specification, including the previously quoted example

$$x := x + 1.$$

Correctness of a program means that every possible observation of any run of the program will yield values which make the specification true; for example, the specification $(x' > x \wedge y' = y)$ is satisfied by the observation $(x = 4 \wedge x' = 5 \wedge y' = y = 7)$. The formal way of defining satisfaction is that the specification is implied by a description of the observation, for example

$$(x = 4 \wedge x' = 5 \wedge y' = y = 7) \Rightarrow (x' > x \wedge y' = y).$$

This implication is true for all values of the observable variables

$$\forall x, y, \ldots, y', z' :: (x = 4 \wedge x' = 5 \wedge y' = y = 7) \Rightarrow (x' > x \wedge y' = y).$$

In future, we will abbreviate such universal quantification by Dijkstra's conventional square brackets, which surround the universally qualified formula thus

$$[(x = 4 \wedge x' = 5 \wedge y = y' = 7) \Rightarrow (x' > x \wedge y' = y)].$$

In fact, the specification is satisfied not just by this single observation but by every possible observation of every possible run of the program $x := x+1$:

$$[(x := x + 1) \Rightarrow x' > x \wedge y' = y].$$

This mixture of programming with mathematical notations may seem unfamiliar; it is justified by the identification of each program with the predicate describing exactly its range of possible behaviours. Both programs and specifications are predicates over the same set of free variables; and that is why the concept of program correctness can be so simply explained as universally quantified logical implication between a program and its specification.

Logical implication is equally interesting as a relation between two programs or between two specifications. If S and T are specifications, $[S \Rightarrow T]$ means that T is a more general or abstract specification than S, and at least as easy to implement. Indeed, by transitivity of implication, any program that correctly implements S will serve as an implementation of T, though not necessarily the other way round. So a logically weaker specification is easier to implement, and the easiest of all is the predicate *true*, which can be implemented by anything.

Similarly, if P and Q are programs, $[P \Rightarrow Q]$ means that P is a more specific or determinate program than Q, and it is (in general) more useful. Indeed, by transitivity of implication, any specification met by Q will be met by P, though not necessarily the other way round. So a logically weaker program is for any given purpose less likely to serve; and the weakest program **true** is the most useless of all.

The initial specification of a complex product is usually separated from its eventual implementation by one or more stages of development. The interface between each stage can in principle be formalised as a design document D. If this is also interpreted as a predicate, the correctness of the design is assured by the implication $[D \Rightarrow S]$ and the correctness of the later implementation P by $[P \Rightarrow D]$. The correctness of P with respect to S (and the validity of the whole idea of stepwise development) follows simply by transitivity of implication:

$$\text{If } [P \Rightarrow D] \text{ and } [D \Rightarrow S] \text{ then } [P \Rightarrow S].$$

When a predicate is used as a specification, there is no reason to restrict the mathematical notations available for its expression. Indeed, any notation with a clear meaning should be pressed into service, because clarity of specification is the only protection we have against misunderstanding of the client's requirements, which can often lead to subsequent disappointment or even rejection of a delivered product.

Particularly important aids to clarity of specification are the simple connectives of Boolean algebra, conjunction (and), disjunction (or), and negation (not). Conjunction is needed to connect individual requirements such as "Temperature must be less than 30° *and* more than 27°". Disjunction is needed to provide useful options for economic implementation: "For mixing, use either the pressure vessel *or* the evaporation tank". And negation is needed for even more important reasons: "It must *not* explode".

The freedom of notation which is appropriate for specification cannot be extended to the programming language in which the ultimate implementation is expressed. Programming notations must be selected to ensure computability, compilability, and reasonable efficiency of execution. In a given programming language, there is a limited collection of combinators available for construction of programs from their primitive components. Typical components include assignments, inputs and outputs; and typical combinations include conditionals, sequential composition, and some form of iteration or recursion. It is for good reason that most programming languages exclude the Boolean combinators and quantifiers of mathematical logic. For example, there is no programming language or compiler that would enable you to protect against disaster by writing a program that *causes* an explosion, and then avoid explosion by just negating the program before execution.

A result of these practical restrictions is that, although we can interpret all programs as predicates, the converse is obviously invalid: not every predicate describes the behaviour of a program. For example, consider the extreme predicate *false*. No observation satisfies this predicate, so the only object that it could correctly describe is one that gives rise to no observation whatsoever.

From a scientific viewpoint, such an object does not exist and could never be constructed. The notations of a programming language must therefore be defined to ensure that they can never express the predicate *false*, or any other wholly unimplementable predicate.

However, we must live with the danger of proposing and accepting an unimplementable predicate for specifications. Indeed, any general notational restriction that ensures computability (or even just satisfiability) could seriously impact clarity and conciseness of specification, and so increase the much greater risk of failure to capture the true requirements and goals of the project. Once these have been successfully formalised, a check on implementability (and on efficiency of implementation) may be made separately with the aid of mathematics or good engineering judgement; and this will be confirmed in the end by successful delivery of an actual product which meets the specification. There is fortunately no danger whatsoever of delivering an implementation of an unimplementable specification.

2.2 The programming language

In this section we shall give a denotational semantics of our simple sequential programming language in terms of predicates describing the behaviour of any program expressed in that language. As explained earlier, the variables $x, y, \ldots, z$ stand for the initial values of the like-named global variables of the program, and $x', y' \ldots, z'$ stand for the final values.

Let $e, f, \ldots, g$ stand for expressions such as $x + 1$, $3 \times y + z, \ldots$ that can feature on the right hand side of an assignment. Clearly, their free variables are confined to the undashed variables of the program; and for simplicity, we assume that all expressions always evaluate successfully to a determinate result. Generalising an example given earlier, we define a simple assignment,

$$(x := e) \quad =_{df} \quad (x' = e \wedge y' = y \wedge \ldots \wedge z' = z).$$

The program which makes no change is just a special case

$$\mathrm{I\!I} \quad =_{df} \quad x := x.$$

A multiple assignment has a list of distinct variables on the left hand side, and a list of the same number of expressions on the right; it is defined

$$(x, y := e, f) \; =_{df} \; (x' = e \wedge y' = f \wedge \ldots \wedge z' = z).$$

A clear consequence of the definition is that an implementation must evaluate all the expressions on the right hand side before assigning any of the resulting values to a variable on the left hand side.

Other consequences can be simply formulated as algebraic laws; they have very simple proofs. For example

$$x := e \quad = \quad x, y := e, y$$
$$x, y := e, f \quad = \quad y, x := f, e.$$

All the definitions and laws extend to lists of more than two variables, for example

$$(z, y := g, f) \quad = \quad (x, y, \ldots, z := x, f, \ldots, g).$$

In fact every assignment may be transformed by such laws to a *total* assignment

$$x, y, \ldots, z \ := \ e, f, \ldots, g$$

where the left hand side is a list of *all* the free variables of the program, in some standard order. In future we will abbreviate this to

$$v := f(v),$$

where v is the vector $(x, y, \ldots, z)$ of program variables, and f is a total function from vectors to vectors. Predicates will be similarly abbreviated

$$P(v, v') \ \text{ instead of } \ P(x, y, \ldots, z, x', y', \ldots, z').$$

Any non-trivial program is composed from its primitive components by the combining notations (combinators) of the programming language. The run-time behaviour of a composite program is obtained by actual execution of its components — all, some, or sometimes even none of them. Consequently, at a more abstract level, a predicate describing this composite behaviour can be defined by an appropriate composition of predicates describing the individual behaviours of the components. So a combinator on programs is defined as a combinator on the corresponding predicates. That is part of what it means for a semantics to be denotational.

The first combinator we consider is the *conditional*. Let b be a program expression, containing only undashed variables and always producing a Boolean result (true or false); and let P and Q be predicates describing two fragments of program. A conditional with these components describes a program which behaves like P if b is initially true, and like Q if b is initially false. It may therefore be defined as a simple truthfunction

$$P \lhd\ b\ \rhd Q \ =_{df} \ (b \wedge P) \vee (\neg b \wedge Q).$$

A more usual notation for a conditional is

$$\textbf{if } b \textbf{ then } P \textbf{ else } Q \ \text{ instead of } \ P \lhd\ b\ \rhd Q.$$

The reason for the change to infix notation is that it simplifies the expression of algebraic laws:

$$\begin{aligned}
P \lhd b \rhd P &= P \\
P \lhd b \rhd Q &= Q \lhd \neg b \rhd P \\
(P \lhd b \rhd Q) \lhd b \rhd R &= P \lhd b \rhd (Q \lhd b \rhd R) \\
&= P \lhd b \rhd R \\
P \lhd b \rhd (Q \lhd c \rhd R) &= (P \lhd b \rhd Q) \lhd c \rhd (P \lhd b \rhd R).
\end{aligned}$$

The first law expresses idempotence, the second gives a form of skew symmetry, the third is an associative law, and the fourth states the distribution of any conditional operator $\lhd b \rhd$ through the conditional $\lhd c \rhd$, for any condition c. All the laws may be proved by propositional calculus; the easiest way is to consider separately the cases when b is true and when it is false. In the first case, replace $P \lhd b \rhd Q$ by P and in the second case by Q. The purpose of the algebraic laws is to help in mathematical reasoning, without such tedious case analyses.

The most characteristic combinator of a sequential programming language is sequential composition, often denoted by semicolon. $(P\,;\,Q)$ may be executed by first executing P and then Q. Its initial state is that of P, and its final state is that of Q. The final state of P is passed on as the initial state of Q; but this is only an intermediate state of $(P\,;\,Q)$, and it cannot be directly observed. All we know is that it exists. The formal definition therefore uses existential quantification to hide the intermediate observation, and to remove the variables which record it from the list of free variables of the predicate.

$$P(v,v')\,;\,Q(v,v') \;=_{df}\; \exists v^0 :: P(v,v^0) \wedge Q(v^0,v').$$

Here, the vector variable v^0 stands for the correspondingly decorated list of bound variables

$$(x^0, y^0, \ldots, z^0).$$

These record the intermediate values of the program variables

$$(x, y, \ldots, z),$$

and so represent the intermediate state as control passes between P and Q. But this operational explanation is far more detailed than necessary. A clever implementation is allowed to achieve the defined effect by more direct means, without ever passing through any of the possible intermediate states. That is the whole purpose of a more abstract definition of the programming language.

In spite of the complexity of its definition, sequential composition obeys some simple, familiar and obvious algebraic laws. For example, it is associative and has $\mathbb{I}$ as its left and right unit. Finally, sequential composition distributes leftward (but not rightward) over the conditional. This asymmetry arises because the condition b is allowed to mention only the initial values of the variables, and not the final (dashed) variables.

$$(P\,;Q)\,;R = P\,;(Q\,;R)$$
$$\mathbb{I}\,;P \;=\; P \;=\; P\,;\mathbb{I}$$
$$(P \lhd b \rhd Q)\,;R \;=\; (P\,;R) \lhd b \rhd (Q\,;R).$$

If e is any expression (only mentioning undashed variables), the assignment

$$x := e$$

changes the value of x so that its final value is the same as the initial value of e, obtained by evaluating e with all its variables taking their initial values. So if $P(x)$ is any predicate mentioning x, P is true of the value of x after the assignment in just the case that P is true of e, i.e.,

$$\begin{aligned} x := e\,;P(x) &= (\exists x^0 : x^0 = e : P(x_0)) \\ &= P(e). \end{aligned}$$

But $P(e)$ is just P with x substituted by e. This substitution effect is defined to generalise to any expression:

$$(x := e\,;f(x)) \;=\; f(e).$$

For example

$$(x := x+1\,;(3 \times x + y < z)) \;=\; (3 \times (x+1) + y < z).$$

Other common notations for substitution are f_e^x, $f[e/x]$ and $f[x/e]$. Substitution permits a rightward distribution law for conditionals:

$$x := e\,;(P \lhd b \rhd Q) \;=\; (x := e\,;P) \lhd x := e\,;b \rhd (x := e\,;Q).$$

Non-determinism is the programming concept that we define next. Let P and Q be predicates describing the behaviour of programs. Their disjunction $(P \vee Q)$ describes the behaviour of a program which may behave like P or like Q, but with no indication which it will be. As an operator of our programming language, disjunction may be easily implemented by arbitrary selection of either of the operands; and the selection may be made at any time, either before or after the program is compiled or even after it starts execution. Disjunction is an extremely simple explanation of the traditionally obscure phenomenon of non-determinism in computing science; and its simplicity provides additional justification for the interpretation and manipulation of programs as predicates.

All the program combinators defined so far distribute through disjunction. This means that separate consideration of each case is adequate for all reasoning about non-determinism. Curiously, disjunction also distributes through itself and through the conditional

$$\begin{array}{lclcl}
P \lhd b \rhd (Q \vee R) & = & (P \lhd b \rhd Q) & \vee & (P \lhd b \rhd R) \\
P\,;(Q \vee R) & = & (P\,;Q) & \vee & (P\,;R) \\
(Q \vee R)\,;P & = & (Q\,;P) & \vee & (R\,;P) \\
P \vee (Q \vee R) & = & (P \vee Q) & \vee & (P \vee R) \\
P \vee (Q \lhd b \rhd R) & = & (P \vee Q) & \lhd b \rhd & (P \vee R).
\end{array}$$

As a consequence of distribution through disjunction, all program combinators also enjoy the property of monotonicity. A function f is said to be *monotonic* if it preserves the relevant ordering, in this case implication. More formally

$$[f.X \Rightarrow f.Y] \text{ whenever } [X \Rightarrow Y].$$

(Here, X and Y are mathematical variables ranging over *predicates* with the same given alphabet, and the line displayed above is true, no matter what predicates take the place of X and Y). All program combinators defined so far are monotonic in all arguments; for example

$$[X\,;Y \Rightarrow X'\,;Y'] \text{ whenever } [X \Rightarrow X'] \text{ and } [Y \Rightarrow Y'].$$

Monotonicity is a very important principle in engineering. Consider an assembly which tolerates a given range of variation in its working environment. Consider also one of its components, which also has a certain tolerance t. The tolerance of the whole assembly can be expressed as some function f of the component tolerance t. The engineer usually assumes that f is a monotonic function, so that if the component is replaced by one with a broader tolerance t', then the tolerance of the whole assembly will in general also be broader, or at worst, the same:

$$[t \leq t' \Rightarrow f(t) \leq f(t')],$$

where $\leq$ is a partial ordering for comparison of breadth of tolerance. Problems arising from violation of monotonicity are in practice the most difficult to diagnose and rectify, because they invalidate the whole theory upon which design of the assembly has been based.

When faced with the task of implementing a complex specification S, it is usual to make an early decision on the general structure of the product, for example as the sequential composition of two program components. To formalise and communicate this decision, each of these components is going to need separate specifications, say D and E. The correctness of these specifications can be checked before implementation by proof of the implication

$$[(D\,;E) \Rightarrow S], \qquad (*)$$

where the sequential composition between specifications has the same definition as between programs considered as predicates. Now what remains are the presumably simpler tasks of finding two programs P and Q which implement the two designs, i.e.,

$$[P \Rightarrow D] \text{ and } [Q \Rightarrow E].$$

Now just deliver the product $(P\,;\,Q)$. By monotonicity of sequential composition

$$[(P\,;\,Q) \Rightarrow (D\,;\,E)],$$

and the fact that

$$[(P\,;\,Q) \Rightarrow S]$$

follows by transivity from a proof of the correctness of the design step $(*)$. What is more, this proof was completed before the start of implementation of P or Q. The technique can be repeated on the components P and Q; and because of monotonicity it extends to all other program combinators. Their monotonicity is essential to the general engineering method of stepwise design decomposition.

But this account of stepwise design requires a simultaneous guess of both the component designs D and E. This is like trying to factorise a given whole number S by simultaneously guessing both the factors D and E, and then checking the guess by multiplication:

$$D \times E = S.$$

If one of the factors E is already known, there is a much more certain way of calculating D by long division

$$D = S \div E,$$

where $\div$ is an approximate inverse of $\times$, in the sense that

$$D \leq S \div E \text{ if and only if } D \times E \leq S.$$

Such an inverse is called a Galois connection, and is very useful when exact inverses are not available.

It turns out that sequential composition also has a Galois connection (which we also write as $\div$). So if E specifies some program already designed or already available in a library, it is possible to calculate $S \div E$ as the weakest specification which satisfies the original intention

$$(S \div E); E \Rightarrow S.$$

Such Galois connections exist for any operator that distributes through disjunction (through empty and infinite disjunctions as well). However, the symbolic calculations required to simplify the predicate $S \div E$ may be heavy; and the result will be unimplementable in the case that there is no way of using E as proposed to implement S. For example

$$(x' \text{ is odd }) \div (x := 2 \times x) \quad = \quad \mathit{false}.$$

All the symbolic calculations described above require that designs are expressed in a mixture of programming notations (for decisions that have already been taken) and more general predicates (for the parts that are specified but still need to be designed). This is yet another advantage of the philosophy of expressing both programs and specifications in the same logical space of predicates.

2.3 Recursion

A final advantage of monotonicity is that it permits a simple treatment of the important programming concept of recursion, and of its important special case, iteration; without this, no program can take significantly longer to execute than to input. Predicates over a given set of observational variables may be regarded as a complete lattice under implication ordering, with universal quantification as meet and existential as join. The bottom of the lattice is the strongest predicate *false* and the top is **true**. Here we will use bold font to distinguish **true** (considered as a program predicate over free variables v, v') from italic *true*, which is a possible value of a Boolean expression b (containing only undashed free variables v).

Moving to a second-order predicate calculus, we introduce a variable X to stand for an arbitrary predicate over the standard set of first-order variables. Fortunately, all the combinators of our programming language are monotonic, and any formula constructed by monotonic functions is monotonic in all its free variables. Let $G.X$ be a predicate constructed solely by monotonic operators and containing X as its only free predicate variable. Tarski's theorem [12] guarantees that the equation

$$X = G.X$$

has a solution for X; and this is called a fixed point of the function G. Indeed, among all the fixed points, there is a weakest one in the implication ordering. This will be denoted by

$$(\mu X :: G.X).$$

It can be implemented as a single non-recursive call of a parameterless procedure with name X and with body $(G.X)$. Occurrences of X within $(G.X)$ are implemented as recursive calls on the same procedure.

The mathematical definition of recursion is given by Tarski's construction:

$$(\mu X :: G.X) =_{df} \bigvee\{X : [X \Rightarrow G.X] : X\}$$

where $\bigvee$ is the lattice join applied to the set of all solutions of $[X \Rightarrow G.X]$. The following laws state that the join is indeed a fixed point of G, and that it is the weakest such:

$$[G.(\mu X :: G.X) \equiv (\mu X :: G.X)]$$
$$[Y \Rightarrow \mu X :: G.X] \text{ whenever } [Y \Rightarrow G.Y].$$

A simple common case of recursion is the iteration or while loop. If b is a condition,

$$\textbf{while } b \textbf{ do } P$$

repeats the program P for as long as b is true before each iteration. More formally, it can be defined as the recursion

$$(\mu X :: (P\,;\,X) \lhd\ b\ \rhd \mathrm{I\!I}).$$

An even simpler example (but hopefully less common) is the infinite recursion which never terminates

$$\mu X.X.$$

This is the weakest solution of the trivial equation

$$X = X;$$

it is therefore the weakest of all predicates, namely **true**. In engineering practice, a non-terminating program is the worst of all programs, and must be carefully avoided by any responsible engineer. That will have to suffice as justification for practical use of a theory which equates any non-terminating program with a totally unpredictable one, which is the weakest in the lattice ordering.

Consider now the program which starts with an infinite loop:

$$(\mu X :: X)\,;\, x, y, \ldots, z := 3, 12, \ldots, 17.$$

In any normal implementation, this would fail to terminate, and so be equal to $(\mu X :: X)$. Unfortunately, our theory gives the unexpected result

$$x' = 3\ \wedge\ y' = 12\ \wedge \ldots \wedge\ z' = 17,$$

the same as if the prior non-terminating program had been omitted. To achieve this result, an implementation would have to execute the program backwards, starting with the assignment, and stopping as soon as the values of the variables are known. While backward execution is not impossible (indeed, it is standard for lazy functional languages), it is certainly not efficient for normal procedural languages. Since we want to allow the conventional forward execution, we are forced to accept the practical consequence that the program

$$(\mu X :: X) ; P$$

will fail to terminate for any program P; and the same is true of

$$P ; (\mu X :: X).$$

Substituting $(\mu X :: X)$ by its value **true** we observe in practice of all programs P that

$$\mathbf{true} ; P = \mathbf{true}$$
$$P ; \mathbf{true} = \mathbf{true}.$$

These laws state that **true** is a zero for sequential composition.

But these laws are certainly not valid for an arbitrary predicate P. As always in science, if a theory makes an incorrect prediction of the behaviour of an actual system, it is the theory that must be adapted; and this usually involves an increase in complication. The violation of plausible simpler laws is what requires and justifies introduction of new concepts and variables, which cannot perhaps be directly observed or controlled, but which are needed to explain what would otherwise be anomalies in more directly observable quantities. All the discoveries of fundamental forces and particles in modern physics have been made in this way. Of course, the old theory is not actually refuted: it would still apply to a lazy implementation. But we are not interested in such an inefficient kind of implementation. It is not the one that we wanted.

2.4 Preconditions and postconditions

In the case of computer programs, the anomaly described in the previous section is resolved by investigating more closely the phenomena of starting and stopping of programs. The collection of free variables describing programs is enlarged to include two new Boolean variables:

ok, which is true of a program that has started in a fully defined state.

ok', which is true of a program that has stopped in a fully defined state.

If ok' is false, the final values of the program variables are unobservable, and the predicate describing the program should make no prediction about these values. Similarly, if ok is false, even the initial values are unobservable. These considerations underlie the validity of the desired zero laws.

The new variables ok and ok' must never appear in the text of any program. However, they *do* appear in the list of existentially quantified variables in the definition of sequential composition, and in the list of universally quantified variables which are abbreviated by the square brackets.

The variables *ok* and *ok'* are useful also in specifications of components of larger programs. The correctness and even the termination of a component with specification Q is often dependent on some assumed properties of the initial values of the variables. This assumption is described by a *precondition* P, which will be true before the program starts. The specification can therefore be written

$$(ok \wedge P) \Rightarrow (ok' \wedge Q),$$

or in words "If the program component starts in a state satisfying P, it stops in a state satisfying Q." For historical reasons, Q is known as a *postcondition*.

The responsibility for ensuring that P is true at the start is thereby delegated to the preceding part of the program. If the assumption is violated, no constraint whatsoever is placed on the behaviour of the program; it may even fail to terminate. Successful teamwork in a large engineering project always depends on appropriately selected assumptions made by the individual designers engaged on a particular sub-task, and the corresponding obligations undertaken by their colleagues. So it is worth while to introduce a special notation

$$(P, Q) =_{df} (ok \wedge P \Rightarrow ok' \wedge Q).$$

This is the primitive notation used by Morgan in [13]. The clear distinction of precondition P from postcondition Q is also an advantageous feature of VDM [14].

In the interpretation of programs as single predicates, the concepts of correctness and refinement are identified with the implication ordering. The same ordering can be defined for pairs of predicates, as shown by the theorem

$$[(P_1, Q_1) \Rightarrow (P_2, Q_2)] \text{ iff } [P_2 \Rightarrow P_1] \text{ and } [P_2 \ \& \ Q_1 \Rightarrow Q_2].$$

Here, (P_1, Q_1) is better because it has a weaker precondition P_1, and so it can be used more widely; furthermore, in all circumstances (P_2) where (P_2, Q_2) can be used, (P_1, Q_1) has a stronger postcondition, so its behaviour can be more readily predicted and controlled.

Equivalence of predicate pairs can be defined by mutual implication:

$$[(P_1, Q_1) \equiv (P_2, Q_2)] \text{ iff } [(P_1, Q_1) \Rightarrow (P_2, Q_2)] \quad \text{and} \quad [(P_1, Q_1) \Rightarrow (P_2, Q_2)].$$

It follows that many equivalent predicate pairs actually denote the same predicate, for example:

$$[(P, Q) \equiv (P, P \wedge Q)] \quad \text{and} \quad [(P, Q) \equiv (P, P \Rightarrow Q)].$$

It is convenient to regard the form $(P, P \Rightarrow Q)$ as a standard form for writing the predicate pair.

The definition of a predicate pair shows how any pair of predicates can be converted to a single predicate by introducing two new variables *ok* and *ok'* into its alphabet. The resulting predicate $R(ok, ok')$ is monotonic in *ok'* and antistrict in *ok*, in the sense that

$$[R(ok, false) \Rightarrow R(ok, true)] \quad \text{and} \quad [R(false, ok')].$$

Conversely, any predicate R with the above properties can be written as a predicate pair (R_0, R'), where

$$R_0 = \neg R(true, false) \quad \text{and} \quad R' = R(true, true).$$

The definition establishes a bijection between predicate pairs and single predicates with the required monotonic and antistrict properties. All the notations of our programming language (including the refinement ordering on programs) can be applied to predicate pairs, as will be shown in the rest of this section. Consequently the bijection is an isomorphism. For all purposes of mathematical calculation, one may use either interpretation of the meaning of programs.

A significant advantage of explicit mention of preconditions and postconditions is a solution of the postponed problem of undefined expressions in assignments. For each expression e of a reasonable programming language, it is possible to calculate a condition $\mathcal{D}e$ which is *true* in just those circumstances in which e can be successfully evaluated. For example

$$\begin{aligned}
\mathcal{D}17 &= \mathcal{D}x = true \\
\mathcal{D}(e+f) &= \mathcal{D}e \wedge \mathcal{D}f \\
\mathcal{D}(e/f) &= \mathcal{D}e \wedge \mathcal{D}f \wedge (f \neq 0).
\end{aligned}$$

Successful execution of an assignment relies on the assumption that the expression will be successfully evaluated, so we formulate a *new* definition of assignment

$$x := e \;=_{df}\; (\mathcal{D}e \;,\; x' = e \;\wedge\; y' = y \;\wedge\; \ldots \wedge z' = z).$$

Expressed in words, this definition states that

either the program never starts ($ok = false$) and nothing can be said about its initial and final values,

or the initial values of the variables are such that evaluation of e fails ($\neg\mathcal{D}e$), and nothing can be said about the final values,

or the program terminates ($ok' = true$), and the value of x' is e, and the final values of all the other variables are the same as their initial values.

The definition of the conditional also needs to be modified to take into account the possibility that evaluation of the condition is undefined

$$P \lhd b \rhd Q = (\mathcal{D}b \Rightarrow (b \land P \lor \neg b \land Q)).$$

However, in future we will maintain the simplifying assumption that all program expressions are everywhere defined.

The normal combinators of the programming language can be defined directly in terms of predicate pairs as shown by the theorems:

$$\begin{array}{lcl} (P_1, Q_1) \lor (P_2, Q_2) & = & (P_1 \land P_2 \, , \, Q_1 \lor Q_2) \\ (P_1, Q_1) \lhd b \rhd (P_2, Q_2) & = & (P_1 \lhd b \rhd P_2 \, , \, Q_1 \lhd b \rhd Q_2) \\ (P_1, Q_1); (P_2, Q_2) & = & (\neg(\neg P_1; \mathit{true}) \ \& \ \neg(Q_1; \neg P_2) \, , \, Q_1; Q_2). \end{array}$$

These theorems show that all the combinators of the programming language map pairs of predicates to pairs of predicates. The functions are all monotonic, and pairs of predicates form a complete lattice under implication ordering:

$$\begin{array}{lcl} \bigvee_i (P_i, Q_i) & = & ((\bigwedge_i P_i), (\bigvee_i Q_i)) \\ \bigwedge_i (P_i, Q_i) & = & ((\bigvee_i P_i), \bigwedge_i (P_i \Rightarrow Q_i)). \end{array}$$

It follows that the least fixed point of such a function is also expressible as a predicate pair: this property of all programs is therefore maintained even in the presence of recursion.

To be more specific, any function of predicate pairs can be analysed as a pair of functions applied to (P, Q), for example

$$(F(P,Q), \quad G(P,Q)).$$

Here, F is monotonic in P and antimonotonic in Q, whereas for G it is the other way round. The least fixed point is calculated over the complete lattice of predicate pairs, and gives the mutually recursive formula

$$\mu(X,Y) :: (F(X,Y), G(X,Y)) = (P(Q), Q))$$

$$\begin{array}{lrcl} \text{where} & P(Y) & = & \nu X :: F(X,Y) \\ & Q & = & \mu Y :: (P(Y) \Rightarrow G(P(Y), Y)). \end{array}$$

Here, νx denotes the strongest fixed point. It may be calculated from the weakest fixed point by the law:

$$(\nu x :: g.x) = (\neg \mu x :: \neg g(\neg x)).$$

The treatment given above applies to arbitrary predicate pairs (P, Q), provided that they do not contain *ok* or *ok'*. In particular, the precondition P is even allowed to mention the dashed final values of the program.

Morgan's refinement calculus has shown that this freedom may be useful in the early stages of specification and design. But programs with such preconditions would be expensive or impossible to execute. In fact, all predicate pairs expressible in the notations of the programming language satisfy the restriction that their preconditions do not mention final values. This permits simplification of the precondition for sequential composition. This fact will be proved and used in the next chapter.

The preference of many researchers is to define the predicate pair as the *meaning* of the program, and use the theorems given above as *definitions* of the combinators of the programming language. This avoids the introduction of the "unobservable" variables *ok* and *ok'*, widely regarded as a distasteful coding trick. However, the coding trick is still useful in revealing the isomorphism between single-predicate and two-predicate definitions of the same language. The programmer may use whichever is most convenient for the immediate purpose, and change freely between them when pursuing different purposes. That is the practical benefit of unifying theories of programming.

For the proof of general algebraic laws which apply equally to specifications as to programs, there is no doubt that the single predicate formulation is the most convenient. However, there are a few important laws which are not valid for general predicates, but only for those that are expressible as predicate pairs. These include the unit laws for composition

$$\mathbb{I};\ (P,Q) = (P,Q) = (P,Q)\ ;\ \mathbb{I}$$

which are valid for the new definition of $\mathbb{I}$ as

$$(true, x' = x \wedge y' = y \wedge \ldots \wedge z' = z).$$

Even more important is the first zero law, which originally motivated introduction of the predicate pairs, which is now trivial

$$\mathbf{true}\ ;\ (P,Q) = \mathbf{true}.$$

The second law zero is

$$P; \mathbf{true} = \mathbf{true}, \qquad \text{for all programs } P.$$

This can be proved easily for assignments; and a simple induction extends the proof to all programs written without recursion. However, proof of the recursive case uses methods developed only in the next chapter.

2.5 Predicate transformers

The definition of our programming language was originally given by Dijkstra in terms of predicate transformers. For any program Q, its weakest precondition $P = wp(Q, R)$ maps a postcondition R describing the final state after its execution onto a predicate P. This is the weakest precondition that describes all initial states in which execution of the program is guaranteed to terminate in a state satisfying the given postcondition. For a *liberal* precondition (*wlp*), termination is not guaranteed; but if it occurs, the postcondition will be satisfied.

Let Q be a predicate describing the behaviour of a program. Its weakest liberal precondition can be defined

$$wlp(Q, R) =_{df} \neg(Q; \neg R).$$

This means that execution of Q cannot end in a state that fails to satisfy R. From this definition, we can derive theorems that characterise the notations of the programming language, for example,

$$\begin{aligned} wlp(Q_1 \vee Q_2, R) &= wlp(Q_1, R) \wedge wlp(Q_2, R) \\ wlp(Q_1 \lhd b \rhd Q_2, R) &= wlp(Q_1, R) \lhd b \rhd wlp(Q_2, R) \\ wlp(Q_1; Q_2, R) &= wlp(Q_1, wlp(Q_2, R)). \end{aligned}$$

The simplicity of the last equation is impressive.

The *wlp* function satisfies the main *healthiness* condition required by Dijkstra, that it distributes through arbitrary universal quantification

$$wlp(Q, \forall i :: R_i) = \forall i :: wlp(Q, R_i).$$

Conversely, let f be any function with this distributive property. Then there is a unique Q such that

$$wlp(Q, R) = f.R, \qquad \text{for all } R.$$

The Q is defined by

$$Q = f.\neg\mathbb{I}.$$

An isomorphism has been established between a language defined in terms of single predicates (containing dashed and undashed variables), and one defined in terms of universally distributive predicate transformers. The original and continuing attraction of predicate transformers is that there is no need to use dashed variables: if the postcondition does not contain them, neither does the precondition. Yet another coding trick is eliminated.

The weakest liberal precondition suffers from the same problem with nontermination as the single-predicate theory of programming described in section 2.3, for example

$$wlp(\mathbf{true}, \mathit{true}) = \mathit{true}$$

To obtain the full strength of Dijkstra weakest precondition in guaranteeing termination, the two-predicate theory is better. Let (P, Q) be a pair of predicates describing the behaviour of a program, as described in section 2.4. This can be converted to a predicate transformer by

$$wp((P,Q),R) =_{df} P \,\&\, wlp(Q,R).$$

Like *wlp*, this transformer is also universally distributive, and satisfies the same laws that characterise the combinators of the programming language. However, non-termination is treated in a more realistic fashion:

$$wp(\mathbf{true}, \mathit{true}) = wp((\mathit{false}, \mathit{true}), \mathit{true}) = \mathit{false}.$$

Chapter 3

The Algebra of Programs.

In this chapter we confine attention to that subset of predicates which are expressible solely in the limited notations of a simple programming language, defined syntactically in table 1. The semantic definitions have been given in the previous section, and provide the basis for proof of a number of algebraic laws. Hitherto, most of these have been valid for arbitrary predicates; but now we can prove additional laws, valid only for predicates which are programs. To emphasize the algebraic orientation, we shall use normal equality between programs in place of the previous universally quantified equivalence

$$P = Q \qquad \text{for} \qquad [P \equiv Q].$$

Such laws are valid for all P and Q ranging over programs. Of course, P and Q themselves are predicates which contain free variables in the appropriate alphabet. Capital letters are used to distinguish these "second order" variables from the lower case variables which they contain as predicates.

Table 1. Syntax.

```
<program> ::= true
   | <variable list> := <expression list>
   | <program> ◁ <Boolean expression> ▷ <program>
   | <program> ; <program>
   | <program> ∨ <program>
   | <recursive identifier>
   | μ <recursive identifier> :: <program>
```

In the form ($\mu X :: P$), X must be the only free recursive identifier in P.

Algebraic laws in the form of equations and inequations have many advantages in practical engineering. As in more traditional forms of calculus, they are useful in calculating parameters and other design details from more general structural decisions made by engineering judgement. There are good prospects of delegating part of the symbolic calculation to a mechanised term rewriting system like OBJ3 [15]. And finally, a theory presented as a set of equations is often easier to teach and to learn than one presented as a mathematical model. Differential calculus is much more widely taught, and more widely used by scientists and engineers, than the foundationary definitions of analysis on which pure mathematicians have shown it to be based.

That is why each of the formal definitions given in the previous section has been followed by a statement of its most important algebraic properties. Proof of these properties is rightly the responsibility of a mathematician; that is the best way of helping engineers, whose skill lies in calculation rather than proof. The goal is to compile a complete collection of laws, so that any other true law that may be needed can be derived by symbolic calculation from the original collection, without ever again expanding the definition of the notations involved.

A valuable aid to the achievement of completeness is the definition of a *normal form*. A normal form uses only a subset of the primitives and combinators of the language, and only in a certain standard order. For example, the conjunctive normal form of Boolean Algebra has conjunction as its outermost combinator, disjunction next, and negation as its innermost operator. The algebraic laws must be sufficient to ensure that every program in the language can be transformed by calculation using just these laws to a program expressed in normal form. There is often a simple test of equality between normal forms; so reduction to normal form generalises this test to arbitrary expressions of the language.

The laws may be classified according to the role that they play in the reduction to normal form.

1. Elimination laws remove operators which are not allowed in the normal form. Such laws contain more occurrences of the operator on one side of the equation than on the other.

2. Distribution laws ensure that the remaining operators can be rearranged to a standard order of nesting.

3. Association and commutation laws are needed to determine equality of normal forms which unavoidably admit variation in their written representation.

For purposes of exposition, we will define a series of normal forms, of increasing complexity and generality, dealing successively with assignment, nondeterminism, non-termination, and recursion.

3.1 Assignment normal form

The first in our series of a normal forms is the total assignment, in which all the variables of the program appear on the left hand side in some standard order:

$$x, y, \ldots, z \;:=\; e, f, \ldots, g.$$

Any non-total assignment can be transformed to a total assignment by vacuous extension of the list, for example:

$$(x, y := e, f) \quad = \quad (x, y, \ldots, z := e, f, \ldots, z).$$

As mentioned before, we abbreviate the entire list of variables $(x, y, \ldots, z)$ by the simple vector variable v, and the entire list of expressions by the vector expressions $g(v)$ or $h(v)$; these will usually be abbreviated to g or h. Thus the normal form will be written

$$v := g \quad \text{or} \quad v := h(v).$$

The law that eliminates sequential composition between normal forms is

$$(v := g\,;\; v := h(v)) \quad = \quad (v := h(g)).$$

The expression $h(g)$ is easily calculated by substituting the expressions in the list g for the corresponding variables in the list v. For example

$$\begin{aligned} &(x\,, y := x+1, y-1 \;\;;\;\; x, y := y, x) \\ &\qquad = (x, y := y-1, x+1). \end{aligned}$$

We now need to assume that our programming language allows conditional expressions on the right hand side of an assignment. Such an expression is defined mathematically

$$\begin{aligned} e \lhd c \rhd f \quad &= \quad e \qquad\qquad \text{if } c \\ &= \quad f \qquad\qquad \text{if } \neg c. \end{aligned}$$

The definition can be extended to lists, for example

$$(e1, e2) \lhd c \rhd (f1, f2) \quad = \quad ((e1 \lhd c \rhd f1), (e2 \lhd c \rhd f2)).$$

Now the elimination law for conditionals is

$$((v := g) \lhd c \rhd (v := h)) \quad = \quad v := (g \lhd c \rhd h).$$

Finally, we need a law that determines when two differently written normal forms are equal. For this, the right hand sides of the two assignments must be equal:

$$(v := g) \quad = \quad (v := h) \quad \text{iff} \quad [g = h].$$

Of course, if g and h are expressions of an undecidable calculus, the algebra of programs will be equally incomplete. This means that a kind of relative completeness has to be accepted as the best that can be achieved in a calculus of programming.

3.2 Non-determinism

Disjunction between two semantically distinct assignments cannot be reduced to a single assignment, which is necessarily deterministic. We therefore move to a more complicated normal form, in which the disjunction operator connects a finite non-empty *set* of total assignments

$$(v := f) \vee (v := g) \ \vee \ldots \vee \ (v := h).$$

Let A and B be such sets; we will write the normal form as $\bigvee A$ and $\bigvee B$. All the previous normal forms can be trivially expressed in the new form as a disjunction over the unit set

$$v := g \quad = \quad \bigvee\{v := g\}.$$

The easiest operator to eliminate is disjunction itself; it just forms the union of the two sets:

$$(\bigvee A) \vee (\bigvee B) \quad = \quad \bigvee(A \cup B).$$

The other operators are eliminated by distribution laws

$$\begin{aligned}
(\bigvee A) \lhd b \rhd (\bigvee B) \quad &= \quad \bigvee\{P, Q : P \in A \wedge Q \in B : (P \lhd b \rhd Q)\} \\
(\bigvee A); (\bigvee B) \quad &= \quad \bigvee\{P, Q : P \in A \wedge Q \in B : (P; Q)\}.
\end{aligned}$$

The right hand sides of these laws are disjunctions of terms formed by applying the relevant operator to total assignments P and Q, which have been selected in all possible ways from A and B. Each of these terms can therefore be reduced to a total assignment, using the laws of 3.1. Thus the occurrences of ; and $\lhd b \rhd$ in the right hand sides of the laws given above are also eliminable.

The laws which permit comparison of disjunctions are

$$\begin{aligned}
&[(\bigvee A) \Rightarrow R] \text{ iff } \forall P : P \in A : [P \Rightarrow R] \\
&[v := f \Rightarrow (v := g \vee \ldots \vee v := h)] \text{ iff } [f \in \{g, \ldots, h\}].
\end{aligned}$$

The first law is a tautology; it enables a disjunction in the antecedent to be split into its component assignments, which are then decided individually by the second law.

3.3 Non-termination

The program constant **true** is not an assignment, and cannot in general be expressed as a finite disjunction of assignments. Its introduction into the language requires a new normal form

$$\mathbf{true} \lhd b \rhd P$$

where P is in the previous normal form. It is more convenient to write this as a disjunction

$$b \vee P.$$

Any unconditional normal form P can be expressed as

$$\mathit{false} \vee P$$

and the constant **true** as

$$\mathit{true} \vee \mathrm{I\!I}.$$

The other operators between the new normal forms can be eliminated by the laws

$$\begin{array}{lll} (b \vee P) \vee (c \vee Q) & = & (b \vee c) \vee (P \vee Q) \\ (b \vee P) \lhd d \rhd (c \vee Q) & = & (b \lhd d \rhd c) \vee (P \lhd d \rhd Q) \\ (b \vee P); (c \vee Q) & = & (b \vee (P; c)) \vee (P; Q). \end{array}$$

The third law relies on the fact that b and c are conditions (not mentioning dashed variables), and P and Q are disjunctions of assignments; from this, it follows that

$$[b; c \Rightarrow b] \text{ and } [b; Q \equiv b]. \qquad (*)$$

We also need a right distribution law

$$x := e; (P \lhd b \rhd Q) = (x := e; P) \lhd x := e; b \rhd (x := e; Q)$$

if P and Q are disjunctions of assignments. The law for testing implication is

$$[(b \vee P) \ \Rightarrow \ (c \vee Q)] \text{ iff } [b \Rightarrow c] \text{ and } [P \Rightarrow c \vee Q].$$

3.4 Recursion

The introduction of recursion into the languages permits construction of a program whose degree of non-determinism depends on the initial state. For example, let n be a non-negative integer variable in

$$\mathbf{while}\ n \text{ is odd } \mathbf{do}\ (n := n \ominus 1 \vee n := n \ominus 2)$$

where $n \ominus k = 0 \lhd k \geq n \rhd n - k$. The effect of this is clearly described by the predicate

$$n' \leq n \ \& \ n' \text{ is even} \ \& \ (n \text{ is even} \Rightarrow n' = n).$$

Informally, this can be expressed as a disjunction of assignments:

$$\begin{array}{rcl} n & := & (n \lhd n \text{ is even } \rhd n-1) \\ \vee\, n & := & (n \lhd n \text{ is even } \rhd n-3) \\ & & \dots \\ \vee\, n & := & (n \lhd n \text{ is even } \rhd 0) \end{array}$$

But there is no *finite* set of assignments whose disjunction can replace the informal ellipses (...) shown above, because the length of the disjunction depends on the initial value of n.

The solution is to represent the behaviour as an *infinite* sequence of expressions

$$S = \{i : i \in \mathcal{N} : S_i\}.$$

Each S_i is a finite normal form, as defined in 3.3; it correctly describes all the possible behaviours of the program, but maybe some impossible ones as well. So we arrange that each S_{i+1} is potentially stronger and therefore a more accurate description than its predecessor S_i:

$$[S_{i+1} \Rightarrow S_i], \qquad \text{for all } i.$$

This is called the descending chain condition. It allows the later members of the sequence to exclude more and more of the impossible behaviours; and in the limit, every impossible behaviour is excluded by some S_i, provided that i is large enough. Thus the exact behaviour of the program is captured by the intersection of the whole sequence, written $(\bigwedge_i S_i)$, or more briefly $(\bigwedge S)$.

For the example shown above, we define the infinite sequence S as follows

$$\begin{array}{rcl} S_0 & = & \textit{true} \\ S_1 & = & n' = n \lhd n \text{ is even } \rhd n \geq 1 \\ S_2 & = & n' = n \lhd n \text{ is even } \rhd (n \geq 3 \vee n' = 0) \\ S_3 & = & n' = n \lhd n \text{ is even } \rhd (n \geq 5 \vee n' \in \{0, 2\}) \\ & \vdots & \\ S_i & = & n' = n \lhd n \text{ is even } \rhd (n \geq 2i - 1 \vee n' < i \ \& \ n' \text{ even}). \end{array}$$

Each S_i is expressible in finite normal form. It describes exactly the behaviour of the program when n is initially less than $2i$, so that the number of iterations is bounded by i. The exact behaviour of the program is described by any S_i with i greater than the initial value of n. It follows that the predicate describing the behaviour of the whole program is equal to the required infinite conjunction $\bigwedge_i S_i$. The laws for recursion given below will provide a general method for calculating the successive approximations S_i describing the behaviour of any particular loop.

The calculation depends critically on the descending chain condition for S, because it is this that permits distribution of all the operators of the language through intersection:

$$\begin{array}{lcll}
(\bigwedge S) \vee P & = & \bigwedge_i (S_i \vee P) & \\
(\bigwedge S) \lhd b \rhd P & = & \bigwedge_i (S_i \lhd b \rhd P) & \\
P \lhd b \rhd (\bigwedge S) & = & \bigwedge_i (P \lhd b \rhd S_i) & \\
(\bigwedge S); P & = & \bigwedge_i (S_i; P) & \\
P; (\bigwedge S) & = & \bigwedge_i (P; S_i) & \text{provided that } P \text{ is in finite normal form.}
\end{array}$$

Operators that distribute through intersections of descending chains are called *continuous*. Every combination of continuous operators is also continuous in *all* its arguments. This permits formulation of a continuity law for recursion:

$$\mu X :: \bigwedge_i S_i.X = \bigwedge_i \mu X :: S_i.X$$

provided that S_i is continuous for all i, and it forms a descending chain for all X, i.e.,

$$[S_{i+1}.X \Rightarrow S_i.X].$$

Another advantage of the descending chain condition is that a descending chain of descending chains can be reduced to a single descending chain by diagonalisation

$$\bigwedge_k (\bigwedge_l S_{kl}) \quad = \quad \bigwedge_i S_{ii},$$

provided that $\forall k, l, i.\ S_{ki} \leq S_{ki+1}$ & $S_{il} \leq S_{i+1l}$. This ensures that a function F, continuous in each of its arguments separately, is also continuous in its arguments taken together

$$F(\bigwedge S, \bigwedge T) = \bigwedge_i F(S_i, T_i).$$

In turn, this gives the required elimination laws for the three operators of the language

$$\begin{array}{lcl}
(\bigwedge S) \vee (\bigwedge T) & = & \bigwedge_i (S_i \vee T_i) \\
(\bigwedge S) \lhd b \rhd (\bigwedge T) & = & \bigwedge_i (S_i \lhd b \rhd T_i) \\
(\bigwedge S); (\bigwedge T) & = & \bigwedge_i (S_i; T_i).
\end{array}$$

The occurrence of the operators on the right hand side of these laws can eliminated by the laws of 3.3, since each S_i and T_i is finite.

The continuity laws ensure that descending chains constitute a valid normal form for all the combinators of the language; and the stage is set for treatment of recursion. Consider first an innermost recursive program (containing no other recursions)

$$\mu X :: F.X,$$

where $F.X$ contains X as its only free recursive identifier. The recursive identifier X is certainly not in normal form, and this makes it impossible to express $F.X$ in normal form. However, all the other components of $F.X$ are expressible in normal form, and all its combinators permit reduction to normal form. So, if X were replaced by a normal form (say **true**), (F.**true**) can be reduced to finite normal form, and so can F.(F.**true**), F.(F.(F.**true**)),... Furthermore, because F is monotonic, this constitutes a descending chain of normal forms. Since F is continuous, by Kleene's famous recursion theorem, the limit of this chain is the least fixed point of F

$$(\mu X :: F.X) \quad = \quad \bigwedge_n F^n.\mathbf{true}.$$

This reduction can be applied first to replace all the innermost recursions in the program by limits of descending chains. The remaining innermost recursions now have the form

$$\mu Y :: H(\bigwedge_m F^m.\mathbf{true}, \bigwedge_m G^m.\mathbf{true}, \ldots, Y).$$

By continuity of H, this transforms to

$$\mu Y :: \bigwedge_m H_m.Y$$

where $H_m.Y = H.(F^m.\mathbf{true}, G^m.\mathbf{true}, \ldots, Y)$, which is (for fixed Y) a descending chain in m. By continuity of μ, this equals

$$\bigwedge_m \mu Y :: H_m.Y.$$

and by Kleene's theorem

$$\bigwedge_m \bigwedge_n (H_m^n.\mathbf{true}).$$

Because this is descending in both n and m, we get

$$\bigwedge_n H_n.\mathbf{true}.$$

Thus the next innermost recursions are converted to normal form; by repeating the process, the whole program can be converted to normal form

$$\bigwedge_n S_n.$$

Another way of describing the same conversion is that S_n is the result of replacing *every* recursion $(\mu X :: F.X)$ in the program by the nth element of its approximating series, i.e, F^n. **true**.

There is no direct algebraic way of testing equality between limits of descending chains. A partial solution to this problem is to relax the descending chain condition, and represent a program as the conjunction of *all* finite normal forms which it implies. For all programs P,

$$P = \bigwedge\{X : X \text{ is a finite normal form and } [P \Rightarrow X] : X\}.$$

This means that if P and Q are different programs, there exists a finite normal form that is implied by one of them and not the other.

The proof of this fact is not trivial, and relies on the continuity of all the operators in the programming language. It also suggests another normal form for programs as intersections of arbitrary sets of finite forms. Each operator F of the language is applied pointwise to the finite forms in their operands

$$F.S = \{P : \ P \in S : \ F.P\}.$$

Recursion is explained as the greatest fixed point in the inclusion ordering for sets of finite forms. The original semantics of the language should be recoverable by applying the $\bigwedge$ operator to all the sets.

But this construction will not work for arbitrary sets of finite forms: they have to satisfy certain additional *closure* properties. These are properties shared by any set generated from a program P by taking all its finite approximations:

$$S = \{p : [P \Rightarrow p] : p\}.$$

1. If $p \in S$ and $p \Rightarrow q$ then $q \in S$.

2. If $T \subseteq S$ and $\bigwedge T$ is expressible in finite form then $\bigwedge T \in S$.

The calculus of programs should be extensible to intersections of arbitrary sets which are closed in this sense.

The finite normal forms play a role similar to that of rational numbers among the reals. Firstly, there is only a countable number of them. A second similarity is that every real is the limit of a descending chain of rationals. Finally, the rationals are dense, in the sense that any two distinct real numbers can be shown to be so by a rational number which separates them. The application of these insights to computer programs is the contribution of Scott's theory of continuous domains.

3.5 Computability

The algebraic laws given in 3.1, 3.2 and 3.3 permit every finite program (one that does not use recursion) to be reduced to finite normal form. The reduction rules are nothing but simple algebraic transformations, of the kind that can be readily mechanised on a computer, and therefore even on a Turing machine. The infinite normal form $(\bigwedge_i S_i)$ of section 3.4 can never be computed in its entirety; however, for each n, the finite normal form S_n can be readily computed; for example by replacing each internal recursion $(\mu X :: F.X)$ by $(F^n.\ \mathbf{true})$.

This suggests a highly impractical method of executing a program, starting in a known initial state s, in which Boolean conditions can be evaluated to *true* or *false*. The machine calculates the series S_n of finite normal forms from the program. Each of these is a disjunction $(b_n \vee P_n)$. If $(s; b_n)$ evaluates to *true*, the machine continues to calculate the next S_{n+1}. If *all* the $(s; b_n)$ are *true*, this machine never terminates; but that is the right answer, because in this case the original program, when started in the given initial state s, contains an infinite recursion or loop. But as soon as a false $(s; b_n)$ is encountered, the corresponding P_n is executed, by selecting and executing an arbitrary one of its constituent assignments. We want to prove that the resulting state will be related to the initial state by the predicate $(\bigwedge_i S_i)$. Unfortunately, this will not be so if the selected assignment is not represented in S_m, for some m greater than n.

The validity of this method of execution depends on an additional property of the normal form, that once n is high enough for b_n to be false, all the assignments P_m remain the same as P_n, for all m greater than n. This can be formalised:

$$[(b_n \vee P_n) \equiv (b_n \vee P_{n+k})], \text{ for all } n, k.$$

Let us therefore define a new ordering relation $\leq$ between normal forms, one that is stronger than the familiar implication ordering. For finite normal forms, this requires that if the condition of the weaker program is false, its effect is exactly the same as that of the stronger program

$$(b \vee P) \leq (c \vee Q) \text{ iff } [b \Rightarrow c] \text{ and } [\neg c \Rightarrow (P \equiv Q)].$$

This is clearly a preorder, with weakest element $(\mathit{true} \vee \mathbf{true})$. What is more, it is respected by all the combinators of the programming language. If $F.X$ is a program, it follows that $\{i :: F^i.\mathbf{true}\}$ is a descending chain in this new ordering. This shows that all innermost recursions enjoy the property that we are trying to prove.

Furthermore, because of monotonicity, any program combinator H preserves this property:

$$H(\bigwedge_i S_i) = \bigwedge_i (H.S_i).$$

For nested recursions, the proof uses the same construction as given at the end of the previous section. All the chains involved are descending in the new ordering as well.

3.6 Completeness

A reduction to normal form gives a method of testing the truth of any proposed implication between any pair of programs: reduce both of them to normal form, and test whether the inequation satisfies the simpler conditions laid down for implication of normal forms. If so, it holds also between the original programs. This is because the reduction laws only substitute equals for equals and each of the tests for implication between normal forms has been proved as a theorem.

For the algebra of programs, the converse conclusion can also be drawn: if the test for implication fails for the normal forms, then the implication does not hold between the original programs. The reason is that the tests give both necessary and sufficient conditions for the validity of implication between normal forms. For this reason, the algebraic laws are said to be *complete*. Of course, since the normal form is infinite, there cannot be any general decision procedure.

Completeness is a significant achievement for a theory of programming. Each of the laws requires a non-trivial proof, involving full expansion of the definitions of all the operators in the formulae, followed by reasoning in the rather weak framework of the predicate calculus. But after a complete set of laws have been proved in this more laborious way, proof of any additional laws can be achieved by purely algebraic reasoning; it will never be necessary again to expand the definitions.

For example, we have to prove the right zero law

$$P\,;\,\mathbf{true} = \mathbf{true}.$$

Since P is a program, it can be reduced to normal form $\bigwedge S$.

$$\begin{aligned}
\bigwedge S;\,\mathbf{true} &= \bigwedge_i (b_i \vee P_i);\mathbf{true}\\
&= \bigwedge_i (b_i \vee (\bigvee_j (v := e_j)\,;\,\mathbf{true}))\\
&= \bigwedge_i b_i \vee (\bigvee_j (v := e_j\,;\,\mathbf{true}))\\
&= \bigwedge_i b_i \vee (\bigvee_j \mathbf{true})\\
&= \mathbf{true}.
\end{aligned}$$

Apart from the practical advantages, completeness of the laws has an important theoretical consequence in characterising the nature of the programming language. For each semantically distinct program there is a normal form with the same meaning, and this can be calculated by application of the laws. It is therefore possible to regard the normal form itself as a definition of the meaning of the program, and to regard the algebraic laws as a definition of the meaning of the programming language, quite independent of the interpretation of programs as predicates describing observations. This is the philosophy of "initial algebra" semantics for abstract data types.

There are many advantages in this purely algebraic approach. Firstly, algebraic reasoning is much easier in practical use than the predicate calculus. It is often quite easy to decide what laws (like the zero laws) are needed or wanted for a programming language; and then it is much easier just to postulate them than to prove them. And there is no need to indulge in curious coding tricks, like the introduction of *ok* and *ok'*. Finally, most algebraic laws are valid for many different programming languages, just as most of conventional schoolroom algebra holds for many different number systems. Even the differences between the systems are most clearly described and understood by examining the relatively simple differences in their algebraic presentations, rather than the widely differing definitions which they are given in the foundations of mathematics.

The real and substantial benefits of algebra are achieved by completely abstracting from the observational meaning of the variables and operators occurring in the formulae. Full advantage should be taken of the benefits of this abstraction, and for as long as possible. But if the algebra is ever going to be applied, either in engineering or in science (or even in mathematics itself), the question arises whether the laws are actually true in the application domain.

To answer this question, it is necessary to give an independent meaning to the variables and operators of the algebra, and then to prove the laws as theorems. It is a matter of personal choice whether the investigation of algebra precedes the search for possible meanings, or the other way round (as in this monograph). The experienced mathematician probably explores both approaches at the same time. When the task is complete, the practising engineer or programmer has a secure intellectual platform for understanding complex phenomena and a full set of calculation methods for the reliable design of useful products. And that is the ultimate, if not the only, goal of the investigations.

Chapter 4

Operational Semantics

The previous chapters have explored mathematical methods of reasoning about specifications and programs and the relationships between them. But the most important feature of a program is that it can be automatically executed by a computing machine, and that the result of the computation will satisfy the specification. It is the purpose of an operational semantics to define the relation between a program and its possible executions by machine. For this we need a concept of execution and a design of machine which are sufficiently realistic to provide guidance for real implementations, but sufficiently abstract for application to the hardware of a variety of real computers. As before, we will try to derive this new kind of semantics in such a way as to guarantee its correctness.

In the most abstract view, a computation consists of a sequence of individual *steps*. Each step takes the machine from one state m to a closely similar one m'; the transition is often denoted $m \rightarrow m'$. Each step is drawn from a very limited repertoire, within the capabilities of a simple machine. A definition of the set of all possible single steps simultaneously defines the machine and all possible execution sequences that it can give rise to in the execution of a program.

The step can be defined as a relation between the machine state before the step and the machine state after. In the case of a stored program computer, the state can be analysed as a pair (s, P), where s is the data part (ascribing actual values to the variables $x, y, \ldots, z$), and P is a representation of the rest of the program that remains to be executed. When this is $\mathbb{I}$, there is no more program to be executed; the state $(t, \mathbb{I})$ is the last state of any execution sequence that contains it, and t defines the final values of the variables.

It is extremely convenient to represent the data part of the state by a total assignment

$$x, y. \ldots, z \ := \ k, l, \ldots, m,$$

where $k, l, \ldots, m$ are *constant* values which the state ascribes to $x, y, \ldots, z$ respectively. If s is an initial data state interpreted as an assignment, and if P is any program interpreted as a predicate, $(s; P)$ is a predicate like P, except that all occurrences of undashed program variables have been replaced by their initial values $(k, l, \ldots, m)$. If this is the identically **true** predicate, execution of P started in s may fail to terminate. Otherwise it is a description of all the possible final data values v' of any execution of P started in s. If t is any other data state,

$$[(t; \mathbb{I}) \Rightarrow (s; P)]$$

means that the final state (t, II) is one of the possible results of starting execution of P in state s. Furthermore,

$$[t; Q \Rightarrow s; P]$$

means that every result (including non-termination) of executing Q starting from data state t is a permitted result of executing P from state s. If this implication holds whenever the machine makes a step from (s, P) to (t, Q), the step will be correct in the sense that it does not increase the set of final states that result from the execution; and if ever a final state (t, II) is reached by a series of such steps, that will be correct too.

In order to derive an operational semantics, let us start by regarding the machine step relation $\rightarrow$ as just another (backwards) way of writing implication:

$$(s, P) \rightarrow (t, Q) \quad \text{instead of} \quad [(t; Q) \Rightarrow (s; P)].$$

The following theorems are now trivial.

1. $\quad (s, v := e) \quad \rightarrow \quad ((v := (s; e)), \text{II}).$

The effect of a total assignment $v := e$ is to end in a final state, in which the variables of the program have constant values $(s; e)$, i.e., the result of evaluating the list of expressions e with all variables in it replaced by their initial values. Here, we return to the simplifying assumption that expressions are everywhere defined.

2. $\quad (s, (\text{II}; Q)) \quad \rightarrow \quad (s, Q).$

A II in front of a program Q is immediately discarded.

3. $\quad (s, (P; R)) \quad \rightarrow \quad (t, (Q; R)), \qquad$ whenever $(s, P) \rightarrow (t, Q)$

The first step of the program $(P; R)$ is the same as the first step of P, with R saved up for execution (by the preceding law) when P has terminated.

4. $\quad (s, P \vee Q) \quad \rightarrow \quad (s, P)$
   $\quad (s, P \vee Q) \quad \rightarrow \quad (s, Q)$

The first step of the program $(P \vee Q)$ is to discard either one of the components P or Q. The criterion for making the choice is completely undetermined.

5. $\quad (s, P \lhd b \rhd Q) \quad \rightarrow \quad (s, P) \qquad$ whenever $s; b$
   $\quad (s, P \lhd b \rhd Q) \quad \rightarrow \quad (s, Q) \qquad$ whenever $s; \neg b$

The first step of the program $(P \lhd b \rhd Q)$ is similarly to select one of P or Q, but the choice is made in accordance with the truth or falsity of $(s; b)$, i.e., the result of evaluating b with all free variables replaced by their initial values.

$$6. \qquad (s, \mu X :: F.X) \quad \rightarrow \quad (s, F.(\mu X :: F.X))$$

Recursion is implemented by the copy rule, whereby each recursive call within the procedure body is replaced by the whole recursive procedure.

$$7. \qquad (s, \mathbf{true}) \quad \rightarrow \quad (s, \mathbf{true}).$$

The worst program **true** engages in an infinite repetition of vacuous steps.

The formulae 1-7 have been presented above as theorems, easily provable from the algebraic laws of the programming language. But it is even easier to present these laws as a *definition* of the way in which the programs are to be executed. In this definition, the components s and P of each state are represented concretely by their texts. The states are identified as pairs (`s`, `P`), where `s` is a *text* describing the data state, and `P` is a program *text*, defining the program state. We use typewriter font to distinguish text from meaning: `P` is the text of a program whose meaning is the predicate P. The symbol $\rightarrow$ is a relation between the texts; it is actually *defined* inductively by the list of laws given above: it is the *smallest* relation satisfying just those laws.

A purely operational presentation of a programming language has many advantages. Firstly, it is surprisingly short and simple; furthermore it closely reflects the programmer's understanding (sometimes called "intuition") of how programs are actually executed. This is especially important in the diagnosis of unexpected behaviour of a program, carelessly written perhaps by someone else. Finally, the clear correspondence to an executing mechanism gives an immediate guarantee against the danger of incomputability, without the long proof of section 3.5. Many researchers into the theory of programming regard an operational semantics as the standard or even the only way of starting research in the subject.

The disadvantages of an operational presentation arise from its extremely low level of abstraction. Mathematical reasoning based on this presentation has to appeal constantly to the completeness clause, i.e., to the absence of any transitions other than those postulated. Small additions or changes in the laws can have unexpectedly gross (or even rather subtle) effects on the theory. What is worse, the assumption that the whole program text is observable means that there can be no non-trivial algebraic equations between programs. Finally, since operational reasoning starts with the text of the program, it is absolutely useless for direct reasoning about specification and design *before* the program is written. In the later sections of this chapter, we will investigate methods of raising the level of abstraction of operational semantics to that of the algebraic or even denotational semantics.

4.1 Total correctness

In the previous section, we used the implication relation $\Rightarrow$ to justify the seven clauses of the operational semantics of the language. The intention was to guarantee the consistency of the operational semantics, in the sense that every final state of an execution satisfies the predicate associated with the program. This is certainly a necessary condition for correctness. But it is not a sufficient condition. There are two kinds of error that it does not guard against:

(i) There may be too few transitions (or even none at all!). An omitted transition would introduce a new and unintended class of terminal state. A more subtle error would be omission of the second of the two laws (4) for $(P \vee Q)$, thereby eliminating non-determinism from the language.

(ii) There may be too many transitions. For example, the transition

$$(s, Q) \rightarrow (s, Q)$$

is entirely consistent with the approach of the previous section, since it just expresses reflexivity of implication. But its inclusion in the operational definition of the language would mean that every execution of every program could result in an infinite iteration of this dumb step.

To guard against these dangers, we need to obtain a clear idea of what it means for an operational semantics to be correct. The purpose of an operational semantics is to define the "machine code" of an abstract machine. One of the best and most common ways of defining a machine code is to write an interpreter for it in a high level programming language, whose meaning is already known; and a language immediately available for this purpose is the one whose observational meaning has been given in Chapter 2. The criterion of total correctness of the interpreter is that its application to a textual representation of the program $\mathtt{P}$ gives rise to exactly the same observations as are described by P, when given its meaning as a predicate.

The alphabet of global variables of our interpreter will be

s to hold the data state (as text)

p to hold the program state (as text).

They will be updated on each step of the interpreter by the assignment

$$s, p := \mathit{next}\,(s, p).$$

The definition of the next function operating on texts is taken directly from the individual clauses of the operational semantics; for example:

$$\mathit{next}\,(s, \mathtt{II}; \mathtt{Q}) = (s, \mathtt{Q}), \quad \text{for all } s.$$

For non-determinism we need two assignments

$$s, p := \mathit{next}1\,(s, p) \vee s, p := \mathit{next}2\,(s, p) \qquad \ldots \text{STEP}$$

where $\mathit{next}1\,(s, \mathtt{P} \vee \mathtt{Q}) = (s, \mathtt{P})$ and $\mathit{next}2\,(s, \mathtt{P} \vee \mathtt{Q}) = (s, \mathtt{Q})$. In all other cases $\mathit{next}1\,(s, \mathtt{P}) = \mathit{next}2\,(s, \mathtt{P}) = \mathit{next}\ (s, \mathtt{P})$. The inner loop of the interpreter repeats this step until the program terminates

$$(p \neq \text{`}\mathbb{I}\text{'}) * \text{ STEP} \qquad \ldots \text{LOOP.}$$

The interpreter defined as LOOP is a predicate describing the relationship between the initial and final values of the program variables s and p. It leaves unchanged the values of the program variables $x, y, \ldots, z$. The predicate P, corresponding to the initial value of p, describes the way that the abstract variables $x, y, \ldots, z$ are updated. The interpreter is correct if the updates correspond with each other in some appropriate sense. The relevant sense is described exactly by the final value of the variable s, produced by the interpreter. This should describe one of the particular final values permitted by the predicate P.

So there are two ways of describing the execution of a program P.

(i) First initialise the values of s and p to $\mathtt{S}$ and $\mathtt{P}$. Then execute the interpreter LOOP. This updates s and p, but not the program variables $x, y, \ldots, z$. If and when the loop terminates, execute the final value of s, which will assign the desired final values to the program variables. The program variables s and p are local to the interpreter, and we are no longer interested in their final values. So they may as well be reinitialised. In summary, the effect of this method of execution is described by the predicate

$$s, p := \mathtt{S}, \mathtt{P}\ ;\ \text{LOOP};\ s\ ;\ s, p := \mathtt{S}, \mathtt{P}$$

where the programming notations have the meaning described in Chapter 2.

(ii) First execute the initial value $\mathtt{S}$, thereby assigning to the program variables the same values as they have in the interpreted execution (i). Then execute the program P, which updates the program variables, hopefully to the same final state as that produced by the interpreter. The variables s and p can then be updated to their same final state as in (i). This execution method is described by the predicate

$$S\ ;\ P\ ;\ s, p := \mathtt{s}, \mathtt{P}.$$

It may be worth a reminder that $\mathtt{S}$ and $\mathtt{P}$ are constant program texts, S and P give their meaning as predicates, whereas s and p are local variables of the interpreter, and their values are changed by the LOOP.

The hope that the two execution methods are the same can be justified by proof of the equation between the two predicates displayed under (i) and (ii). If this can be proved entirely from the algebraic laws, it establishes the desired correspondence between the operational and the algebraic semantics. If the theorem is weakened to an implication, the operational semantics is still correct, but may be more deterministic than the denotational.

Note that both the statement of the theorem and its proof depend utterly on acceptance of the priority of the denotational or algebraic definition as a *specification* of the language. Indeed, unless there is such an independent specification, the question of correctness of an operational semantics cannot arise.

In the absence of an independent meaning for the programming language, the operational semantics may be the only one acceptable or available as a foundation for a theory of programming. In that case, other methods are needed for escaping from its deplorably low level of abstraction. Such methods are discussed in the remaining sections of this chapter. They will be of only secondary interest to readers who recognise that the essential task of proving equivalence of denotational, algebraic and operational semantics has already been completed.

4.2 Bisimulation

As mentioned above, the operational semantics uses the actual text of programs to control the progress of the computation; in principle, two programs are equal only if they are written in exactly the same way, so there cannot be any non-trivial algebraic equations. Instead, we have to define and use some reasonable *equivalence* relation (conventionally denoted $\sim$) between program texts. In fact, it is customary to define this relation between complete machine states, including the data part. Two programs P and Q will then be regarded as equivalent if they are equivalent whenever they are paired with the same data state:

$$\mathtt{P} \sim \mathtt{Q} =_{df} \forall \mathtt{s} :: (\mathtt{s}, \mathtt{P}) \sim (\mathtt{s}, \mathtt{Q}).$$

Now the basic question is: what is meant by a "reasonable" equivalence between states? The weakest equivalence is the universal relation, and the strongest is textual equality; clearly we need something between these two extremes. There are a great many possible answers to these questions; but we shall concentrate on two of the first and most influential of them, which are due to Milner and Park [16].

To exclude the universal relation, it is sufficient to impose an obligation on a proposed equivalence relation $\sim$ that it should preserve the distinctness of a certain minimum of "obviously" distinguishable states. For example, the terminal states of each computation are intended to be recognisable as such, and their data parts are intended to be directly observable. We define the terminal states:

$$(\mathtt{s},\mathtt{P}) \not\rightarrow =_{df} \neg\exists(\mathtt{t},\mathtt{Q}) :: (\mathtt{s},\mathtt{P}) \rightarrow (\mathtt{t},\mathtt{Q}),$$

and require that

(i) if $(\mathtt{s},\mathtt{P}) \sim (\mathtt{t},\mathtt{Q})$ and $(\mathtt{t},\mathtt{Q}) \not\rightarrow$, then $\mathtt{s} = \mathtt{t}$ and $(\mathtt{s},\mathtt{P}) \not\rightarrow$.

The second condition on a reasonable equivalence relation between states is that it should respect the transition rules of the operational semantics; or more formally

(ii) If $(\mathtt{s}, \mathtt{P}) \sim (\mathtt{t}, \mathtt{Q})$ and $(\mathtt{t}, \mathtt{Q}) \rightarrow (\mathtt{v}, \mathtt{S})$ then there is a state $(\mathtt{u}, \mathtt{R})$ such that

$$(\mathtt{s}, \mathtt{P}) \rightarrow (\mathtt{u}, \mathtt{R}) \text{ and } (\mathtt{u}, \mathtt{R}) \sim (\mathtt{v}, \mathtt{S}).$$

(iii) Since $\sim$ is an equivalence relation, the same must hold for the converse of $\sim$.

The condition (ii) may be expanded to a weak commuting diagram:

$$\begin{array}{ccc} (\mathtt{u},\mathtt{R}) & \sim & (\mathtt{v},\mathtt{S}) \\ \uparrow & \supseteq & \uparrow \\ (\mathtt{s},\mathtt{P}) & \sim & (\mathtt{t},\mathtt{Q}) \end{array}$$

Or it may be contracted to a simple inequation in the calculus of relations:

$$(\sim;\rightarrow) \subseteq (\rightarrow;\sim).$$

A relation $\sim$ which satisfies these three reasonable conditions is called a strong *bisimulation*. Two states are defined to be strongly *bisimilar* if there exists *any* strong bisimulation between them. So bisimilarity is a relation defined as the union of *all* bisimulations. Fortunately, distribution of relational composition through such unions means that bisimilarity is itself a bisimulation, in fact the weakest relation satisfying the three reasonable conditions listed above.

An additional most important property of an equivalence relation is that it should be respected by all of the operators of the programming language, for example:

If $\mathtt{P} \sim \mathtt{P}'$ and $\mathtt{Q} \sim \mathtt{Q}'$ then $(\mathtt{P};\mathtt{Q}) \sim (\mathtt{P}';\mathtt{Q}')$.

An equivalence relation with this property is called a *congruence*. It justifies the principle of substitution which underlies all algebraic calculation and reasoning: without it, the algebraic laws would be quite useless. Fortunately, the bisimilarity relation happens to be a congruence for all the operators mentioned in our operational semantics; this claim needs to be checked by mathematical proof.

As an example of the use of bisimulation, we will prove the commutative law for disjunction. The trick is to define a relation $\sim$ which makes the law true, and then prove that it has the properties of a bisimulation. So let us define a reflexive relation $\sim$ that relates every state of the form $(\mathtt{s}, \mathtt{P} \vee \mathtt{Q})$ with itself and with the state $(\mathtt{s}, \mathtt{Q} \vee \mathtt{P})$, and relates every other state only to itself. This is clearly an equivalence relation. It vacuously satisfies bisimilarity condition (i). Further if $(\mathtt{s}, \mathtt{P} \vee \mathtt{Q}) \rightarrow (\mathtt{t}, \mathtt{X})$ then inspection of the two transitions for $\vee$ reveals that $\mathtt{t} = \mathtt{s}$ and $\mathtt{X}$ is either $\mathtt{P}$ or $\mathtt{Q}$. In either case, inspection of the laws shows that $(\mathtt{s}, \mathtt{Q} \vee \mathtt{P}) \rightarrow (\mathtt{s}, \mathtt{X})$. The mere existence of this bisimulation proves the bisimilarity of $(\mathtt{P} \vee \mathtt{Q})$ with $(\mathtt{Q} \vee \mathtt{P})$.

But there is no bisimulation that would enable one to prove the idempotence laws for disjunction. For example, let $\sim$ be a relation such that

$$(\mathtt{s}, \mathbb{I}) \sim (\mathtt{s}, \mathbb{I} \vee \mathbb{I}).$$

Clearly $(\mathtt{s}, \mathbb{I} \vee \mathbb{I}) \rightarrow (\mathtt{s}, \mathbb{I})$. However, the operational semantics deliberately excludes any $(\mathtt{t}, \mathtt{X})$ such that $(\mathtt{s}, \mathbb{I}) \rightarrow (\mathtt{t}, \mathtt{X})$; condition (ii) for bisimulation is therefore violated. In fact, this condition is so strong that it requires any two bisimilar programs to terminate in exactly the same number of steps. Since one of the main motives for exploring equivalence of programs is to replace a program by one that can be executed in fewer steps, strong bisimilarity is far too strong a relation for this purpose.

Milner's solution to this problem is to define a weak form of bisimilarity (which we denote $\approx$) for which the three conditions are weakened to

(i) if $(\mathtt{s}, \mathtt{P}) \approx (\mathtt{t}, \mathtt{Q})$ and $(\mathtt{t}, \mathtt{Q}) \not\rightarrow$

then there is a state $(\mathtt{t}, \mathtt{R})$ such that $(\mathtt{s}, \mathtt{P}) \stackrel{*}{\rightarrow} (\mathtt{t}, \mathtt{R})$ and $(\mathtt{t}, \mathtt{R}) \not\rightarrow$.

(ii) $(\approx; \rightarrow) \subseteq (\stackrel{*}{\rightarrow}; \approx)$,

(iii) $\approx$ is symmetric

where $\stackrel{*}{\rightarrow}$ is the reflexive transitive closure of $\rightarrow$. (This is very similar to the confluence condition, used in the proof of the Church-Rosser property of a set of algebraic transformations). Weak bisimilarity is defined from weak bisimulation in the same way as for strong bisimilarity; and a similar check has to be made for the congruence property.

Now we can prove idempotence of disjunction. Let $\approx$ relate every $(\mathtt{s}, \mathtt{P})$ just to itself and to $(\mathtt{s}, \mathtt{P} \vee \mathtt{P})$ and vice versa. In the operational semantics $(\mathtt{s}, \mathtt{P} \vee \mathtt{P})$ is related by $\rightarrow$ only to $(\mathtt{s}, \mathtt{P})$; fortunately $(\mathtt{s}, \mathtt{P}) \stackrel{*}{\rightarrow} (\mathtt{s}, \mathtt{P})$ since $\stackrel{*}{\rightarrow}$ is reflexive. Conversely, whenever $(\mathtt{s}, \mathtt{P}) \rightarrow (\mathtt{s}', \mathtt{P}')$ then $(\mathtt{s}, \mathtt{P} \vee \mathtt{P}) \rightarrow (\mathtt{s}, \mathtt{P}) \rightarrow (\mathtt{s}', \mathtt{P}')$, so equality is restored after two steps. This therefore is the weak bisimulation that shows the bisimilarity

$$(P \vee P) \approx P.$$

Unfortunately, we still cannot prove the associative law for disjunction. The three simple states $(s, x := 1), (s, x := 2)$ and $(s, x := 3)$ end in three distinct final states; and by condition (i), none of them is bisimilar to any other. The state $(s, (x := 1 \vee x := 2))$ and $(s, (x := 2 \vee x := 3))$ are also distinct, because each of them has a transition to a state (i.e., $(s, x := 1)$ and $(s, x := 3)$ respectively) which cannot be reached in any number of steps by the other. For the same reason $(s, (x := 1 \vee x := 2) \vee x := 3)$ is necessarily distinct from $(s, x := 1 \vee (x := 2 \vee x := 3))$. The associative law for disjunction is thereby violated.

In fact, there is a perfectly reasonable sense in which it is possible to observe the operational distinctness between the two sides of an associative equation

$$(P \vee Q) \vee R = P \vee (Q \vee R).$$

Just take *two* copies of the result of executing the first step. In the case of the left hand side, this will give either of the two pairs.

$$(R, R) \text{ and } ((P \vee Q), (P \vee Q)).$$

Now each copy is run independently. There are now five possibilities for the subsequent behaviours of the pair:

$$(R, R), (P, P), (P, Q), (Q, P) \text{ and } (Q, Q).$$

But if the same procedure is applied to the right hand side of the associative equation, the five possibilities are

$$(P, P), (Q, Q), (Q, R), (R, Q), \text{ and } (R, R).$$

The third and fourth possibilities in each case are sufficient to distinguish the two different bracketings. If copying the state is allowed after any step of the operational semantics, it is possible to make a pair of observations of the two copied programs that may tell which way the original program was bracketed. Such an observation would invalidate the associative law.

So even weak bisimilarity is not weak enough to give one of the laws that we quite reasonably require. Unfortunately, it is also too weak for our purposes: it gives rise to algebraic laws that we definitely do *not* want. For example, consider the program

$$\mu X :: (\mathrm{II} \vee X).$$

This could lead to an infinite computation (if the second disjunct X is always selected); or it could terminate (if the first disjunct II is ever selected, even only once). Weak bisimilarity ignores the non-terminating case, and equates the program to $\mathbb{I}$. However, in our theory it is equated to **true**, the weakest fixed point of the equation

$$\mathtt{X} = (\mathbb{I} \vee \mathtt{X}).$$

Our weaker interpretation **true** permits a wider range of implementations: for example, the "wrong" choice may be infinitely often selected at run time; indeed, the "right" choice can even be eliminated at compile time! For a theory based on bisimilarity, neither of these implementations is allowed. A non-deterministic construction $(\mathtt{P} \vee \mathtt{Q})$ is expected to be implemented *fairly*: in any infinite sequence of choices, each alternative must be chosen infinitely often. Weak bisimilarity is a very neat way of imposing this obligation, which at one time was thought essential to the successful use of non-determinism.

Unfortunately, the requirement of fairness violates the basic principle of monotonicity, on which so much of engineering design depends. The program $(\mathtt{X} \vee \mathbb{I})$ is necessarily less deterministic than X, so $(\mu\mathtt{X} :: \mathtt{X} \vee \mathbb{I})$ should (by monotonicity) be less deterministic than $(\mu\mathtt{X} :: \mathtt{X})$, which is the least deterministic program of all. However, weak bisimulation identifies it with the completely deterministic program $\mathbb{I}$. It would therefore be unwise to base a calculus of design on weak bisimulation.

But that was never the intention; bisimulation was originally designed by Milner as the strongest equivalence that can reasonably be postulated between programs, and one that could be efficiently tested by computer, without any consideration of any possible meaning of the texts being manipulated. It was used primarily to explore the algebra of communication and concurrency. It was not designed for application to a non-deterministic sequential programming language; and the problems discussed in this section suggest it would be a mistake to do so.

A great many alternative definitions of program equivalence based on operational semantics have been explored by subsequent research. One of them is described in the next section. In the final section of this chapter there is yet another solution: to derive an observational semantics from the operational. The algebraic laws can then be proved from the observations, as described in Chapter 3.

4.3 From operations to algebra

In this section we will define a concept of simulation which succeeds in reconstructing the algebraic semantics of our chosen language on the basis of its operational semantics. First we define $\overset{*}{\rightarrow}$ as the reflexive transitive closure of $\rightarrow$:

$$\begin{array}{rll} (\mathtt{s},\mathtt{P}) \rightarrow^{0} (\mathtt{t},\mathtt{Q}) & \text{iff} & \mathtt{s} = \mathtt{t} \text{ and } \mathtt{P} = \mathtt{Q} \\ (\mathtt{s},\mathtt{P}) \rightarrow^{n+1} (\mathtt{u},\mathtt{R}) & \text{iff} & \exists \mathtt{t},\mathtt{Q} :: (\mathtt{s},\mathtt{P}) \rightarrow (\mathtt{t},\mathtt{Q}) \text{ and } (\mathtt{t},\mathtt{Q}) \rightarrow^{n} (\mathtt{u},\mathtt{R}) \end{array}$$

$$(\mathtt{s},\mathtt{P}) \rightarrow^{*} (\mathtt{t},\mathtt{Q}) \quad \text{iff} \quad \exists n :: (\mathtt{s},\mathtt{P}) \rightarrow^{n} (\mathtt{t},\mathtt{Q}).$$

Secondly, we define the concept of *divergence*, being a state that can lead to an infinite execution:

$$(\mathtt{s},\mathtt{P}) \uparrow =_{df} \forall n :: \exists \mathtt{t},\mathtt{Q} :: (\mathtt{s},\mathtt{P}) \rightarrow^{n} (\mathtt{t},\mathtt{Q}).$$

Thirdly, we define an ordering relation $\sqsubseteq$ between states. One state is *better* in this ordering than another if any result given by the better state is also possibly given by the worse; and furthermore, a state that can fail to terminate is worse than any other:

$$\begin{aligned}(\mathtt{s},\mathtt{P}) \sqsubseteq (\mathtt{t},\mathtt{Q}) \equiv_{df} \quad & (\mathtt{s},\mathtt{P}) \uparrow \text{ or } \neg(\mathtt{t},\mathtt{Q}) \uparrow \\ & \text{and } (\forall \mathtt{u} :: (\mathtt{t},\mathtt{Q}) \xrightarrow{*} (\mathtt{u},\mathbb{I}) \Rightarrow (\mathtt{s},\mathtt{P}) \xrightarrow{*} (\mathtt{u},\mathbb{I})).\end{aligned}$$

One program is better than another if it is better in all data states

$$\mathtt{P} \sqsubseteq \mathtt{Q} \quad \text{iff} \quad \forall \mathtt{s} :: (\mathtt{s},\mathtt{P}) \sqsubseteq (\mathtt{s},\mathtt{Q}).$$

The $\sqsubseteq$ relation is often called *refinement* or *simulation* of the worse program by the better.

It is easy to see that the syntactically defined $\sqsubseteq$ relation is transitive and reflexive, i.e., a preorder. As a result the relation

$$\mathtt{P} \sim \mathtt{Q} =_{df} \quad \mathtt{P} \sqsubseteq \mathtt{Q} \text{ and } \mathtt{Q} \sqsubseteq \mathtt{P}$$

is an equivalence. Since it is also a congruence, it can be used in exactly the same way as proposed for the bisimilarity relation in the previous section: one can thereby derive a full collection of algebraic laws for the programming language from its operational semantics. For example, associativity of $\vee$ follows from the two lemmas

$$\begin{aligned}&(s, P \vee (Q \vee R)) \xrightarrow{*} (t,\mathbb{I}) \\ &\qquad \text{iff } (s,P) \xrightarrow{*} (t,\mathbb{I}) \text{ or } (s,Q) \xrightarrow{*} (t,\mathbb{I}) \text{ or } (s,R) \xrightarrow{*} (t,\mathbb{I}) \\ &(s, P \vee (Q \vee R)) \uparrow \text{ iff } (s,P) \uparrow \text{ or } (s,Q) \uparrow \text{ or } (s,R) \uparrow .\end{aligned}$$

The same holds for the other bracketing as well.

But this is rather a laborious way of proving the rather large collection of laws. Each of these laws of Chapter 3 is a theorem of the form $[P \equiv Q]$. They can all be automatically lifted to simulation laws by the single theorem:

$$\mathtt{P} \sim \mathtt{Q} \quad \text{if} \quad [P \equiv Q].$$

In fact, the above theorem can be strengthened to an equivalence, so that the laws proved by simulation are exactly those of the algebraic semantics. The algebraic semantics is isomorphic to the operational, when abstracted by this particular notion of simulation.

4.4 From operations to observations

An operational semantics is in essence an inductive definition of all possible sequences of states that can arise from any execution of any program expressed in the notations of the language. This can be turned directly into an isomorphic observational semantics by just assuming that the whole execution sequence generated by each program can be observed. We will use the free variable e to stand for such an execution sequence. Of course, the level of abstraction is exactly the same as that of the operational semantics. To hide the excessive degree of detail, we need to define a predicate which relates each execution sequence onto an observation just of its initial and final states. Using this relation as a coordinate transformation, the definitions of the observational semantics given in Chapter 2 can be proved as theorems. The proofs are wholly based on the definition of the operational semantics, thereby completing the cyclic proof of mutual consistency of all three theories of programming.

An execution is formally defined as an empty, finite or infinite sequence of states in which every adjacent pair of states is related by the operational transition $\rightarrow$

$$E = \{e \mid \forall i : 0 < i < \sharp e : e_{i-1} \rightarrow e_i\},$$

where $\sharp$ gives the length of a finite sequence, or ∞ for an infinite one. The execution sequences generated by an initial state $(\mathtt{s}, \mathtt{P})$ are those that begin with this state

$$E(\mathtt{s}, \mathtt{P}) = \{e \mid e \in E \ \& \ (e = \langle\rangle \vee e_0 = (\mathtt{s}, \mathtt{P}))\}$$

The observations of a program $\mathtt{P}$ are those in which $\mathtt{P}$ is the initial state of the stored program

$$E(\mathtt{P}) = \bigcup_{\mathtt{s}} E(\mathtt{s}, \mathtt{P}).$$

The function E defines an observational semantics of each program text $\mathtt{P}$. The definition can be rewritten into an equivalent recursive form

$$E(\mathtt{s}, \mathtt{P}) = (\mathtt{s}, \mathtt{P})^\wedge\{e \mid \exists \mathtt{t}, \mathtt{Q} : (\mathtt{s}, \mathtt{P}) \rightarrow (\mathtt{t}, \mathtt{Q}) \ \& \ e \in E(\mathtt{t}, \mathtt{Q})\},$$

where $x^\wedge X =_{df} \{\langle\rangle\} \cup \{\langle x \rangle e \mid e \in X\}$. As in Chapter 2, the recursion here is understood to define the weakest (i.e. largest) fixed point, which will include all the infinite sequences as well.

But the definition lacks the important denotational property, requiring that the semantics of each combinator is defined in terms of the semantics of its components. Fortunately, the problem is solved by proof of the following theorems, which closely follow the structure of the operational semantics

$$
\begin{array}{lcll}
E(\mathtt{s},\mathbb{I}) & = & \{<(\mathtt{s},\mathbb{I})>,<>\} & \\
E(\mathtt{s},\mathtt{v}:=\mathtt{f}) & = & (\mathtt{s},\mathtt{v}:=\mathtt{f})^\wedge\{<>,(\mathtt{v}:=(s;\mathtt{f}),\mathbb{I})\} & \\
E(\mathtt{s},\mathtt{P}\vee\mathtt{Q}) & = & E(\mathtt{s},\mathtt{P})\cup E(\mathtt{s},\mathtt{Q}) & \\
E(\mathtt{s},\mathtt{P}\lhd\ \mathtt{b}\ \rhd\mathtt{Q}) & = & (\mathtt{s},\mathtt{P}\lhd\ \mathtt{b}\ \rhd\mathtt{Q})^\wedge E(\mathtt{s},\mathtt{P}) & \text{if } s;\mathtt{b} \\
 & = & (\mathtt{s},\mathtt{P}\lhd\ \mathtt{b}\ \rhd\mathtt{Q})^\wedge E(\mathtt{s},\mathtt{Q}) & \text{if } \neg s;\mathtt{b} \\
E(\mathtt{s},\mathtt{P};\mathtt{Q}) & = & \ \mathrm{add}Q(E(\mathtt{s},\mathtt{P})) & \\
 & & \cup\{ef \mid e\in\ \mathrm{add}Q(E(\mathtt{s},\mathtt{P}))\ \&\ f\in E(\mathtt{Q}) & \\
 & & \&\ \exists\mathtt{t}: e \text{ ends in } (\mathtt{t},\mathbb{I};\mathtt{Q}) \text{ and } f \text{ begins with } (\mathtt{t},\mathtt{Q})\} &
\end{array}
$$

where $\mathrm{add}Q(e)_i = (\mathtt{s},\mathtt{P};\mathtt{Q})$ whenever $e_i = (\mathtt{s},\mathtt{P})$,

$$E(\mathtt{s},\mu\mathtt{X}::\mathtt{F.X}) = (\mathtt{s},\mu\mathtt{X}::\mathtt{F.X})^\wedge E(\mathtt{s},\mathtt{F}.(\mu\mathtt{X}::\mathtt{F.X})).$$

If desired, these equations could be presented as the definition of the operational semantics of the language. Although they do not mention the step relation $\rightarrow$, they define exactly the same execution sequences. In fact, the step relation may be defined afterwards as that which holds between the first and second members of any execution sequence

$$(\mathtt{s},\mathtt{P})\rightarrow(\mathtt{t},\mathtt{Q}) \ \text{ iff } \ \exists e :: e\in E(\mathtt{s},\mathtt{P})\ \&\ e_2 = (\mathtt{t},\mathtt{Q}).$$

The proof of this theorem establishes an isomorphism between the traditional presentation of the operational semantics, given at the beginning of this chapter, and its denotational presentation, given by the above definition of E.

Our original presentation of a denotational semantics in Chapter 2 mapped each program text onto a predicate describing its observations. The definition of E maps each program text onto a *set* of observations, i.e., its execution sequences. Such a set can easily be timed into a predicate $e\in E(\mathtt{P})$, which uses the free variable e to stand for an observed value of the execution sequence generated by P. Similarly, any predicate P, with free variables ranging over known sets, can be turned into a set of observations by simple comprehension, for example $\{e \mid P\}$. The distinction between these two presentations as sets and as predicates is entirely ignorable. So the equations shown above could equally well be regarded as an ***observational*** semantics of the programming language, in exactly the style of Chapter 2. In future, let us use the italic $P(e)$ to stand for the predicate $e\in E(\mathtt{P})$.

But of course, the level of abstraction of this new observational semantics is identical to that of the operational semantics. To raise the level, we use a standard technique of data refinement. This requires us to specify which of the more concrete observations we are interested in, and which parts of them we want to observe. These decisions can be expressed as a predicate containing *both* sets of free variables, ones standing for the concrete observations, as well as ones standing for the abstract.

Our original abstract observations were selected on the assumption that we want to observe only a very small part of the information contained in each execution, namely the initial and final data states. Furthermore, we choose never to observe a finite non-empty execution that has not yet terminated. So we are interested only in minimal or maximal executions – those which cannot be reduced or extended

$$e = <> \lor e \text{ ends in } \mathbb{I} \lor e \text{ is infinite.}$$

The distinction between empty, terminated and infinite execution sequences is captured in the Boolean variables *ok* and *ok'*:

$$ok = (e \neq <>) \text{ and } ok \Rightarrow (ok' = (e \text{ ends in } \mathbb{I})).$$

For non-empty sequences, we can observe the initial data state v

$$ok \Rightarrow \exists \mathtt{s}, \mathtt{P} :: e_0 = (\mathtt{s}, \mathtt{P}) \;\&\; init(\mathtt{s})$$

where *init* is a function that maps the text $\mathbf{v} := \mathbf{k}$ to the predicate $v = k$. Similarly, for terminating executions, we can observe the final states

$$ok' \Rightarrow \exists \mathtt{s} :: e \text{ ends in } (\mathtt{s}, \mathbb{I}) \;\&\; final(\mathtt{s})$$

where *final* is a function that maps the text $\mathbf{v} := \mathbf{k}$ to the predicate $v' = k$.

Let ABS be the conjunction of the predicates displayed above. It has free variables ok, ok', v, v' and e. It describes all the ways in which an execution e can give rise to particular values of the other more abstract variables. So the abstract observations obtainable from any possible execution of P are described by the predicate

$$abs(P(e)) =_{df} \exists e : ABS \land P(e).$$

The function *abs* maps predicates describing execution sequences (denoted by the free variable e) to predicates describing observations of the initial and final states of a subset of those sequences (denoted by the free variables ok, ok', v, v'). This latter class of predicates is exactly the one used to give the original denotational semantics of the programming language in Chapter 2. We wish to claim that the semantics defined above by the functions *abs* is the same as that of Chapter 2. The claim is substantiated by showing that *abs* is an isomorphism between the relevant subsets of the two classes of predicate, namely those predicates that can be expressed as programs.

The predicate ABS is often called a linking invariant, because it relates observations of the same system at two levels of abstraction. It is used to define the function *abs*, which translates a low level predicate, whose free variables denote details of an implementation, to a predicate describing the same system at a higher level of abstraction. The predicate *abs*(P) is the strongest specification expressible at this higher level which is met by any concrete implementation described by P. But in a top-down design, we would prefer to translate in the opposite direction; given a specification S with free variables v, v', ok, ok', what is the weakest description of the low-level observations that will satisfy the specification? The answer is given by the definition

$$abs^{-1}(S) = (\forall v, v', ok, ok' : \mathrm{ABS} \Rightarrow S).$$

The transformations in each direction are linked by the Galois connection

$$[abs(P) \Rightarrow S] \quad \text{iff} \quad [P \Rightarrow abs^{-1}(S)].$$

From this it follows that *abs* is monotonic with respect to implication, and

$$P \Rightarrow abs^{-1}(abs(P)) \text{ and } abs(abs^{-1}(S)) \Rightarrow S.$$

Even more important are the equations

$$abs(P) = abs(abs^{-1}(abs(P))) \quad \text{and} \quad abs^{-1}(S) = abs^{-1}(abs(abs^{-1}(S))).$$

This means that $abs \circ abs^{-1}$ is the identity function over the range of *abs*, and similarly for $abs^{-1} \circ abs$. In our case, this range includes *all* the predicates expressible in the programming language. So *abs* is a bijection between predicates with alphabet $\{e\}$ describing execution sequences and predicates with alphabet $\{v, v', ok, ok'\}$ describing initial and final states. In both cases the predicates both arise from program texts.

Our last task is to show that *abs* is a homomorphism in the usual algebraic sense, that it is respected by all the combinators of the programming language. The required theorems look very like a denotational definition of the *abs* function.

$$abs(\mathbf{true}) = \mathbf{true}$$

$$abs(\mathrm{I\!I}) = \neg ok \vee (ok' \wedge v' = v)$$

$$abs(v := f) = \neg ok \vee (ok' \wedge v' = f)$$

$$abs(P \vee Q) = abs(P) \vee abs(Q)$$

$$abs(P \lhd b \rhd Q) = (abs(P) \wedge b) \vee (abs(Q) \wedge \neg b)$$

$$abs(P; Q) = abs(P); abs(Q)$$

$$abs(\mu X :: F.X) = \mu Y :: F'.Y$$

$$\text{where } F'.Y = abs(\mathtt{F}(abs^{-1}Y)).$$

In the last clause, the functions *abs* and abs^{-1} are needed to translate F from a function over predicates with e as their only free variable to a function F' over predicates with the more abstract alphabet.

The form of these definitions is exactly the same as those of the original observational semantics of the language in Section 2.2. Indeed, on omission of occurrences of the function *abs* and abs^{-1}, the two definitions are the same. More formally, the theorems show that *abs* is an isomorphism between two different interpretations of the notations of the same programming language: one as a description of execution sequences derived from the operational semantics, and one as a description of a relationship between initial and final values of variables v, ok.

This completes the task of unifying observational, algebraic and operational theories of the same programming language. In each case the basic definitions or axioms of each theory have been derived as theorems in its preceding theory, in a cyclic fashion. It is therefore a matter of choice which theory is presented first. My preference is to start with the most abstract, because this gives the most help in specification, design and development of programs. Furthermore, derivation of algebraic laws is quite straightforward, using standard mathematical proof methods. Finally, proof of the properties of the operational semantics can take advantage of previously established theorems. In general, a top-down approach seems easier than starting at a low level of abstraction and working upwards. But the operational semantics has considerable attraction, and is currently quite fashionable among theorists investigating the foundations of Computing Science.

Chapter 5

Conclusion

This monograph has recommended three distinct approaches to the construction of theories relevant to computing — the operational, the algebraic, and the observational. They have each an important distinctive role, which can and should be studied independently by specialists. But the full benefits of theory are obtained by a clear and consistent combination of the benefits of all three approaches. The method of consistent combination has been illustrated by application to a very simple programming language for expression of sequential algorithms with possible non-determinism. This is only a small part of the total task of clarifying the foundations of Computing Science.

We will need to build up a large collection of models and algebras, covering a wide range of computational paradigms, appropriate for implementation either in hardware or in software, either of the present day or of some possible future. But even this is not enough. What is needed is a deep understanding of the relationships between the different models and theories, and a sound judgment of the most appropriate area of application of each of them. Of particular importance are the methods by which one abstract theory may be embedded by translation or interpretation in another theory at a lower level of abstraction. In traditional mathematics, the relations between the various branches of the subject have been well understood for over a century, and the division of the subject into its branches is based on the depth of this understanding. When the mathematics of computation is equally well understood, it is very unlikely that its branches will have the same labels that they have today. Quoting from Andreski [17], "the contours of truth never coincide with the frontiers between embattled parties and cliques". So we must hope that the investigations by various schools of programming theory will contribute to the understanding which leads to their own demise.

The establishment of a proper structure of branches and sub-branches is essential to the progress of science. Firstly, it is essential to the efficient education of a new generation of scientists, who will push forward the frontiers in new directions with new methods unimagined by those who taught them. Secondly, it enables individual scientists to select a narrow specialisation for intensive study in a manner which assists the work of other scientists in related branches, rather than just competing with them. It is only the small but complementary and cumulative contributions made by many thousands of scientists that has led to the achievements of the established branches of modern science. But until the framework of complementarity is well understood, it is impossible to avoid gaps and duplication, and achieve rational collaboration in place of unscientific competition and strife.

Quoting again from Andreski

> "...the reason why human understanding has been able to advance in the past, and may do so in the future, is that true insights are cumulative and retain their value regardless of what happens to their discoverers; while fads and stunts may bring an immediate profit to the impresarios, but lead nowhere in the long run, cancel each other out, and are dropped as soon as their promoters are no longer there (or have lost the power) to direct the show. Anyway let us not despair."

References

[1] E. W. Dijkstra. *A Discipline of Programming.* Prentice-Hall Series in Automatic Computation. Prentice-Hall, 1976.

[2] Robin Milner. *A Calculus of Communicating Systems.* Springer, 1980. Lecture Notes in Computer Science, Volume 92.

[3] J.A. Bergstra and J.W. Klop. Algebra of communicating processes with abstraction. *Theoretical Computer Science*, 37(1):77–121, 1985.

[4] S.D. Brookes, C.A.R. Hoare, and A.W. Roscoe. A theory of communicating sequential processes. *Journal of the ACM*, 31:560–599, 1984.

[5] Dana Scott and Christopher Strachey. Towards a mathematical semantics for computer languages. In *Proceedings, 21st Symposium on Computers and Automata*, pages 19–46. Polytechnic Institute of Brooklyn, 1971. Also Technical Monograph PRG 6, Oxford University, Programming Research Group.

[6] M.B. Smyth. Power domains. *JCSS*, 16:23–26, 1978.

[7] E.C.R. Hehner. Predicative programming Parts I and II. *Communications of the ACM*, 27(2):134–151, Feb 1984.

[8] A. Tarski. On the calculus of relations. *The Journal of Symbolic Logic*, 6(3):73–89, Sept 1941.

[9] Gordon D. Plotkin. *A structural approach to Operational Semantics.* DAIMI FN-1. Computing Science Department, Aarhus, 1981.

[10] C.A.R. Hoare, I.J. Hayes, Jifeng He, C.C. Morgan, A.W. Roscoe, J.W. Sanders, I.H. Sorensen, J.M. Spivey, and B.A. Sufrin. The laws of programming. *Communications of the ACM*, 30(8):672-687, Aug 1987; see Corrigenda in Communications of the ACM, 30(9):770.

[11] C.A.R. Hoare and P.E. Lauer. Consistent and complementary formal theories of the semantics of programming languages. *Acta Informatica*, 3(2):135–153, 1974.

[12] A. Tarski. A lattice-theoretical fixpoint theorem and its applications. *Pacific J. of Math.*, 5:285–309, 1955.

[13] C.C. Morgan. *Programming from specifications.* Prentice-Hall International Series in Computer Science; London, 1990.

[14] Cliff B. Jones. *Systematic Software Development Using VDM.* Series in Computer Science. Prentice-Hall International Series in Computer Science, 1986.

[15] Joseph Goguen and Timothy Winkler. Introducing OBJ3. Technical Report SRI-CSL-88, SRI International, Computer Science Lab, Aug 1988. Revised version to appear with additional authors José Meseguer, Kokichi Futatsugi and Jean-Pierre Jouannaud, in *Applications of Algebraic Specification using OBJ*, edited by Joseph Goguen, Derek Coleman and Robin Gallimore, Cambridge.

[16] R. Milner. *Communication and Concurrency.* Prentice-Hall International Series in Computer Science, 1989.

[17] S. Andreski. *Social Science as Sorcery.* Pelican Books.

Part 3

Programming Paradigms

Not only programming notations vary to a large extent, but also the styles which are advocated and used for programming. Different programming paradigms base on different computational models. In order to come to grips with a programming paradigm, one must put it on a solid mathematical foundation. It can then be investigated what laws are valid and how valuable they are for program design. Hence a mathematical formalization is the prerequisite for the comprehension and scientifically based judgement of a programming paradigm. Mathematical models of different programming paradigms help to compare them.

John C. Mitchell, Gert Smolka, Jean-Raymond Abrial

On the relationship between classes, objects and data abstraction

Kathleen Fisher[1] and John C. Mitchell[2] *

[1] AT&T Research, 600 Mountain Ave., Murray Hill, NJ 07974
kfisher@research.att.com

[2] Computer Science Department, Stanford University, Stanford, CA 94305
mitchell@cs.stanford.edu

Abstract. While most object-oriented programming is done in class-based languages, the trend in theoretical study has been to develop formal systems that are object-based, without classes and often without explicit inheritance mechanisms. This paper studies the correspondence between object primitives and class constructs of the form found in C++, Eiffel, and Java. The main qualitative insight is that such classes require both an extensible aggregate, to serve as the basis for inheritance, and a non-extensible form of object to support subtyping. We compare three approaches to modeling classes, the first using records of object components called "premethods" and the latter two using an extensible form of object called a "prototype." While the first approach uses fewer primitive operations on objects, it does not provide several important features of class-based languages. In the latter two approaches, we overcome these deficiencies by combining prototypes with standard abstraction mechanisms. All three treatments of classes use translations into provably sound object calculi.

1 Introduction

There are several forms of object-oriented programming languages, including class-based languages such as C++ [ES90, Str94], Eiffel [Mey92], Java [AG96] Simula [BDMN73], and Smalltalk [GR83], prototype-based languages such as Self [US91] and Obliq [Car95], and multi-method-based approaches such as CommonLisp [Ste84] and Cecil [Cha95]. This paper is concerned with the study of class-based languages and the relationship between three language constructs: classes, prototype-based inheritance primitives, and traditional data abstraction of the form found in languages such as CLU [LSAS77, L$^+$81], Ada [US 80], and ML [MTH90]. We begin by identifying some essential properties of the class constructs that appear in class-based object-oriented languages. This analysis gives rise to evaluation criteria that we will later use to assess proposed class constructs. We then present three ways of representing classes in terms of basic operations on objects. These representations are evaluated using both our design

* Supported in part by NSF Grant CCR-9303099.

criteria and comparisons between the kinds of object operations involved. The differences between the three translations from classes into object operations provide some insight into the suitability of prototype-based object systems for carrying out traditional class-based programming. To the best of our knowledge, the translations also provide the first type soundness proofs for standard forms of class constructs.

The first treatment of classes is based on an encoding into record operations. A version of this encoding appears in [AC96c, AC96b], where it is used to show how a pure object calculus can be used to represent simple class constructs. In this model, each class has three parts: a type for the instances of the class, a constructor function to produce instances, and a record of object-part implementations. One deficiency of this model is that it is not possible to inherit initialization code. There are also limitations in the way that changes may be propagated through a class hierarchy. Finally, the translation uses structural subtyping on object types to account for the similarities between classes related via inheritance; consequently, it is not possible to explicitly control the inheritance hierarchy or the implementation of objects with any object type. An advantage of this approach is that this form of class may be expressed using a relatively straightforward and understandable object system, without cloning or object extension. While we believe the lack of constructor inheritance and the inability to control the class hierarchy are serious limitations (as elaborated in this paper), others may disagree [AC96b, RR96].

The second approach to classes, which addresses the deficiencies in the first approach, provides the central features found in Java, Eiffel, and C++. The fundamental improvements arise from the use of extensible objects to define extensible constructors and a form of extensible abstract datatype to provide extensible types with implementation guarantees. In the process, we gain explicit control over the subtyping hierarchy without invalidating ordinary structural subtyping. One appeal of this translation is that it clearly shows how classes may be viewed as an orthogonal combination of pure operations on objects (providing aggregation but no encapsulation) and data abstraction (providing encapsulation but no aggregation).

The third translation of classes into object operations is an elaboration of the second, allowing the type of base-class methods to be refined in derived classes. This ability to refine the types of methods has been called "method specialization" or "mytype specialization" in the literature [Mit90, BSv95]. Although method specialization complicates the typing issues considerably, we are again able to show that classes may be viewed as an orthogonal combination of object and data abstraction primitives. The third translation requires a target language with extensible objects, specific forms of higher-order polymorphism, and so-called variance analysis.

All of the translations yield type soundness results for their classes as a consequence of the type soundness of their target object calculi. The soundness proof for the first target calculus appears in [AC96b]; the analysis of extensible object calculi is developed in full in [Fis96]. For brevity and to present the major design issues as clearly as possible, we do not repeat the technical details in this paper. The interested reader may consult our previous publications [FHM94, FM95a] for discussions of type soundness proofs for extensible object calculi, and [FM95b] for a general discussion of type systems, object calculi, and object-oriented programming. An alternate view of many of these issues may be found in the excellent book by Abadi and Cardelli [AC96b].

The remainder of this paper is structured as follows. In the next section, we discuss the use of classes in traditional object-oriented programming, developing criteria for evaluating the properties of various class constructs. Section 3 gives the syntax of a sample class construct and two examples to be used in comparing different class mechanisms. In Section 4, we describe our notation and discuss the background material necessary to understand the first class encoding. Sections 5 to 7 describe and analyze three translations of the sample class syntax into specific formal calculi, each inducing a different set of restrictions on the use of objects and inheritance. Sections 8 and 9 describe related work and conclude.

2 Classes

In contemporary class-based object-oriented languages such as Eiffel, Java, and C++, classes are the main tool for program organization. Since program organization is a complex task, classes are designed to serve a number of purposes. In brief, a class declaration provides a mechanism for creating a family of objects, all sharing some or all of their implementation. Inheritance, which enables the reuse of object implementations, is structured along class lines, with derived classes incorporating some or all of the implementation of their base classes. Classes typically provide access control mechanisms, which allow programmers to hide implementation details. A class declaration also defines a type, namely, the type that all objects created by this class will share. Finally, in contemporary languages, class declarations have explicitly determined the subtyping relation between object types. We discuss each of these issues in turn.

2.1 Defining Objects

A class declaration provides a mechanism for creating a family of objects. Typically, all objects created from a single class will have the same representation, with the values of some data field differing from instance to instance. For example, the following class declaration, written in a Java-like pseudo-code, defines a family of `Point` objects, created (or *instantiated*) by calling the *constructor* function of the class. For simplicity, we illustrate the main ideas using one-dimensional points.

$$
\begin{array}{rl}
\textbf{class}\ \texttt{Point}\ \{ & \\
\texttt{x} & : \ int \\
\texttt{setX} & : \ int \rightarrow Point \\
\texttt{getX} & : \ int \\
\texttt{move} & : \ int \rightarrow Point \\
\textbf{constructor}\ \texttt{newPoint} & : \ int \rightarrow Point \\
\} &
\end{array}
$$

Each object constructed from this class has an `x` component, representing the point's geometric location. The `setX` and `getX` components write and read the `x` coordinate, respectively, while `move` changes the location by a given offset. For simplicity, we omit the implementations of these components. We will see the need for `setX` and `getX` when we discuss access controls in Section 2.3.

For efficiency, components that are the same for all objects instantiated from a given class are typically shared. In Smalltalk for example, data, called *instance variables*, are stored within each object, while functions, called *methods*, are shared by all objects instantiated from the same class. The distinction between components allocated per object and per class does not always fall between data and functions, but that division is a common one. In the `Point` example, the `x` component would probably be an instance variable, set to different values in different `Point` objects, while `setX`, `getX`, and `move` would be shared among all instances of the class.

Programmers create *instances* of the `Point` class (*i.e.*, `Point` objects) by calling the `Point` class constructor function `newPoint`, here distinguished by the keyword **constructor**. We again omit the implementing code. Since objects are created by calling constructor functions, we must be able to call the constructors of a class before any objects have been created. Hence constructor functions are not components of the objects they create. Despite this independence, constructor definition is an important part of class design since each object must be initialized in accordance with any intended object invariants.

2.2 Defining Related Objects Via Inheritance

Inheritance allows a new family of objects to be defined as an incremental modification to another. For example, we may use inheritance to define the class `ColorPoint` from `Point` as follows:

$$
\begin{array}{rl}
\textbf{class}\ \texttt{ColorPoint}\ :\ \texttt{Point}\ \{ & \\
\texttt{c} & : \ color \\
\texttt{turnRed} & : \ ColorPoint \\
\texttt{getC} & : \ color \\
\textbf{constructor}\ \texttt{newColorPoint} & : \ int \rightarrow color \rightarrow ColorPoint \\
\} &
\end{array}
$$

The `ColorPoint` class *inherits* the `Point` `x`, `setX`, `getX`, and `move` methods and adds three new ones: `c`, `turnRed`, and `getC`. The `Point` class is said to be a *base*, *parent* or *super* class and `ColorPoint` a *derived*, *child* or *sub* class. In addition to adding components, derived classes may also modify existing methods during inheritance. For example, the `ColorPoint` class may redefine the `Point` `move` method so that `ColorPoint` objects change color whenever they are moved. The constructor function `newColorPoint` is responsible for creating fully initialized `ColorPoint` objects.

As this example illustrates, inheritance is a code-reuse mechanism. In principle, for every object or class of objects defined via inheritance, there is an equivalent definition that does not use inheritance, obtained by expanding the definition so that the inherited code is duplicated. The importance of inheritance is that it saves the effort of duplicating (or reading duplicated) code, and that when one class is implemented by inheriting from another, changes to the parent are automatically reflected in its descendants. This mechanism has a significant impact on program maintenance and modification.

A benefit of class-based inheritance is that classes are essentially static structures. Class-based inheritance may be implemented as a compile-time operation, greatly simplifying the static checking involved in compiling class declarations. In particular, the correctness of a derived class declaration depends on both the presence and absence of certain components in its base class. This information is available at compile time if the base class declaration is known. If we want to statically check *delegation*, a run-time form of inheritance based on individual objects instead of classes, then it is necessary to include both presence and absence of components as part of the static type system. This complicates subtyping, as discussed in [FHM94, FM95b, FM95a], for example.

2.3 Access Restrictions

Classes typically provide access controls, which are used to restrict the visibility of various object components. Some common visibility levels are *private*, for use only within a given class implementation; *protected*, for use only by derived classes; and *public*, for use by anyone. As an example, we may add the annotations **private**, **protected**, and **public** to the `Point` class above to obtain the class

class `Point` {

private	`x`	:	*int*
protected	`setX`	:	$int \rightarrow Point$
public	`getX`	:	*int*
	`move`	:	$int \rightarrow Point$
constructor	`newPoint`	:	$int \rightarrow Point$

}

This modified class permits the `x` component to be used only within the `Point` class declaration, *i.e.*, within `setX`, `getX`, `move`, and `newPoint`. Similarly, `setX` is visible only to the implementation of the `Point` class and all classes derived from it, *e.g.*, the `ColorPoint` class defined above. We may similarly annotate the `ColorPoint` class to make its `c` component private.

Visibility levels help insure that separate blocks of code have no hidden dependencies and hence may be separately maintained. For example, if the designers of the `Point` class decide to change its underlying representation (say to a polar representation), they are free to modify private components without fear of breaking either client code or derived classes. Similarly, they may change protected components without worrying about breaking client code; however, in this case, they may have to fix derived classes. Throughout this paper, we will use the term "client code" for code restricted to public-level access and "derived classes" for code granted protected-level access.

There is a subtle interaction between visibility and constructors, which we will explain by example. Consider the `ColorPoint` constructor `newColorPoint`. To create an initialized `ColorPoint` object, `newColorPoint` must initialize the inherited portion of the object, in particular, the `x` component. However, the `ColorPoint` class has no access to its parent's private components. Hence there must be a way for base classes to provide initialization code for use within the constructors of derived classes. Typically, derived-class constructors invoke their parent's constructor to initialize these portions.

2.4 Defining Object Types

The fourth function of a class declaration is to define an object type. Each class associates a type with the objects it creates; this type will be the return type of the class constructor functions. In principle, there are at least two forms of object type that a language may use: *interface* types and *implementation* types. As the name suggests, interface types constrain only the interface of objects, specifying the set of messages each object must understand and the types of these messages. This form of object type has been extensively studied in the theoretical literature, *e.g.*, [AC96c, Bru93, FHM94, PT94] and the earlier papers appearing in [GM94], but it is not used in current class-based languages. In contrast, implementation types are a form of abstract type. An implementation type specifies both the interface and a portion of the implementation of an object. Although interface types offer increased flexibility, existing languages such as C++, Java, and Eiffel ([ES90, AG96, Mey92]) typically use implementation types by identifying the type of each object type with the class that created the object. To avoid confusion, we will distinguish between classes and the object types they define by writing class names in teletype font (`ClassName`) and object types in italics ($TypeName$). We discuss interface and implementation types in more detail in the remainder of this section.

Interface types An interface type specifies a list of operations, generally as method names and return types. If an object is declared to implement all of the operations with the appropriate types, it is considered an element of an interface type. For example, objects instantiated from the `Point` class have the following interface type, written using a notation defined in Section 4.2.

$$Point_{Inter} \stackrel{def}{=} \mathbf{obj}\, u.\langle\!\langle \mathtt{getX} : int \rightarrow u,\ \mathtt{move} : int \rightarrow u \rangle\!\rangle$$

An expression with this type is guaranteed to be an object that has at least `getX` and `move` methods. Furthermore, when either of these messages is sent with an integer parameter to such an object, the result is an object with the same type as the receiver, *i.e.*, $Point_{Inter}$. In more detail, to reflect the fact that the `Point` class is recursive (the return type of `getX` and `move` methods is *Point*), we use a form of recursive type for the `Point` class interface type. To that end, we have replaced the *Point* return type of the `getX` and `move` methods with a type variable u, which is bound by the keyword **obj** to the type $Point_{Inter}$. We omit the `x` and `setX` methods from the interface type because they are not public components and are therefore not visible to clients of the class. We also omit the constructor function `newPoint` because constructors are not part of the objects they create.

Similarly, we may write the interface type for the access-controlled version of the `ColorPoint` class as follows:

$$ColorPoint_{Inter} \stackrel{def}{=} \mathbf{obj}\, u.\langle\!\langle\ \texttt{getX} : int \rightarrow u,\ \texttt{move} : int \rightarrow u, \\ \texttt{turnRed} : u,\ \texttt{getC} : color \rangle\!\rangle$$

Because interface types specify only the names and types of operations, objects with the same interface may have significantly different internal representations. In particular, objects from many different classes may all be given the same interface type. For example, objects instantiated from any class with public `getX` and `move` methods of the appropriate types will have type $Point_{Inter}$, not just objects instantiated from the `Point` class.

A significant advantage of interface types is the flexibility they provide. For example, using interface types we can define a single type of matrix object and then represent dense matrices with one form of object and sparse matrices with another. Both matrix representations may be given the same type and therefore be used interchangeably in any program. This kind of flexibility is particularly useful when we write library operations on matrices without assuming any particular implementation. Such library functions may be written using a standard interface type without concern for how matrices might be implemented in later (or earlier) development of a software system.

Implementation Types Intuitively, implementation types are in a one-to-one correspondence with classes. More specifically, an implementation type is a form of object type that guarantees both a specific public interface and some aspects of the implementation of objects, in a fashion similar to abstract datatypes. For example, $ColorPoint_{Imp}$, the implementation type associated with the `ColorPoint` class, imposes two requirements: if an object has $ColorPoint_{Imp}$, then it must (i) support the public interface $ColorPoint_{Inter}$ and (ii) have the private and protected components defined in the `ColorPoint` class. Thus $ColorPoint_{Imp}$ objects must have private `x` and `c` components, with `x` occurring before `c` in the representation if the language requires it. This implementation constraint that is *not* reflected in the public interface.

The extra information supplied by an implementation type is useful for several reasons, including optimizing component lookup, guaranteeing behavioral similarity, and allowing class-level protection mechanisms. We discuss these issues in more detail in the following paragraphs.

Optimizing Lookup. This point may be explained by example. If we know that all *Point* objects inherit a specific representation of x and y coordinates, then a program may be optimized to take advantage of this static guarantee. The usual implementations of C++, for example, use type information to statically calculate the offset of member fields relative to the starting address of the object. A similar calculation is used to find the offset of methods in the virtual function table at compile time [ES90, Section10.7c]. (Virtual function tables store the methods associated with all objects of a given class.) Such optimizations are not possible in an untyped language such as Smalltalk [GR83] and would not be possible in a typed language where objects of a single type could have arbitrarily dissimilar implementations. In particular, the Java interface mechanism allows us to specify that several classes that do not have similar implementations all satisfy a common interface. When all that is known about an object (say, passed as an actual parameter to a function) is an interface it supports, then basic facts about the implementation of the object are not known at compile time.

Behavioral Guarantees. A second, more methodological reason that programmers may be interested in implementation types is that they convey greater behavioral guarantees. For example, since the only non-trivial way to create a new object with a given implementation type is to call the constructor function for the associated class, we can be sure that an object with a given implementation type has been properly initialized and satisfies the expected invariants for its class.

Class-Level Protection. A more subtle reason to use types that restrict the implementations of objects has to do with the implementation of binary operations. In an object-oriented context, a binary operation on type A is realized as a method that requires another A object as a parameter. In a language where all objects of type A share some common representation, it is possible for an A method to safely access part of the private internal representation of another A object. A simple example arises with *Set* objects that have only a membership test and a union operation in their public interfaces. With interface types, some objects of type *Set* might be represented internally using bit vectors, while others might use linked lists. In this case, there is no type-safe way to implement union, since no single operation will access both a bit vector and a linked list correctly. With only interface types, it is necessary to extend the public interface of both kinds of sets with some sort of "`elementsAsList`" method to make this operation possible. In contrast, if the type of an object conveys implementation information, then a less flexible but type-safe implementation of set union is possible without polluting the public interfaces. In this case, all *Set* objects would have one representation and a union operation could be implemented by taking advantage of this uniformity. See [BCC$^+$96] for a detailed discussion of binary methods.

Interface vs. Implementation Types In summary, interface types and implementation types have complementary strengths:

	Implementation types	Interface types
Flexibility	-	+
Efficiency	+	-
Behavioral Guarantees	+	-
Binary Methods	+	-

Current languages use implementation types, essentially identifying classes and their associated object types, but interface types have been widely studied in the theoretical literature [AC96c, Bru93, FHM94, PT94, GM94]. It is possible to associate both an interface and an implementation type with each classes, since, as we will see in the next section, each implementation type is typically a subtype of the interface type obtained by "forgetting" its implementation constraints. This technique combines the benefits of both forms of object type to be realized in a single language. To some extent, the Java class and interface mechanisms provide this combination of features, although since it is necessary to specifically declare membership in each Java interface, Java interfaces are actually a form of abstract data type in which the implementation constraint is vacuous, not a pure interface type of the form we discuss in this paper.

2.5 Defining a Subtyping Hierarchy

Like inheritance, *subtyping* is a code reuse mechanism. Instead of supporting the reuse of object definitions, however, subtyping supports the reuse of client code. The basic principle associated with subtyping is *substitutivity*: if A is a subtype of B, then any expression of type A may be used without type error in any context that requires an expression of type B. To see the importance of this principle, suppose we have written a large body of code to manipulate B objects. If A is a subtype of B, then it follows from the basic substitutivity principle of subtyping that the same code may also be able used to manipulate A objects (at least without type errors). The main significance of subtyping is that it permits uniform operations over various types of data that share some basic structure. Subtyping therefore makes it possible to have heterogeneous data structures containing objects that belong to different subtypes of some common base type. We write "$A <: B$" to indicate that A is a subtype of B.

It is important to keep in mind that subtyping is a relation on types, not classes. This distinction is easily understood in languages where objects are either untyped (in which case subtyping is an external notion that can be used to describe the differences between interfaces of objects) or where objects are created individually, without using classes. However, in class-based languages, as we have already noted, it is typical to associate a type with each class. This may either be an interface type or the form of abstract type we call an implementation type. The standard subtype relation on interface types is a form of structural subtyping, while the standard form of subtyping on implementation types requires some further correspondence between hidden implementation constraints. The only practical way to guarantee subtyping between implementation types, without violating principles of data abstraction by exposing hidden components of objects, is to make the subtyping hierarchy a subset of the inheritance hierarchy, *i.e.*, if

$$ColorPoint_{Imp} <: Point_{Imp}$$

then `ColorPoint` must inherit from `Point`. See [?, Fis96] for more details. Thus with implementation types, the class construct plays a substantial role in defining the subtyping hierarchy.

Since the implementation subtyping and inheritance hierarchies are closely related, providing one mechanism (class hierarchies) to declare both the subtyping and inheritance relations can be a significant conceptual simplification. This simplification does not mean, however, that the two hierarchies are necessarily the same. For example, C++ private inheritance allows two classes to be related in the inheritance hierarchy without inducing an associated subtype relationship. As mentioned briefly earlier, Java *interfaces* allow implementation types to have a common supertype, without sharing implementations.

With interface types, the subtype relation may be inferred by comparing the conditions imposed by each interface. For example,

$$ColorPoint_{Inter} <: Point_{Inter}$$

may be inferred simply by looking at the structure of these two types. While classes are fundamental to subtyping between implementation types, interface types allow subtyping and inheritance to be unrelated.

There is a third language design possibility, arising because implementation types are subtypes of interface types [Fis96]. In this scenario, each class has an associated implementation type that is a subtype of the interface type obtained by "forgetting" the extra implementation constraints. For example,

$$Point_{Imp} <: Point_{Inter}$$

The class hierarchy is responsible for determining subtyping relationships between implementation types, while subtype inference rules are used to infer relationships between interface types. Hence in this case, the class hierarchy defines the "bottom" portion of the subtyping hierarchy, while structural subtyping rules define the "top."

2.6 Evaluation Criteria

Sections 2.1 through 2.5 reveal that classes play a fundamental role in structuring code, providing both reuse (subtyping and inheritance) and code separation (access restrictions, initialization code) mechanisms. Since the goal of incorporating code reuse and code separation mechanisms into a programming language is to support large-scale programming, any class construct should be evaluated with large-scale programming in mind. Based on this consideration and the discussions in the preceding sections, we believe that a class mechanism should provide the features outlined in the following paragraphs. Where appropriate, we annotate these paragraphs with the sections that contain relevant discussion.

Coherent, Extensible Collection [Sections 2.1 and 2.2]. A class construct should provide an extensible collection of object components. Such components may include methods, data, local constants, specifications of communication protocols, *etc.* These parts should be *coherent*, in the sense that if a class is well-formed, then it should be guaranteed that all objects instantiated from the class will themselves be well-formed. For example, if one of the methods of a class requires an integer `f` method, then the class must contain an integer `f` method. The consistency of the components should be checked when the class is formed, not when objects are instantiated from the class, a timeliness which facilitates debugging. The *extensibility* of the collection means that derived classes may reuse object components from their parents, possibly adding additional coherent components and replacing others. This inheritance should be statically checked as well, insuring that derived classes are also well-formed.

Guaranteed Initialization [Sections 2.1 and 2.2]. Classes should provide code to initialize their collection of components when objects are instantiated. A class-based language should insure that this initialization code is executed both for objects instantiated directly from a given class and for those instantiated from its descendants. Running this initialization code is essential for establishing invariants for the components, particularly in light of the private level of visibility. Since derived classes cannot see private components and are not supposed to be aware of their parents' implementation details, they should not (and indeed cannot) be responsible for initializing and setting up invariants for inherited components.

Enforced Access Restrictions [Section 2.3]. A class construct should allow programmers to specify a level of visibility for each of the components defined in a class. These specifications should be enforced by the language. For example, if a programmer specifies that a given method is private, the type system should insure that only the implementation of the given class may access the method in any way. This restriction guarantees that the implementor of a class may change its private implementation without fear of introducing type errors into clients or descendants of the class. Access controls should clearly indicate the visibility of each of the parts for both the current class and all derived classes, so if a method is private in a given class, that method is guaranteed to be private in all derived classes as well.

Explicit Type Hierarchy [Sections 2.4 and 2.5]. Another important point of comparison for class constructs is whether or not they provide implementation types and explicit control over a portion of the subtyping hierarchy. Because implementation types are subtypes of interface types, classes that provide implementation types give programmers more flexibility. They may use implementation types where the extra information is useful and interface types where more generality is required.

Automatic Propagation of Base Class Changes. Because classes are intended to support large-scale programs and because such programs are inevitably modified, classes should support code maintenance. In particular, incremental changes to a base class should automatically be propagated to all derived classes. If any type errors are introduced in derived classes because of such changes, these errors should be reported immediately. For example, if a programmer adds a new method to a parent class, that method should then be automatically added to all derived classes. If the new method conflicts with a method defined in a descendent, the conflict should be reported immediately, not when objects are instantiated from the given derived class.

3 Class Construct

In this brief section, we give the syntax of the class construct we will translate into formal calculi in Sections 5 to 7.

class ⟨*class-name*⟩ { : ⟨*superclass* ⟩ }
  constructor ⟨*name*⟩ : ⟨*type*⟩ → ⟨*class-name*⟩ ;
  private {⟨*name*⟩ : ⟨*type*⟩; }*
  public {⟨*name*⟩ : ⟨*type*⟩; }*
end;

This declaration of class *class-name* may specify a parent class *superclass*, gives a constructor function returning instances of the class with the type *class-name* (using the pun between class names and object types) and lists public and private methods. For the sake of simplicity, we leave the implementing code implicit, specifying an implementation in each object calculus, and omit the **protected** level of visibility. We also restrict our attention to the special case of exactly one constructor per class.

3.1 `Point` and `ColorPoint` Examples

To simplify the exposition in the following sections, we will use the familiar example of points and color points, slightly streamlined from the versions described in Section 2. Using the syntax above these classes may be written as follows:

class `Point`
  constructor `newPoint` : *int* → *Point*;
  private `x` : *int*;
  public `setX` : *int* → *Point*;
  `getX` : *int*;
end;

$$\begin{array}{lll} \textbf{class}\ \texttt{ColorPoint} : \texttt{Point} & & \\ \quad \textbf{constructor} & \texttt{newCPoint} & : \textit{color} \rightarrow \textit{int} \rightarrow \textit{ColorPoint}; \\ \quad \textbf{private} & \texttt{c} & : \textit{color}; \\ \quad \textbf{public} & \texttt{setC} & : \textit{color} \rightarrow \textit{ColorPoint}; \\ & \texttt{getC} & : \textit{color}; \\ \textbf{end}; & & \end{array}$$

Intuitively, the `set` methods are used to assign to private location or color data while the `get` methods read these private values. The constructor functions `newPoint` and `newCPoint` return new `Point` and `ColorPoint` objects, respectively, initialized to store the values passed to them as parameters.

4 Preliminaries: Objects, Object Types and Existential Types

Before proceeding to the first translation, we introduce some background material, including basic object calculi terminology and notation, the object types we will use, and a form of existential type that will be useful in hiding private components.

4.1 Objects, Fields and Methods

Many of the basic object operations that we will need may be presented using Cardelli's Obliq [Car95], a simple dynamically-typed language for distributed programming that has been used for a variety of purposes, including graphics and web "oblets." Obliq has a pure object system which is not based on classes. Although we refer to Obliq as an example of a "real" programming language based on pure objects without classes, we will use a simplified lambda-calculus-style notation that has fewer syntactic distinctions than Obliq, but retains the same essential expressiveness.

An Obliq object may be written in the form

$$\{\mathtt{m}_1 = e_1, \ldots, \mathtt{m}_k = e_k\}$$

where each e_i defines a *component* of the object. For clarity, Obliq distinguishes between two forms of components, *data fields* and *methods*. The distinction is not based on the types of components, since functions are considered a particular kind of data. Instead, the distinction is based on how the component is accessed. Specifically, simple selection, as in records, is used for fields, while method invocation is used for methods. The distinction may be illustrated using the object

$$\mathtt{a} \stackrel{def}{=} \{\mathtt{x} = 6,\ \mathtt{setX} = \lambda(\mathtt{self},\mathtt{n})\,\mathtt{self.x} := \mathtt{n}\}$$

We have $\mathtt{a.x} = 6$ simply by extracting the value of x, and

$$\mathtt{a.setX(8)} \quad = \quad \{\mathtt{x} = 8,\ \mathtt{setX} = \lambda(\mathtt{self},\mathtt{n})\,\mathtt{self.x} := \mathtt{n}\}$$

by a function call that involves passing the receiver object a to the `setX` method as the first actual parameter. The important point is that method invocation is considered an atomic operation; if a component is a method, it is not possible to extract the method and apply it (as a function) to any object other than the object to which it belongs. An object with only fields may be called a *record.*

Several formal calculi that provide Obliq operations, and more, have been developed [AC96c, AC95, AC96a, Mit90, FHM94, FM95a]. As a foundational point, it is possible to eliminate the distinction between fields and methods since any field may be represented as a method which does not use the `self` argument. For example, we could replace $\mathtt{x} = 6$ by $\mathtt{x} = \lambda(\mathtt{self})\,6$. The only differences are that method invocation is less efficient than field selection in practice and that objects of a given class typically have different data but the same methods. This second distinction allows objects to share method code and method lookup tables.

For syntactic simplicity, we will assume every object component is a method and write objects in the form

$$\langle \mathtt{m_1} = e_1, \ldots, \mathtt{m_k} = e_k \rangle$$

where each expression $e_1, \ldots, e_k$ is assumed to be a function that expects its host object as its first actual parameter when invoked. Using the symbol $\Leftarrow$ for method invocation, we have

$$\langle \mathtt{m_1} = e_1, \ldots, \mathtt{m_k} = e_k \rangle \Leftarrow \mathtt{m_i} \quad = \quad e_i \langle \mathtt{m_1} = e_1, \ldots, \mathtt{m_k} = e_k \rangle$$

A useful operation on objects is to *override* a method, creating a new object with a different value at that method. Using the symbol $\leftarrow$ for override, we have

$$\langle \mathtt{m_1} = e_1, \ldots, \mathtt{m_i} = e_i, \ldots, \mathtt{m_k} = e_k \rangle \leftarrow \mathtt{m_i} = e_i' \quad = \quad \langle \mathtt{m_1} = e_1, \ldots, \mathtt{m_i} = e_i', \ldots, \mathtt{m_k} = e_k \rangle$$

These object operations are shared by the calculi defined in [AC96c, AC95, AC96a, Mit90, FHM94, FM95a] and subsequent variants.

4.2 Object Types

We will use the type expression

$$\mathbf{obj}\, u.\langle\!\langle \mathtt{m_1} : \tau_1, \ldots, \mathtt{m_k} : \tau_k \rangle\!\rangle$$

to type objects that have methods $\mathtt{m_1}, \ldots, \mathtt{m_k}$ with associated types $\tau_1, \ldots, \tau_k$. In this expression, the bound type variable u may appear in the type of a method, in which case it refers to the type of the host object. For example, if method `setX` is invoked on an object of type $\mathbf{obj}\, u.\langle\!\langle \mathtt{x} : int, \mathtt{setX} : int \rightarrow u \rangle\!\rangle$, the result is a function which, given an integer, returns an object of type $\mathbf{obj}\, u.\langle\!\langle \mathtt{x} : int, \mathtt{setX} : int \rightarrow u \rangle\!\rangle$. When u does not appear in the types $\tau_1, \ldots, \tau_k$, we will use the shorthand

$$\mathbf{obj.}\langle\!\langle \mathtt{m_1} : \tau_1, \ldots, \mathtt{m_k} : \tau_k \rangle\!\rangle$$

As described elsewhere [FHM94, FM95a, AC96c, AC95] **obj** types are essentially a specialized form of recursive type, with "folding" combined with object formation and "unfolding" combined with method invocation.

4.3 Existential quantification and data abstraction

Since we will use existential types in several places, we review a general form of datatype declaration based on existential types. For further information on data abstraction, the reader is referred to [LSAS77, MP88, Mor73, Rey83, Set89]; the connection between existential types and data abstraction is elaborated in [MP88, Mit93].

The declaration form

$$\texttt{Abstype } t \texttt{ with } x_1 : \sigma_1, \ldots, x_k : \sigma_k \texttt{ is } \langle \tau; M_1, \ldots, M_k \rangle \texttt{ in } N$$

declares an abstract type t with "operations" $x_1, \ldots, x_k$ and implementation $\langle \tau; M_1, \ldots, M_k \rangle$. The scope of this declaration is N. For example, the expression

$$
\begin{array}{ll}
\texttt{Abstype} & \textit{stream} \texttt{ with} \\
 & s : \textit{stream}, \\
 & \textit{first} : \textit{stream} \rightarrow \textit{nat}, \\
 & \textit{rest} : \textit{stream} \rightarrow \textit{stream} \\
\texttt{is} & \\
 & \langle \tau, M_1, M_2, M_3 \rangle \\
\texttt{in} & \\
 & N
\end{array}
$$

declares an abstract datatype *stream* with distinguished element $s : \textit{stream}$ and functions *first* and *rest* for operating on streams. Within the scope N of the declaration, the stream s and operations *first* and *rest* may be used to compute natural numbers or other results. However, the type of N may not be *stream,* since this type is local to the expression. In computational terms, the elements of the abstract type *stream* are represented by values of the type τ given in the implementation. Operations s, *first* and *rest* are implemented by expressions M_1, M_2 and M_3. Since the value of s must be a stream, the expression M_1 must have type τ, the type of values used to represent streams. Similarly, we must have $M_2 : \tau \rightarrow \textit{nat}$ and $M_3 : \tau \rightarrow \tau$. Using cartesian products, we may put any abstract datatype declaration in the form $\texttt{Abstype } t \texttt{ with } x{:}\sigma \texttt{ is } \langle \tau, M \rangle \texttt{ in } N$. For example, the *stream* declaration may be put in this form by combining the three operations s, *first* and *rest* into a single operation of type $\textit{stream} \times (\textit{stream} \rightarrow \textit{nat}) \times (\textit{stream} \rightarrow \textit{stream})$. There is no loss in doing so, since we may recover s, *first* and *rest* using projection functions.

Some useful flexibility is gained by considering abstract datatype declarations and datatype implementations separately. An implementation for the abstract type *stream* mentioned above consists of a type τ, used to represent *stream*'s, together with expressions for the specified stream s and the stream operations *first* and *rest.* If we want to describe implementations of streams in general, we might say that in any implementation, "there exists a type t with elements of types t, $t \rightarrow \textit{nat}$, and $t \rightarrow t$." This description would give just enough information about an arbitrary implementation to determine that an `Abstype` declaration makes sense, without giving any information about how streams are represented. This fits the general goals of data abstraction, as discussed in [Mor73], for example.

We may add abstract datatype implementations and `Abstype` declarations to a type system using *existential types* of the form $\exists t.\sigma$. Intuitively, each element of an existential type $\exists t.\sigma$ consists of a type τ and an element of $[\tau/t]\sigma$. Using products to combine s, *first*, and *rest*, an implementation of *stream* would have type $\exists t.[t \times (t \rightarrow nat) \times (t \rightarrow t)]$, for example.

We will write an implementation of an abstract type in the form $\langle t = \tau, M{:}\sigma \rangle$, where $t = \tau$ binds t in the remainder of the expression. The reader may think of $\langle t = \tau, M{:}\sigma \rangle$ as a "pair" $\langle \tau, M \rangle$ in which access to the representation type τ has been restricted. The bound type variable t and the type expression σ serve to disambiguate the type of the expression. The type of a well-formed expression $\langle t = \tau, M : \sigma \rangle$ is $\exists t.\sigma$. For notational simplicity, we write $\langle t = \tau, M \rangle$ instead of $\langle t = \tau, M : \sigma \rangle$ when the type σ is clear from context.

The `Abstype` declaration form allows us to bind names to the type and value components of a datatype implementation. More specifically, if expression e_1 has type $\exists t.\tau$ and, under the assumption that $x{:}\tau$ the expression e_2 has type σ, then

$$\texttt{Abstype}\ t\ \texttt{with}\ x:\tau\ \texttt{is}\ e_1\ \texttt{in}\ e_2\ :\ \sigma$$

Informally, this rule binds type variable t and ordinary variable x to the type and value part of the implementation e_1, respectively, with scope e_2. A technical-sounding condition is that in $\texttt{Abstype}\ t\ \texttt{with}\ x:\tau\ \texttt{is}\ e_1\ \texttt{in}\ e_2\ :\ \sigma$, the type variable t cannot occur free in the type σ of the entire expression. This is actually a straightforward scoping condition: the bound variable t should not be allowed to escape from its scope.

Existential types and objects Existential types are useful in the context of modeling objects because they can hide private components [PT94, PT93]. To develop a feel for this idea, we briefly explain how point objects may be modeled as elements of existential type. The type of an object with private `x` and public `getX` and `setX` components could be written as follows

$$
\begin{array}{l}
\mathit{Point} \stackrel{def}{=} \exists\, \mathit{Rep}.\ \mathit{record} \\
\qquad\qquad\qquad \texttt{state} \ \ : \mathit{Rep}; \\
\qquad\qquad\qquad \texttt{methods} : \mathit{record} \\
\qquad\qquad\qquad\qquad\qquad \texttt{getX} : \mathit{Rep} \rightarrow \mathit{int}; \\
\qquad\qquad\qquad\qquad\qquad \texttt{setX} : \mathit{Rep} \rightarrow \mathit{int} \rightarrow \mathit{unit}; \\
\qquad\qquad\qquad\qquad\quad \mathit{end}; \\
\qquad\qquad \mathit{end};
\end{array}
$$

Each object has an internal `state` component and a record of methods that operate on that state. The type of the internal state is hidden by the existential quantifier, so external functions cannot operate on it. The types of an object's methods are written in terms of the type of the hidden state.

Now that we have defined the operations and types that we need, we may proceed with the first class encoding, translating the class construct we saw in Section 3 into a simple object calculus.

5 Record-of-Premethods Model

In Cook's important early work on the foundations of object-oriented languages [Coo87, Coo89], each class is represented by a function, called an *object generator*. The fixed point of a generator is a record that represents an object, with recursion used to resolve references to *self*. The reason for object generators is that inheritance cannot work directly on objects that encapsulate (or hide) the dependence of one method on another. However, inheritance may be formulated as an ordinary operation on method bodies if we explicitly treat methods as functions of *self* [Mit90].

In the context of object calculi, it seems natural to define inheritance using premethods, functions that are written with the intent of becoming object methods, but which are not yet installed in any object. Like Cook's generators, premethods are functions that explicitly depend on the "object itself," typically assumed to be the first parameter to the function. Following this idea, Abadi and Cardelli have proposed encoding classes in a pure object system using records of premethods [AC96c]. In this model, a class is an object that contains a record of premethods and a constructor function used to package these premethods into objects. We illustrate this proposal by example in the next subsection. A similar approach is used in [RR96].

5.1 Example Classes

Figure 1 shows the six declarations that result from translating the classes given in Section 3 into a simple object calculus along the lines outlined above. The first three declarations give (1) the object type associated with the `Point` class:

$$\mathbf{obj.}\langle\!\langle \mathtt{x} : int,\ \mathtt{getX} : int,\ \mathtt{setX} : int \rightarrow unit \rangle\!\rangle,$$

(2) the record of `Point` premethods `PointImpl`, and (3) the `Point` constructor `newPoint`. For the sake of simplicity, we assume we are working with an imperative semantics and `setX` works by changing the value of `x`. This assumption allows us to make *unit* the return type of `setX`, side-stepping the problem of *method specialization*. Method specialization occurs when the types of methods are "improved" during inheritance to reflect their new, more specialized host object. For example, if the type of `setX` were *Point* in `Point` objects and *ColorPoint* in `ColorPoint` objects, we would say the type of `setX` was *specialized* during inheritance. Although this general approach can support method specialization, significantly more complex types are required to do so [AC96b].

The next three declarations give the corresponding representation for the `ColorPoint` class. Note that the implementation `CPointImpl` of `ColorPoint` is defined from the implementation `PointImpl` of `Point`. This reuse is type correct because $ColorPoint <: Point$, and therefore, by the contravariance of function types, we have $Point \rightarrow \tau <: ColorPoint \rightarrow \tau$ for each type τ. However, this dependency appears to be the extent to which inheritance can be realized. It is not possible to reuse the `Point` constructor because objects are not extensible in this model. Thus, we cannot use a `Point` object to create a `ColorPoint` object (except trivially).

```
type Point          = obj.⟪x : int, getX : int, setX : int → unit⟫;

val PointImpl       : record
                               x : Point → int;
                            getX : Point → int;
                            setX : Point → int → unit;
                      end;
                    = record
                               x = λ(self) 0;
                            getX = λ(self) self ⇐ x;
                            setX = λ(self) λ(newX) self.x: = newX;
                      end;

fun newPoint        : int → Point
                    = λ xi : int.⟨     x = PointImpl.x,
                                    getX = PointImpl.getX,
                                    setX = PointImpl.setX  ⟩ ⇐ setX xi


type ColorPoint     = obj.⟪ x : int, getX : int, setX : int → unit,
                            c : color, getC : color, setC : color → unit⟫

val CPointImpl      : record
                               x : ColorPoint → int;
                            getX : ColorPoint → int;
                            setX : ColorPoint → int → unit;
                               c : ColorPoint → color;
                            getC : ColorPoint → color;
                            setC : ColorPoint → color → unit;
                      end;
                    = record
                               x = PointImpl.x;
                            getX = PointImpl.getX;
                            setX = PointImpl.setX;
                               c = λ(self) red;
                            getC = λ(self) self ⇐ c;
                            setC = λ(self) λ(newC) self.c: = newC;
                      end

fun newColorPoint   : int → color → ColorPoint
                    = λ xi : int. λ ci : color.
                      ((getX = CImpl.getX, ... ⟩ ⇐ setX xi) ⇐ setC ci
```

Fig. 1. Type and constructor declarations for `Point` and `ColorPoint` classes.

5.2 Adding Protection

Notice that in the basic record-of-premethods encoding, the `x` and `c` components of an object are not private; they are just as accessible as the public methods. There are three ways of incorporating private fields into this approach, each with advantages and disadvantages. The first approach uses closures to create private fields whenever a constructor is called. The second uses subtyping to restrict the return type of constructor functions to reflect only public methods. The third uses a variation of the existential-type object model.

5.3 Protection Via Closures

$$
\begin{array}{lll}
\texttt{type}\ Point & = \textbf{obj}\texttt{.}\langle\!\langle \texttt{getX} : int,\ \texttt{setX} : int \rightarrow unit \rangle\!\rangle; \\
\texttt{val PointImpl} & : record \\
 & \qquad \texttt{getX} : int \rightarrow Point \rightarrow int; \\
 & \qquad \texttt{setX} : int \rightarrow Point \rightarrow int \rightarrow unit; \\
 & \quad end; \\
 & = \texttt{record} \\
 & \qquad \texttt{getX} = \lambda(\texttt{x})\,\lambda(\texttt{self})\ \texttt{x}; \\
 & \qquad \texttt{setX} = \lambda(\texttt{x})\,\lambda(\texttt{self})\,\lambda(\texttt{newX})\ \texttt{x} := \texttt{newX}; \\
 & \quad \texttt{end}; \\
\texttt{fun newPoint} & : int \rightarrow Point \\
 & = \lambda\,\texttt{xi} : int.\,\texttt{let x = xi in} \\
 & \qquad\qquad \langle\, \texttt{getX} = \texttt{PointImpl.getX x}, \\
 & \qquad\qquad\ \ \texttt{setX} = \texttt{PointImpl.setX x}\ \rangle
\end{array}
$$

Fig. 2. Declarations for the `Point` class with modifications to support closure-style private fields.

Figure 2 shows a version of the `Point` class, modified to incorporate closure-style private fields. The necessary changes include:

- removing the `x` field from type *Point* to indicate that `x` is not publicly available,
- removing the `x` field from `PointImpl`, since the `x` field will be allocated within the constructor function on a per object basis,
- rewriting the remaining premethods (and their types) to take an extra parameter, a placeholder for the `x` field that will be supplied whenever a `Point` object is created, and
- modifying the constructor function to allocate the private `x` field and pass it to each of the premethods during their installation in the new object.

This mechanism succeeds in hiding the `x` field from general clients; only the `setX` and `getX` methods of a particular object may access its `x` field. The approach does not succeed, however, at hiding private fields from derived classes. Notice that because inheritance manipulates records of premethods instead of constructors, the private x field is not inherited by the `ColorPoint` class. Instead, the constructor for the derived class `ColorPoint` has to allocate its own `x` field for the inherited methods, illustrating the fact that the supposedly private `x` component has been exposed to derived classes. This exposure of private `x` is also reflected in the new types for the `setX` and `getX` premethods. Another problem with this approach is its lack of support for private *methods*. Because methods need to reside within objects to allow access to the *self* parameter, they cannot be stored in closures.

5.4 Protection Via Restricting Constructor Return Type

In this version, which appears in Figure 3, there are two object types associated with the `Point` class. Type *Point* lists the public components, while $Point_{Priv}$

$$
\begin{array}{lll}
\texttt{type } \mathit{Point} & = & \texttt{obj.}\langle\!\langle \texttt{getX} : \mathit{int}, \texttt{setX} : \mathit{int} \rightarrow \mathit{unit} \rangle\!\rangle; \\
\texttt{type } \mathit{Point}_{Priv} & = & \texttt{obj.}\langle\!\langle \texttt{x} : \mathit{int}, \texttt{getX} : \mathit{int}, \texttt{setX} : \mathit{int} \rightarrow \mathit{unit} \rangle\!\rangle; \\
\texttt{val PointImpl} & : & \mathit{record} \\
 & & \quad \texttt{x} : \mathit{Point}_{Priv} \rightarrow \mathit{int}; \\
 & & \quad \texttt{getX} : \mathit{Point}_{Priv} \rightarrow \mathit{int}; \\
 & & \quad \texttt{setX} : \mathit{Point}_{Priv} \rightarrow \mathit{int} \rightarrow \mathit{unit}; \\
 & & \mathit{end}; \\
 & = & \texttt{record} \\
 & & \quad \texttt{x} = \lambda(\texttt{self})\ 0; \\
 & & \quad \texttt{getX} = \lambda(\texttt{self})\ \texttt{self} \Leftarrow \texttt{x}; \\
 & & \quad \texttt{setX} = \lambda(\texttt{self})\,\lambda(\texttt{newX})\ \texttt{self.x} := \texttt{newX}; \\
 & & \texttt{end}; \\
\texttt{fun newPoint} & : & \mathit{int} \rightarrow \mathit{Point} \\
 & = & \lambda\, \texttt{xi} : \mathit{int}.\langle \quad \texttt{x} = \texttt{PointImpl.x}, \\
 & & \quad\quad \texttt{getX} = \texttt{PointImpl.getX}, \\
 & & \quad\quad \texttt{setX} = \texttt{PointImpl.setX}\ \rangle \Leftarrow \texttt{setX xi}
\end{array}
$$

Fig. 3. Declarations for the `Point` class with modifications to support private components via restricting the return type of `Point`'s constructor function.

lists the public and the private components. The premethods assume the *self* object has the private as well as the public components, but the constructor function exposes only the public methods by making its return type *Point* instead of $Point_{Priv}$. This technique is type correct because $Point_{Priv}$ is a subtype of *Point* ($Point_{Priv} <: Point$).

Although this approach succeeds in hiding the x component from clients who invoke the newPoint constructor function, it fails to hide the x premethod. Consequently, if the Point class designers wish to change their private representation, they must worry about breaking client code that has come to rely on a supposedly private premethod. A second encapsulation failure stems from the fact that the ColorPoint class must explicitly extract the private components of PointImpl to inherit them. Hence in this approach, derived classes must again be aware of the private components of their parents. On the positive side, this technique does support private methods in addition to private fields.

5.5 A Refinement

In response to our criticism, Martin Abadi has suggested a refinement to the record-of-premethods approach that involves using a form of existential type to hide the types of private components [Aba96]. Figure 4 shows the necessary declarations for the Point class. In this refinement, the x premethod is still exposed in PointImpl, but its type is hidden by an existential type. Since only

```
type Point          = obj.⟪getX : int, setX : int → unit⟫;
type Point_Priv(R)  = obj.⟪x : R, getX : int, setX : int → unit⟫;
 val PointImpl      : ∃R.record
                              x : Point_Priv(R) → int;
                           getX : Point_Priv(R) → int;
                           setX : Point_Priv(R) → int → unit;
                           newP : int → Point;
                           end;
                    = ⟨ R  =  int,
                        record
                              x = λ(self) 0;
                           getX = λ(self) self ⇐ x;
                           setX = λ(self) λ(newX) self.x:= newX;
                           newP = λ xi : int.⟨     x = PointImpl.x,
                                                getX = PointImpl.getX,
                                                setX = PointImpl.setX ⟩ ⇐ setX xi
                        end⟩;
 fun newPoint       : int → Point
                    = λ xi : int. PointImpl.newP xi
```

Fig. 4. Declarations for the Point class with modifications to support private components via restricting the return type of Point's constructor function and hiding the types of private methods via data abstraction.

the other premethods of the `Point` class are within the scope of this existential, only they can see `x`'s true type; hence only they can use the `x` premethod in any non-trivial way. The main changes from Figure 3 include

- parameterizing $Point_{Priv}$ by the representation type for `x`,
- giving `PointImpl` existential type, to hide the private representation for `x`,
- changing the argument type for each of the premethods to be the private type $Point_{Priv}(R)$ instantiated to the hidden representation type,
- implementing `PointImpl` as an element of existential type and binding the representation type variable R to the selected representation type, *int*, and
- moving the code for the constructor function into `PointImpl` so that it will be within the scope of the hidden representation type.

This approach succeeds in hiding the private `x` component from general use. Furthermore, since the `x` premethod has an unknown type, clients cannot usefully access it, despite its public availability, because its hidden type means it can only be used in trivial ways. Derived classes still must explicitly name the private component to inherit it, which is somewhat problematic. For private fields, there is at least a partial solution to this naming problem [Aba96]. In particular, a convention could be adopted in which each class has a special component named "`private_fields`" that stores a record of all the private fields. Instead of having to name each of the fields individually, the derived class can just name `private_fields`. At the price of this one name being globally known, we can insure that the actual private names do not leak out.

Since this approach is the most successful at modeling the private level of visibility, we will adopt this version as the "record-of-premethods" class interpretation for the purposes of the analysis in the next section.

5.6 Evaluation

The primary advantage of the record-of-premethods encoding of classes is that it does not require a complicated form of object. All that is needed is a way of forming an object from a list of component definitions. However, looking back at the list of desiderata in Section 2.6, it is clear that this approach has some serious drawbacks. We discuss each of the criteria in order.

Coherent, Extensible Collection. The combination of a record of premethods and a constructor function may be thought of as a coherent, extensible collection. Because premethods are simply fields in a record, nothing requires that they be coherent until a constructor function is supplied. Since the constructor function installs the premethods into an object, however, the fact that a given constructor is typable implies that the premethods it uses are coherent. Notice, however, that nothing requires a given constructor to mention all of the premethods in a given premethod record.

Guaranteed Initialization. In this model, class designers *can* guarantee that objects instantiated directly from their class are initialized if they insure that each of the classes that they inherit from behave properly; however, they *cannot* guarantee that classes deriving from theirs will inherit properly initialized components. The problem arises from the fact that inheritance manipulates records of premethods instead of constructor functions. Because inherited private components have unknown type, derived classes cannot initialize them directly. Constructor functions return non-extensible objects, so they cannot be used for this initialization either, except trivially. Instead, parent classes must supply separate initialization functions for this purpose. Nothing requires derived classes to invoke the supplied initialization code, however. Hence the designers of parent classes cannot assume that derived classes will always properly initialize the components they inherit. Furthermore, if a class (or one of its parents) does not initialize its inherited components properly, then the objects instantiated from that class may not be properly initialized either.

Enforced Access Restrictions. As we saw above, the final version of the record-of-premethod approach largely succeeds in providing enforced access restrictions, at the cost of complicating the object primitives with existential types. The only (small) leak is the name of private methods. A subtle point is that premethods are available to derived classes, not just methods. In other words, a derived class has access to a record of functions, and these functions may be used in any way. They need not be used as methods of an object, but could be used to define functions or could be tested, for example to determine whether a base class method depends on another methods of the object.

Explicit Type Hierarchy. Another issue that we believe is equally significant is the lack of explicit control over the subtyping hierarchy. In many existing class-based languages, it is possible to restrict the subtypes of an object type to classes that inherit all or part of the class implementation. As we discussed in Section 2.4, this restriction may be useful for optimizing operations on objects, allowing object access in binary methods and guaranteeing semantic consistency beyond type considerations (some discussion of these points appears in [KLM94]). A special case of this capability is the ability to define *final* classes, as recognized in work on Rapide [KLM94] and incorporated (presumably independently) as a language feature in Java. This ability is lacking in the record-of-premethods approach since any object whose type is a structural subtype can be used as an object of that type.

Automatic Propagation of Base Class Changes. Because derived class constructors must explicitly name the components that they wish to inherit, the record-of-premethods approach does not automatically propagate base class changes. In particular, if a derived class `D` is defined from a base class `B` in Java or related languages, then adding a method to `B` will result in an additional method of `D`, and similarly for every other class derived from `B` (and there may be many). With the approach shown here, derived class constructors must be explicitly rewritten each time their base classes change. Since object-oriented programs are typically quite large and maintenance may be distributed across many people, the person who maintains a base class may fail to inform those maintaining its derived classes of its change, causing unpredictable errors. There is no mechanism in the language to detect such errors automatically. Another way of stating this point is to say that in this approach, the translation from classes to object operations is not local – the definition of the constructor of a class depends on all of the definitions of classes from which components are inherited.

While reasonable people may disagree about the potential for records of premethods in large-scale object-oriented system design, it seems clear to us that this approach involves a significant loss of language support, in favor of very simple object primitives. While we readily admit that simplicity is a virtue, it also seems worthwhile to carefully examine the features that are lost.

6 Classes as Extensible Objects + Abstraction

Extensible objects provide a rich alternative to generators or premethods. They obviate the need for premethods, since collections of methods that are already installed in objects may be extended. Because of this fact, we may impose static constraints on the ways in which one method may be combined with others. For example, if an object contains two mutually recursive methods, then we cannot replace one with another of a different type. In contrast, in the record-of-premethods approach, it is possible to form a record of premethods without a "covering" constructor that checks to be sure that all of the premethods are coherent. A second advantage of extensible objects is that class constructors and initialization code can be inherited, *i.e.*, reused in derived classes. For example, to create `ColorPoint` objects, we may invoke the `Point` class constructor and add color methods to the resulting extensible object. This process guarantees that the `Point` class has the opportunity to properly initialize the inherited components. Another advantage arises with private (or protected) methods. In the extensible-object formulation, methods always remain within an object, even when it is extended. These hidden methods exist in all future extensions, but they can only be accessed by methods that were defined before the method became hidden. Furthermore, these private methods need not be explicitly manipulated by derived class constructors to insure that they are treated properly.

In this section and the next, we show how to obtain classes by combining extensible objects and a form of data abstraction. As a first step, we describe a suitable object calculus and briefly discuss the type system necessary to support it. We then introduce a refined form of existential type to provide the necessary data abstraction. Finally, we put these two mechanisms together to get the desired class construct.

6.1 Extensible Objects

Extensible objects support the operation of *object extension* in addition to method invocation and method override. Object extension, which we denote by the symbol $\leftarrow\!\!+$, adds new components to objects. Although it seems appealing to work in a calculus with a single form of object that is both overridable and extensible, there is unfortunately a fundamental conflict between these operations and subtyping, another crucial aspect of object types. In particular, we lose *width subtyping* with extension and *depth subtyping* with override, as illustrated in the following table.

Object Inheritance Operations	Sound Subtyping
none	width and depth
override	width only
extension	depth only
override and extension	none

Width subtyping implies that longer object types (ones with more methods) are subtypes of narrower ones. Depth subtyping is the orthogonal notion; it specifies that object types with specialized components are subtypes of object types with more general components. For example, assuming that $ColorPoint <: Point$, then $\mathbf{obj}.\langle\!\langle \mathtt{x} : ColorPoint \rangle\!\rangle$ is a depth subtype of $\mathbf{obj}.\langle\!\langle \mathtt{x} : Point \rangle\!\rangle$. The conflict between subtyping and object-based inheritance is described in [FM95b].

Because of this conflict, we need two forms of object type to construct classes properly: first, an extensible form for use within class declarations to support inheritance and second, a non-extensible form that provides rich subtyping for use as the objects instantiated from classes. Ideally, there would be an easy way to obtain the non-extensible form from the extensible one. This conversion process should occur whenever programmers ask for instances of the class. We shall refer to extensible objects as *prototypes* and non-extensible ones as proper objects.

Type checking prototypes is significantly more challenging than type checking proper objects. Intuitively, the difficulties arise from two sources. The first is that we need to insure that a given method name is *not* used in a prototype before we add a new method with that name. This guarantee insures that the new method body cannot violate any typing assumptions the old prototype may have made for methods of that name. To provide this kind of information, we have equipped our type system with a kinding system that tracks method *absences*. In principle, a type system that tracks potential method *presences* may also suffice.

The second source of difficulty arises from the need to insure that each method body works correctly for all objects, extensible and otherwise, that it may eventually live in. Since we may extend prototypes, we must show that a candidate method has the necessary type for all possible extensions of its original host. To express this form of polymorphism, we need to be able to say formally "all future extensions" of a prototype. To that end, we introduce a version of Wand's *rows* [Wan87] into our type system. Intuitively, a *row* is a list of methods and associated types; it has the form $\langle\!\langle \mathtt{m}_1 : \tau_1, \ldots, \mathtt{m}_k : \tau_k \rangle\!\rangle$. *Row variables* range over row expressions. Such variables, when implicitly universally quantified, stand for any list of method-type pairs. When we bound such a row variable with a "supertype" listing the methods of a prototype, we obtain (a representation of) the collection of all possible future extensions of that prototype.

Since we will use both object extension and method override for inheritance, we lump the two inheritance operations, method override and object extension, together in this paper. We will use types of the form $\mathbf{pro}.\langle\!\langle \mathtt{m}_1 : \tau_1, \ldots, \mathtt{m}_k : \tau_k \rangle\!\rangle$

to type our extensible, overridable objects. It turns out that we may obtain non-extensible objects from these extensible ones via subsumption [FM95a]:

$$\mathbf{pro.}\langle\!\langle \mathtt{m}_1 : \tau_1, \ldots, \mathtt{m}_k : \tau_k \rangle\!\rangle <: \mathbf{obj.}\langle\!\langle \mathtt{m}_1 : \tau_1, \ldots, \mathtt{m}_k : \tau_k \rangle\!\rangle$$

6.2 Data Abstraction

To provide the encapsulation needed for classes, we introduce a fairly standard form of bounded existential quantification [CW85, MP88]. Bounded existential types are similar to the existential types we described in Section 4.3, except that they provide a supertype bound for the hidden type, revealing partial information about it. In more detail, a declaration of the form

$$\texttt{Abstype}\ p <:_w P :: \kappa\ \texttt{with}\ \mathit{Ops}\ \texttt{is}\ \mathit{Impl}\ \texttt{in}\ \mathit{Client}$$

introduces a new row name, p, with upper bound P and kind κ. When we use this construct to model classes, p will play the role of a class name. Intuitively, the upper bound P lists the methods from p that clients will be able to use, while the kind information indicates how derived classes may extend p. The functions listed in *Ops* are the only operations that can access the private representation of p. Typically, class constructors and friend functions appear in this list. *Impl* gives the actual implementation of the abstraction, which *Client* uses as allowed by P, κ, and *Ops*.

Typically, a class implementation specifies an object layout and a set of method bodies (code for the methods of its objects). In our approach, object layouts are given by row expressions and method bodies as parts of constructor functions. Formally, the implementation of each class (*Impl*, above) is given by an expression of the form

$$\{\!|p <:_w P :: \kappa = P_{priv},\ \mathit{ConsImp}|\!\}$$

As above, the row variable p may be thought of as the name of the class implemented by this expression. Row P lists the class's publicly supported methods and their types. Kind κ describes how the class may be extended via inheritance. Thus P and κ essentially form the interface of the class implementation construct. The remaining components, P_{priv} and *ConsImp*, give the actual implementation of the class. P_{priv} describes the "layout" of the class's instances by listing the names of all of the methods defined within the class and specifying their types. The constructor function *ConsImp* provides the method bodies and initializes instantiated objects as necessary.

Following general principles of data abstraction, method bodies may rely on aspects of their host's representation that are hidden from other parts of the program. This fact is reflected in the above construct in that the constructor function *ConsImp* is within the scope of the private interface P_{priv}, and hence it may use private methods omitted from P. Our framework allows any number of constructor functions or other "non-virtual" operations to be provided in the *ConsImp* slot. However, for simplicity, we discuss only the special case of one constructor per class.

6.3 Example Classes

$$
\begin{array}{l}
\texttt{Abstype}\; p <:_w P_{pub} :: \{\texttt{c}, \texttt{getC}, \texttt{setC}\} \\
\quad \texttt{with newPoint} : int \rightarrow \textbf{pro.}p \\
\quad\quad \texttt{is}\; \{\!|\, p <:_w P_{pub} :: \{\texttt{c}, \texttt{getC}, \texttt{setC}\} = P_{priv},\; \texttt{Imp}_p \,|\!\} \\
\quad\quad \texttt{in} \\
\quad\quad\quad \texttt{Abstype}\; cp <:_w CP_{pub} :: \emptyset \\
\quad\quad\quad\quad \texttt{with newCPoint} : int \rightarrow color \rightarrow \textbf{pro.}cp \\
\quad\quad\quad\quad\quad \texttt{is}\; \{\!|\, cp <:_w CP_{pub} :: \emptyset = CP_{priv},\; \texttt{Imp}_{cp} \,|\!\} \\
\quad\quad\quad\quad\quad \texttt{in} \\
\quad\quad\quad\quad\quad\quad \texttt{let}\;\; \texttt{newPoint} : int \rightarrow \textbf{obj.}p = \texttt{newPoint} \\
\quad\quad\quad\quad\quad\quad \texttt{let}\;\; \texttt{newCPoint} : int \rightarrow color \rightarrow \textbf{obj.}cp = \texttt{newCPoint} \\
\quad\quad\quad\quad\quad\quad \texttt{in}\;\;\; \langle \texttt{Program} \rangle
\end{array}
$$

where

$$
\begin{array}{lcl}
P_{pub} & \stackrel{def}{=} & \langle\!\langle \texttt{getX} : int,\; \texttt{setX} : int \rightarrow unit \rangle\!\rangle \\
P_{priv} & \stackrel{def}{=} & \langle\!\langle \texttt{x} : int,\; \texttt{getX} : int,\; \texttt{setX} : int \rightarrow unit \rangle\!\rangle \\
CP_{pub} & \stackrel{def}{=} & \langle\!\langle p \,|\, \texttt{getC} : color,\; \texttt{setC} : color \rightarrow unit \rangle\!\rangle \\
CP_{priv} & \stackrel{def}{=} & \langle\!\langle p \,|\, \texttt{c} : color,\; \texttt{getC} : color,\; \texttt{setC} : color \rightarrow unit \rangle\!\rangle
\end{array}
$$

and

$$
\begin{array}{l}
\texttt{Imp}_p \stackrel{def}{=} \lambda\, \texttt{xi} : int. \langle\;\; \texttt{x} = \lambda(\texttt{self})\;\; 0; \\
\qquad\qquad\qquad\quad \texttt{getX} = \lambda(\texttt{self})\;\; \texttt{self} \Leftarrow \texttt{x}; \\
\qquad\qquad\qquad\quad \texttt{setX} = \lambda(\texttt{self})\;\; \lambda(\texttt{newX})\;\; \texttt{self.x} := \texttt{newX} \rangle \Leftarrow \texttt{setX xi};
\end{array}
$$

$$
\begin{array}{l}
\texttt{Imp}_{cp} \stackrel{def}{=} \lambda\, \texttt{xi} : int. \lambda\, \texttt{ci} : color. \\
\qquad\qquad\quad \langle\, \texttt{Imp}_p\; \texttt{xi} \;\; \longleftarrow\!\!\!+ \;\; \texttt{c} = \lambda(\texttt{self})\;\; \texttt{red}; \\
\qquad\qquad\qquad\qquad\quad \longleftarrow\!\!\!+\; \texttt{getC} = \lambda(\texttt{self})\;\; \texttt{self} \Leftarrow \texttt{c}; \\
\qquad\qquad\qquad\qquad\quad \longleftarrow\!\!\!+\; \texttt{setC} = \lambda(\texttt{self})\;\; \lambda(\texttt{newC}) \\
\qquad\qquad\qquad\qquad\qquad\qquad\qquad \texttt{self.c} := \texttt{newC} \rangle \Leftarrow \texttt{setC ic};
\end{array}
$$

Fig. 5. Nested abstract datatypes for `Point` and `ColorPoint` classes.

Using extensible objects and the abstract datatype mechanism from Section 6.2, we may encode our example class hierarchy as shown in Figure 5. As in the previous section, we work with an imperative version of the underlying object calculus to avoid complications arising from method specialization[3]. The class interpretation presented in the next section provides enough machinery to allow

[3] The most relevant soundness proof for an imperative version of the underlying object calculus appears in [DF96], which also includes support for concurrency.

us to address the method specialization problem directly; it does not require an imperative semantics. However, as this additional machinery is somewhat complex, we first present the basic ideas behind the "Classes = Extensible Objects + Data Abstraction" model without support for method specialization.

In order, the two abstract type declarations in Figure 5 define the `Point` and `ColorPoint` classes. The `Abstype` clause of the `Point` class declaration introduces row variable p, which formally serves as the "name" of the `Point` class. Row p is bounded by row P_{pub}, `Point`'s public interface. This row, defined in Figure 5, lists the methods available to clients and descendants of the `Point` class, namely `getX` and `setX`. The kind $\{\mathtt{c}, \mathtt{getC}, \mathtt{setC}\}$ indicates that descendant classes, *e.g.*, `ColorPoint`, may add methods with these names.

The `with` clause of the declaration specifies that the constructor for the `Point` class, named `newPoint`, will return expressions with **pro** type, specifically with type $\mathbf{pro}.p$, so that descendant classes may define their objects via extension from the `Point` class constructor. The row variable p in this return type signals that objects returned by the constructor are defined in the `Point` class. We call such types *partially abstract* because a portion of their structure is hidden.

The `is` clause of the abstraction describes the implementation of the `Point` class. Within this implementation, the row P_{priv} describes the full representation of `Point` class objects by listing all of the methods defined in the class and giving their associated types. The expression $\mathtt{Imp}_p$, defined towards the bottom of Figure 5, gives the implementation of the constructor `newPoint`. The type system insures that $\mathtt{Imp}_p$'s type is $\mathbf{pro}.p$ with p replaced by `Point`'s private interface, P_{priv}.

The structure of the `ColorPoint` class is similar. There are several aspects of this declaration that are worth mentioning however. In particular, notice that the `ColorPoint` constructor, implemented via the $\mathtt{Imp}_{cp}$ function defined at the bottom of Figure 5, invokes `newPoint` to inherit `Point`'s implementation and initialization code. Notice also that the `ColorPoint` class publicly names the `Point` class as its parent by listing p in its public interface CP_{pub}. Without thus naming its parent, the `ColorPoint` object type ($\mathbf{obj}.cp$) would not be a subtype of `Point`'s object type ($\mathbf{obj}.p$) because the only way to be a subtype of a partially abstract type is to extend that type. The row variable p in CP_{pub} signals that `ColorPoint` objects were formed via extension from `Point` prototypes, and hence their partially abstract types may be related via subtyping. This fact is the theoretical counterpart to the observation we made in Section 2 that the only way for implementation types to be related via subtyping is for them to be related via inheritance.

Since the `ColorPoint` class could have omitted p from CP_{pub}, this formulation of classes gives programmers explicit control over whether or not they wish a derived class's implementation object type to be a subtype of its parent's. As described in Section 2, such explicit control is useful for insuring that all objects of a certain type satisfy particular invariants (by guaranteeing they share a particular implementation). It is also useful for improving the efficiency of message sending and for typing binary methods. Additionally, when we combine this encapsulation mechanism with constructors that return the non-extensible form of object, we get *final* classes, classes that have no descendants. Such classes are used in Java as a security measure and in optimizing compilers to speed message sending.

Finally, the two `let` declarations in Figure 5 give constructors for the two classes that return proper objects, instead of prototypes. Technically, this shadowing is type correct because **pro** types are subtypes of **obj** types [FM95a].

6.4 Evaluation

The example in this section illustrates that a principled way to think about class-based object-oriented languages is as the combination of two orthogonal components: one, an object system that supports inheritance and message sending and two, an encapsulation mechanism that provides hiding. Referring to the class-evaluation checklist in Section 2.6, we can see that this approach successfully addresses each of the points listed there.

Coherent, Extensible Collection. The type checking of our object calculus insures that the methods defined in each class are compatible with each other: the methods are typed at class-creation time, not at object-instantiation time.

Guaranteed Initialization. The data abstraction mechanism provided by this framework allows each class to control the initialization of its objects by insuring that the only way to get an object from a particular class is to call its constructor. Hence class-writers can be sure that any invariants necessary for their objects will hold for all non-trivial expressions with the object type defined in their class. For example, the only way to get a nontrivial expression with type **obj**.p is to call the constructor for the `Point` class, insuring that such an expression has been properly initialized. This property holds with respect to derived classes as well, since the only way for the `ColorPoint` class to inherit from `Point` is to invoke the `Point` class constructor, again guaranteeing that `Point`'s invariants will hold.

Enforced Access Restrictions. The mechanism presented here also allows programmers to explicitly state which methods are public and which are private. Private methods are guaranteed to be used only within the class implementation itself; hence they may be changed with great flexibility.

Explicit Type Hierarchy. An additional strength of this model is that it provides explicit control over the bottom portion of the subtyping hierarchy, as discussed in Section 2.5. In particular, object types that include existentially bound row variables correspond to implementation types; the row variable signals the class which defined the type, putting such types in one-to-one correspondence with classes. The only subtyping that holds between these types is that declared via the inheritance hierarchy, *e.g.*, for **obj**.cp to be a subtype of **obj**.p, the `ColorPoint` class had to inherit from `Point`. Furthermore, to enable this subtyping relationship, the `ColorPoint` class had to explicit list `Point` as its parent by including p in CP_{pub}, the public interface for `ColorPoint` objects. Thus the bottom portion of the inheritance hierarchy is explicitly under programmer control. In contrast, object types without row variables correspond to interface types. For these types, subtyping is determined by inference rules that examine their structure. Finally, since we have subtyping relationships between these two kinds of types, *e.g.*,

$$\mathbf{obj.}p <: \mathbf{obj.}\langle\!\langle \texttt{getX} : int, \texttt{setX} : int \rightarrow unit \rangle\!\rangle$$

we get the rich, hybrid subtyping hierarchy described in Section 2.5. In particular, we get the benefits of implementation types for the bottom portion of the hierarchy, while maintaining the flexibility of interface types in the top portion.

Automatic Propagation of Base Class Changes. The class structure described in this section provides explicit support for inheritance. In particular, when a derived class inherits from its parent, it must do so by calling its parent's constructor since the full representation of the parent is hidden. Hence changes to the parent class are automatically reflected in all derived classes. If such a change causes a type error in a derived class, that error is detected as soon as the derived class is rechecked.

Problems with this approach. A weakness of the current formulation of our idea that "Classes = Objects + Data Abstraction" is that subclasses must be declared within the scope of their parent classes. We believe this shortcoming may be overcome by replacing our block-structured construct with modules and a dot notation in the style of [CL90, HL94]. A further weakness of our interpretation is the need for negative information. In effect, parent classes must currently "guess" what methods their descendants will want to add. We believe a variant of the kinding system used here that tracks method *presences* instead of absences combined with some form of α-binding to hide private method names may solve this problem. However, we leave these tasks to future work.

A problem that we do know how to solve, however, is that of finding a way for method types to refer to the type of their host object. Such a type, frequently called "mytype," is useful both in automatically specializing method types during inheritance and in supporting binary methods. As an example of the first of these roles, we would like the `setX` and `setC` methods from our example to have return types indicating that the expressions they return have the same type as the object to which they were sent. (Recall that we adopted an imperative semantics earlier to side-step this problem.) Because such types change during inheritance *e.g.*, `setX` returns *Point* objects when it is sent to *Points* and *ColorPoint* objects when sent to *ColorPoints*, the ability to name "mytype" is said to support *method specialization.* It also allows us to write binary methods. In the next section, we present a framework that combines the strengths (and the above weaknesses) of the current system with additional support for method specialization and binary methods.

7 Classes with "Mytype" Specialization

To support binary methods and method specialization, we need to have a way for method types to refer to the type of their host object. To that end, we use type variables and recall that both **pro** and **obj** are type binding operators. In particular, when a type variable u appears within one of the types τ_i in

$$\mathbf{pro}\, u.\langle\!\langle \mathtt{m_1} : \tau_1, \ldots, \mathtt{m_n} : \tau_n \rangle\!\rangle$$

u refers to the whole type (*i.e.*, to $\mathbf{pro}\, u.\langle\!\langle \mathtt{m_1} : \tau_1, \ldots, \mathtt{m_n} : \tau_n \rangle\!\rangle$), and similarly for **obj** types.

The row variables that we used to express unknown lists of methods in the previous section must now be parametric in the type of the host object. Hence in this section, row variables will range over *row functions*, functions from the type of the self object to the list of method-type pairs it supports. Row functions are written using the notation $\lambda t.\langle\!\langle \mathtt{m_1} : \tau_1, \ldots, \mathtt{m_k} : \tau_k \rangle\!\rangle$.

The final change necessary is that we need to be able to infer subtyping relationships between types of the form $\mathbf{obj}\, u.cp\, u$ and $\mathbf{obj}\, u.p\, u$. Now simply knowing that the *cp* row variable must stand for a row that extends whatever row p represents is not sufficient to insure the corresponding types are subtypes, because of the possible contravariance of the u type variable in the type $\mathbf{obj}\, u.p\, u$. Nor can we simply look at the type $\mathbf{obj}\, u.p\, u$ to determine that u appears covariantly, since the type contains only type and row variables. Instead, we need to add a variance analysis system to our kind system to keep track of the variance of type variables that appear in our type and row expressions. For example, if we may give a row function R kind $T^+ \rightarrow (M;\, \emptyset)$, then the variance annotation $+$ on the T indicates that R's argument type appears covariantly. The $\emptyset$ in the kind indicates that R contains no free type variables.

7.1 Example Classes with "Mytype"

Figure 6 shows our simple class hierarchy written in the full system. This version has the same structure as the one in the previous section, differing only in the addition of the "mytype" type variables and the variance annotations. Because of the richer type system, we may write the constructors for these two classes without using an imperative semantics, and we could write binary methods such as `eq` methods for the two classes.

7.2 Evaluation

The type system used to type check the above example is presented in full and proved type sound in [Fis96]. Because this version has the same class structure as in the previous section, it provides the same rich support for classes. Also as before, it provides a rich subtyping hierarchy, the bottom portion of which is under explicit programmer control. In addition, the system described here supports method specialization and binary methods. The remaining problems to be solved are to change the kinding system to track positive, instead of negative, information about method existence and to replace our block-structured abstraction mechanism with modules and a dot notation in the style of [CL90, HL94]. We leave these tasks to future work.

$$
\begin{array}{l}
\texttt{Abstype}\ p <:_w P_{pub} :: T^+ \rightarrow (\{\texttt{c}, \texttt{getC}, \texttt{setC}\};\ \emptyset) \\
\quad \texttt{with newPoint}\ :\ int \rightarrow \mathbf{pro}\, u\texttt{.}p\,u \\
\quad\quad \texttt{is}\ \{\!| p <:_w P_{pub} :: (T^+ \rightarrow (\{\texttt{c}, \texttt{getC}, \texttt{setC}\};\ \emptyset)) = P_{priv},\ \texttt{Imp}_p |\!\} \\
\quad\quad \texttt{in} \\
\quad\quad\quad \texttt{Abstype}\ cp <:_w CP_{pub} :: T^+ \rightarrow (\emptyset;\ \emptyset) \\
\quad\quad\quad\quad \texttt{with newCPoint}\ :\ int \rightarrow color \rightarrow \mathbf{pro}\, u\texttt{.}cp\,u \\
\quad\quad\quad\quad\quad \texttt{is}\ \{\!| cp <:_w CP_{pub} :: (T^+ \rightarrow (\emptyset;\ \emptyset)) = CP_{priv},\ \texttt{Imp}_{cp} |\!\} \\
\quad\quad\quad\quad\quad \texttt{in} \\
\quad\quad\quad\quad\quad\quad \texttt{let}\quad \texttt{newPoint} : int \rightarrow \boldsymbol{obj}\, u\texttt{.}p\, u = \texttt{newPoint} \\
\quad\quad\quad\quad\quad\quad \texttt{let}\quad \texttt{newCPoint} : int \rightarrow color \rightarrow \boldsymbol{obj}\, u\texttt{.}\, cp\, u = \texttt{newCPoint} \\
\quad\quad\quad\quad\quad\quad \texttt{in}\quad \langle \texttt{Program} \rangle
\end{array}
$$

where

$$
\begin{array}{ll}
P_{pub} & = \lambda u.\langle\!\langle \texttt{getX} : int,\ \texttt{setX} : int \rightarrow u \rangle\!\rangle \\
P_{priv} & = \lambda u.\langle\!\langle \texttt{x} : int,\ \texttt{getX} : int,\ \texttt{setX} : int \rightarrow u \rangle\!\rangle \\
CP_{pub} & = \lambda u.\langle\!\langle p\,u \,|\, \texttt{getC} : color,\ \texttt{setC} : color \rightarrow u \rangle\!\rangle \\
CP_{priv} & = \lambda u.\langle\!\langle p\,u \,|\, \texttt{c} : color,\ \texttt{getC} : color,\ \texttt{setC} : color \rightarrow u \rangle\!\rangle
\end{array}
$$

and

$$
\begin{array}{lrl}
\texttt{Imp}_p \stackrel{def}{=} \lambda\, \texttt{xi} : int.\langle & \texttt{x} = & \lambda(\texttt{self})\ \ 0; \\
& \texttt{getX} = & \lambda(\texttt{self})\ \ \texttt{self} \Leftarrow \texttt{x}; \\
& \texttt{setX} = & \lambda(\texttt{self})\ \ \lambda(\texttt{newX})\ \ \texttt{self} \leftarrow \lambda(\texttt{s})\texttt{newX}\rangle \Leftarrow \texttt{setX xi};
\end{array}
$$

$$
\begin{array}{l}
\texttt{Imp}_{cp} \stackrel{def}{=} \lambda\, \texttt{xi} : int.\lambda\, \texttt{ci} : color. \\
\qquad\qquad \langle\, \texttt{Imp}_p\ \texttt{xi}\ \ \longleftarrow\!\!+\ \ \ \texttt{c} = \lambda(\texttt{self})\ \ \texttt{red}; \\
\qquad\qquad\qquad\qquad\quad \longleftarrow\!\!+\ \texttt{getC} = \lambda(\texttt{self})\ \ \texttt{self} \Leftarrow \texttt{c}; \\
\qquad\qquad\qquad\qquad\quad \longleftarrow\!\!+\ \texttt{setC} = \lambda(\texttt{self})\ \ \lambda(\texttt{newC}) \\
\qquad\qquad\qquad\qquad\qquad\qquad\qquad \texttt{self} \leftarrow \lambda(\texttt{s})\texttt{newC}\rangle \Leftarrow \texttt{setC ic};
\end{array}
$$

Fig. 6. `Point` and `ColorPoint` classes encoded as nested abstract datatypes with support for method specialization.

8 Other Approaches

8.1 Existential Model

In [PT94], Pierce and Turner interpret classes as object-generating functions. In their interpretation, inheritance is interpreted as modifications to the object-generating functions that model classes. This encoding is somewhat cumbersome, since it requires programmers to explicitly manipulate `get` and `put` functions, which intuitively convert between the hidden state of parent class objects and derived class objects. Also, because this model provides protection at the object-level, as opposed to the class-level, binary methods require extra machinery. One such solution appears in [PT93].

In [HP96], Hofmann and Pierce introduce a refined version of $F_{<:}$ that permits only positive subtyping. With this restriction, get and put functions are both guaranteed to exist and hence may be handled in a more automatic fashion in class encodings. A disadvantage of this approach is that no subtyping is possible between existential types because the put functions for existential types are not meaningful. Since objects have existential type, this model does not currently support subtyping between object types; some form of explicit coercions are necessary. Extensions that combine positive subtyping with normal subtyping, addressing this lack of object subtyping, seem possible.

8.2 Direct Models

Kim Bruce has developed a family of type-safe formal languages that model classes directly instead of via an interpretation as the combination of more basic primitives. In [Bru93], Bruce describes TOOPL, a functional object-oriented language. PolyTOIL, presented in [BSv95], incorporates imperative features and introduces the notion of *matching*, a relationship between object types that holds whenever the first is an extension of the second, regardless of the variance of the "mytype" type variable. In these languages, the type of an object reflects only its public interface; it cannot convey implementation information.

Scott Smith and the Hopkins Object Group have designed a type-safe class-based object-oriented language with a rich feature set called I-LOOP, [EST95]. Their type system is based on polymorphic recursively constrained types, for which they have a sound type inferencing algorithm. The main advantage of this approach is the extreme flexibility afforded by recursively constrained types. Currently, the main problem is that it returns large, difficult-to-read types. Some form of simplification may be required. Work in this area is in progress.

9 Conclusion

Pure object systems are simpler than class-based systems; however, most of the empirical evidence for the utility of object-oriented languages is based on class-based languages, and certain features of classes seem critical to this success. These features include:

- Static checking that guarantees that the object "parts" defined in a class are consistent with each other (for example, all premethods must have consistent assumptions about the type of "self").
- Control over the initialization of objects (both in a given class and in all of its descendants). This control is essential for establishing inheritable invariants, for example.
- The ability to specify which methods and fields of a given class are private and which are public. These specifications should be meaningful both within the class itself and within its descendants.
- Explicit control over a portion of the subtyping hierarchy.

- Preservation of the relationships between classes when private or public components are added. This property implies that when changes are made to a parent class, those changes are automatically reflected in its descendants.

We believe it is important to evaluate pure object systems in light of their ability to support class-based programming styles. The fundamental constructs that seem necessary are (i) an extensible form of object, or perhaps pre-object, to support inheritance, (ii) a non-extensible form of object that can be created easily from an extensible form, and (iii) subtyping on the types of proper objects. Given structural subtyping on object types, we have shown in this paper how to gain more precise control over the class hierarchy using standard data abstraction, enhanced with subtype constraints as described in [CW85].

A further insight from the analysis in this paper is that the Abadi/Cardelli and Fisher/Mitchell calculi are incomparable. The Fisher/Mitchell calculi were originally formulated with method specialization in mind, which has led to typing rules for object formation that enforce a high degree of polymorphism. This polymorphism is necessary to make each method that is added to a prototype suitable for inclusion in any future extension of that prototype. However, it precludes the formation of certain objects that would behave sensibly only if not extended. The record-of-premethod class objects in Section 5 are an instance of this phenomenon. In its current form, the Fisher/Mitchell calculi cannot type these objects. Explicit universal quantification over rows must be added to remedy this deficiency. On the other hand, the Abadi/Cardelli family of calculi do not provide any form of extensible object. This lack makes it difficult to support class-based programming, for the reasons enumerated in Section 5. A key technical issue for further exploration is to devise a system with two forms of objects: an extensible form (similar to our prototypes) for representing class implementations and second form that supports method override and width subtyping (like the Abadi/Cardelli objects) to behave as proper objects, with a natural form of conversion between the first and the second forms. We do not see any fundamental impediment to completing a study of this additional part of the design space. Recent work by Luigi Liquori [Liq96] addresses this issue.

References

[Aba96] M. Abadi. A few thoughts on objects (ML2000). Private Email, August 1996.

[AC95] M. Abadi and L. Cardelli. A theory of primitive objects: Second-order systems. *Science of Computer Programming*, 25(2-3):81–116, December 1995. A preliminary version appeared in the 1994 Proc. of European Symposium on Programming.

[AC96a] M. Abadi and L. Cardelli. An imperative object calculus. *Theory and Practice of Object Systems*, 1(3):151–166, 1996. Earlier version appeared in TAPSOFT '95 proceedings.

[AC96b] M. Abadi and L. Cardelli. *A Theory of Objects.* Springer, 1996.

[AC96c] M. Abadi and L. Cardelli. A theory of primitive objects: Untyped and first-order systems. *Information and Computation*, 125(2):78–102, 1996. Earlier version appeared in TACS '94 proceedings, LNCS 789.

[AG96] K. Arnold and J. Gosling. *The Java Programming Language*. Addison Wesley, 1996.

[BCC+96] K. Bruce, L. Cardelli, G. Castagna, The Hopkins Object Group, G. Leavens, and B. Pierce. On binary methods. *Theory and Practice of Object Systems*, 1(3):221–242, 1996.

[BDMN73] G.M. Birtwistle, O.-J. Dahl, B. Myhrhaug, and K. Nygaard. *Simula Begin*. Studentlitteratur, Box 1717, S-222 01 Lund, Sweden; Auerbach, Philadelphia, 1973.

[Bru93] K. Bruce. Safe type checking in a statically-typed object-oriented programming language. In *Proc. 20th ACM Symp. Principles of Programming Languages*, pages 285–298, 1993.

[BSv95] K. Bruce, A. Schuett, and R. van Gent. PolyTOIL: A type-safe polymorphic object-oriented language. In *Proc. 9th European Conference on Object-Oriented Programming*, pages 26–51, Aarhus, Denmark, 1995. Springer LNCS 952.

[Car95] L. Cardelli. A language with distributed scope. *Computing Systems*, 8(1):27–59, January 1995.

[Cha95] C. Chambers. The Cecil language: Specification and rationale. Available electronically from http://www.cs.washington.edu/research/projects/cecil/www/Papers/cecil-spec.html, December 1995.

[CL90] L. Cardelli and X. Leroy. *Abstract types and the dot notation*, pages 479–504. IFIP State of the Art Reports. North Holland, March 1990. Also appeared as SRC Research Report 56.

[Coo87] W.R. Cook. A *self*-ish model of inheritance. Manuscript, 1987.

[Coo89] W.R. Cook. *A Denotational Semantics of Inheritance*. PhD thesis, Brown University, 1989.

[CW85] L. Cardelli and P. Wegner. On understanding types, data abstraction, and polymorphism. *Computing Surveys*, 17(4):471–522, 1985.

[DF96] P. DiBlasio and K. Fisher. A concurrent object calculus. In *CONCUR '96 Proc.*, pages 655–670, Pisa, 1996. Springer LNCS 1119.

[ES90] M. Ellis and B. Stroustrup. *The Annotated C^{++} Reference Manual*. Addison-Wesley, 1990.

[EST95] J. Eifrig, S. Smith, and V. Trifonov. Sound polymorphic type inference for objects. In *ACM Conf. Object-Oriented Programming: Systems, Languages and Applications*, pages 169–184, October 1995.

[FHM94] K. Fisher, F. Honsell, and J.C. Mitchell. A lambda calculus of objects and method specialization. *Nordic J. Computing (*formerly *BIT)*, 1:3–37, 1994. Preliminary version appeared in *Proc. IEEE Symp. on Logic in Computer Science,* 1993, 26–38.

[Fis96] K. Fisher. *Type Systems for object-oriented programming languages*. PhD thesis, Stanford University, 1996.

[FM95a] K. Fisher and J.C. Mitchell. A delegation-based object calculus with subtyping. In *Proc. 10th Int'l Conf. Fundamentals of Computation Theory (FCT'95)*, pages 42–61. Springer LNCS 965, 1995.

[FM95b] K. Fisher and J.C. Mitchell. The development of type systems for object-oriented languages. *Theory and Practice of Object Systems*, 1(3):189–220,

1995. Preliminary version appeared in TACS '94 proceedings.

[GM94] C.A. Gunter and J.C. Mitchell, editors. *Theoretical Aspects of Object-Oriented Programming.* MIT Press, Cambridge, MA, 1994.

[GR83] A. Goldberg and D. Robson. *Smalltalk–80: The Language and its Implementation.* Addison Wesley, 1983.

[HL94] R. Harper and M. Lillibridge. A type-theoretic approach to higher-order modules with sharing. In *Proc. 21st ACM Symp. on Principles of Programming Languages*, 1994.

[HP96] M. Hofmann and B. Pierce. Positive subtyping. *Information and Computation*, 126(1):11–33, 1996. Preliminary version appeared in *Proc. 22nd ACM Symp. on Principles of Programming Languages*, 1995.

[KLM94] D. Katiyar, D. Luckham, and J.C. Mitchell. A type system for prototyping languages. In *Proc. 21-st ACM Symp. on Principles of Programming Languages*, 1994.

[L+81] B. Liskov et al. *CLU Reference Manual.* Springer LNCS 114, Berlin, 1981.

[Liq96] L. Liquori. An extended theory of primitive objects: First and second order systems. Technical Report CS-23-96, Dipartimento di Informatica, Universitá di Torino, 1996.

[LSAS77] B. Liskov, A. Snyder, R. Atkinson, and C. Schaffert. Abstraction mechanisms in CLU. *Comm. ACM*, 20:564–576, 1977.

[Mey92] B. Meyer. *Eiffel: The Language.* Prentice-Hall, 1992.

[Mit90] J.C. Mitchell. Toward a typed foundation for method specialization and inheritance. In *Proc. 17th ACM Symp. on Principles of Programming Languages*, pages 109–124, January 1990.

[Mit93] J.C. Mitchell. *Introduction to Programming Language Theory.* MIT Press, to be published 1993.

[Mor73] J.H. Morris. Types are not sets. In 1st *ACM Symp. on Principles of Programming Languages*, pages 120–124, 1973.

[MP88] J.C. Mitchell and G.D. Plotkin. Abstract types have existential types. *ACM Trans. on Programming Languages and Systems*, 10(3):470–502, 1988. Preliminary version appeared in *Proc. 12th ACM Symp. on Principles of Programming Languages,* 1985.

[MTH90] R. Milner, M. Tofte, and R. Harper. *The Definition of Standard ML.* MIT Press, 1990.

[PT93] B.C. Pierce and D.N. Turner. Statically typed friendly functions via partially abstract types. Technical Report ECS-LFCS-93-256, University of Edinburgh, LFCS, April 1993. Also available as INRIA-Rocquencourt Rapport de Recherche No. 1899.

[PT94] B.C. Pierce and D.N. Turner. Simple type-theoretic foundations for object-oriented programming. *Journal of Functional Programming*, 4(2):207–248, 1994. Preliminary version appeared in *Proc. 20th ACM Symp. on Principles of Programming Languages*, 1993, under the title "Object-oriented programming without recursive types".

[Rey83] J.C. Reynolds. Types, abstraction, and parametric polymorphism. In *Information Processing '83*, pages 513–523. North-Holland, Amsterdam, 1983.

[RR96] J.H. Reppy and J.G. Riecke. Classes in Object ML via modules, 1996. Presented at FOOL3 workshop.

[Set89] R. Sethi. *Programming Languages: Concepts and Constructs.* Addison-Wesley, 1989.

[Ste84] G.L. Steele. *Common Lisp: The Language.* Digital Press, 1984.

[Str86] B. Stroustrup. *The C++ Programming Language*. Addison-Wesley, 1986.

[Str94] B. Stroustrup. *The Design and Evolution of* C^{++}. Addison-Wesley, 1994. Chapter 3: The birth of C^{++}.

[US 80] US Dept. of Defense. *Reference Manual for the Ada Programming Language*. GPO 008-000-00354-8, 1980.

[US91] D. Ungar and R.B. Smith. Self: The power of simplicity. *Lisp and Symbolic Computation*, 4(3):187–206, 1991. Preliminary version appeared in *Proc. ACM Symp. on Object-Oriented Programming: Systems, Languages, and Applications,* 1987, 227-241.

[Wan87] M. Wand. Complete type inference for simple objects. In *Proc. IEEE Symp. on Logic in Computer Science*, pages 37–44, 1987. Corrigendum in *Proc. IEEE Symp. on Logic in Computer Science,* page 132, 1988.

The Oz Programming Model

Gert Smolka
Programming Systems Lab
German Research Center for Artificial Intelligence (DFKI)
Stuhlsatzenhausweg 3, 66123 Saarbrücken, Germany
email: `smolka@dfki.uni-sb.de`

Abstract

The Oz Programming Model (OPM) is a concurrent programming model subsuming higher-order functional and object-oriented programming as facets of a general model. This is particularly interesting for concurrent object-oriented programming, for which no comprehensive formal model existed until now. The model can be extended so that it can express encapsulated problem solvers generalizing the problem solving capabilities of constraint logic programming. OPM has been developed together with a concomitant programming language Oz, which is designed for applications that require complex symbolic computations, organization into multiple agents, and soft real-time control. An efficient, robust, and interactive implementation of Oz is freely available.

This paper has appeared in: Computer Science Today, Jan van Leeuwen, editor, Lecture Notes in Computer Science, Volume 1000, pages 324–343, Springer-Verlag, Berlin, 1995.

1 Introduction

Computer systems are undergoing a revolution. Twenty years ago, they were centralized, isolated, and expensive. Today, they are parallel, distributed, networked, and inexpensive. However, advances in software construction have failed to keep pace with advances in hardware. To a large extent, this is a consequence of the fact that current programming languages were conceived for sequential and centralized programming.

A basic problem with existing programming languages is that they delegate the creation and coordination of concurrent computational activities to the underlying operating system and network protocols. This has the severe disadvantage that the data abstractions of the programming language cannot be shared between communicating computational agents. Thus the benefits of existing programming languages do not extend to the central concerns of concurrent and distributed software systems.

Given this state of affairs, the development of concurrent programming models is an important research issue in Computer Science. A concurrent programming model must support the creation and coordination of multiple computational activities. Simple concurrent programming models can be obtained by accommodating concurrency in the basic control structure of the model. This way concurrency appears as a generalization rather than an additional feature.

The development of simple, practical, high-level, and well-founded concurrent programming models turned out to be difficult. The main problem was the lack of a methodology and formal machinery for designing and defining such models. In the 1980's, significant progress has been made on this issue. This includes the development of abstract syntax and structural operational semantics [17, 27]; functional and logic programming, two declarative programming models building on the work of logicians (lambda calculus and predicate logic); CCS [12] and the π-calculus [13], two well-founded concurrent programming models developed by Milner and others; and the concurrent constraint model [10, 18], a concurrent programming model that originated from application-driven research in concurrent logic programming [21] and constraint logic programming [6].

This paper reports on the Oz Programming Model, OPM for short, which has been developed together with the concurrent high-level programming language Oz. OPM is an extension of the basic concurrent constraint model, adding first-class procedures and stateful data structures. OPM is a concurrent programming model that subsumes higher-order functional and object-oriented programming as facets of a general model. This is particularly interesting for concurrent object-oriented programming, for which no comprehensive formal model existed until now. There is a conservative extension of OPM providing the problem-solving capabilities of constraint logic programming. The resulting problem solvers appear as concurrent agents encapsulating search and speculative computation with constraints.

Oz and OPM have been developed at the DFKI since 1991. Oz [25, 23, 22] is designed as a concurrent high-level language that can replace sequential high-level languages such as Lisp, Prolog and Smalltalk. There is no other concurrent language combining a rich object system with advanced features for symbolic processing and problem solving. First applications of Oz include simulations, multi-agent systems, natural language processing, virtual reality, graphical user interfaces, scheduling, time tabling, placement problems, and configuration. The design and implementation of Oz took ideas from AKL [7], the first concurrent constraint language with encapsulated search.

An efficient, robust, and interactive implementation of Oz, DFKI Oz, is freely available for many Unix-based platforms (see remark at the end of this paper). DFKI Oz features a programming interface based on GNU Emacs, a concurrent browser, an object-oriented interface to Tcl/Tk for building graphical user interfaces, powerful interoperability features, an incremental compiler, and a run-time system with an emulator and a garbage collector.

DFKI Oz proves that an inherently concurrent language can be implemented efficiently on sequential hardware. Research on a portable parallel implementation for shared memory machines has started. More ambitiously, we have also begun work towards a distributed version of Oz supporting the construction of open systems.

This paper describes OPM in an informal manner. Calculi formalizing the major aspects of OPM can be found in [23, 22]. The Oz Primer [25] is an introduction to programming in Oz. Basic implementation techniques for Oz are reported in [11].

2 Computation Spaces

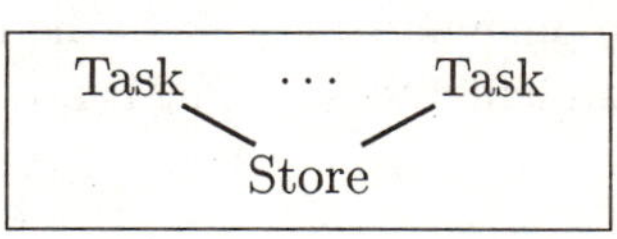

Computation in OPM takes place in a *computation space* hosting a number of *tasks* connected to a shared *store*. Computation advances by *reduction of tasks*. The reduction of a task can manipulate the store and create new tasks. When a task is reduced it disappears. Reduction of tasks is an atomic operation, and tasks are reduced one by one. Thus there is no parallelism at the abstraction level of OPM.

Tasks can *synchronize* on the store in that they become reducible only once the store satisfies certain conditions. A key property of OPM is that *task synchronization is monotonic*, that is, a reducible task stays reducible if other tasks are reduced before it.

Typically, many tasks are reducible in a given state of a computation space. To obtain fairness, reactivity, and efficiency, a *reduction strategy* is needed to select the reducible tasks qualifying for the next reduction step. *Fairness* ensures that several groups of tasks can advance simultaneously. *Reactivity* means that one can create computations that react to outside events within foreseeable time bounds. The following is an example of a fair and reactive reduction strategy:

> All tasks are maintained in a queue, where the first task of the queue is the one to be considered next for reduction. If it is not reducible, it is moved to the end of the queue. If it is reducible, it is reduced and the newly created tasks are appended at the end of the queue.

We will see later that this strategy is inefficient since its degree of fairness is too fine-grained for OPM. A practical reduction strategy will be given in Section 16.

3 Concurrency and Parallelism

OPM is a concurrent and nonparallel programming model. Concurrency means that one can create several simultaneously advancing computations, possibly synchronizing and communicating. Parallelism means that the execution of several hardware operations overlaps in time. Concurrency can be obtained in a nonparallel setting by interleaving reduction steps. This is typically the case in operating systems that advance several concurrent processes on single processor machines. We can see concurrency as a programming abstraction and parallelism as a physical phenomenon.

The fact that OPM is nonparallel does not exclude a parallel implementation, however. The reason for making OPM concurrent but not parallel is the desire to make things as simple as possible for programmers. In OPM, the semantics of programs does not depend on whether they run on a sequential or parallel implementation. Thus the complexities of parallelism need only concern the implementors of OPM, not the programmers.

4 Synchronization as Logic Entailment

We will now see how OPM realizes monotonic task synchronization. The basic idea is very simple. We assume that a set of logic formulas, called *constraints*, is given. The set of constraints is closed under conjunction, and for constraints a logic entailment relation ("C implies D") is defined. We also assume that the store of a computation space holds a constraint in a special compartment, called the *constraint store*. The only way the constraint store can be updated is by *telling* it a constraint C, which means that the constraint store advances from S to the conjunction $S \wedge C$. Finally, we assume that it is possible to synchronize a task on a constraint, called its *guard.* A synchronized task becomes reducible if its guard is entailed by the constraint store.

It is easy to see that this synchronization mechanism is monotonic. At any point in time, the constraint store can be seen as a conjunction

$$\texttt{true} \wedge C_1 \wedge C_2 \wedge \ldots \wedge C_n$$

where $C_1, \ldots, C_n$ are the constraints told so far. The beauty of this arrangement is that the information in the constraint store increases monotonically with every further constraint told, and that the order in which constraints are told is insignificant as far as the information in the store is concerned (conjunction is an associative and commutative operation).

We assume that the constraint store is always satisfiable. Consequently, it is impossible to tell a constraint store S a constraint C if the conjunction $S \wedge C$ is unsatisfiable.

It suffices to represent the constraint store modulo logic equivalence. This means that the synchronization mechanism is completely declarative. It turns out that there are constraint systems for which synchronization as entailment is both expressive and efficient.

Synchronization on a constraint store appeared first in Prolog II [4] in the primitive form of the so-called freeze construct. The idea to synchronize on entailment of constraints is due to Maher [10].

5 Constraint Structures

We now make precise the notions of constraint and entailment. We will also see that the constraint store is the place where information about the values participating in a computation is stored. An important property of the constraint store is the fact that it can store partial (i.e., incomplete) information about the values of variables.

A *constraint structure* is a structure of first-order predicate logic. The elements of a constraint structure are called *values*, and the first-order formulas over the signature of a constraint structure are called *constraints.* We assume that constraints are built over a fixed infinite alphabet of variables. A constraint C *entails* a constraint D if the implication $C \rightarrow D$ is valid in the constraint structure. A constraint C *disentails* a constraint D if C entails $\neg D$. Two constraints C and D are *equivalent* if C entails D and D entails C.

The constraint structure must be chosen such that its elements are the values we want to compute with. The values will typically include numbers, ordered pairs of values, and additional primitive entities called names. Values can be thought of as stateless data structures. Note that this set-up requires that values are defined as mathematical entities, and that operations on values are described as mathematical functions and relations.

To ensure that checking entailment between the constraint store and guards is computationally inexpensive, one must carefully restrict the constraints that can be written in the constraint store and that can be used as guards.

We now outline a concrete constraint structure INP. As values of INP we take the integers, an infinite set of primitive entities called *names*, and all ordered pairs that can be obtained over integers and names. We write $v_1|v_2$ for the ordered pair whose left component is the value v_1 and whose right component is the value v_2. Moreover, we assume that the signature of INP provides the following *primitive constraints*:

- $x = n$ says that the value of the variable x is the integer n.
- $x = \xi$ says that the value of the variable x is the name ξ.
- $x = y|z$ says that the value of the variable x is the pair having the value of the variable y as left and the value of the variable z as right component.
- $x = y$ says that the variables x and y have the same value.

An example of a constraint store over INP is

$$x = y \ \wedge \ y = z|u \ \wedge \ z = 3.$$

This constraint store asserts that the value of z is 3, that the value of y is a pair whose left component is 3, and that x and y have the same value. While this constraint store has total information about the value of the variable z, it has only partial information about the values of the other variables. In fact, it has no information about any variable other than x, y and z.

The constraint store above entails the constraint $x = 3|u$ and disentails the constraint $x = 3$. It neither entails nor disentails the constraint $y = 3|5$.

In practice, one uses more expressive constraint structures than INP. The constraint structure CFT [26, 2] offers constraints over possibly infinite records called feature trees. Oz employs an extension of CFT.

6 A Simple Concurrent Constraint Language

We now present a sublanguage OCC of OPM that is also a sublanguage of Saraswat's concurrent constraint model [18]. OCC cannot yet express indeterministic choice, which we will accommodate later (see Section 8).

The store of an OCC computation space consists only of the constraint store. As constraint structure we take INP to be concrete. As tasks we take expressions according to the abstract syntax

E	$::=$	C	*constraint*
	$\mid$	$E_1 \wedge E_2$	*composition*
	$\mid$	**if** C **then** E_1 **else** E_2	*conditional*
	$\mid$	**local** x **in** E	*declaration*

where C ranges over a suitably restricted class of constraints, and where x ranges over the variables used in constraints. A declaration **local** x **in** E binds the variable x with scope E. *Free* and *bound* variables of expressions are defined accordingly.

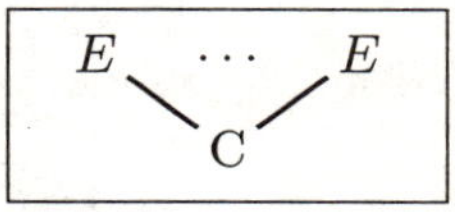

An OCC computation space consists of tasks which are expressions as defined above and a store which is a satisfiable constraint. Tasks which are constraints, compositions or declarations are unsynchronized. Conditional tasks synchronize on the constraint store and become reducible only once their guard is entailed or disentailed by the constraint store.

The reduction of a constraint task C *tells* the constraint store the constraint C. We say that such a reduction performs a *tell operation*. If the conjunction $S \wedge C$ of the present constraint store S and C is satisfiable, the reduction of the task C will advance the constraint store to $S \wedge C$. If the conjunction $S \wedge C$ of the present constraint store S and C is unsatisfiable, the reduction of the task C will not change the constraint store and announce *failure*. A concrete language has three possibilities to handle the announcement of failure: to ignore it, to abort computation, or to handle it by an exception handling mechanism.

Reduction of a composition $E_1 \wedge E_2$ creates two tasks E_1 and E_2. Reduction of a conditional **if** C **then** E_1 **else** E_2 creates the task E_1 if C is entailed and the task E_2 if C is disentailed by the constraint store. Reduction of a declaration **local** x **in** E chooses a fresh variable y and creates the task $E[y/x]$ obtained from E by replacing all free occurrences of x with y. A variable is *fresh* if it does not occur in the current state of the computation space.

The expressions of OCC provide basic operations for concurrent programming. Compositions make it possible to obtain several concurrent tasks from a single task. Conditionals make it possible to synchronize tasks on the constraint store. Telling constraints makes it possible to fire synchronized tasks. Declarations make it possible to obtain fresh variables. This will become significant as soon as we introduce procedures. For now observe that two identical tasks **local** x **in** E will reduce to two different tasks $E[y/x]$ and $E[z/x]$, where y and z are distinct fresh variables.

Telling constraints makes it possible to assert information about the values of variables (e.g., $x = 7$). The combination of conditionals and telling makes it possible to access the constituents of nonprimitive values. The task

$$\textbf{if } \exists y \exists z (x = y|z) \textbf{ then } x = u|v \textbf{ else } E$$

will equate the variables u and v to the left and right component of x if x turns out to be pair, and reduce to the task E otherwise. We call this construction a *synchronized decomposition.* To have a convenient notation, we will write

$$\textbf{if } x_1 \ldots x_n \textbf{ in } C \textbf{ then } E_1 \textbf{ else } E_2$$

as an abbreviation for

$$\textbf{if } \exists x_1 \ \ldots \ \exists x_n \, C \textbf{ then local } x_1 \textbf{ in } \ldots \textbf{ local } x_n \textbf{ in } (C \wedge E_1) \textbf{ else } E_2$$

With that we can write the above task as

$$\textbf{if } y \ z \textbf{ in } x = y|z \textbf{ then } u = y \wedge v = z \textbf{ else } E$$

The reason for having the conditional synchronize symmetrically on entailment and disentailment is that the incremental algorithms for checking entailment automatically also check for disentailment [1, 26]. These algorithms have in fact three outcomes: entailed, disentailed, or neither. The symmetric form of the conditional also has the nice property that it makes negated guards unnecessary since **if** $\neg C$ **then** E_1 **else** E_2 is equivalent to **if** C **then** E_2 **else** E_1.

Given a state of a computation space, we say that a variable x is *bound to* an integer n [a name ξ, a pair] if the constraint store entails the constraint $x = n$ [$x = \xi$, $\exists y \exists z (x = y|z)$].

7 First-class Procedures

Every programming language has procedures. Procedures are the basic mechanism for expressing programming abstractions. If provided in full generality, procedures have spectacular expressivity. As is well-known from the lambda calculus, creation and application of nonrecursive functional procedures alone can express all computable functions.

A programming language provides *first-class procedures* if

- procedures can create new procedures,
- procedures can have lexically scoped global variables,
- procedures are referred to by first-class values.

First-class procedures are available in functional programming languages such as Scheme, SML or Haskell. They are typically not available in today's concurrent programming languages although they can provide crucial functionality for concurrent and distributed programming (see the later sections of this paper and also [3]).

In OPM, a *procedure* is a triple

$$\xi : z/E$$

consisting of a name ξ (see Section 5), a *formal argument* z (a variable), and a *body* E (an expression). A procedure binds its formal argument z with scope E. The *free* or *global* variables of a procedure are defined accordingly. Procedures can actually have any number of formal arguments, but for now we consider only one argument to ease our presentation.

Besides the constraint store, OPM's store has a second compartment called the *procedure store*. The procedure store contains finitely many procedures such that for one name there is at most one procedure. Once a procedure has been entered into the procedure store, it cannot be retracted. Information about the values of the global variables of a procedure is kept in the constraint store. What we call a procedure is often called a closure in the literature.

There are two new expressions for creating and applying procedures:

E	$::=$	**proc** $\{x\ z\}\ E$	*definition*
	$\mid$	$\{x\ y\}$	*application*

A definition **proc** $\{x\ z\}\ E$ binds its *formal argument* z (a variable) with scope E. Definitions are always reducible. The reduction of a definition **proc** $\{x\ z\}\ E$ chooses a fresh name ξ, tells the constraint store the constraint $x = \xi$, and writes the new procedure $\xi : z/E$ into the procedure store.

An application $\{x\ y\}$ must wait until the procedure store contains a procedure $\xi : z/E$ such that the constraint store entails $x = \xi$. If this is the case, the application task $\{x\ y\}$ can reduce to the task $E[y/z]$, which is obtained from the body of the procedure by replacing all free occurrences of the formal argument z with the actual argument y, avoiding capturing.

The sublanguage of OPM introduced so far can express both eager and lazy higher-order functional programming [23]. For instance, a higher-order function

$$\textit{MkMap}: (\textit{Value} \rightarrow \textit{Value}) \rightarrow (\textit{List} \rightarrow \textit{List})$$

returning a list mapping function can be expressed as a binary procedure

```
proc {MkMap F Map}
   proc {Map Xs Ys}
      if X Xr in Xs=X|Xr then
         local Y Yr in Ys=Y|Yr  {F X Y}  {Map Xr Yr} end
```

```
      else Ys=Nil fi
   end
end
```

We are now using concrete Oz syntax, where a composition $E_1 \wedge E_2$ is written as a juxtaposition $E_1\ E_2$. A list $v_1, \ldots, v_n$ is represented as a nested pair $(v_1|(\ldots(v_n|\nu))\ldots)$, where ν is a name representing the empty list. We assume that the variable `Nil` is bound to ν. The procedure `MkMap` takes a binary procedure `F` as input and creates a binary procedure `Map` mapping lists elementwise according to `F`.

Since our model employs logic variables, there is no static distinction between input and output arguments. The functionality offered by a procedure $\xi{:}\,z/E$ is simply the ability to spawn any number of tasks $E[y/z]$, where the variable y replacing the formal argument z can be chosen freely each time.

To ease our notation, we will suppress auxiliary variables by means of *nesting*. For instance, we will write

```
{{MkMap F} 1|2|Nil X}
```

as an abbreviation for

```
local Map One Two A B in
   {MkMap F Map}  One=1  Two=2  A=One|B  B=Two|Nil  {Map A X}
end
```

The procedure `MkMap` actually implements a *concurrent function*. For instance, the task

```
{{MkMap F} A|B|C X}
```

will tell the constraint `X=U|V|W`, where `U`, `V`, and `W` are fresh variables. It will also create tasks that automatically synchronize on the variables `F`, `A`, `B`, and `C` and that will compute the values of `U`, `V`, and `W` when the necessary information is available.

The representation of functional computation as concurrent computation has been studied carefully for calculi formalizing the relevant aspects of OPM [23, 16, 15]. The main results include the identification of confluent subcalculi, embeddings of the eager and the lazy lambda calculus, and a correctness proof for the eager embedding. Lazy functional programming can be embedded such that argument computations are shared, a crucial feature of implementations that cannot be modeled with the lambda calculus [9].

OPM combines higher-order programming with first-order constraints. The idea to interface variables and procedures through freshly chosen names appeared first in Fresh [24].

8 Cells

Besides the constraint and the procedure store, OPM's store has a third and final compartment called the *cell store.* A cell is a mutable binding of a name to a variable. Cells make it possible to express stateful and concurrent data structures, which can serve as a communication medium between concurrent agents. There is an exchange operation on cells that combines reading and writing into a single atomic operation, thus providing mutual exclusion and indeterminism as needed for many-to-one communication.

The cell store contains finitely many *cells* $\xi\colon x$ representing mutable bindings of names to variables. Similar to the procedure store, the cell store contains at most one cell per name. Given a cell $\xi\colon x$ in the cell store, we say that the cell ξ *hosts* the variable x. The task

```
{NewCell X Y}
```

chooses a fresh name ξ, tells the constraint store the constraint $\mathtt{Y} = \xi$, and writes the new cell $\xi\colon\mathtt{X}$ into the cell store. Once a cell has been entered into the cell store, it cannot be retracted. The task

```
{Exchange X Y Z}
```

must wait until the cell store contains a cell $\xi\colon u$ such that the constraint store entails $\mathtt{X} = \xi$. The task can then be reduced by updating the cell to host the variable Z and telling the constraint store the constraint $\mathtt{Y} = u$.

Cells introduce indeterminism into OPM since the order in which multiple exchange tasks for the same cell are reduced is unspecified.

Cells are different from assignable variables in multi-threaded imperative languages. For one thing, OPM ensures mutual exclusion for concurrent exchange tasks for the same cell (since OPM is nonparallel and task reduction is an atomic operation). Moreover, an exchange task combines reading and writing of a cell into a single atomic operation. In the presence of logic variables, this atomic combination turns out to be expressive since one can write a new variable into a cell whose value will be computed only afterwards from the value of the old variable in the cell. This cannot be obtained in an imperative setting since it requires that consumers of a variable are automatically synchronized on the event that the value of the variable becomes known.

9 Ports

Building on cells, we can express complex concurrent data structures with state. The internal structure of such data structures can be hidden by means of procedural abstraction and lexical scoping of variables. We can thus obtain abstract concurrent data types with state.

As a first example we consider ports [8], which can serve as message queues for agents. A *port* is a procedure connected to a stream. A *stream* is a variable S that is incrementally constrained to a list by telling a constraint for every element of the list:

$S=X_1|S_1,\ S_1=X_2|S_2,\ S_2=X_3|S_3,\ S_3=X_4|S_4,\ \ldots$

It is assumed that nobody but the procedure P writes on the stream. An application {P X} will tell a constraint $S_i=X|S_{i+1}$, where S_i is the current tail of the stream and S_{i+1} is a new variable serving as the new tail of the stream. A port has state because it must remember the current tail of its stream. A port is a concurrent data structure since it allows several concurrent computations to write consistently on a single stream.

The procedure

```
proc {NewPort Stream Port}
   local Cell in
      {NewCell Stream Cell}
      proc {Port Message}
         local Old New in
            {Exchange Cell Old New}  Old=Message|New
         end
      end
   end
end
```

creates a new port Port connected to a stream Stream. The port holds the current tail of its stream in a private cell Cell. Note how lexical scoping ensures that no one but the port can see the cell. Also note that NewPort is a higher-order procedure in that it creates and returns a new procedure Port.

How can we enter two messages A and B to a port such that A appears before B on the associated stream? To make things more interesting, we are looking for a solution making it possible that other concurrently sent messages can be received between A and B (it may take a long time before B is sent).

One possible solution makes assumptions about the employed reduction strategy (see Section 16). Here we will give a solution that will work for every reduction strategy. The basic idea is to model a port as a binary procedure

```
{Port Message Continuation}
```

that will tell the constraint Continuation=Port after Message has been put on the stream [8]. Two messages A and B can then be sequentialized by writing

```
local Continuation Dummy in
   {Port A Continuation} {Continuation B Dummy}
end
```

Such synchronizing ports can be created with

```
proc {NewSyncPort Stream Port}
   local Cell in
      {NewCell Port|Stream Cell}
      proc {Port Message Continuation}
         local New in
            {Exchange Cell Continuation|Message|New Port|New}
         end
      end
   end
end
```

10 Names

Names serve as dynamically created capabilities that cannot be faked. It is often useful to be able to obtain reference to fresh names that do not designate procedures or cells. For this purpose we introduce a primitive task

```
{NewName X}
```

which can be reduced by choosing a fresh name ξ and telling the constraint X=ξ. Referring to names by means of variables has the advantage of lexical scoping and also avoids the need for a concrete syntax for names. Using names, lexical scoping, and procedures, sophisticated access control schemes can be expressed.

11 Agents

An agent is a computational abstraction processing messages received through a port. It maintains an internal state and may send messages to other agents. An example of an agent is a queue that can handle concurrent enqueue and dequeue requests.

We assume that the functionality of an agent is given by a procedure

$$\textit{Serve}\colon\ \textit{State} \times \textit{Message} \ \rightarrow\ \textit{NewState}$$

describing how the agent serves a message and how it advances its state. The procedure

```
proc {NewAgent Serve Init Port}
   local Stream Feed in
      {NewPort Stream Port}
      {Feed Stream Init}
      proc {Feed Ms State}
```

```
            if Message Mr NewState in Ms=Message|Mr then
               {Serve State Message NewState}
               {Feed Mr NewState}
            else true fi
         end
      end
   end
```

creates a new agent that receives messages through `Port` and operates as specified by the procedure `Serve` and the initial state `Init`. Note that an agent hides the stream queueing its messages.

A queue agent receiving messages through a port `Q` can be created with

```
local Xs in {NewAgent QueueServe Xs|Xs Q} end
```

where the procedure `QueueServe` is defined as follows:

```
{NewName Enqueue}
{NewName Dequeue}
proc {QueueServe State Message NewState}
   if First Last in State=First|Last then
      if X NewLast in Message=Enqueue|X then
         Last=X|NewLast  NewState=First|NewLast
      else
         if X NewFirst in Message=Dequeue|X then
            First=X|NewFirst  NewState=NewFirst|Last
         else true fi
      fi
   else true fi
end
```

Messages are represented as pairs `Enqueue|X` and `Dequeue|X`, where the variables `Enqueue` and `Dequeue` are bound to names identifying the corresponding operations. Using lexical scoping, one can construct contexts in which none or only one of the two operations is visible.

A message `Enqueue|X` will enqueue `X`, and a message `Dequeue|X` will dequeue an item and bind it to `X`. In case the queue is empty, a dequeue request will wait in a queue of unserved dequeue requests, which is served as soon as an item is entered into the queue. The procedure `QueueServe` shows that this synchronization idea can be expressed elegantly by means of logic variables.

12 Objects

Objects are a modular programming abstraction for concurrent data structures with state. We model objects as procedures `{Object Message}` that are applied to messages. A message is a pair `MethodName|Argument`. When an object is applied to a message, it invokes the requested method with the given argument and advances to a new state. Similar to agents, we assume that the functionality of an object is specified by a procedure

$$\mathit{Serve}\colon\ \mathit{State} \times \mathit{Message} \times \mathit{Self} \ \rightarrow\ \mathit{NewState}$$

describing how the agent serves a message and how it advances its state. The argument *Self* is a reference to the object invoking *Serve* making it possible to have a self reference within *Serve* and still share *Serve* between several objects. The procedure

```
proc {NewObject Serve Init Object}
   local Cell in
      {NewCell Init Cell}
      proc {Object Message}
         local State NewState in
            {Exchange Cell State NewState}
            {Serve State Message Object NewState}
         end
      end
   end
end
```

creates a new object `Object` from a procedure `Serve` and an initial state `Init`.

It is straightforward to express classes defining serve procedures in a modular fashion by means of named methods. Methods are modeled as procedures similar to serve procedures. Objects can then be obtained as instances of classes. The states of objects are modeled as finite mappings from attributes to variables, where attributes are modeled as names. Methods can then construct new states from given states by "assigning" variables to attributes. One can also provide for *inheritance*, that is, the ability to construct new classes by inheriting methods and attributes from existing classes. All this is a matter of straightforward higher-order programming. Exploiting the power of lexical scoping and names, it is straightforward to express private attributes and methods.

OPM is a simple and powerful base for expressing concurrent object-oriented programming abstractions. It was in fact designed for this purpose. Concrete programming languages will of course sweeten frequently used programming abstractions with a convenient notation. For a concrete system of object-oriented abstractions and notations we refer the reader to the Oz object system [5, 25].

The reader will have noticed the similarity between agents and objects. We can see agents as active objects. An object can easily be turned into an agent by interfacing it through a port.

13 Distribution

OPM can be extended to serve as a model for distributed programming. Distribution means that a program can spread computations over a network of computers. At the abstraction level of OPM, this can be modeled by assigning a *site* to every task and by assuming that the store is distributed transparently. Moreover, we assume that new tasks inherit the site of the creating task.

We can now see a clear difference between agents and objects. When we send a message to an agent, the message is served at the site where the agent was created (there is a task waiting for the next message sent). When we apply an object to a message, the message is served at the site where the object is applied. In other words, agents are stationary and objects are mobile.

Since OPM has first-class procedures, it is straightforward to express compute servers. Cardelli [3] gives an excellent exposition of distributed programming techniques available in a lexically-scoped language with first-class procedures and concurrent state.

The assumption of a transparently distributed store is not realistic for many applications. It conflicts with the ability to model fault-tolerance, for instance. We have started work on a less abstract model where the store appears as a directed graph whose nodes are situated similar to tasks.

14 Incremental Tell

The tell operation of OCC (see Section 6) is not suitable for a parallel implementation. The reason is that a constraint must be told in a single reduction step. Since telling a constraint (e.g., $x = y$) may involve scanning the entire store, other tell tasks may be blocked for a long time. The problem can be resolved by telling a constraint piecewise. The basic idea is to reduce a constraint task T by keeping the task T as is and by advancing the constraint store from S to a slightly stronger constraint store S' entailed by $S \wedge T$. This amplifying reduction step is repeated until the constraint store entails T, in which case the task T is discarded. Since the constraint store must always be satisfiable, the case where $S \wedge T$ is unsatisfiable needs special care.

To make the incremental tell operation precise, we introduce the notion of a constraint system. A *constraint system* consists of a constraint structure, a set of constraints called *basic constraints*, and, for every basic constraint T, a binary relation $\rightarrow_T$ on basic constraints such that:

1. The basic constraints are closed under conjunction and contain $\perp$ (i.e., false).

2. For every basic constraint T, the relation $\rightarrow_T$ is well-founded, that is, there exists no infinite chain $S_1 \rightarrow_T S_2 \rightarrow_T S_3 \rightarrow_T \cdots$.

3. If $S \rightarrow_T S'$, then (i) S' entails S, (ii) $S \wedge T$ entails S', (iii) S is satisfiable, and (iv) S' is unsatisfiable if and only if $S' = \bot$.

4. If T is not entailed by S and both are basic constraints, then there exists S' such that $S \rightarrow_T S'$.

The *tell reductions* $\rightarrow_T$ correspond to the visible simplification steps of the incremental algorithms implementing the necessary operations on constraint stores. Such algorithms can be found, for instance, in [1, 26]. Note that the tell reductions may be nondeterministic; that is, for given S and T, there may be different S_1 and S_2 such that $S \rightarrow_T S_1$ and $S \rightarrow_T S_2$.

Let S be a satisfiable basic constraint and T a basic constraint. Then the tell reduction $\rightarrow_T$ satisfies the following properties:

1. S entails T if and only if S is irreducible with respect to $\rightarrow_T$.

2. S disentails T if and only if every maximal chain $S \rightarrow_T \cdots$ ends with $\bot$.

3. Let $S \rightarrow_T \cdots \rightarrow_T S'$ be a chain such that S' is irreducible with respect to T. Then (i) $S \wedge T$ is equivalent to S' and (ii) $S \wedge T$ is unsatisfiable if and only if $S' = \bot$.

Given a constraint system, we assume that the constraints appearing in expressions and the constraint store are all basic, where the constraints appearing in guards may be existentially quantified. Given a constraint store S and a constraint task T, the *incremental tell operation* is defined as follows: if S is irreducible with respect to $\rightarrow_T$, then the task T is discarded. Otherwise, choose some basic constraint S' such that $S \rightarrow_T S'$. If $S' = \bot$, then announce failure and discard the task T; if $S' \neq \bot$, then advance the constraint store to S' and keep the task T.

The canonical constraint system for the constraint structure INP comes with the basic constraints

$$C \ ::= \ \bot \mid \top \mid \langle \textit{primitive constraint} \rangle \mid C_1 \wedge C_2.$$

Primitive constraints were defined in Section 5.

As long as failure does not occur, it is not important to know which tell reductions are used. However, if $S \wedge T$ is unsatisfiable and computation can continue after failure (e.g., since there is exception handling), all chains $S \rightarrow_T \cdots \rightarrow_T S'$ should only add local information. The notion of "local information" cannot be made precise in general. However, there are straightforward definitions for INP and other practically relevant constraint systems. Here we will just give an example for INP. Given $S \equiv (x = 1|2 \ \wedge \ y = u|3)$ and $T \equiv (x = y)$, the tell reduction $\rightarrow_T$ should only permit two maximal chains issuing from S: $S \ \rightarrow_T \ S \wedge u = 1 \ \rightarrow_T \ \bot$ and $S \ \rightarrow_T \ \bot$.

15 Propagators

The algorithms for telling and checking entailment and disentailment of basic constraints must be efficient. The typical complexity should be constant time, and the worst-case complexity should be quadratic or better in the size of the guard and the constraint store. Consequently, expressive constraints such as $x+y=z$ and $x*y=z$ cannot be written into the constraint store and hence cannot be accommodated as basic constraints. (For nonlinear constraints over integers satisfiability is undecidable (Hilbert's Tenth Problem).)

Nonbasic constraints can be accommodated as tasks that wait until the constraint store contains enough information so that they can be equivalently replaced with basic constraints. For instance, a task $x+y=z$ may wait until there exist two integers n and m such that the constraint store entails $x=n \wedge y=m$. If this is the case, the task can be reduced to the basic constraint $z=k$, where k is the sum of n and m. Nonbasic constraints that are accommodated in this way are called *propagators*.

Another example of a propagator is a Boolean order test for integers:

$$\textit{less}(x,y,z) \;\equiv\; (x<y \leftrightarrow z=\texttt{True}) \;\wedge\; (z=\texttt{True} \vee z=\texttt{False}).$$

`True` and `False` are variables bound to distinct names. This propagator can reduce to $z=$ `True` or $z=$ `False` as soon as the constraint store contains sufficient information about the values of x and y.

16 Threads

We now give an efficient reduction strategy for OPM that is fair and reactive (see Section 2). An efficient reduction strategy must make it possible to write programs that create only a moderate amount of concurrency, which can be implemented efficiently on both single and multi-processor architectures.

The example of the Fibonacci function (we use a sugared notation suppressing auxiliary variables)

```
proc {Fib N M}
   local B X Y in
      {Less 2 N B}
      if B=True
      then {Fib N-1 X}  {Fib N-2 Y}  X+Y=M
      else M=1 fi
   end
end
```

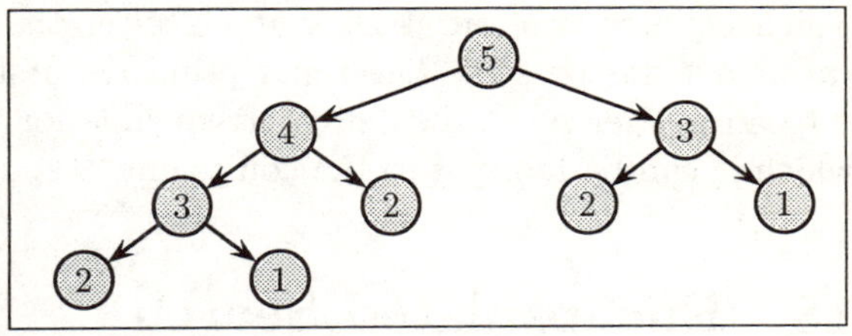

shows that the naive reduction strategy in Section 2 is impractical: it will traverse the recursion tree of `{Fib 5 M}`, say, in breadth-first manner, thus requiring exponential space. On the other hand, sequential execution will traverse the recursion tree in depth-first manner from left to right and will thus only need linear space. This difference clearly matters.

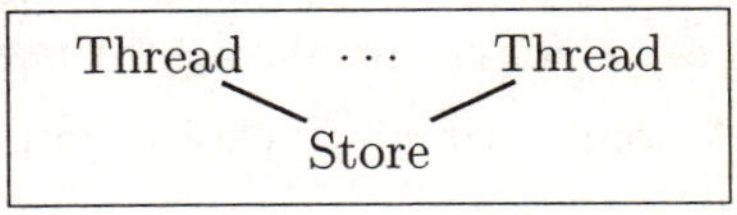

The efficient reduction strategy organizes tasks into threads, where every thread is guaranteed to make progress. Thus fairness is guaranteed at the level of threads. A *thread* is a nonempty stack of tasks, where only the topmost task of a thread can be reduced. If the topmost task of a thread is reduced, it is replaced with the newly created tasks, if there are any. If a composition $E_1 \wedge E_2$ is reduced, the left expression E_1 goes on top of the right expression E_2, which means that E_1 is considered before E_2. If the topmost task of a thread is irreducible over the current store and the thread contains further tasks, the topmost task is moved to a newly created thread. A thread disappears as soon as it has no task left.

The outlined reduction strategy tries to be as sequential as possible and as concurrent as necessary. It will execute the task `{Fib 5 M}` sequentially in a single thread, thus requiring only linear space.

The concurrent execution of an expression E in a separate thread can be forced by writing

local `Fire` **in if** `Fire=1` **then** E **else true** `fi Fire=1` **end**

With threads it is straightforward to send messages sequentially through a port. If ports are defined as in Section 9, the simple composition

```
{Port A}  {Port B}
```

will send `A` before `B` provided the store binds `Port` already to a port.

17 Time

It is straightforward to extend OPM such that tasks can be synchronized on time points. Here we specify a timer primitive

```
{Sleep T X Y}
```

which equates two variables X and Y after the time span specified by the variable T has passed. The timer primitive first waits until the store binds T to an integer n. It then stays irreducible for further n milliseconds, after which it can be reduced to the constraint X=Y.

18 Encapsulated Search

Since OPM has constraints and logic variables, it will subsume the problem solving capabilities of constraint logic programming when extended with a nondeterministic choice combinator. However, a completely new idea is needed for encapsulating the resulting problem solvers into concurrent agents.

A nondeterministic choice combinator can be provided as an expression

$$E_1 \textbf{ or } E_2$$

called a *choice*. Choice tasks can only be reduced if no other task is reducible. If this is the case, a choice can be reduced by *distributing* the computation space into two spaces obtained by replacing the choice with its left and right alternative, respectively. The resulting search tree of computation spaces can be explored with a suitable strategy. If a tell operation announces failure, computation in the corresponding computation space is aborted. The leaves of the search tree are either failed or unfailed computation spaces. Unfailed leaves will contain the solutions of the problem being solved as bindings to certain variables.

While the outlined semantics for nondeterministic choice provides the expressivity of constraint logic programming, distributing the top level computation space is not compatible with the idea of concurrent computation. What we would like to have are concurrent agents to which we can present a search strategy and a problem to be solved and from which we can request the solutions of the problem one by one. This means that the search agent should encapsulate search. It turns out that such a search agent can be programmed with a single further primitive, called a *search combinator*. The search combinator spawns a subordinate computation space and reduces in case the subordinate space fails, becomes irreducible, or is distributed. In the case of distribution, the two alternative local spaces are frozen and returned as first-class citizens represented as procedures. The details of this elaborate construction are reported in [20, 19, 22].

The resulting model is realized in Oz together with further concurrent constraint combinators [20, 19, 14]. Oz gets constraint logic programming out of its problem solving ghetto and integrates it into a concurrent and lexically scoped language with first-class procedures and state. This integration eliminates the need for Prolog's ad hoc constructs and also increases the expressivity of the problem solving constructs.

19 Summary

We have presented a simple and expressive model OPM for high-level concurrent programming. The model is lexically scoped and consists of the concurrent constraint kernel OCC, first-class procedures, and cells providing for concurrent state. It computes with logic variables and constraints and monotonically synchronizes on a declarative constraint store. The constraint store is the exclusive place where information about the values of variables is stored. Dynamically created values called names interface the constraint store with the procedure and the cell store. This way OPM realizes an orthogonal combination of first-order constraints with first-class procedures and stateful cells. We have shown how OPM can express higher-order functions, agents and objects. We have added an incremental tell operation to improve the potential for parallelism. We have also added propagators, threads, and a timer primitive as needed for a practical language. Finally, we have outlined how the model can be extended so that it can express encapsulated problem solvers generalizing the problem solving capabilities of constraint logic programming. Oz translates the presented ideas into an exciting new programming language.

Acknowledgements

The development of OPM and Oz would have been impossible without the combined contributions of the members of the Programming Systems Lab at DFKI. Much inspiration and technical knowledge came from the developers of AKL at SICS, the developers of LIFE at Digital PRL, and the other partners of the Esprit basic research action ACCLAIM.

I'm grateful to Seif Haridi, Martin Henz, Michael Mehl, Joachim Niehren, Andreas Podelski, and Christian Schulte who read and commented on drafts of this paper.

The research reported in this paper has been supported by the BMBF (contract ITW 9105), the Esprit Basic Research Project ACCLAIM (contract EP 7195), and the Esprit Working Group CCL (contract EP 6028).

Remark

The DFKI Oz system and related papers are available at `http://www.ps.uni-sb.de/`.

References

[1] H. Aït-Kaci, A. Podelski, and G. Smolka. A feature-based constraint system for logic programming with entailment. *Theoretical Computer Science*, 122(1–2):263–283, 1994.

[2] R. Backofen. A complete axiomatization of a theory with feature and arity constraints. *Journal of Logic Programming*, vol 24, pages 37–71, 1995.

[3] L. Cardelli. Obliq: A Language with Distributed Scope. In *Proc. 22nd Ann. ACM Symposium on Principles of Programming Languages (POPL'95)*, pages 286–297, 1995.

[4] A. Colmerauer, H. Kanoui, and M. V. Caneghem. Prolog, theoretical principles and current trends. *Technology and Science of Informatics*, 2(4):255–292, 1983.

[5] M. Henz, G. Smolka, and J. Würtz. Object-oriented concurrent constraint programming in Oz. In V. Saraswat and P. V. Hentenryck, editors, *Principles and Practice of Constraint Programming*, pages 27–48. The MIT Press, Cambridge, MA, 1995.

[6] J. Jaffar and M. J. Maher. Constraint logic programming: A survey. *The Journal of Logic Programming*, 19/20:503–582, 1994.

[7] S. Janson and S. Haridi. Programming paradigms of the Andorra kernel language. In V. Saraswat and K. Ueda, editors, *Logic Programming, Proc. 1991 Int. Symposium*, pages 167–186. The MIT Press, Cambridge, MA, 1991.

[8] S. Janson, J. Montelius, and S. Haridi. Ports for objects. In *Research Directions in Concurrent Object-Oriented Programming.* The MIT Press, Cambridge, MA, 1993.

[9] J. Launchbury. A natural semantics for lazy evaluation. In *Proc. 20th Ann. ACM Symposium on Principles of Programming Languages (POPL'93)*, pages 144–154, 1993.

[10] M. J. Maher. Logic semantics for a class of committed-choice programs. In J.-L. Lassez, editor, *Logic Programming, Proc. 4th Int. Conference*, pages 858–876. The MIT Press, Cambridge, MA, 1987.

[11] M. Mehl, R. Scheidhauer, and C. Schulte. An abstract machine for Oz. In *Proc. 7th Int. Symposium on Programming Languages, Implementations, Logics and Programs (PLILP'95).* Lecture Notes in Computer Science, vol. 982, pages 151–168, Springer-Verlag, Berlin, 1995.

[12] R. Milner. *A Calculus of Communicating Systems.* Lecture Notes in Computer Science, Vol. 92, Springer-Verlag, Berlin, 1980.

[13] R. Milner. Functions as processes. *Journal of Mathematical Structures in Computer Science*, 2(2):119–141, 1992.

[14] T. Müller, K. Popow, C. Schulte, and J. Würtz. Constraint programming in Oz. DFKI Oz documentation series, DFKI, Saarbrücken, Germany, 1994.

[15] J. Niehren. *Funktionale Berechnung in einem uniform nebenläufigen Kalkül mit logischen Variablen.* Doctoral Dissertation. Universität des Saarlandes, Saarbrücken, Germany, December 1994.

[16] J. Niehren and G. Smolka. A confluent relational calculus for higher-order programming with constraints. In J.-P. Jouannaud, editor, *Proc. 1st Int. Conference on Constraints in Computational Logics (CCL'94)*, pages 89–104. Lecture Notes in Computer Science, Vol. 845, Springer-Verlag, Berlin, 1994.

[17] G. D. Plotkin. A structural approach to operational semantics. DAIMI FN-19, Dept. of Computer Science, Aarhus University, Denmark, 1981. Reprinted 1991.

[18] V. A. Saraswat. *Concurrent Constraint Programming.* The MIT Press, Cambridge, MA, 1993.

[19] C. Schulte and G. Smolka. Encapsulated search in higher-order concurrent constraint programming. In M. Bruynooghe, editor, *Logic Programming, Proc. 1994 Int. Symposium*, pages 505–520. The MIT Press, Cambridge, MA, 1994.

[20] C. Schulte, G. Smolka, and J. Würtz. Encapsulated search and constraint programming in Oz. In A. Borning, editor, *Proc. 2nd Int. Workshop on Principles and Practice of Constraint Programming (PPCP'94)*, pages 134–150. Lecture Notes in Computer Science, Vol. 874, Springer-Verlag, Berlin, 1994.

[21] E. Shapiro. The family of concurrent logic programming languages. *ACM Computing Surveys*, 21(3):413–511, 1989.

[22] G. Smolka. The definition of Kernel Oz. In A. Podelski, editor, *Constraints: Basics and Trends*, pages 251–292. Lecture Notes in Computer Science, Vol. 910, Springer-Verlag, Berlin, 1995.

[23] G. Smolka. A foundation for higher-order concurrent constraint programming. In J.-P. Jouannaud, editor, *Proc. 1st Int. Conference on Constraints in Computational Logics (CCL'94)*, pages 50–72. Lecture Notes in Computer Science, Vol. 845, Springer-Verlag, Berlin, 1994.

[24] G. Smolka. Fresh: A higher-order language with unification and multiple results. In D. DeGroot and G. Lindstrom, editors, *Logic Programming: Relations, Functions, and Equations*, pages 469–524. Prentice-Hall, Englewood Cliffs, NJ, 1986.

[25] G. Smolka. An Oz primer. DFKI Oz documentation series, DFKI, Saarbrücken, Germany, 1995.

[26] G. Smolka and R. Treinen. Records for logic programming. *Journal of Logic Programming*, 18(3):229–258, 1994.

[27] G. Winskel. *The Formal Semantics of Programming Languages.* Foundations of Computing. The MIT Press, Cambridge, MA, 1993.

A Foundation for Higher-order Concurrent Constraint Programming

Gert Smolka
Programming Systems Lab
German Research Center for Artificial Intelligence (DFKI)
Stuhlsatzenhausweg 3, 66123 Saarbrücken, Germany
email: smolka@dfki.uni-sb.de

Abstract

We present the γ-calculus, a computational calculus for higher-order concurrent programming. The calculus can elegantly express higher-order functions (both eager and lazy) and concurrent objects with encapsulated state and multiple inheritance. The primitives of the γ-calculus are logic variables, names, procedural abstraction, and cells. Cells provide a notion of state that is fully compatible with concurrency and constraints. Although it does not have a dedicated communication primitive, the γ-calculus can elegantly express one-to-many and many-to-one communication.

There is an interesting relationship between the γ-calculus and the π-calculus: The γ-calculus is subsumed by a calculus obtained by extending the asynchronous and polyadic π-calculus with logic variables.

The γ-calculus can be extended with primitives providing for constraint-based problem solving in the style of logic programming. A such extended γ-calculus has the remarkable property that it combines first-order constraints with higher-order programming.

This paper first appeared in: J.-P. Jouannaud, editor, Proc. 1st Int. Conference on Constraints in Computational Logics (CCL'94), *pages 50–72. Lecture Notes in Computer Science, Vol. 845, Springer-Verlag, Berlin, 1994.*

1 Introduction

Concurrent constraint programming [21] is a research direction aiming at a unified framework for high-level concurrent programming and constraint-based problem solving. Its roots are concurrent logic programming [23] and constraint logic programming [3, 10]. Although concurrent programming and constraint-based problem solving have different structure and applications, they do have significant commonalities:

- both come in a relational and concurrent setting
- constraint propagation is a concurrent activity
- logic variables are the canonical form of reference for constraints and concurrent computation.

This paper presents the γ-calculus, a computational calculus for higher-order concurrent programming. As is, the calculus can elegantly express higher-order functions (both eager and lazy) and concurrent objects with encapsulated state and multiple inheritance. Constraint-based problem solving in the style of logic programming requires additional primitives, which can be chosen such that one obtains a combination of higher-order programming with first-order constraints. This is in sharp contrast to approaches based on higher-order logic [18], where higher-order programming comes with the operational burden of higher-order constraints.

An extension [24, 22] of the γ-calculus providing for constraint-based problem solving serves as the foundation of Oz [8], a full-fledged programming language and system under development at the Programming Systems Lab of DFKI.[1]

The primitives of the γ-calculus are logic variables, names, procedural abstraction, and cells. Cells provide a notion of state that is fully compatible with concurrency and constraints. Although it does not have a dedicated communication primitive, the γ-calculus can elegantly express one-to-many and many-to-one communication.

It is illuminating to compare the γ-calculus with the π-calculus [17, 16, 15]. Both are concurrent systems with first-class names. While the γ-calculus has logic variables, the π-calculus has formal input arguments only (as in functional programming). As is well-known from logic programming, logic variables do not necessitate a static distinction between input and output, thus providing for a free data flow combining smoothly with concurrent control. While the π-calculus has communication as its principal primitive, the

[1]The Oz programming system and its documentation are available at `http://www.ps.uni-sb.de/`.

γ-calculus has logic variables, procedural abstraction, and cells as its principal primitives. The primitives of the γ-calculus were chosen with the consideration that programming abstractions such as higher-order functions and concurrent objects be easily expressible. If we extend the π-calculus with logic variables, it can express procedural abstraction and cells. Logic variables increase the expressivity of the π-calculus in two crucial aspects: They allow to equate communication links, and they provide the possibility to express procedures with input *and* output arguments (recall that a function is a procedure with input and output).

The paper is organized as follows. Section 2 gives the formal definition of the γ-calculus. Sections 3–5 provide important intuitions and examples for the expressivity of the γ-calculus. Sections 6 and 7 show how the eager and the lazy λ-calculus can be embedded into the γ-calculus. Section 8 shows how the γ-calculus can express records. Sections 9 and 10 show how the γ-calculus can express concurrent objects with encapsulated state and multiple inheritance. Section 11 discusses communication issues. Section 12 presents a possible execution strategy for the γ-calculus. Section 13 shows how the γ-calculus can be extended with general first-order constraints. Section 14 clarifies the relationship between the γ-calculus and the π-calculus.

2 The Gamma Calculus

Figure 1 shows the syntax of the γ-calculus. It assumes that an infinite alphabet of *variables* and a disjoint and infinite alphabet of *names* are given. Variables and names are jointly referred to as *references*. Variables are placeholders for names. There are no other values but names.

The expressions of the γ-calculus are relational, as in logic programming or the π-calculus. Seen from the perspective of predicate logic, expressions play the role of formulas and references play the role of terms. Composition is like conjunction in logic programming and parallel composition in the π-calculus. A declaration $\exists uE$ introduces a new reference u with scope E. Declaration of variables is like existential quantification in logic programming; declaration of names is like restriction in the π-calculus. Equations are like equations in logic. Names stand for themselves and thus are different if they are syntactically different (so-called unique name assumption). A (named) abstraction $a{:}\overline{x}/E$ consists of a name a, formal arguments $\overline{x}$ ($\overline{x}$ stands for a possibly empty sequence of variables), and a body E (the expression being abstracted from). There is the side condition that the sequence of formal arguments $\overline{x}$ be linear (i.e., consist of pairwise distinct variables). Abstractions can be seen as procedure or predicate definitions. An application $u\overline{v}$ consists of a reference u designating the abstraction to be applied and the actual arguments $\overline{v}$. Applications can be seen as procedure or predicate calls. A conditional **if** $u = v$ **then** E **else** F reduces to either E or F, depending on whether u

Symbols

x, y, z	:		variables
a, b, c	:		names
u, v, w	$::=$	$x \mid a$	references

Expressions

E, F, G	$::=$	$\top$	null
	$\mid$	$E \wedge F$	composition
	$\mid$	$\exists u E$	declaration
	$\mid$	$u = v$	equation
	$\mid$	$a{:}\,\overline{x}/E$	abstraction
	$\mid$	$u\overline{v}$	application
	$\mid$	**if** $u = v$ **then** E **else** F	conditional
	$\mid$	$a{:}u$	cell

Figure 1: Syntax of the γ-calculus

and v turn out to be equal or different. A cell $a{:}u$ has the name a and holds the reference u; reduction with an application avw will impose the equation $u = v$ and update the cell to hold w.

From the above it is clear that the γ-calculus has one binder for names ($\exists aE$) and two binders for variables ($\exists xE$ and $a{:}\,\overline{x}/E$). *Free and bound references* of expressions are defined accordingly.

The γ-calculus is an expressive computational system. We will show that it can elegantly express higher-order functional programming, data structures, and concurrent objects with encapsulated state and multiple inheritance.

A distinctive feature the γ-calculus shares with logic programming is that variables can be used without explicitly saying how their values are obtained (so-called logic variables). Information about the values of variables can be stated through equations, which can be seen as constraints. Equations can express partial (e.g, $x = y$) and total (e.g., $x = a$) information. Recall that names are the only values variables can take in the γ-calculus.

The computational intuitions expressed above are formalized by rules rewriting the expressions of the calculus. This is a common setup also found in the λ-calculus (functional computation) and SLD-resolution (relational computation). For the γ-calculus, this setup is refined in that the rules are applied modulo a structural congruence, and in that the rules can only be applied to specific positions.[2]

Applying rewrite rules modulo a structural congruence is actually quite common, although it is often not made explicit. In the λ-calculus, it is common practise to "identify" expressions that are equal up to α-conversion (consistent renaming of bound variables). In logic programming and unification, one typically rewrites multisets of atomic formulas, where the multisets are obtained by making conjunction associative and commutative. First-order rewriting modulo equations is an established topic [5], serving as a foundation for the specification language OBJ [6] and Meseguer's rewriting logic [14].

The *structural congruence* of the γ-calculus is the least congruence "$E \equiv F$" on the set of expressions satisfying the following laws:

- composition $E \wedge F$ of expressions is associative, commutative, and satisfies $E \wedge \top \equiv E$ (thus we can see composition as multiset union and $\top$ as the empty multiset)
- declaration $\exists u E$ of references allows for consistent renaming of the declared reference u and satisfies

$$\begin{aligned} \exists u E \wedge F &\equiv \exists u(E \wedge F) \qquad \text{if } u \text{ not free in } F \\ \exists u \exists v E &\equiv \exists v \exists u E \\ \exists u \top &\equiv \top \end{aligned}$$

  (thus declarations can always be moved above compositions, and declarations of references not being used can be deleted)
- abstractions $a{:}\overline{x}/E$ allow for consistent renaming of the formal arguments $\overline{x}$
- equations $u = v$ are symmetric.

Reduction in the γ-calculus is defined in Figure 2 by a system of inference rules. Only the structure rules have premises, all other rules are axioms. The structure rules say that reduction is modulo structural congruence, and that reductions of subexpressions not appearing beneath abstractions and conditionals can be taken as reductions of the entire expression. A reduction $E \rightarrow F$ is possible if and only if it can be derived with the structure rules from exactly one instance of an axiom. The Application Rule comes with the side condition that the number $|\overline{u}|$ of actual arguments in the application equals the number $|\overline{x}|$ of formal arguments in the abstraction.

[2] A similar setup is used in a recent presentation of the π-calculus devised by Milner [16, 15].

Structure

$$\frac{E \equiv E' \quad E' \to F' \quad F' \equiv F}{E \to F} \qquad \frac{E \to E'}{E \wedge F \to E' \wedge F} \qquad \frac{E \to E'}{\exists u E \to \exists u E'}$$

Elimination

$$\exists x(x = u \ \wedge\ E) \ \to\ E[u/x] \qquad \text{if } x \neq u \text{ and } u \text{ free for } x \text{ in } E$$

Application

$$a\overline{u} \ \wedge\ a{:}\overline{x}/E \ \to\ E[\overline{u}/\overline{x}] \ \wedge\ a{:}\overline{x}/E \qquad \text{if } \overline{u} \text{ free for } \overline{x} \text{ in } E \text{ and } |\overline{u}| = |\overline{x}|$$

Conditional

$$\textbf{if } u = u \textbf{ then } E \textbf{ else } F \ \to\ E \qquad \textbf{if } a = b \textbf{ then } E \textbf{ else } F \ \to\ F \quad \text{if } a \neq b$$

Exchange

$$a{:}u \ \wedge\ avw \ \to\ a{:}w \ \wedge\ v = u$$

Figure 2: Reduction in the γ-calculus

Proposition 1 *Let* contexts *be defined as* $C \ ::= \ \bullet \mid C \wedge E \mid E \wedge C \mid \exists u C$. *Then* $E \to E'$ *is a reduction in the* γ*-calculus if and only if there exists a context* C *and an instance* $G \to G'$ *of an axiom in Figure 2 such that* $E \equiv C[G]$, *and* $C[G'] \equiv E'$.

3 The Chemical Metaphor

Reduction in the γ-calculus can be seen as evolution of a *computation space* containing a multiset of freely floating *molecules*.[3] The molecules are equations, abstractions, applications, conditionals, and cells. The structural congruence of the γ-calculus is defined such that every expression can be seen as a computation space: After pushing all declarations to the top (possibly involving α-conversion), we are left with a conjunction of molecules. The expression $\top$ describes the empty computation space. Expressions appearing as the constituents of abstractions and conditionals do not yet contribute to the computation space.

A computation space evolves by reduction with the rules given in Figure 2. The Application and Exchange Rules describe reactions between two molecules sharing a name. The rules for the conditional describe transformations of a single molecule. The Elimination Rule deletes an equational

[3]The metaphor of seeing concurrent computation as chemical reaction appeared with Berry and Boudol's chemical abstract machine [4].

molecule and eliminates a variable by replacing all its occurrences with another reference.

When a conditional reduces, it injects one of its constituent expressions into the computation space, thus possibly contributing new molecules and new references (the operational reading of a declaration $\exists u E$ is: Create a new reference u). Similarly, when an application reacts with an abstraction, a copy of the body of the abstraction is injected into the computation space, where the actual arguments of the application replace the formal arguments of the abstraction. The Application Rule is the only rule that copies expressions. As the space evolves, the number of molecules and the number of connecting references can increase and decrease. Every infinite reduction chain $E_1 \rightarrow E_2 \rightarrow \ldots$ must involve the Application Rule.

The Elimination Rule provides all the constraint handling needed in the γ-calculus. If a computation space contains a molecule $x = u$, then x can be eliminated by replacing it with u, provided u is different from x. We assume that a computation space does not have free variables. Injecting an equation $x = a$ into a computation space amounts to an attempt to fix the value of the variable x to a. There might be competing such attempts, as in

$$\exists x(x = a \ \wedge \ x = b \ \wedge \ E).$$

Which value is taken for x is an indeterministic choice: The space can either reduce to $a = b \wedge E[a/x]$ or to $a = b \wedge E[b/x]$, where the choice being made cannot be retracted. Note that all occurrences of x will be replaced with only one of the two names. The fact that there were conflicting attempts to fix the value of x remains partly visible since the "inconsistent" equation $a = b$ remains in the space.[4] There are three possibilities to handle such a conflict: consider it a regular event (the choice taken in the γ-calculus), consider it a run-time error, or consider it a failure in the sense of logic programming (we will say more about failure in Section 13).

The expression (a is not free in E)

$$\exists a(\textbf{if } x = a \textbf{ then } \top \textbf{ else } E)$$

has an interesting operational reading: inject the expression E in the computation space once the variable x has been assigned a value (i.e., has been replaced by a name). Put more informally, the above expression synchronizes E upon the event that the value of x becomes known.

The Exchange Rule describes a reaction of a cell $a{:}u$ with an application avw. The reaction updates the reference hold by the cell to w and equates the references u and v (exploiting logic variables). Thus reading and writing of a cell are merged into one atomic operation. Cells yield a notion of state

[4]Equations of the form $u = u$ and $a = b$ do not have a computational effect. Hence they can be deleted in an implementation of the γ-calculus.

that is fully compatible with concurrency and constraints. Cells are essential for expressing objects.

The Application and the Exchange Rule have in common that they describe reactions between two molecules that agree on the same name (i.e., a). As computation proceeds, new abstractions and cells may be created. This necessitates the creation of fresh names, an operation elegantly expressible in the γ-calculus.

4 Creating Fresh Names

The operational reading of $\exists a(x = a)$ is: Create a fresh name and make it the value of the variable x. To see why this is so, consider the expression

$$\exists x \exists y(\exists a(x = a) \ \wedge\ \exists a(y = a) \ \wedge\ \textbf{if } x = y \textbf{ then } E \textbf{ else } F)$$

and suppose that x and y are distinct variables that do not occur free in E and F. Moreover, assume that a is a name not occurring free in E and F. We will show that the expression reduces to F.

First, we move the left declaration of a to the outside of the expression using the laws for declarations and compositions and exploiting the assumption that a does not occur free in E and F.

$$\equiv\ \exists a \exists x \exists y(x = a \ \wedge\ \exists a(y = a) \ \wedge\ \textbf{if } x = y \textbf{ then } E \textbf{ else } F)$$

Next we exchange the declarations of x and y and eliminate x with the Elimination Rule.

$$\rightarrow\ \exists a \exists y(\exists a(y = a) \ \wedge\ \textbf{if } a = y \textbf{ then } E \textbf{ else } F)$$

Next we rename the inner name a to b, where b is assumed to be different and to not occur free in E and F.

$$\equiv\ \exists a \exists y(\exists b(y = b) \ \wedge\ \textbf{if } a = y \textbf{ then } E \textbf{ else } F)$$

This brings us in a position where we can eliminate y in the same way we did it for x before.

$$\rightarrow\ \exists a \exists b(\textbf{if } a = b \textbf{ then } E \textbf{ else } F)$$

Now, since a and b are different, we obtain

$$\rightarrow\ \exists a \exists b F$$

using the appropriate rule for the conditional. It remains to get rid of the declarations of the names a and b. This can be done using the congruence laws:

$$\equiv\ \exists a \exists b(\top \wedge F)\ \equiv\ (\exists a \exists b \top)\ \wedge\ F\ \equiv\ \top\ \wedge\ F\ \equiv\ F.$$

5 Possible Indeterminisms

The γ-calculus involves several indeterminisms:

1. if there are two applications for the same cell, the order of their reduction is indeterministic

2. if there are two equations $x = a$ and $x = b$ for the same variable, the choice of the name replacing x is indeterministic

3. if an application matches more than one abstraction or cell, the choice of the abstraction or cell it reacts with is indeterministic.

The first indeterminism is essential for concurrent computation (see the section on objects). The other indeterminisms should not occur with well-written programs.

The third form of indeterminism can be excluded with a straightforward syntactic condition: extend the γ-calculus with the syntactic variants

$$\begin{aligned} x{:}\overline{y}/E \quad &:\equiv \quad \exists a(x = a \wedge a{:}\overline{y}/E) \\ x{:}u \quad &:\equiv \quad \exists a(x = a \wedge a{:}u) \end{aligned}$$

and admit only initial expressions not containing the primitive forms $a{:}\overline{y}/E$ and $a{:}u$. One can show that reduction sequences issuing from such expressions cannot involve the third form of indeterminism.

Provided one excludes cells, there is a syntactic condition excluding all remaining indeterminisms; a thus restricted version of the γ-calculus is the δ-calculus studied and proven confluent in [20]. The δ-calculus seems to be a promising alternative to the λ-calculus for the foundation of functional programming.

Remark. The syntactic extensions $x{:}\overline{y}/E$ and $x{:}u$ defined above are *static*; that is, they must be expanded *before* a reduction rule is applied. This is since $x{:}\overline{y}/E$ changes its meaning when the elimination rule replaces x with a name a.

6 Embedding of the Eager Lambda Calculus

To embed the eager λ-calculus (see [26]) into the γ-calculus, we extend the expressions of the γ-calculus such that one can write λ-terms in equations:

$$\begin{aligned} E, F, G \quad &::= \quad \ldots \mid x = M \\ M, N \quad &::= \quad x \mid \lambda x M \mid MN. \end{aligned}$$

The semantics of the new equations is given by the congruences

$$x = \lambda y M \quad :\equiv \quad x{:}\, yz/z = M$$

$$x = MN \quad :\equiv \quad \exists y \exists z (y = M \;\wedge\; z = N \;\wedge\; yzx)$$

providing a translation from the extended syntax to the base syntax (the syntactic extension $x{:}\, yz/E$ was defined in the previous section). As one would expect, functional abstractions translate into relational abstractions with an input and an output argument. It is instructive to consider the translation of the identity function:

$$x = \lambda yy \quad \equiv \quad \exists a(x = a \;\wedge\; a{:}\, yz/z = y).$$

The translation of functional applications exploits that functional nesting can be expressed by composition and declaration of auxiliary variables.

The soundness of the embedding is established by the following theorem [19].

Theorem 2 *Let M be a closed λ-term. Then M converges in the eager λ-calculus if and only if $\exists x(x = M)$ converges in the γ-calculus.*

In contrast to the λ-calculus, the γ-calculus can express (mutual) recursion directly. For instance, $\exists x \exists y\, (x = \lambda u M \;\wedge\; y = \lambda v N \;\wedge\; E)$ defines two possibly mutually recursive functions x and y that can be used in E.

Eager functional programming with mutual recursion can in fact be expressed in a confluent subcalculus of the γ-calculus, called the δ-calculus [20].

7 Embedding of the Lazy Lambda Calculus

The embedding of the lazy λ-calculus (see [26]) into the γ-calculus is more subtle than the embedding of the eager λ-calculus. The basic idea is to represent a lazy function by an abstraction with three arguments: one argument for the input of the function, one argument for the output of the function, and one argument for requesting that the input of the function be computed.

In the following we will use r and s to denote variables used to request subcomputations. We extend the syntax of the γ-calculus as follows:

$$E, F, G \quad ::= \quad \ldots \mid x{\cdot}r = K$$

$$K, L \quad ::= \quad x \mid \lambda x K \mid KL \mid x{\cdot}r.$$

An equation $xr = K$ equates x to the result of the λ-term K, where evaluation of K must be requested explicitly through the variable r.

The semantics of the new expressions is given by the congruences

$$
\begin{array}{rcl}
x{\cdot}r = y & :\equiv & x = y \\
x{\cdot}r = y{\cdot}s & :\equiv & x = y \wedge r = s \\
x{\cdot}r = \lambda yK & :\equiv & x{:}\, ysz/z{\cdot}r = K[y{\cdot}s/y] \\
x{\cdot}r = KL & :\equiv & \exists y \exists y' \exists z \exists s\, (y{\cdot}r = K \;\wedge\; ryy' \;\wedge\; y'zsx \;\wedge\; z{\cdot}s = L)
\end{array}
$$

providing a translation from the extended syntax to the base syntax of the γ-calculus. The translation of an equation $x{\cdot}r = K$ will admit no other rule but the Elimination Rule, eliminating unnecessary auxiliary variables (e.g., the translation of $x{\cdot}r = y(z{\cdot}s)$ will reduce to $\exists y'(ryy' \wedge y'zsx)$). Evaluation of $x{\cdot}r = K$ must be requested explicitly by composing it with $r = \lambda xx$ (the "eager" equation $r = \lambda xx$ was defined in the previous section). Evaluation is made lazy by switching the connection between abstractions and applications only when the result of the application is needed. The switch is realized by an application ryy', which is fired by equating r to the identity function.

Concerning the correctness of the embedding of the lazy λ-calculus, we conjecture the following theorem to hold.

Theorem 3 *Let M be a closed λ-term. Then M converges in the lazy λ-calculus if and only if $\exists x \exists r(r = \lambda yy \wedge x{\cdot}r = M)$ converges in the γ-calculus.*

Reduction in the lazy λ-calculus is not a fully satisfactory model of reduction in lazy functional programming languages [13]. The problem is that β-reduction possibly copies the arguments of applications, which will duplicate reductions to be done if the arguments are reducible terms. For instance, $(\lambda x(xx))M$ will reduce to MM containing two copies of the possibly reducible term M. The γ-calculus avoids this problem completely since it copies the bodies of abstractions rather than the actual arguments of functional applications. Launchbury [13] carefully analyses sharing in lazy functional programming and provides an operational semantics providing an accurate model for sharing.

The following facts provide evidence that the γ-calculus is superior to the λ-calculus as an operational model of functional programming languages:

- The γ-calculus can directly express (mutual) recursion;
- the γ-calculus can express sharing;
- the γ-calculus can mix lazy with eager functions;
- the γ-calculus provides a unified framework for functional and concurrent programming.

8 Records

Records can be expressed in the γ-calculus as functions mapping field names to their associated values. For instance, the record

```
[A:U B:V C:W]
```

can be expressed as the function

```
fun {F}
   if F=C then W
   elseif F=B then V
   elseif F=A then U
   else undefined fi
end
```

returning the name `undefined` in case the argument is not equal to one of the field names `A`, `B`, `C`. We have now switched to a concrete syntax for the γ-calculus. Variables are written as identifiers starting with capital letters, and names are written as identifiers starting with lower case letters (e.g., `undefined`). Functional notation translates as in the section on the embedding of the eager λ-calculus.

Note that the field names of the above record are given as variables. In case two or more field names turn out to be equal, the rightmost value specification wins.

Record adjunction takes the union of two records, where conflicts are resolved by giving priority to the right record; for instance,

```
[a:1 b:2 c:3] * [b:8 d:6] = [a:1 b:8 c:3 d:6]
```

In the γ-calculus, record adjunction can be expressed as the higher-order function

```
Adjoin = fun {R S}
            fun {F}
               local V = {S F} in
                  if V=undefined then {R F} else V fi
               end
            end
         end
```

9 Procedures with Encapsulated State

The following defines a procedure `{Num X}` maintaining an internal counter initialized with `0`.

```
local C = {NewCell 0} in
   proc {Num X}
      local Y in {C X Y} {Plus X 1 Y} end
   end
end
```

An application {Num X} will equate X with the current value of the counter and then increment the counter. It is straightforward to represent numbers in the γ-calculus. The procedure NewCell is defined as

$$\exists a(\texttt{NewCell} = a \wedge a{:}\, xy/\exists c(x = c \wedge c{:}y)).$$

Now suppose the computation space contains the applications

```
{Num X}  {Num Y}  {Num Z}
```

Then the variables X, Y, and Z will be equated to different numbers and the internal counter of Num will be incremented three times. One possible outcome is X=0, Y=1, Z=2. Another possible outcome is X=1, Y=0, Z=2. However, X=3, Y=0, Z=2 is impossible, provided there are no other applications of Num but the ones above.

The procedure Num builds a state sequence

$$u_1,\ u_2,\ u_3,\ \ldots,\ u_k$$

whose members are linked by constraints {Plus u_i 1 u_{i+1}}, and whose respective last member is hold in the cell C. Concurrent applications of Num create concurrent exchange requests for the cell C, which are serialized indeterministically. Reduction of an application {C X Y} will equate X to the current end of the sequence and make Y the new end of the sequence. Note that this construction makes crucial use of logic variables, and that mutual exclusion of the competing state accesses is obtained for free.

The procedure Num is unsafe in so far that an application {Num 567}, say, may set the counter to 568, due to the indeterministic choice of the equation to be used with the Elimination Rule. A safe version of Num is

```
local C = {NewCell 0} in
   proc {Num A}
      local X Y in {C X Y} {Plus X 1 Y} {Wait X A} end
   end
end
```

where {Wait X A} is defined as $\exists a(\textbf{if}\ \texttt{X} = a\ \textbf{then}\ \top\ \textbf{else}\ X = A)$.

```
proc {Create MethodTable O}
   local C = {NewCell EmptyRecord} in
      proc {O Message}
         local Method State NewState in
            Method = {MethodTable {Message methodName}}
            {C State NewState}
            {Method State Message O NewState}
         end
      end
   end
end
```

Figure 3: Object creation

10 Objects

Objects are procedures with encapsulated state. They are specified by a collection of methods, possibly obtained by inheritance from other objects. Objects are applied to messages. A message is a record `[methodName:M ... ]` specifying the name `M` of the method to be applied, possibly together with input and output arguments. A method is a possibly indeterministic function

$$\textit{method}\colon \textit{state} \times \textit{message} \times \textit{object} \;\rightarrow\; \textit{state}$$

evolving the state of the object according to the message and the object itself (the so-called self reference).

When an object is applied to a message, the method requested by the message

```
Method = {MethodTable {Message methodName}}
```

is obtained from the method table of the object (represented as a record). Next a request

```
{C State NewState}
```

to extend the state sequence of the object is issued (`C` is the encapsulated cell holding the end of the state sequence) and the selected method is applied

```
{Method State Message O NewState}
```

to link the new state with the old state.

Figure 3 shows a procedure `{Create MethodTable O}` creating a new object `O` from a method table given as argument. States are represented as records, and the initial state is the empty record represented as follows:

```
EmptyRecord = fun {F} undefined end
```

The procedure Create is oversimplified in that it does not

- handle the case where the requested method is undefined
- provide a possibility to initialize the state of the newly created object (which is a must in a concurrent setting)
- provide more sophisticated synchronization, for instance, state access only after the method to be applied is known
- provide a possibility to close an object.

All these features can be incorporated easily [8]. Initialization can be taken care of by giving Create an initial message as extra argument.

Using the syntax of Oz [8], a simple counter object C can be created as follows:

```
create C
   meth init(X)  val <- X end
   meth inc(X)  val <- @val+X end
   meth read(X)  X=@val end
end
```

This translates in an application of the procedure Create in Figure 3 to the method table shown in Figure 4. The state of the counter is represented as a one field record [val:_]. The methods init and inc "update" the attribute val by means of record adjunction.[5] A message requesting that the counter be incremented by 67, say, takes the form [methodName:inc arg:67]. The generality obtained by representing states as records and attribute updates as adjunctions is needed when the methods of the counter are inherited to objects with additional attributes.

Creating an object O by inheritance from objects $O_1, \ldots, O_n$ means to obtain the method table of O by combining the method tables of $O_1, \ldots, O_n$, possibly by record adjunction. To enable inheritance, the method table of an object must be made accessible. One straightforward way to do this is to equip an object with a pseudo-method returning its method table.

From our discussion it should be clear that there is more than one style of object-orientation the γ-calculus can express. A fully developed style of object-orientation based on the ideas outlined here is realized in Oz [8].

The observation that objects are procedures with encapsulated state is well-known in the Lisp community [1]. Our contribution here is to show that this idea carries over smoothly to the concurrent setting of the γ-calculus.

[5] Attributes are the field names of states and represent what is called an instance variable in Smalltalk.

```
[init: proc {InState Message Self OutState}
          OutState = {Adjoin InState [val: {Message arg}]}
       end
 inc:  proc {InState Message Self OutState}
          OutState = {Adjoin InState
                           [val: {Plus {InState val}
                                       {Message arg}}]}
       end
 read: proc {InState Message Self OutState}
          OutState = InState
          {Message arg} = {InState val}
       end
]
```

Figure 4: Method table of a simple counter object

Our object model can express private methods and private attributes by restricting the visibility of method and attribute names exploiting the statically scoped setting of the γ-calculus. Although attributes are not directly accessible, they may be visible to methods added by inheritance.

11 Communication

We have seen that we can express communicating concurrent objects as procedures with encapsulated state. This model is different from the established model, where a concurrent object is an agent reading messages from a communication medium (e.g., streams in concurrent logic programming [23], mail boxes in the actor model [9], and ports [12] in AKL). Moreover, the principal notion of process algebras and the π-calculus is communication through channels. So, how is it that the γ-calculus can express communicating concurrent objects without a dedicated communication prmitive?

The answer is simple: Explicit communication is unnecessary if procedures can be applied concurrently and can have encapsulated state. State is obtained from cells, which can be seen as a primitive and standardized form of procedures with state. Thus, communication and state turn out to be different sides of the same coin. This observation is fundamental, but certainly not new.

Our object model provides for straightforward many-to-one communication. In contrast, streams in concurrent logic programming [23] provide for easy one-to-many communication, but have severe problems with many-to-one communication (see [12] for a discussion of this issue).

```
proc {NewPort Port Stream}
   local C = {NewCell Stream} in
      proc {Port Message}
         local S in {C [token:Message next:S] S} end
      end
   end
end
```

Figure 5: Creating ports

Ports [12] are a communication structure well-suited for both many-to-one and one-to-many communication. Ports can be easily expressed in the γ-calculus. The procedure `{NewPort Port Stream}` in Figure 5 creates a new port (a procedure) and connects it to a stream (a logic variable to be constrained incrementally to a list). An application `{Port Message}` extends the stream associated with the port with the reference `Message`. One easily obtains many-to-many communication since the port can be shared by many message senders and the stream can be shared by many message receivers.

12 An Execution Strategy

A programming language based on the γ-calculus must make some assumptions about the order in which possible reduction steps are to be carried out. Such assumptions are needed so that the programmer can write fair[6] and efficient programs. We will outline one possible execution strategy below.

Our execution strategy organizes a computation space into a blackboard and a collection of threads.

Blackboard

Thread $\cdots$ Thread

The *blackboard* is a composition of abstractions, cells, and redundant equations of the form $u = u$ or $a = b$.[7] A *thread* is a nonempty stack of expressions. The execution strategy considers the threads of a computation space in a round-robin fashion making sure that every reducible thread will make progress. As computation proceeds, existing threads may terminate and new threads may be created.

[6] Fairness roughly means that reduction steps that could be done will be done eventually.

[7] Equations of the form $u = u$ or $a = b$ have no computational significance and can be dropped in an implementation.

A thread is reduced by considering its topmost expression. The reduction rules for threads are derived from the rules of the γ-calculus. A thread is not reducible if it consists of a single expression E and E is either a conditional whose guard does not have the form $u = u$ or $a = b$, or an application that does not match an abstraction or a cell on the blackboard. In all other cases, a thread is reduced by popping its topmost expression and if it is

1. $E \wedge F$: push first F and then E
2. $\exists x E$: create a fresh variable y and push $E[y/x]$
3. $\exists a E$: create a fresh name b and push $E[b/a]$
4. $x = u$ or $u = x$, where $x \neq u$: replace all occurrences of x with u
5. $u = u$, $a = b$, $a{:}\overline{x}/E$, or $a{:}x$: write it on the blackboard
6. $a\overline{u}$ and the blackboard contains a matching abstraction $a{:}\overline{x}/E$: push $E[\overline{u}/\overline{x}]$
7. **if** $u = u$ **then** E **else** F: push E
8. **if** $a = b$ **then** E **else** F, where $a \neq b$: push F
9. avw and the blackboard contains a matching cell $a{:}u$: push $v = u$ and replace $a{:}u$ with $a{:}w$ on the blackboard
10. an application or conditional that cannot reduce yet with one of the above rules: make it the single expression of a new thread (**Suspension Rule**).

The congruence laws must not be applied. We assume that computation starts with a computation space where no variable is free and no free name is declared. These assumptions ensure that capturing of references cannot occur. The rules have the remarkable property that a reducible thread stays reducible if other threads are reduced before it.

The idea is to start with a computation space with an empty blackboard and a single thread containing a single expression. If the top of a nonsingleton thread is not yet reducible, it is *suspended* by moving it to a newly created thread. This way the thread is not blocked and the next expression can be reduced. One can force the creation of a new thread executing E by writing

$$\exists x\,(\textbf{if } x = a \textbf{ then } E \textbf{ else } \top \;\wedge\; x = a).$$

An expression is called *sequential* if it will execute with a single thread; that is, if we start with a computation space consisting just of one singleton thread containing the expression, it cannot evolve into a space with more than one thread. An expression is called *quasi-sequential* if it is congruent to a sequential expression. If E_1 and E_2 are sequential, then

$$\exists x\,(\textbf{if } x = a \textbf{ then } E_1 \textbf{ else } \top \;\wedge\; \textbf{if } x = a \textbf{ then } E_2 \textbf{ else } \top \;\wedge\; x = a)$$

is quasi-sequential but not sequential.

A implementation may execute several threads in parallel. Our execution strategy has the interesting property that a sequential expression may be easily rewritten such that it executes with several possibly parallel threads.

Let M be a closed λ-term. Then the expression $\exists x(x = M)$ obtained with the translation embedding the eager λ-calculus into the γ-calculus is sequential. Expressions obtained with the translation embedding the lazy λ-calculus are in general not even quasi-sequential.

13 First-order Constraints and Search

We will now extend the γ-calculus with general first-order constraints. The extension to general constraints will confront us with the problem of failure, which we could circumvent nicely for the simple constraints of the γ-calculus.

In the following we can only present some basic ideas concerning the extension of the γ-calculus to general constraints and search. For a deeper investigation of these issues we refer the reader to [24, 22].

We base our notion of constraint system on first-order predicate logic with equality. A *constraint system* consists of

1. a signature Σ (a set of constant, function and predicate symbols)
2. a consistent theory Δ (a set of sentences over Σ having a model)
3. an infinite set of constants in Σ called *names* satisfying two conditions:
   (a) $\Delta \models \neg(a \doteq b)$ for every two distinct names a, b
   (b) $\Delta \models \phi \leftrightarrow \psi$ for every two sentences ϕ, ψ over Σ such that ψ can be obtained from ϕ by permutation of names.

Given a constraint system, we will call every first-order formula over its signature a *constraint*. We use $\bot$ for the constraint that is always false, and $\top$ for the constraint that is always true.

The minimal constraint system has no other symbols but names in its signature. The usual tree constraint systems (finite or rational constructor trees) can be made into constraint systems in our sense by simply distinguishing infinitely many constants as names.

We now extend the γ-calculus with three new forms

$$E, F, G \quad ::= \quad \ldots \mid \phi \mid \textbf{if } \phi \textbf{ then } E \textbf{ else } F \mid E \nabla F$$
$$\phi, \psi \quad : \quad \text{constraints}$$

called *constraints*, *conditionals*, and *disjunctions*, respectively. We assume that all constraints are taken from some fixed constraint system. Recall that a constraint is simply a first-order formula over the constraint signature. A real programming language will of course carefully restrict the constraints a programmer can actually write (see, for instance, Oz [8]). The new expressions subsume the expressions $\top$, $u = v$, and **if** $u = v$ **then** E **else** F of the γ-calculus.

13.1 Constraints

The semantics of constraints in the extended γ-calculus is given by four congruence laws:

1. conjunction of constraints is congruent to composition of constraints
2. existential quantification $\exists x\phi$ of constraints is congruent to variable declaration $\exists x\phi$ over constraints
3. $\phi \equiv \psi \quad$ if $\Delta \models \phi \leftrightarrow \psi$
4. $x = u \wedge E \;\equiv\; x = u \wedge E[u/x] \quad$ if u free for x in E.

The first three laws provide for constraint simplification. Law (4) extends the equality imposed by constraints to all expressions. The Elimination Rule of the γ-calculus is subsumed by the new congruence laws and is thus not present in the extended calculus.

Proposition 4 *If $\Delta \models \phi \wedge \psi \leftrightarrow \psi'$, then $\phi \wedge \psi \wedge E \;\equiv\; \psi' \wedge E$. If $\Delta \models \phi \rightarrow \psi$, then $\phi \wedge E \;\equiv\; \psi \wedge \phi \wedge E$.*

13.2 Conditionals

The semantics of the conditional is given by the congruence law

$$\phi \wedge \textbf{if } \psi \textbf{ then } E \textbf{ else } F \quad \equiv \quad \phi \wedge \textbf{if } \phi \wedge \psi \textbf{ then } E \textbf{ else } F$$

providing for relative simplification of conditional guards (see [2, 25]) and two reduction rules

$$\textbf{if } \top \textbf{ then } E \textbf{ else } F \;\rightarrow\; E \qquad\qquad \textbf{if } \bot \textbf{ then } E \textbf{ else } F \;\rightarrow\; F$$

subsuming the corresponding rules of the γ-calculus.

Proposition 5 *If* $\Delta \models \phi \rightarrow \psi$*, then* $\phi \wedge \textbf{if } \psi \textbf{ then } E \textbf{ else } F \rightarrow \phi \wedge E$*. If* $\Delta \models \phi \rightarrow \neg\psi$*, then* $\phi \wedge \textbf{if } \psi \textbf{ then } E \textbf{ else } F \rightarrow \phi \wedge F$*.*

A useful generalization of the conditional is obtained by allowing for multiple clauses

$$\textbf{if } \phi_1 \textbf{ then } E_1 \,[]\, \ldots \,[]\, \phi_n \textbf{ then } E_n \textbf{ else } F$$

where the conditional can reduce with any clause whose guard is entailed. This introduces a new form of indeterminism known as committed choice. If the guards of all clauses are disentailed, then the generalized conditional can reduce to the else constituent.

13.3 Disjunctions

The semantics of disjunctions is given by the congruence laws

$$E \nabla F \equiv F \nabla E \qquad\qquad \phi \wedge (E \nabla F) \equiv (\phi \wedge E) \nabla (\phi \wedge F)$$

and the reduction rules

$$(\bot \wedge E) \nabla F \rightarrow F \qquad\qquad \top \nabla F \rightarrow \top.$$

Note that disjunctions do not introduce any form of backtracking. Read from right to left, the second congruence law allows to lift shared constraints (an idea also realized in the constructive disjunction of [7]). For instance,

$$(x = 1 \wedge y = 1) \nabla (x = 1 \wedge y = 2) \equiv x = 1 \wedge (y = 1 \nabla y = 2).$$

13.4 Failure

A expression E is called *failed* if $E \equiv E \wedge \bot$. In a failed expression, all conditionals and disjunctions become trivially reducible. Thus computation must be stopped as soon as failure occurs. Note that this is in contrast to the situation in the pure γ-calculus, where computation can proceed orderly in the presence of inconsistent equations $a = b$.

13.5 Search

The extension of the γ-calculus to first-order constraints is of practical use only in conjunction with a facility for search.

Search in the style of Prolog can be provided as follows: Computation proceeds as long as reduction rules are applicable and failure does not occur. If computation arrives at an unfailed and irreducible expression, a disjunctive molecule $E \nabla F$ is selected (if there is any) and two don't know alternatives are created by replacing $E \nabla F$ with E and F, respectively. The alternatives are reduced as before and may be explored following a backtracking strategy. Unfailed and irreducible expressions not containing disjunctive molecules are taken as solutions.

Prolog-style search suffers from many problems. For one thing, it is not obtained within the computational calculus but formulated at the meta-level. Moreover, the idea of backtracking is incompatible with the idea of concurrent and reactive computation.

Combining reactive computation with search has been one of the (unsolved) challenges of the Japanese Fifth Generation Project. A computational calculus solving the problem through encapsulation of search into deep guard combinators has been devised with the concurrent constraint language AKL [11]. Oz realizes a more flexible scheme based on the γ-calculus and a higher-order search combinator spawning a local computation space [22].

13.6 Higher-order Programming and First-order Constraints

The extended γ-calculus has the remarkable property that it combines first-order constraints with higher-order programming. The only requirement on constraints imposed by higher-order programming is the accommodation of names. This is in sharp contrast to approaches based on higher-order logic [18], where higher-order programming comes with the operational burden of higher-order constraints. Although we do not doubt the usefulness of higher-order constraints for some applications (e.g., reasoning about programs), we feel that higher-order programming and higher-order constraints are two separate issues that should be decoupled as much as possible.

14 Relationship with the π-calculus

It is illuminating to compare the γ-calculus with the π-calculus [17, 16, 15], a calculus of concurrent computation that evolved from research on algebraic process calculi. Although the γ-calculus and the π-calculus were conceived with very different goals and intuitions—a unified model of computation in the case of the γ-calculus and a model of communicating processes in the case of the π-calculus—they are strikingly close technically. In fact, both calculi can be obtained as specializations of a slightly more general calculus, which is obtained from the polyadic π-calculus [15] by distinguishing between names and variables and making variables logical. Logic variables increase the expressivity of the π-calculus in two crucial aspects: They allow to equate communication links, and they provide the possibility to express procedures with input *and* output arguments (recall that a function is a procedure with input and output).

While the γ-calculus has logic variables, the π-calculus has formal arguments only (as in functional programming). While the π-calculus has communication as its principal primitive, the γ-calculus has logic variables, abstraction, and cells as principal primitives. We shall show below that the π-calculus can be extended with logic variables, and that the thus extended asynchronous π-calculus can express abstractions and cells.

To put the comparison of the two calculi on solid ground, we introduce yet another calculus, called the *κ-calculus*. The κ-calculus is an asynchronous and polyadic version of the π-calculus in [16] extended with equations. Its abstract syntax is given by

$$A, B \ ::= \ \top \mid A \wedge B \mid \exists x A \mid x{::}\overline{y}/A \mid x\overline{y} \mid x{:}\overline{y}/A \mid x = y$$

where $\top$ is null, $A \wedge B$ is composition, $\exists x A$ is restriction, $x :: \overline{y}/A$ is an input agent, $x\overline{y}$ is an asynchronous output agent, and $x{:}\overline{y}/A$ is a replicating input agent (i.e, $!\,x :: \overline{y}/A$). The only form not present in the π-calculus are equations $x = y$. In contrast to the π-calculus, where x and y would be called names, they are called variables in the κ-calculus.

Seen from the perspective of the γ-calculus, we have dropped conditionals and the distinction between names and variables, and we have added the form $x{::}\overline{y}/A$, which will turn out to be a once-only abstraction.

The structural congruence of the κ-calculus is given by the usual laws for composition and restriction, α-conversion for both input agents, symmetry for equations, and replication for replicating input agents:

$$x{:}\overline{y}/A \ \equiv \ x{::}\overline{y}/A \ \wedge \ x{:}\overline{y}/A.$$

The reduction axioms are the Communication Rule

$$x\overline{y} \ \wedge \ x{::}\overline{z}/A \ \rightarrow \ A[\overline{y}/\overline{z}] \qquad \text{if } \overline{y} \text{ free for } \overline{z} \text{ in } A$$

and the Elimination Rule

$$\exists x(x = y \ \wedge \ A) \ \rightarrow \ A[y/x] \qquad \text{if } x \neq y \text{ and } y \text{ free for } x \text{ in } E.$$

The structural reduction rules are the usual ones.

Seen from the perspective of the γ-calculus, an output agent is an application and a replicating input agent is an abstraction. Ordinary input agents are once-only abstractions providing extra expressivity. In fact, cells can be expressed using once-only abstractions:

$$x{:}y \quad :\equiv \quad \exists z\,(x{::}\,uv/(u = y \wedge zv) \ \wedge \ z{:}\,w/\,x{::}\,uv/(u = w \wedge zv)).$$

The κ-calculus does not make a distinction between variables and names. Without this distinction, there is nothing that can make two variables different. Hence the symmetric conditional of the γ-calculus does not carry over to the κ-calculus. However, we could still have an asymmetric conditional just testing for equality.

One easily verifies that our embeddings of the eager and lazy λ-calculus into the γ-calculus carry over to the κ-calculus. Due to the presence of logic variables, they are simpler than the ones for the π-calculus given by Milner [16]. In contrast to Milner's encoding, our embedding of the lazy λ-calculus shares reductions of arguments (as in implementations of lazy functional programming).

It seems that the κ-calculus cannot express record adjunction and, consequently, inheritance with method overwriting. The problem is that two variables cannot be established as different. Thus names and a corresponding symmetric conditional seem to be crucial for modeling inheritance.

15 Future Research

Our investigations of the γ-calculus are at an early stage. So far, they have mainly been driven by considerations concerning the design and implementation of the programming language Oz, of which it formalizes important aspects. Directions for future research are type disciplines and reasoning about programs. In particular, a declarative characterization of program equivalence is desirable, the investigation of which may start from the techniques developed for the π-calculus. Another interesting topic are extensions of the γ-calculus so that it can model distributed computation and mobility.

Acknowledgements

I'm thankful to Martin Müller and Joachim Niehren for continued discussions accompanying the development of the γ-calculus. Martin Henz, Andreas Podelski, Ralf Treinen and Jörg Würtz helped by commenting on a draft version of this paper.

References

[1] H. Abelson, G. J. Sussman, and J. Sussman. *Structure and Interpretation of Computer Programs.* The MIT Press, Cambridge, Mass., 1985.

[2] H. Aït-Kaci, A. Podelski, and G. Smolka. A feature-based constraint system for logic programming with entailment. *Theoretical Computer Science*, 122(1–2):263–283, January 1994.

[3] F. Benhamou and A. Colmerauer, editors. *Constraint Logic Programming: Selected Research.* The MIT Press, Cambridge, Mass., 1993.

[4] G. Berry and G. Boudol. The chemical abstract machine. In *Proceedings of the 17th ACM Conference on Principles of Programming Languages*, pages 81–94, 1990.

[5] N. Dershowitz and J.-P. Jouannaud. Rewrite systems. In *Handbook of Theoretical Computer Science*, volume B, chapter 15. North–Holland, Amsterdam, Holland, 1990.

[6] K. Futatsugi, J. Goguen, J.-P. Jouannaud, and J. Meseguer. Principles of OBJ2. In *Proceedings of the 12th ACM Conference on Principles of Programming Languages*, pages 52–66, 1985.

[7] P. v. Hentenryck, V. Saraswat, and Y. Deville. Design, implementations, and evaluation of the constraint language cc(FD). Technical Report CS-93-02, Brown University, Box 1910, Providence, RI 02912, 1993.

[8] M. Henz, M. Mehl, M. Müller, T. Müller, J. Niehren, R. Scheidhauer, C. Schulte, G. Smolka, R. Treinen, and J. Würtz. The Oz Handbook. Research Report RR-94-09, Deutsches Forschungszentrum für Künstliche Intelligenz, Stuhlsatzenhausweg 3, D-66123 Saarbrücken, Germany, 1994.

[9] C. Hewitt, P. Bishop, and R.Steiger. A universal modular ACTOR formalism for artificial intelligence. In *Proceedings of the International Joint Conference on Artificial Intelligence*, pages 235–245, 1973.

[10] J. Jaffar and M. J. Maher. Constraint Logic Programming: A Survey. *The Journal of Logic Programming*, to appear, 1994.

[11] S. Janson and S. Haridi. Programming paradigms of the Andorra kernel language. In V. Saraswat and K. Ueda, editors, *Logic Programming, Proceedings of the 1991 International Symposium*, pages 167–186, San Diego, USA, 1991. The MIT Press.

[12] S. Janson, J. Montelius, and S. Haridi. Ports for objects. In *Research Directions in Concurrent Object-Oriented Programming.* The MIT Press, Cambridge, Mass., 1993.

[13] J. Launchbury. A natural semantics for lazy evaluation. In *Proceedings of the 20th ACM Conference on Principles of Programming Languages*, pages 144–154, 1993.

[14] J. Meseguer. Conditional rewriting logic as a unified model of concurrency. *Theoretical Computer Science*, 96:73–155, 1992.

[15] R. Milner. The polyadic π-calculus: A tutorial. ECS-LFCS Report Series 91-180, Laboratory for Foundations of Computer Science, University of Edinburgh, Edinburgh EH9 3JZ, October 1991.

[16] R. Milner. Functions as processes. *Journal of Mathematical Structures in Computer Science*, 2(2):119–141, 1992.

[17] R. Milner, J. Parrow, and D. Walker. A calculus of mobile processes, I. *Information and Computation*, 100(1):1–40, September 1992.

[18] G. Nadathur and D. Miller. An overview of λProlog. In R. A. Kowalski and K. A. Bowen, editors, *Proceedings of the Fifth International Conference and Symposium on Logic Programming*, pages 810–827, Seattle, Wash., 1988. The MIT Press.

[19] J. Niehren. Functional Computation as Concurrent Computation. *Proceedings of the 23rd ACM Symposium on Principles of Programming Languages*, pages 333–343, St. Petersburg Beach, Florida, 1996. The ACM Press.

[20] J. Niehren and G. Smolka. A confluent relational calculus for higher-order programming with constraints. In J.-P. Jouannaud, editor, *Proc. 1st Int. Conference on Constraints in Computational Logics (CCL'94)*, pages 89–104. Lecture Notes in Computer Science, Vol. 845, Springer-Verlag, Berlin, 1994.

[21] V. A. Saraswat. *Concurrent Constraint Programming.* The MIT Press, Cambridge, Mass., 1993.

[22] C. Schulte and G. Smolka. Encapsulated search in higher-order concurrent constraint programming. In M. Bruynooghe, editor, *Logic Programming, Proc. 1994 Int. Symposium*, pages 505–520. The MIT Press, Cambridge, MA, 1994.

[23] E. Shapiro. The family of concurrent logic programming languages. *ACM Computing Surveys*, 21(3):413–511, September 1989.

[24] G. Smolka. A calculus for higher-order concurrent constraint programming with deep guards. Research Report RR-94-03, Deutsches Forschungszentrum für Künstliche Intelligenz, Stuhlsatzenhausweg 3, D-66123 Saarbrücken, Germany, February 1994.

[25] G. Smolka and R. Treinen. Records for logic programming. *The Journal of Logic Programming*, 18(3):229–258, April 1994.

[26] G. Winskel. *The Formal Semantics of Programming Languages.* The MIT Press, Cambridge, Mass., 1993.

Part 4

Mechanical Proof Assistance

Part 0 has shown that proof is a major ingredient of careful program design. This part deals with the question how theorem proving can be supported by machine. Mechanical assistance is particularly welcome if it relieves the programmer of routine steps. Given a proof strategy, a proof tool should carry out the technical details of the proof. So, the programmer can concentrate on the essentials. In addition, proof tools should help to cope with large axiomatizations, such as arise in hardware design. In the following, two instances of proof assistants are presented.

Lawrence C. Paulson

N. Shankar, C.A.R. Hoare, Gert Smolka

Tool Support for Logics of Programs

Lawrence C. Paulson

Computer Laboratory, University of Cambridge

Key words: proof tools / generic reasoning / logic programming / logical variables / unification / Isabelle / higher-order logic / set theory / inductive definitions / cryptographic protocols

Summary. Proof tools must be well designed if they are to be more effective than pen and paper. Isabelle supports a range of formalisms, two of which are described (higher-order logic and set theory). Isabelle's representation of logic is influenced by logic programming: its "logical variables" can be used to implement step-wise refinement. Its automatic proof procedures are based on search primitives that are directly available to users. While emphasizing basic concepts, the article also discusses applications such as an approach to the analysis of security protocols.

1. What Should Proof Tools Do For Us?

Computer scientists often use pen and paper for proofs, as mathematicians have always done. Informal proofs leave big gaps, gaps that minds can bridge but machines often cannot. Proof tools require formal calculi, comprising a rigid syntax of formulas and rules for transforming them. Tools and formal calculi can be hard to use. They must give us something in return.

- *Soundness*: we can trust the result
- *Transparency*: we can follow the reasoning
- *Power*: the tool is mightier than the pen

There are trade-offs among these benefits. A tool that puts all the emphasis on soundness may sacrifice power, and vice versa. Transparency involves a combination of soundness (the reasoning is correct) and power (the reasoning is expressed at a high level). Even unsound tools can be valuable: consider floating-point arithmetic. If soundness is not guaranteed then we need transparency, in order to check for ourselves that a derivation is sound.

Soundness can be obtained by recording proofs down to the level of inference rules, and checking them with a separate tool. But this requires considerable storage, and does not aid transparency: detailed proofs are too big to understand. If the proof tool is allowed to invoke external reasoners, such as model checkers or computer algebra systems, then it could record all dependencies on such reasoners.

The tool must let us prove things that we cannot using pen and paper. This is the most important requirement of all, and is perhaps the hardest to attain. Tools are mainly valuable for proofs about objects, such as hardware or formal specifications, that are too big to manage on paper.

Marktoberdorf is traditionally devoted to pen-and-paper calculations. This year is an exception, with two lecture courses on automatic proof tools. Shankar presents some of the impressive applications of PVS. The present article complements his by discussing the principles underlying Isabelle [36].

Isabelle uses a unique representation of logic. The benefits include support for many formalisms and powerful tools for logical reasoning. "Logical variables" in goals serve as placeholders for as-yet-unknown subterms, to support refinement of specifications into implementations. Soundness comes from Milner's abstract type approach,[1] transparency from the use of high-level rules, and power through unification and search.

The article continues by covering generic reasoning (Sect. 2), extending it to quantifiers (Sect. 3). It then presents specialized tools: automatic proof (Sect. 4), mechanized set theory (Sect. 5), and induction (Sect. 6). We see how Isabelle has been applied to proofs about cryptographic protocols (Sect. 7). The conclusions (Sect. 8) briefly survey other applications, such as refinement of functional programs.

2. Generic Reasoning: The Propositional Case

A generic proof tool supports reasoning in a variety of logics. It provides independence from formalization details, a flexible treatment of notation, and ease of extension.

There are many formal systems that differ only slightly. Changing just a few rules can make a logic classical instead of constructive, higher-order instead of first-order. A proof tool should let us share the implementation effort for the common part of such logics.

Users should not have to know which of the facts they use are axioms as opposed to theorems. They should not have to know which operators are primitive as opposed to derived. Proof tools should not make it harder to use non-primitive concepts. Minor changes to a formalization should not force us to redo existing proofs.

Good notation matters. The pen can draw any symbols and figures. Our tools cannot match that, but they should be built to be as flexible as possible. We must not dismiss this question as mere syntax.

Most of us do not switch between formal systems, but any proof development requires extending the formal system. Each definition may involve new notation, new laws to be proved and new reasoning methods for those laws. Over time, the effect is to create a new formal system.

Isabelle's approach to generic reasoning is based on a unique interpretation of inference rules. Combining rules in various ways yields different forms of reasoning on the level of theorems.

[1] I have described his approach elsewhere [32, Chapter 7].

2.1 Two Readings of Proof Rules

Most approaches to formal logic take rules of the form

$$\frac{X_1 \quad \ldots \quad X_m}{X}$$

as primitive. We call $X_1, \ldots, X_m$ the *premises* and X the *conclusion*. There are two ways of reading such rules:

- *forward*: if $X_1, \ldots, X_m$ then X
- *backward*: to show X it suffices to show $X_1, \ldots, X_m$

The forward reading is sometimes used in logic texts, especially in proofs organized as a numbered list of formulas. Forward proof is useful when applying a general law to a specific case, and when simplifying the instance so obtained.

The backward reading is better for proof discovery. It concentrates on the given problem, analysing it to simpler subproblems.

Hand proofs consist of a mixture of backward and forward proof. Backward proof forms the main structure of the argument (such as induction followed by case analysis), while forward proof may be used at lower levels. A proof tool should support both styles.

2.1.1 Proof Trees. Here is a typical proof tree, for a simple theorem about sets: if $C \subseteq A$ and $z \in C \cap B$ then $z \in A$.

$$\frac{C \subseteq A \qquad \dfrac{z \in C \cap B}{z \in C}\,\cap\text{-elim1}}{z \in A}\,\subseteq\text{-elim}$$

Proof trees are constructed by composing rules. The tree above consists of two rules joined together. It simply assumes the formulas $C \subseteq A$ and $z \in C \cap B$, which perhaps we could prove given more information about A, C and z.

This example presupposes a proof system for set theory with rules such as these:

$$\frac{x \in A \cap B}{x \in A}\,\cap\text{-elim1} \qquad \frac{x \in A \cap B}{x \in B}\,\cap\text{-elim2} \qquad \frac{A \subseteq B \quad x \in A}{x \in B}\,\subseteq\text{-elim}$$

A common misconception is that proof trees can only use the rules for the standard connectives $\wedge$, $\vee$, etc. It is better to reason using rules that directly express the subject matter. The definition of $\cap$ in set theory is surprisingly complicated, and best forgotten as soon as possible.

In Isabelle, we can derive new rules and use them as if they were primitive rules. Proving X from the assumptions $X_1, \ldots, X_m$ derives the rule

$$\frac{X_1 \quad \ldots \quad X_m}{X.}$$

If $m = 0$ we might call this a theorem.

2.1.2 Forward versus Backward Proof. Proof trees can be built from the root upwards (backward reasoning) or from the leaves downwards (forward reasoning). Refer again to the proof tree shown above. In the forward style, we begin with the assumption $z \in C \cap B$. We apply rule $\cap$-elim1 to obtain $z \in C$, etc., finally concluding $z \in A$.

In the backward style, we begin with the desired conclusion, $z \in A$. Call this the *main goal*. We observe that rule $\subseteq$-elim can reduce it to $z \in C$, where we must prove $C \subseteq A$. We then apply rule $\cap$-elim1 to obtain the subgoal $z \in C \cap B$. At this point, we have reduced the main goal to the two subgoals $C \subseteq A$ and $z \in C \cap B$. This is the derived rule

$$\frac{C \subseteq A \quad z \in C \cap B}{z \in A.}$$

It is of no permanent interest. But it perfectly captures the state of the backward proof. If we know enough about A, C and z to prove the two subgoals then we shall eventually be left with a proof state, $z \in A$, with no subgoals. This proof state is the theorem we intended to prove.

Proof tools usually derive theorems. Perhaps they should instead derive rules. The operation of joining two rules would then implement both forward and backward proof. Isabelle is designed to operate on rules.

2.2 A Simple Rule Calculus

Let us replace the traditional two-dimensional notation for inference rules by $[\![X_1; \ldots ; X_m]\!] \Longrightarrow X$. The brackets $[\![\ ]\!]$ are optional if there is only one premise, and $[\![X_1; \ldots ; X_m]\!] \Longrightarrow X$ abbreviates

$$X_1 \Longrightarrow (\cdots X_m \Longrightarrow X \cdots).$$

Here are the rules $\cap$-elim1 and $\subseteq$-elim in the notation:

$$x \in A \cap B \Longrightarrow x \in A \qquad [\![A \subseteq B; x \in A]\!] \Longrightarrow x \in B$$

Isabelle implements a calculus of inference rules. Its most basic primitives are the *trivial rule* and *resolution*.

The trivial rule has the form $X \Longrightarrow X$. It supports our use of rules as representing proof states, serving the same role as zero does in arithmetic. At the very start of the backward proof, before we have applied any rules, there is one subgoal to be proved, namely the main goal itself. If the main goal is X then the initial proof state is $X \Longrightarrow X$.

Rules of the form $[\![X_1; \ldots ; X_m]\!] \Longrightarrow X$ are called Horn clauses in Prolog terminology. Resolution on Horn clauses involves matching the conclusion of one rule with a premise of another rule. In this example, the conclusion of $[\![X_1; X_2]\!] \Longrightarrow X$ matches Y_2.

$$[\![X_1; X_2]\!] \Longrightarrow X \quad + \quad [\![Y_1; \underline{Y_2}; Y_3]\!] \Longrightarrow Y$$
$$= \text{instance of } [\![Y_1; \underline{X_1; X_2}; Y_3]\!] \Longrightarrow Y$$

Underlining indicates the affected parts. In general we *unify* the formulas X and Y_2, applying the unifying substitution to the result — hence the words "instance of" above.

In general, resolution unifies the conclusion of one rule $[\![X_1; \ldots; X_m]\!] \Longrightarrow X$ with the ith premise of another, $[\![Y_1; \ldots; Y_n]\!] \Longrightarrow Y$. In the resulting rule, Y_i is replaced by $X_1; \ldots; X_m$.

Prolog resolution, extended slightly, is all we need to build proof trees. It automatically matches rules and axioms to the goal being proved. We can even allow some variables in the goal to be updated. Such variables stand for currently unknown parts of the goal. They let us extract information from proofs, say for interactive program derivation. They also make it easier to implement proof procedures for quantifiers.

2.3 Assumptions in Natural Deduction

Natural deduction, due to Gentzen, is based upon three principles.

1. Proof takes place within a varying context of assumptions.
2. Each connective is defined independently of the others.
3. Each connective is defined by *introduction* and *elimination* rules.

In the case of $\wedge$, the *introduction* rule describes how to infer $P \wedge Q$ while the *elimination* rules for $\wedge$ describe what to infer from $P \wedge Q$:

$$\frac{P \quad Q}{P \wedge Q}\,{\scriptstyle \wedge\text{-intr}} \qquad \frac{P \wedge Q}{P}\,{\scriptstyle \wedge\text{-elim1}} \qquad \frac{P \wedge Q}{Q}\,{\scriptstyle \wedge\text{-elim2}}$$

The elimination rule for $\rightarrow$ says what to deduce from $P \rightarrow Q$. It is sometimes called Modus Ponens.

$$\frac{P \rightarrow Q \quad P}{Q}\,{\scriptstyle \rightarrow\text{-elim}}$$

The introduction rule for $\rightarrow$ is characteristic of natural deduction. It says that $P \rightarrow Q$ is proved by assuming P and deriving Q:

$$\frac{\begin{matrix}[P]\\ Q\end{matrix}}{P \rightarrow Q} \qquad (\rightarrow\text{-intr})$$

The key point is that rule $\rightarrow$-intr *discharges* its assumption: even if the proof of Q requires the assumption P, the conclusion $P \rightarrow Q$ does not depend upon that assumption. The notation $[P]$ indicates that any uses of the assumption P are discharged.

Here is a little example. Assuming P and Q, we may prove $P \land Q$ using rule $\land$-intr. By applying rule $\rightarrow$-intr, we discharge the assumption Q to obtain $Q \rightarrow (P \land Q)$, still assuming P. By applying rule $\rightarrow$-intr again, we discharge the assumption P and get $P \rightarrow (Q \rightarrow (P \land Q))$. The conclusion no longer depends upon any assumption, and is clearly a tautology. The proof is shown as follows:

$$\dfrac{\dfrac{\dfrac{[P] \quad [Q]}{P \land Q}\,\land\text{-intr}}{Q \rightarrow (P \land Q)}\,\rightarrow\text{-intr}}{P \rightarrow (Q \rightarrow (P \land Q))}\,\rightarrow\text{-intr}$$

The introduction rules for $\lor$ are straightforward. The elimination rule says that to show some R from $P \lor Q$ there are two cases to consider, one assuming P and one assuming Q.

$$\frac{P}{P \lor Q}\,\lor\text{-intr1} \qquad \frac{Q}{P \lor Q}\,\lor\text{-intr2} \qquad \frac{P \lor Q \quad \begin{array}{c}[P]\\ R\end{array} \quad \begin{array}{c}[Q]\\ R\end{array}}{R}\,\lor\text{-elim}$$

Horn clauses can accommodate natural deduction and assumptions. We allow them to be nested, extend resolution to nested clauses, and introduce a notion of proof by assumption.

2.4 Generalized Horn Clauses

Natural deduction requires a simple generalization of our rule calculus. We can formalize the rule $\rightarrow$-intr as $(P \Longrightarrow Q) \Longrightarrow P \rightarrow Q$. Thus, we regard assumption discharge as the same sort of entailment as that from premises to conclusion. It is implication: not the implication of first-order logic, but implication at the meta-level. We can regard $\rightarrow$-intr as a rule whose premise is itself a rule, namely $\frac{P}{Q}$ [48].

We must augment resolution to allow for nesting of $\Longrightarrow$. Let us consider why. To prove $P \rightarrow (Q \rightarrow (P \land Q))$, resolution with $\rightarrow$-intr yields the subgoal $P \Longrightarrow Q \rightarrow (P \land Q)$; as expected, the step adds P to the assumptions. Now we need to apply $\rightarrow$-intr again, to add Q to the assumptions. But the subgoal has the form $\cdots \Longrightarrow \cdots \rightarrow \cdots$ instead of just $\cdots \rightarrow \cdots$. Lifting in resolution allows a rule to be applied in any context. Lifting the rule $[\![X_1; \ldots ; X_m]\!] \Longrightarrow X$ over the assumption P transforms it to

$$[\![P \Longrightarrow X_1; \ldots ; P \Longrightarrow X_m]\!] \Longrightarrow (P \Longrightarrow X).$$

Lifting and matching to our subgoal transforms $\rightarrow$-intr into the Horn clause

$$(P \Longrightarrow (Q \Longrightarrow P \land Q)) \Longrightarrow (P \Longrightarrow Q \rightarrow (P \land Q)).$$

Resolving this with the proof state replaces our subgoal by $P \Longrightarrow (Q \Longrightarrow P \wedge Q)$, which may be written more concisely as $[\![P; Q]\!] \Longrightarrow P \wedge Q$. Finally, we must apply the rule $\wedge$-intr. Lifting it over the assumptions P and Q, and matching it to the subgoal, transforms $\wedge$-intr into the Horn clause

$$[\![\, [\![P; Q]\!] \Longrightarrow P;\ [\![P; Q]\!] \Longrightarrow Q \,]\!] \Longrightarrow ([\![P; Q]\!] \Longrightarrow P \wedge Q).$$

Resolution using this yields a new proof state having two subgoals,

$$[\![P; Q]\!] \Longrightarrow P \quad \text{and} \quad [\![P; Q]\!] \Longrightarrow Q.$$

We may generalize our notion of trivial rule from $X \Longrightarrow X$ to include subgoals of the form above. Proof by assumption involves deleting a subgoal of the form $[\![X_1; \ldots ; X_m]\!] \Longrightarrow X$ where X matches X_i for some i between 1 and m; the matching substitution must be applied to all other subgoals.

Resolution is normally combined with lifting. For example, the rule

$$[\![Y_1;\ H \Longrightarrow Y_2]\!] \Longrightarrow Y$$

may be seen as a proof state with two subgoals Y_1 and Y_2. The second subgoal has an assumption, H. Resolving the rule $[\![X_1;\ X_2]\!] \Longrightarrow X$ against this proof state replaces the second subgoal by two new ones, each with H as an assumption.

$$[\![X_1;\ X_2]\!] \Longrightarrow X \quad + \quad [\![Y_1;\ H \Longrightarrow Y_2]\!] \Longrightarrow Y$$

$$= \text{instance of } [\![Y_1;\ H \Longrightarrow X_1;\ H \Longrightarrow X_2]\!] \Longrightarrow Y$$

2.5 The Role of Resolution

Isabelle uses resolution to provide proof checking in the forward and backward styles. Unlike classical resolution theorem provers such as Otter [22], Isabelle does not seek contradictions, but derives rules in positive form. Horn clause resolution is a special case of the sophisticated forms of resolution used in Otter. On the other hand, Isabelle generalizes Horn clause resolution in unusual ways. It allows clauses having nested implication, and resolves them using lifting.

Isabelle matches rules and assumptions automatically. Asked to prove a subgoal by assumption, Isabelle searches for suitable assumptions: we do not have to specify one by number. Isabelle considers all matching assumptions, not just the first. Similarly, if we supply a list of rules to match against a subgoal, Isabelle considers all possible matches. We can apply known facts with minimal effort.

Exercise 2.1. Express the rules $\vee$-intr1, $\vee$-intr2 and $\vee$-elim as (possibly generalized) Horn clauses.

Exercise 2.2. Comment on these alternative introduction rules for ∧ and ∨. Why are they correct? What are they good for?

$$\frac{P \quad \begin{matrix}[P] \\ Q\end{matrix}}{P \wedge Q} \qquad\qquad \frac{\begin{matrix}[\neg Q] \\ P\end{matrix}}{P \vee Q}$$

Exercise 2.3. Give a natural deduction proof of $P \to (Q \to P)$.

3. A General Approach to Quantifiers

Quantifiers require a radical generalization of resolution. Isabelle uses the typed λ-calculus to represent the syntax of terms, formulas and rules. Ordinary unification, such as Prolog uses, is unsuitable. Isabelle bases resolution on higher-order unification. Our approach handles not only the quantifiers $\forall$ and $\exists$, but other binding operators such as indexed summations, unions or Cartesian products. It supports sound inference, not just a means of expressing the syntax.

3.1 The Typed λ-Calculus

Here is a capsule review of the typed λ-calculus.[2] The notation $M[L/y]$ stands for the result of substituting the term L for all free occurrences of the variable y in a term M. We regard $((\lambda x.M)N)$ as equal to $M[N/x]$: a β-conversion substitutes the argument, N, into the abstraction's body, M. Bound variables in M are renamed to prevent clashes with variables in N.

The application MN has type τ if M has the function type $\sigma \to \tau$ and N has type σ. The abstraction $\lambda x.M$ has type $\sigma \to \tau$ if M has type τ given that x has type σ. The function type $[\sigma_1, \ldots, \sigma_k] \to \tau$ abbreviates $\sigma_1 \to \cdots (\sigma_k \to \tau) \cdots$. By the standard treatment of curried functions, if M has this type and $N_1, \ldots, N_k$ have types $\sigma_1, \ldots, \sigma_k$, respectively, then $MN_1 \ldots N_K$ has type τ.

Isabelle uses a polymorphic type system. For example, the identity function $\lambda x.x$ gets type $\alpha \to \alpha$, where α is a *type variable*. Type variables can be replaced by any types. The identity function may be regarded as having any type of the form $\tau \to \tau$. Each occurrence of it in an expression may have a different type.

The type prop (short for proposition) is built-in. It is the type of inference rules, which include theorems and axioms as special cases. Rules are sometimes called *meta-level theorems* because Isabelle provides an inference system for them: the *meta-logic*. The meta-logic represents other inference systems, the *object-logics*: HOL, ZF, etc.

[2] The easiest way to learn about the λ-calculus and polymorphism is to learn a programming language based on them, like Haskell or ML. My ML book includes material about the λ-calculus itself [39, Chapter 9].

3.2 Declaring Types and Connectives

To represent an object-logic in Isabelle we extend the meta-logic with types, constants, and axioms. For example, take the predicate calculus.

To represent predicate calculus syntax, we introduce the type bool for meanings of formulas. It is preferable to avoid identifying bool with prop, the type of rules. So we use the constant isTrue to convert one to the other; if A has type bool then isTrue A has type prop.

The logical constants True, False, And, Or, etc., are declared in the obvious manner. With Isabelle such declarations are made in a *theory file*:

```
types
  bool
consts
  isTrue                :: bool => prop
  True, False           :: bool
  And, Or, Implies      :: [bool, bool] => bool
```

The theory file may also specify the constants as having special syntax (such as infix) and describe pretty printing. Let us ignore such matters for now; assume that And, Or, Implies represent the infix operators $\wedge$, $\vee$, $\rightarrow$ with the usual precedences, and associating to the right. Thus the formula $P \wedge Q \wedge R \rightarrow Q$ can be represented by the λ-term

$$\mathsf{Implies}(\mathsf{And}\, P\, (\mathsf{And}\, Q R))Q.$$

We may continue to use the conventional syntax, keeping this representation hidden underneath.

The theory file declares the conjunction rules as follows:

```
rules
  conjI          "[| P;  Q |] ==> P&Q"
  conjunct1      "P&Q ==> P"
  conjunct2      "P&Q ==> Q"
```

Strictly speaking, a rule such as $\wedge$-elim1 should be written

$$\mathsf{isTrue}(P \wedge Q) \Longrightarrow \mathsf{isTrue}(P).$$

Usually isTrue is left implicit, to avoid clutter:

$$P \wedge Q \Longrightarrow P.$$

The need for isTrue can be inferred from the context. Above, both $P \wedge Q$ and P must have type bool since conjunction has type $[\mathsf{bool}, \mathsf{bool}] \rightarrow \mathsf{bool}$. Both operands of $\Longrightarrow$ must have type prop.

3.3 Declaring Quantifiers

Variable-binding notations exist throughout mathematics. Consider the integral $\int_a^b f(x)dx$, where x is bound, and the product $\Pi_{k=0}^n p(k)$, where k is bound. The quantifiers $\forall$ and $\exists$ also bind variables. Isabelle uses λ-notation to represent all such operators uniformly.

To handle the standard quantifiers we could use a theory file containing the following declarations:

```
types
  ind
consts
  All  :: (ind => bool) => bool          (binder "ALL " 10)
  Ex   :: (ind => bool) => bool          (binder "EX " 10)
```

Here ind is some arbitrary type of individuals. For now it does not matter what these individuals are — numbers, sets, etc.[3] The *binder* specification causes Isabelle to interpret ALL $x.P$ as $\mathsf{All}(\lambda x.P)$, and similarly for EX; the 10 concerns operator precedence. Let us use the usual $\forall$ and $\exists$ symbols whenever possible.

Isabelle hides the representation from the user. Let us examine the internal form of an example:

$$\forall x.\, Px \land Qx \quad \mapsto \quad \mathsf{All}(\lambda x.\, Px \land Qx) \quad \mapsto \quad \mathsf{All}(\lambda x.\, \mathsf{And}(Px)(Qx))$$

How do we define a new quantifier, say $\forall_{x \in A} P$ to mean $\forall x.\, x \in A \to P$? For the semantics, we simply define a new constant abbreviating the appropriate expression. For the variable-binding notation, we must add four more lines to the theory file; no programming is required.

3.4 Quantifier Rules Involving Substitution

Of the four natural deduction rules for the quantifiers, two of them involve substituting a term for a bound variable:

$$\frac{\forall x.P}{P[t/x]}\ \forall\text{-elim} \qquad \frac{P[t/x]}{\exists x.P}\ \exists\text{-intr}$$

Isabelle's typed λ-calculus handles the substitution in these rules:

$$(\forall x.P\,x) \Longrightarrow P\,t \qquad (\forall\text{-elim})$$

$$P\,t \Longrightarrow (\exists x.P\,x) \qquad (\exists\text{-intr})$$

---

[3] Isabelle actually uses polymorphic declarations. Using type classes, we can specify whether to allow quantifications over booleans and functions. If we allow them, we get higher-order logic; otherwise we get many-sorted first-order logic [29].

In both of these, P has type $\mathsf{ind} \to \mathsf{bool}$ and stands for a formula with a hole. We may replace it by the λ-abstraction of any formula over x, say $\lambda x.\, Qx \wedge Rxy$. The corresponding instance of $\forall$-elim is then

$$(\forall x.(\lambda x.\, Qx \wedge Rxy)x) \Longrightarrow (\lambda x.\, Qx \wedge Rxy)t$$

or equivalently, by β-reduction,

$$(\forall x.\, Qx \wedge Rxy) \Longrightarrow Qt \wedge Rty.$$

We thus obtain all instances of the traditional rule.

Isabelle hides this machinery: we do not see the β-reductions . In backward proof, $\forall$-elim generalizes a formula of the form $P[t]$, yielding the subgoal $\forall x.\, P[x]$. There are usually countless ways of doing so; to constrain the choices, Isabelle lets you specify P or t.

A rule like $\forall$-elim is normally applied in the forward direction, mapping a theorem such as $\forall x\,.\, 0 + x = x$ to the new theorem $0 + ?t = ?t$. Here $?t$ is an *unknown*: a variable that can be replaced automatically. In Prolog terminology, it is called a "logical variable".

Isabelle provides both free variables x, y, ... , and unknowns $?x$, $?y$, ... From a logical point of view they are all free variables. The difference between the two kinds of variables is pragmatic. Unknowns may be replaced during unification; free variables remain fixed. This article often omits the question marks to avoid clutter; they should be present in all the Isabelle rules shown.

The rule $\exists$-intr is represented in precisely the same way. In backward proof it replaces the goal $\exists x.\, Qx \wedge Rxy$ by the subgoal $Q\,?t \wedge R\,?t\,y$. We need not specify $?t$, but may leave it as an unknown. Then we can split the subgoal in two (by applying $\wedge$-intr). Proving $Q\,?t$ will probably replace $?t$ by something, say 3. The other subgoal will become $R\,3\,y$.

Thus we can strip quantifiers without specifying terms for the bound variables. During the proof, the variables may be filled in automatically, which is valuable in both interactive and automatic proofs.

3.5 Quantifier Rules Involving Parameters

The other two quantifier rules involve provisos expressed in English:

$$\frac{P}{\forall x.P}\ \forall\text{-intr} \qquad \frac{\exists x.P \quad \begin{matrix}[P] \\ Q\end{matrix}}{Q}\ \exists\text{-elim}$$

Rule $\forall$-intr holds provided x is not free in the assumptions, while $\exists$-intr holds provided x is not free in Q or the assumptions. Most quantifier provisos are typical of these. They ensure that the premise makes no assumptions about the value of x: it holds for all x. Isabelle expresses them using !!, its inbuilt notion of 'for all'. We formalize $\forall$-intr by

$$(!!x.\,\mathsf{isTrue}(P\,x)) \Longrightarrow \mathsf{isTrue}(\forall x.P\,x).$$

This means, 'if Px is true for all x, then $\forall x.Px$ is true'. Hiding isTrue simplifies it to

$$(!!x.\,P\,x) \Longrightarrow (\forall x.\,P\,x). \qquad (\forall\text{-intr})$$

Applying this rule in backwards proof creates a subgoal prefixed by !!; the bound variable is called a *parameter* or *eigenvariable*. We have discussed (Sect. 2.4) how rules are lifted over assumptions; they are analogously lifted over parameters.[4] A subgoal's parameters and assumptions form a context; all subgoals resulting from it will have the same context, or one derived from it.

We have defined the object-level universal quantifier ($\forall$) using !!. But we do not require meta-level counterparts of all the connectives of the object-logic. The existential quantifier rule can also be formalized using !!:

$$[\![\exists x.P\,x;\ !!x.P\,x \Longrightarrow Q]\!] \Longrightarrow Q \qquad (\exists\text{-elim})$$

For another example of !!, consider the rule of mathematical induction:

$$\frac{P(0) \quad \begin{array}{c}[P(x)]\\ P(x+1)\end{array}}{P(n)}$$

There is the usual x not free ... proviso. In Isabelle, the rule becomes

$$[\![P(0);\ !!x.\,P(x) \Longrightarrow P(x+1)]\!] \Longrightarrow P(n)$$

In higher-order logic (HOL), we can express induction using predicate variables:

$$P(0) \wedge (\forall x.\,P(x) \to P(x+1)) \to P(n).$$

Isabelle provides the meta-level connectives !! and $\Longrightarrow$ so that users are not forced to work in HOL. Isabelle's treatment of rules recognizes !! and $\Longrightarrow$ but not $\forall$ and $\to$.

3.6 Higher-Order Unification

Isabelle resolves rules by unifying typed λ-terms. This process is called *higher-order unification.* To handle β-conversion, it reduces $(\lambda x.t)u$ to $t[u/x]$: this is easy. But sometimes Isabelle must solve equations like

[4] The rule's premises and conclusion receive the additional quantification $!!x$. All variables in the rule are given x as an additional argument. Their types are changed accordingly [24, 33].

$$?f(t) \equiv g\, u_1 \ldots u_k.$$

This task involves making guesses for the unknown function $?f$. Isabelle uses a refinement of Huet's [14] search procedure. It solves equations by guessing the leading symbol of $?f$, simplifying, then recursively unifying the result.

In the general case, higher-order unification is undecidable. Fortunately, we can usually recognize the problematical cases beforehand: they involve *function unknowns.* Terms such as $?f\,?x\,?y$ and $?f(?g\,x)$ match anything in countless ways. There may be infinitely many unifiers, and the search need not terminate. Isabelle lets you specify some unknowns before attempting unification.

Some uses of function unknowns are harmless. The term $?f\,a$ matches $a + a$ in four ways. Isabelle generates them lazily. Solutions that use the function's argument appear first, as they are usually preferable:

$$?f \equiv \lambda x.\, x + x \quad ?f \equiv \lambda x.\, a + x \quad ?f \equiv \lambda x.\, x + a \quad ?f \equiv \lambda x.\, a + a$$

Terms like $?f\,x\,y\,z$, where the arguments are distinct bound variables, cause no difficulties. They can match another term in at most one way. If the other term is $x + y \times z$ then the only unifier is

$$?f \equiv \lambda xyz.\, x + y \times z$$

The approach can be implemented without full higher-order unification. Pattern unification [28] is much easier to implement because it does not attempt to invent functions. The cost is just a little more user intervention in the problematic cases.

3.7 A Close Look at Quantifier Reasoning

To see how quantifier reasoning works, let us examine two tiny proofs in detail. Both involve stripping quantifiers from the initial goal. Because of the order of the quantifiers, one goal is provable and the other is not. (I have modified Isabelle's output by using special symbols, etc., to improve readability.)

3.7.1 The Good Proof. We start with the goal $\forall x.\, \exists y.\, x = y$.

```
goal thy "∀x. ∃y. x = y";
 Level 0
 ∀x. ∃y. x = y
  1. ∀x. ∃y. x = y
```

The initial proof state is an instance of the trivial rule: the main goal and the only subgoal are identical. For our first inference, we apply the rule ∀-intr.

```
by (resolve_tac [allI] 1);
 Level 1
 ∀x. ∃y. x = y
  1. !!x. ∃y. x = y
```

This yields the subgoal $\exists y. x = y$, where x is bound (by !!) in that subgoal. Remember, x is called a parameter. Next, we remove the existential quantifier by applying $\exists$-intr. We get a subgoal containing an unknown.

```
by (resolve_tac [exI] 1);
 Level 2
 ∀x. ∃y. x = y
  1. !!x. x = ?y1(x)
```

Lifting makes the unknowns depend on the subgoal's parameters. Our subgoal has the form $x = ?y1\,x$, where $?y1$ is a *function* unknown. The unknown $?y1\,x$ may be replaced by any term containing x. The term we require is x itself, with $?y1$ updated to $\lambda x.x$. The last step proves the goal by reflexivity.

```
by (resolve_tac [refl] 1);
 Level 3
 ∀x. ∃y. x = y
 No subgoals!
```

3.7.2 The Bad Proof. We start with the invalid goal $\exists y. \forall x. x = y$.

```
goal thy "∃y. ∀x. x = y";
 Level 0
 ∃y. ∀x. x = y
  1. ∃y. ∀x. x = y
```

The first inference applies $\exists$-intr, yielding the subgoal $\forall x. x = ?y$. The unknown is just $?y$; it may not be updated to terms containing x, since x is a bound variable.

```
by (resolve_tac [exI] 1);
 Level 1
 ∃y. ∀x. x = y
  1. ∀x. x = ?y
```

Removing the universal quantifier, by $\forall$-intr, yields the subgoal $x = ?y$. Still x is bound, though by !! instead of $\forall$.

```
by (resolve_tac [allI] 1);
 Level 2
 ∃y. ∀x. x = y
  1. !!x. x = ?y
```

When we attempt to use reflexivity, unification fails.

```
by (resolve_tac [refl] 1);
 by: tactic failed
```

Quantifier proofs frequently produce unknowns of function type, applied to bound variables. This assures correct treatment of alternating quantifiers. Informally, it is best to regard a term like $?y_1(x)$ as standing for any term, including terms containing occurrences of x. Isabelle takes care of the function unknowns automatically.

The workings of lifting and higher-order unification are complicated, but take place out of sight. The method can be proved correct [33]. It is reasonably fast: Isabelle can do hundreds of such steps per second. It works just as well for user-defined quantifiers like $\forall_{x\in A} Px$ or $\bigcup_{x\in A} Bx$.

Isabelle's original representation of rules was inspired by Schroeder-Heister's [48] 'rules of higher level' in natural deduction. The current approach is essentially identical to that of Felty and Miller [9, 10]. Pfenning [43] surveys other work on logical frameworks.

Exercise 3.1. Express this substitution rule, where P serves as a template for substitution, in Isabelle form:

$$\frac{t = u \quad P[t/x]}{P[u/x]}$$

Exercise 3.2. Suggest a representation of $\forall_{x\in A} P$ in the λ-calculus, where occurrences of x in P are bound.

Exercise 3.3. Give the Isabelle form of this bounded quantifier rule:

$$\frac{\begin{matrix}[x \in A] \\ P\end{matrix}}{\forall_{x\in A} P} \quad (x \text{ not free in assumptions other than } x \in A)$$

4. Generic, Automatic Proof

Tactics and tacticals form the basis for semi-automatic theorem proving in any Isabelle object-logic. The built-in tactics support proof checking using the rules you have declared. Complex tactics can be built using tacticals, which implement several search strategies.

Isabelle provides two tools that work at a higher level. The *classical reasoner* works with analytic rules, such as are found in natural deduction formalisms. The *simplifier* performs rewriting and can be extended to exploit the properties of new connectives.

4.1 Tactics, for Backward Proof

Tactics are the basic unit of backward proof. They can be expressed by combining primitive tactics, using operators called *tacticals*. Milner introduced these concepts in Edinburgh LCF. Regarding automatic proof as impossible, he sought to provide a language in which users could express proof methods, which might be variations on the built-in tactics, or complicated strategies. Although Milner had developed an elegant high-level language for this purpose (ML), he sought an even higher level language just for proof.

Isabelle tactics differ substantially from those of LCF, HOL, etc. A tactic looks at the entire proof state: at all outstanding subgoals instead of just one. While this sacrifices the proof's tree structure, it is essential in order to support unification, where proving one goal may affect others. An Isabelle tactic can update unknowns in other subgoals or the main goal. It can search for proofs of all remaining subgoals.

An Isabelle tactic is a function from a state to a sequence (lazy list) of next states. The sequence enumerates alternatives for backtracking and other kinds of searches [39, Chapter 5]. Because of all this, Isabelle's tactics and tacticals are considerably more powerful than HOL's. We shall discuss applications below — especially the classical reasoner, which is a general tool for quantifier reasoning.

4.2 Proof Checking Tactics

The primitive tactics provide proof checking: one rule at a time. Section 2. describes the concepts behind `assume_tac` and `resolve_tac`; other tactics are variations on those themes.

Tactic `assume_tac` proves a subgoal by assumption. Shouldn't Isabelle do such trivial steps without being told to? But proof by assumption is less trivial than it looks. Consider this proof state:

```
(P(a) & P(b) → P(?x)) & (P(b) & P(c) → P(?x))
 1. [| P(a); P(b) |] ⟹ P(?x)
 2. [| P(b); P(c) |] ⟹ P(?x)
```

Both subgoals have two proofs, but only one of the four combinations proves both goals simultaneously. If backtracking occurs, `assume_tac` searches for another matching assumption. Assumptions are not referred to by number.

Tactic `resolve_tac` applies rules, which may be primitive or user-derived, searching for those that match the subgoal. Backtracking searches for other matches and other rules.

Tactic `eresolve_tac` uses and then deletes an assumption. It is suitable when a rule might make an assumption obsolete. Its effect with the rule ∨-elim (given in Sect. 2.3) is to replace a subgoal of the form

$$\cdots P \vee Q \cdots \Longrightarrow X$$

by two new subgoals of the form $\cdots P \cdots \Longrightarrow X$ and $\cdots Q \cdots \Longrightarrow X$. Backtracking makes `eresolve_tac` search for another matching assumption.

Tactic `res_inst_tac` lets us partly instantiate a rule explicitly during refinement, specifying some unknowns. It is needed when applying rules like ∀-elim, whose conclusion is too general for automatic matching.

Resolution performs the most basic step in mathematics: appeal to a previous result. Many proof checkers make this difficult.

4.3 Tacticals: Control for Tactics

Proofs expressed entirely in terms of primitive tactics would be too long. Isabelle's *tacticals* `THEN`, `ORELSE`, etc., combine existing tactics to yield new ones, providing a rich control language for tactics. They achieve the desired behaviour by operating on lazy lists. Recall that a tactic maps a proof state to a lazy list of possible next states. If *tac*, *tac1*, etc., are tactics then so are the following:

tac1 `THEN` *tac2* returns all states reachable by applying *tac1* then *tac2*
tac1 `ORELSE` *tac2* tries *tac1*; if this fails, it uses *tac2*
tac1 `APPEND` *tac2* calls both *tac1* and *tac2*, appending their results
`DETERM` *tac* returns the first state reachable by applying *tac*
`REPEAT` *tac* returns all states reachable by applying *tac* as long as possible
`DEPTH_FIRST` *satp* *tac* returns all states satisfying *satp* reachable by applying *tac* in depth-first search

Explicit control of backtracking can help keep the search space small. Using `DETERM` prevents backtracking inside its argument. The difference between `ORELSE` and `APPEND` is that `ORELSE` forbids backtracking from its first argument to its second. It is the user's responsibility to ensure that the eliminated alternatives are not needed.

There are tacticals for several other search strategies: iterative deepening [16], best-first, etc. The argument *satp* is a boolean-valued function specifying what kind of state to search for, typically in terms of how many subgoals are left unsolved. Artificial Intelligence textbooks [47] discuss these strategies. Depth-first search is fastest but often gets stuck down an infinite path; iterative deepening is safe but much slower; best-first search can be fast, but must be guided by an accurate heuristic function.

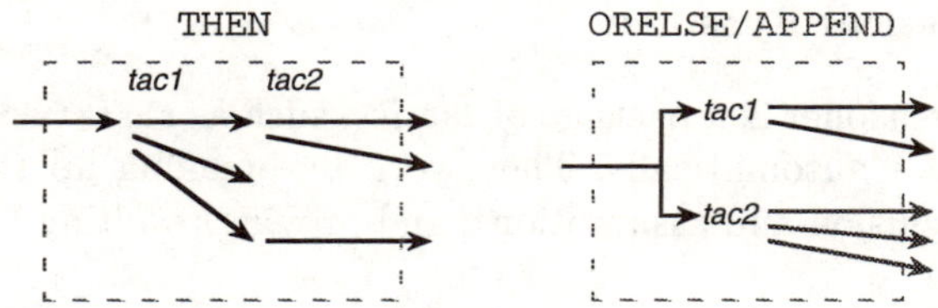

Fig. 4.1. The tacticals `THEN`, `ORELSE` and `APPEND`

Figure 4.1 illustrates some of the possibilities. Consider *tac*1 `THEN` *tac*2. Here *tac*1 returns a sequence of three possible next states. Given the first of these, *tac*2 returns two next states. Given the second of these, *tac*2 returns no next states; thus this possibility contributes nothing to the final sequence of outcomes. Given the third of these, *tac*2 returns one next state. A total of three states can arise from this call to *tac*1 `THEN` *tac*2.

Consider *tac*1 `ORELSE` *tac*2. Here *tac*1 returns a sequence of two possible next states. As this sequence is nonempty, it becomes the full output of *tac*1 `ORELSE` *tac*2; had it been empty, then the output would have been that of *tac*2. Similar is *tac*1 APPEND *tac*2, but its output comprises those of both *tac*1 and *tac*2.

Alternative versions of THEN and APPEND could interleave sequence elements instead of putting all possibilities from one sequence first. That would give more of a depth-first flavour.

Now let us consider a simple example of tacticals. Trying to prove that disjunction is associative has yielded three subgoals, each to prove $P \vee (Q \vee R)$ from the assumption P, Q or R:

```
(P ∨ Q) ∨ R → P ∨ Q ∨ R
 1. P ⟹ P ∨ Q ∨ R
 2. Q ⟹ P ∨ Q ∨ R
 3. R ⟹ P ∨ Q ∨ R
```

At our disposal are the disjunction rules $P \Longrightarrow P \vee Q$ and $Q \Longrightarrow P \vee Q$. We command Isabelle to perform a brute-force search:

```
by (DEPTH_SOLVE
     (assume_tac 1 ORELSE resolve_tac [disjI1, disjI2] 1));
 (P ∨ Q) ∨ R → P ∨ Q ∨ R
 No subgoals!
```

The command attempts proof by assumption and the disjunction rules, with backtracking. Tactical DEPTH_SOLVE uses DEPTH_FIRST to search for a fully solved proof state: no subgoals.

We often reach a point where the result clearly follows by repeated application of certain rules. We can then compose a command like the one above. We could have used APPEND instead of ORELSE, or used other search tacticals.

4.4 The Classical Reasoner

The classical reasoner is a package of tactics, such as `fast_tac` and `best_tac`, that prove goals automatically. They work by breaking up the formulas in the goal's conclusion and assumptions, and proving resulting trivial subgoals by assumption.

Here are some examples of what the classical reasoner can prove. We begin with #40 from Pelletier's problem set [42]. It is rather easy; its proof requires only 0.5 seconds on a fast SPARCstation.

$$(\exists y \forall x.\, Pxy \leftrightarrow Pxx) \to \neg\forall x \exists y \forall z.\, Pzy \leftrightarrow \neg Pzx$$

The classical reasoner can prove many set-theoretic identities. For this task, `fast_tac` uses rules proved specifically about the primitives of set theory, rather than expanding the definitions to primitive logical concepts. High-level rules promote transparency (short, clear proofs) as well as efficiency. This distributive law is proved in 0.3 seconds.

$$\Big(\bigcup_{i\in I} A_i \cup B_i\Big) = \Big(\bigcup_{i\in I} A_i\Big) \cup \Big(\bigcup_{i\in I} B_i\Big)$$

The third example comes from a proof of the soundness and completeness of propositional logic, by Tobias Nipkow and myself [37]. The relation $H \vdash p$ is defined inductively to express that proposition p is deducible from H. The deduction theorem is expressed as follows:

$$\{p\} \cup H \vdash q \Longrightarrow H \vdash \text{implies}\, p\, q$$

To prove it, the first step is induction on $\{p\} \cup H \vdash q$; this yields five subgoals. The second step is application of the classical reasoner, equipped with basic rules of the embedded logic. All five subgoals are proved in under 0.2 seconds.

Analogous but much harder is a proof of the Church-Rosser theorem for combinators. A key lemma is the diamond property for parallel reduction [38]. Again it consists of induction followed by classical reasoning. But this time `fast_tac` (equipped with rules about the behaviour of reductions) needs nearly 50 seconds to prove the four subgoals.

Now let us consider how the classical reasoner works. *Analytic* rules are those that break up a formula. We must distinguish between *safe* and *unsafe* rules. Safe rules do not require backtracking; they represent logical equivalences, and are analogous to rewrite rules such as $x \in A \cup B \leftrightarrow x \in A \vee x \in B$. Unsafe rules may require backtracking.

Safe rules lose no information; they may be attempted on any subgoal. For predicate calculus they include the following:

$$\frac{P \quad Q}{P \wedge Q} \qquad \frac{\begin{matrix}[P] \\ Q\end{matrix}}{P \to Q} \qquad \frac{P}{\forall x.P}$$

The following rules are unsafe because their premises are stronger than their conclusion. (They are sound, but in backward proof they discard information.) The latter rule is also unsafe in the operational sense that repeated application of it could run forever.

$$\frac{m < n}{m < n+1} \qquad \frac{x < y \quad y < z}{x < z}$$

The sequent calculus is better than natural deduction for reasoning about $\vee$ and $\exists$. The classical reasoner simulates the sequent calculus, representing the sequent $P_1, \ldots, P_m \vdash Q_1, \ldots, Q_n$ by

$$[\![P_1; \ldots; P_m; \neg Q_2; \ldots; \neg Q_n]\!] \Longrightarrow Q_1$$

The classical reasoner is coded in terms of tactics and tacticals. It is less efficient than hard-wired tautology checkers, but is more flexible. We may apply it in the predicate calculus, set theory, and in specialized theories. It is especially good at reasoning about inductively defined relations. When it fails, it often fails quickly, and can be single-stepped to locate the trouble spot.

4.5 A Generic Simplifier

Virtually all proof tools provide a simplifier that performs rewriting. Most of them handle conditional write rules such as $x \neq 0 \Longrightarrow x/x = 1$. Some of them even do permutative rewriting, to handle rules like $x + y = y + x$ without looping; this can be used to sort the terms in expressions of the form $e_1 + \cdots + e_n$. But few of them provide generic support to adapt rewriting to new connectives.

Consider the question of using context. For the occurrence of Q in $P \wedge Q$, we may assume that P is true. Simplification can take advantage of this. To rewrite $P \wedge Q$, first rewrite P, yielding say P', and then assume P' while rewriting Q. Here is a case where contextual information assists the simplifier:

$$f(1) = 2 \wedge x \neq 0 \wedge f(x/x) = 2.$$

The first conjunct yields a rewrite rule for f; the second allows conditional rewriting about division, so that the third conjunct can be simplified to true.

If information about context is built in, then users cannot extend it. Isabelle's simplifier is supplied such information as inference rules. To make it treat conjunction as described above, we give it the following rule:

$$\frac{P \leftrightarrow P' \quad \begin{matrix}[P'] \\ Q \leftrightarrow Q'\end{matrix}}{(P \wedge Q) \leftrightarrow (P' \wedge Q')}$$

We can prove an alternative rule that works from right to left, and dual rules for disjunction. If we define bounded quantification over sets, we can prove and install the analogous rule for it:

$$\frac{A = A' \quad \begin{matrix}[x \in A'] \\ P(x) \leftrightarrow P'(x)\end{matrix}}{(\forall_{x \in A} P(x)) \leftrightarrow (\forall_{x \in A'} P'(x))}$$

This says, rewrite $\forall_{x \in A} P(x)$ by first rewriting A to A' and then assuming $x \in A'$ when rewriting $P(x)$. Rules of this form are known as *congruence rules.*

The simplifier can perform automatic case splits and similar operations, also described by inference rules. The simplifier also lets users describe how to extract rewrite rules from general formulas. For example, a formula of the form $\forall_{x \in A} P(x)$ might yield rewrite rules conditional upon $x \in A$.

These extensions are sound because they work by building proofs. A user error cannot make the simplifer replace everything by true, or make the classical reasoner prove assertions that do not follow from the axioms.

5. Mechanized Set Theories

We need set theory whether or not we have types. Simple types (as found in the HOL system [12]) are too rigid: there is no type of n-element lists.

Predicate subtyping (as in PVS) can help. But even in highly expressive type systems, one usually defines some sort of set theory. Set-theoretic primitives can express complex assertions, which is why they are found in specification languages such as Z and B.

Isabelle implements two set theories. Isabelle/ZF [34] is built upon first-order logic using the standard Zermelo-Fraenkel axioms [50]. Isabelle's higher-order logic (Isabelle/HOL) includes a polymorphically typed set theory, with sets represented by predicates. The two theories are similar but not identical; below we shall consider only ZF set theory.

Isabelle/ZF includes the standard operators such as $a \in A$, $A \subseteq B$ and $A \cup B$. Comprehensions are written $\{x \in A.\, P(x)\}$ or $\{b(x).\, x \in A\}$. Finite sets are written $\{a_1, \ldots, a_n\}$, tuples $\langle a_1, \ldots, a_n \rangle$ and functions $\lambda x \in A.\, b(x)$. There are disjoint sums $A+B$, Cartesian products $A \times B$ and function spaces $A \to B$. There are bounded quantifiers $\forall_{x \in A} P(x)$ and $\exists_{x \in A} P(x)$, indexed intersections $\bigcap_{x \in A} B(x)$ and unions $\bigcup_{x \in A} B(x)$, etc.

The definite description $\iota x.\, P(x)$ provides a means of naming an object that is characterized uniquely by P. It is like Hilbert's ϵ-operator, found in the HOL system, but does not assume the axiom of choice.

Note: Isabelle provides a choice of syntaxes for its underlying λ-calculus. In Isabelle/HOL, the application of term M to arguments $N_1, \ldots, N_k$ is written $M\, N_1\, N_k$; in Isabelle/ZF, it is written $M(N_1, \ldots, N_k)$. The former syntax has the advantage of (λ-calculus) tradition; it also facilitates partial application of curried functions, omitting arguments from the right. The latter syntax has the advantage of (set theory) tradition; it also forbids partial application, which does not make sense in first-order logic. The only ambiguity arises in terms of the form $M(a, b)$: in HOL there is one argument, an ordered pair; in ZF there are two arguments.

5.1 High-Level Rules

Proofs are conducted in terms of high-level rules, such as these for the subset relation:[5]

$$\frac{\begin{matrix}[x \in A] \\ \vdots \\ x \in B\end{matrix}}{A \subseteq B}\ \subseteq\text{-intr} \qquad \frac{A \subseteq B \quad c \in A}{c \in B}\ \subseteq\text{-elim}$$

This is how the rules are expressed Isabelle:

$$(!!x.\, x \in A \Longrightarrow x \in B) \Longrightarrow A \subseteq B \qquad (\subseteq\text{-intr})$$

$$[\![A \subseteq B;\ a \in A]\!] \Longrightarrow a \in B \qquad (\subseteq\text{-elim})$$

[5] The first rule has a standard proviso on x: it must not occur free elsewhere in the inference.

They are proved from the Zermelo-Fraenkel axioms when Isabelle/ZF is built. In use, there is no perceptible difference between primitive and derived rules.

The $\{x \in A. P(x)\}$ notation deserves special mention. The axiom of separation is traditionally written $a \in \{x \in A. P(x)\} \leftrightarrow a \in A \wedge P(a)$. Strictly speaking this is an axiom scheme, with a distinct instance for each first-order predicate P. Here is the Isabelle version; for once, let us stop hiding the question marks:

$$?a \in \{x \in ?A\,.\,?P(x)\} \leftrightarrow ?a \in ?A \wedge ?P(?a)$$

This formula behaves as a scheme, with $?P$ automatically replaced by the matching λ-expression and the result β-reduced. Derived rules, incidentally, are just a form of theorem scheme. We can derive convenient rules such as the following:

$$\frac{a \in A \quad P(a)}{a \in \{x \in A.P(x)\}} \qquad \frac{a \in \{x \in A.P(x)\}}{a \in A} \qquad \frac{a \in \{x \in A.P(x)\}}{P(a)}$$

5.2 A Proof about Unions

For a demonstration, let us prove a simple theorem in small steps using a tactic called `step_tac`. This tactic is normally called via `fast_tac`. Calling it directly allows for single-step reasoning and is especially useful when the proof fails.

Calling `step_tac` either applies safe inferences to the entire proof state, or applies one unsafe inference to the specified subgoal. Safe inferences are like rewriting by logical equivalences such as

$$\begin{aligned} x \in A \cap B &\iff x \in A \wedge x \in B \\ \langle x, y\rangle \in A \times B &\iff x \in A \wedge y \in B \\ \{x \in A.\, P(x)\} &\iff x \in A \wedge P(x) \\ x \in \bigcup(A) &\iff \exists y.\, x \in y \wedge y \in A \end{aligned}$$

Natural deduction rules express each direction of an equivalence separately.

The last one above concerns $\bigcup(A)$, the union of a set of sets, which might also be written $\bigcup_{X \in A} X$. Our sample theorem is $\bigcup(A \cap B) \subseteq \bigcup(A)$.

Isabelle's response to the initial `goal` command is to print an initial proof state, represented internally by an instance of the trivial clause $X \Longrightarrow X$.

```
goal ZF.thy "Union(A Int B) ⊆ Union(A)";
 Union(A Int B) ⊆ Union(A)
  1. Union(A Int B) ⊆ Union(A)
```

Calling `step_tac` performs the safe steps of assuming $x \in \bigcup(A \cap B)$, then $x \in y$ and $y \in A \cap B$, replacing the latter by $y \in A$ and $y \in B$, where x and y are arbitrary. In this context we must show $x \in \bigcup(A)$.

```
by (step_tac ZF_cs 1);
 Union(A Int B) ⊆ Union(A)
  1. !!x y. [| x∈y; y∈A; y∈B |] ⟹ x∈Union(A)
```

The next two step_tac calls apply unsafe inferences to subgoal 1. The first of these is the rule $\bigcup$-intr:

$$[\![B \in C;\ A \in B]\!] \Longrightarrow A \in \bigcup(C)$$

To show $x \in \bigcup(A)$ we can exhibit some set U such at $U \in A$ and $x \in U$.

```
by (step_tac ZF_cs 1);
 Union(A Int B) ⊆ Union(A)
  1. !!x y. [| x∈y; y∈A; y∈B |] ⟹ ?B3(x, y)∈A
  2. !!x y. [| x∈y; y∈A; y∈B |] ⟹ x∈?B3(x, y)
```

Isabelle displays the unknown set U as ?B3(x,y). It appears in two goals. In general, different proofs of a goal update its unknowns differently, and we may have to search for one that lets the other subgoals be proved.

The second unsafe step proves subgoal 1 by assumption; this amounts to guessing that ?B3(x,y) is y. The remaining subgoal (now numbered 1) has changed to $x \in y$, and is trivially provable as there is an identical assumption.

```
by (step_tac ZF_cs 1);
 Union(A Int B) ⊆ Union(A)
  1. !!x y. [| x∈y; y∈A; y∈B |] ⟹ x∈y
```

Few examples are simple enough to present and hard enough to need more than one call to step_tac. More interesting and slightly harder to prove is the monotonicity rule for $\bigcup$:

$$A \subseteq B \Longrightarrow \bigcup(A) \subseteq \bigcup(B)$$

Its fast_tac proof takes about 0.2 seconds. Such rules, once proved, can themselves be used in later proofs as atomic inferences.

5.3 A Proof About Functions and Injections

Isabelle/ZF defines functions and the sets of functions and injections, essentially as follows:

$$\begin{aligned}
\mathsf{func}(f) &\equiv \forall x\, y.\ \langle x,y\rangle \in f \longrightarrow (\forall z.\ \langle x,z\rangle \in f \longrightarrow y = z)\\
A \to B &\equiv \{f \subseteq A \times B.\ A \subseteq \mathsf{dom}(f) \wedge \mathsf{func}(f)\}\\
\mathsf{inj}(A,B) &\equiv \{f \in A \to B.\ \forall w, x \in A.\ f`w = f`x \longrightarrow w = x\}
\end{aligned}$$

Here $f`x$ denotes application of the function f to the argument x. This object-level application differs from meta-level application, which is written $f(x)$. There is a real distinction between the two levels. With object-level application, the function is a set of pairs; with meta-level application, it is an Isabelle term of function type. Many meta-level functions, such as $\bigcup$, inj and range, are defined over the entire universe of sets; they are not sets themselves.

Because treating functions as sets of pairs is too low-level, most reasoning is done in terms of λ-abstraction and function application. Users need only know the definition of $\mathsf{inj}(A, B)$, the set of injections from A to B.

Injections, surjections and bijections are fundamental to discrete mathematics and computer science. Most databases label each record with a unique primary key, giving an injection from records to primary keys. The inverse of that injection is a function whose domain is the set of primary keys actually used. This fact is an application of our next theorem,

$$f \in \mathsf{inj}(A, B) \Longrightarrow f^{-1} \in \mathsf{range}(f) \to A.$$

Here is the initial proof state after expanding the definition[6] of $\mathsf{inj}(A, B)$.

```
goalw Perm.thy [inj_def]
    "!!f. f∈inj(A,B) ⟹ f⁻¹∈range(f)→A";
 !!f. f∈inj(A, B) ⟹ f⁻¹∈range(f) → A
  1. !!f. f ∈
          {f ∈ A → B .
           ∀ w∈A. ∀ x∈A. f ‘ w = f ‘ x ⟶ w = x}
          ⟹ f⁻¹∈range(f) → A
```

Unfolding the definitions of function space and $\mathsf{func}(f)$ yields a new proof state. Here, rewriting affects only the subgoal's conclusion, not its assumptions; an occurrence of the function space operator remains in the assumptions as $A \to B$.

```
by (asm_simp_tac (ZF_ss addsimps [Pi_iff, function_def]) 1  THEN
    eresolve_tac [CollectE] 1);
 !!f. f∈inj(A, B) ⟹ f⁻¹∈range(f) → A
  1. !!f. [| f ∈ A → B;
             ∀ w∈A. ∀ x∈A. f ‘ w = f ‘ x ⟶ w = x |]
          ⟹ (∀ x y. <y,x>∈f ⟶
                      (∀ z. <z,x>∈f ⟶ y = z)) &
             f⁻¹ ⊆ range(f) × A
```

The second part of the command applies the rule `CollectE` to break down an assumption of the form $\{f \in A \to B. \ldots\}$.

Default rules (stored in `ZF_ss`) perform most of the reasoning. Furthermore we specify `Pi_iff` and `function_def` as rewrite rules, to expose the representation of functions. Such low-level rules are not included by default.

Rewriting has proved part of the unfolded subgoal, namely $\mathsf{dom}(f^{-1}) \subseteq \mathsf{range}(f)$. We still have to show that f^{-1} is a function and is included in $\mathsf{range}(f) \times A$. The following command rewrites with `apply_iff`, which replaces $\langle a, b\rangle \in f$ by $f`a = b$ provided f is a function.

[6] via the `goalw` command

```
by (asm_simp_tac (ZF_ss addsimps [apply_iff]) 1);
  !!f. f∈inj(A, B) ⟹ f⁻¹∈range(f) → A
   1. !!f. [| f ∈ A → B;
             ∀ w∈A. ∀ x∈A. f ` w = f ` x ⟶ w = x |]
          ⟹ (∀ x y.
                 y∈A & f ` y = x ⟶
                 (∀ z. z∈A & f ` z = x ⟶ y = z)) &
             f⁻¹ ⊆ range(f) × A
```

The equivalence between the quantified assumption and the first conjunct can now be discerned. We still have to prove it, as well as the second conjunct, $f^{-1} \subseteq \mathsf{range}(f) \times A$. Isabelle finds this task to be trivial.

```
by (fast_tac (ZF_cs addDs [fun_is_rel]) 1);
  !!f. f∈inj(A, B) ⟹ f⁻¹∈range(f) → A
  No subgoals!
```

The rule `fun_is_rel` stands for $f \in A \to B \Longrightarrow f \subseteq A \times B$. Again, it concerns how functions are represented as sets of pairs. With the present subgoal, `fast_tac` needs only 1.3 seconds, of which one second is devoted to the first conjunct. This example illustrates how Isabelle's simplifier and classical reasoner can prove nontrivial theorems with a few commands.

Exercise 5.1. Using high-level rules such as those demonstrated above, prove the monotonicity of $\bigcup$, namely $A \subseteq B \Longrightarrow \bigcup(A) \subseteq \bigcup(B)$.

Exercise 5.2. Use the rule $[\![f \in \mathsf{inj}(A, B);\ b \in \mathsf{range}(f)]\!] \Longrightarrow f`(f^{-1}`b) = b$ to strengthen our result to $f \in \mathsf{inj}(A, B) \Longrightarrow f^{-1} \in \mathsf{inj}(\mathsf{range}(f), A)$. The conclusion could even be that f^{-1} is a bijection between $\mathsf{range}(f)$ and A.

6. Induction and Recursion

Most proof tools allow the definition of recursive types and functions. More unusually, Isabelle also supports inductive and co-inductive relations. Such relations are commonly found in theoretical computer science.

6.1 Recursive Types and Functions

A datatype is a disjoint sum and may be recursive. The syntax is based upon Standard ML's [39]. Datatypes can model lists, trees and finite enumerations.[7]

For example, here is a specification of binary trees. A binary tree may be a leaf (Lf) or a branch node (Br) carrying a label and two subtrees. Instead of specifying a fixed type of labels, we use polymorphism and specify the type variable `'a`.

```
datatype 'a bt = Lf  |  Br 'a ('a bt) ('a bt)
```

[7] Here we switch back from ZF to HOL. ZF has similar facilities (more general in fact) but HOL's type checking makes the declarations more concise.

Datatype declarations may be placed in theory files along with declarations of other types, constants, etc. Isabelle makes appropriate definitions and derives the properties required of the datatype.

Isabelle (with ZF or HOL) supports function definition by well-founded recursion. Any relation that can be proved to be well-founded may be used to show termination of recursive calls. We may even interleave the proofs of termination and those of other correctness properties; this is essential for reasoning about certain nested recursive functions, such as the unification algorithm. Slind has recently written a tool to automate much of this process; its accepts function definitions expressed using pattern-matching [49].

Primitive recursion is a simpler form of function definition. It allows recursive calls only to immediate subparts of the argument. Here are two functions to count the branch nodes and leaves, respectively, of a binary tree.

```
consts nodes, leaves :: 'a bt => nat

primrec nodes bt
  nodes Lf            = 0
  nodes(Br a t1 t2)   = Suc(nodes t1 + nodes t2)

primrec leaves bt
  leaves Lf           = 1
  leaves(Br a t1 t2)  = leaves t1 + leaves t2
```

Reasoning about recursive functions is often easy. A classic theorem states that each binary tree has one more leaf node than branch node. Let us set up the induction in Isabelle.

```
goal BT.thy "leaves(t) = Suc(nodes(t))";

by (bt.induct_tac "t" 1);
 leaves t = Suc (nodes t)
  1. leaves Lf = Suc (nodes Lf)
  2. !!a t1 t2.
        [| leaves t1 = Suc (nodes t1);
           leaves t2 = Suc (nodes t2) |]
        ==> leaves (Br a t1 t2) = Suc (nodes (Br a t1 t2))
```

The simplifier trivially proves the base case and inductive step shown above.

6.2 Inductive Definitions

An *inductive definition* specifies the least set closed under a given collection of rules [1]. The set of theorems in a logic is inductively defined. A structural operational semantics [13] inductively defines an evaluation relation on programs. Dually, a *coinductive definition* specifies the greatest set closed under given rules. Equivalence of concurrent processes is often defined coinductively, in terms of bisimulation relations [25].

Figure 6.1 gives an inductive definition of the "permutation of" relation for lists. Here $x\#l$ stands for the list with head x and tail l. The upper two rules say that [] is a permutation of itself and that exchanging the first two elements of a list creates a permutation. The lower left rule says that adding identical elements to both lists preserves the "permutation of" relation. The final rule says that the relation is transitive.

$$[] \leftrightsquigarrow [] \qquad y\#x\#l \leftrightsquigarrow x\#y\#l$$

$$\frac{xs \leftrightsquigarrow ys}{z\#xs \leftrightsquigarrow z\#ys} \qquad \frac{xs \leftrightsquigarrow ys \quad ys \leftrightsquigarrow zs}{xs \leftrightsquigarrow zs}$$

Fig. 6.1. Inductive Definition: Permutations of Lists

A desired collection of rules may be given to Isabelle (ZF and HOL) to specify a (co)inductive definition. Isabelle reduces it to a least fixedpoint (greatest fixedpoint for a coinductive definition). A broad class of definitions is acceptable [35].

Rule induction [52] is a powerful inference rule for proving consequences of $xs \leftrightsquigarrow ys$. Recall that $\leftrightsquigarrow$ is the least set closed under the rules given in Figure 6.1. If some predicate P is also closed under those rules then $xs \leftrightsquigarrow ys$ implies $P\,xs\,ys$ for all x and y. The corresponding subgoals are as follows:

$$P\,[]\,[]$$
$$P\,(y\#x\#l)\,(x\#y\#l)$$
$$\text{if } P\,xs\,ys \text{ then } P\,(z\#xs)\,(z\#ys)$$
$$\text{if } P\,xs\,ys \text{ and } P\,ys\,zs \text{ then } P\,xs\,zs$$

Let us use rule induction to prove that $xs \leftrightsquigarrow ys$ implies $\mathsf{length}\,xs = \mathsf{length}\,ys$. The four subgoals are easily proved:

$\mathsf{length}[] = 0 = \mathsf{length}[]$
$\mathsf{length}\,(y\#x\#l) = 2 + \mathsf{length}\,l = \mathsf{length}\,(x\#y\#l)$
If $\mathsf{length}\,xs = \mathsf{length}\,ys$ then

$$\mathsf{length}(z\#xs) = 1 + \mathsf{length}\,xs = 1 + \mathsf{length}\,ys = \mathsf{length}(z\#ys)$$

If $\mathsf{length}\,xs = \mathsf{length}\,ys$ and $\mathsf{length}\,ys = \mathsf{length}\,zs$ then

$$\mathsf{length}\,xs = \mathsf{length}\,zs \quad \text{(by transitivity of equality)}$$

Symmetry of $\leftrightsquigarrow$ is another example. We can show that $xs \leftrightsquigarrow ys$ implies $ys \leftrightsquigarrow xs$ by simple rule induction: just put $ys \leftrightsquigarrow xs$ for $P\,xs\,ys$. Equivalently, observe that all the rules are preserved if we exchange the operands of $\leftrightsquigarrow$. The Isabelle versions of the declaration of $\leftrightsquigarrow$ and the proof of symmetry appear below.

6.3 Declaring Inductive Sets to Isabelle

This example illustrates the syntax for inductive definitions. Details have been simplified. The theory file declares the constant perm to stand for the relation. Three omitted lines inform Isabelle to use `xs <~~> ys` as notation for `(xs,ys):perm`.

```
consts  perm :: ('a list × 'a list) set

inductive perm
 Nil        [] ↭ []
 swap   y#x#l ↭ x#y#l
 Cons      xs ↭ ys ⟹ z#xs ↭ z#ys
 trans  [| xs ↭ ys; ys ↭ zs |] ⟹ xs ↭ zs
```

If the recursion involves other operators then they must be monotonic; we must supply the corresponding monotonicity theorems.

(Co)inductive definitions are conservative. Isabelle derives basic properties of the newly defined set. These include the rules specified in the definition: the set's *introduction* rules. They include a rule for case analysis over the definition: the *elimination* rule. Finally, they include an induction or coinduction rule. These rules are proved, not merely asserted; for most definitions, this takes only seconds.

Let us prove that ↭ is symmetric. After stating the goal we apply rule induction, which is called `perm.induct`.

```
goal Perm.thy "!!xs. xs ↭ ys ⟹ ys ↭ xs";
by (etac perm.induct 1);
  1. !!xs. [] ↭ []
  2. !!xs l x y. x # y # l ↭ y # x # l
  3. !!xs ws ys z.
        [| ws ↭ ys; ys ↭ ws |] ⟹ z # ys ↭ z # ws
  4. !!xs ws ys zs.
        [| ws ↭ ys; ys ↭ zs;
           ys ↭ ws; zs ↭ ys |] ⟹ zs ↭ ws
```

We can prove all four subgoals trivially using a single command:

```
by (REPEAT (assume_tac 1 ORELSE resolve_tac perm.intrs 1));
```

This repeatedly attempts proof by assumption or one of the introduction rules (denoted by `perm.intrs`). We need only mention subgoal 1, as other subgoals are moved up when the first subgoal is proved.

6.4 Applications of (Co)Inductive Definitions

Several large studies use inductive definitions. Lötzbeyer et al. [18, 31] have related the operational and denotational semantics of Winskel's toy programming language IMP [52]. Using different techniques, Nipkow [30] and Rasmussen [45] have both proved the Church-Rosser theorem. A datatype specifies the set of λ-terms, while inductive definitions specify several reduction relations.

To demonstrate coinductive definitions, Frost [11] has proved the consistency of the dynamic and static semantics for a small functional language. The example, by Milner and Tofte [26], concerns a coinductively defined typing relation. Isabelle/ZF supports *codatatypes*, which are like datatypes but admit infinitely deep nesting. (Constructing non-well-founded trees in the presence of the foundation axiom requires variant pairs and function [40].) Frost defines a codatatype of values and value environments in mutual recursion. Non-well-founded values represent recursive functions; value environments are functions from variables into values.

The Ackermann's function proof [35] demonstrates the flexibility of inductive definitions in Isabelle. The set of primitive recursive functions is difficult to define formally — the composition operator combines a function with a list of functions. The "list of" operator is monotonic, however, and Isabelle allows monotonic operators to appear in inductive definitions.

Pusch [44] is proving the correctness of a compiling algorithm from Prolog to the Warren Abstract Machine (WAM). She uses datatypes to formalize Prolog's syntax and data structures involved in the interpretation, and inductive definitions to formalize the semantics of Prolog and the WAM. The proof involves around ten refinement steps from Prolog to the WAM; five of these steps have been verified using Isabelle. Each step introduces some low-level feature, such as pointers or optimizations of backtracking, and proves semantic equivalence.

Exercise 6.1. Using recursion and an if-then-else construct, define a function $\mathsf{count}\, l\, z$ to count how many times z occurs in the list l.

Exercise 6.2. Using rule induction, prove that $xs \leadsto ys$ implies $\mathsf{count}\, xs\, z = \mathsf{count}\, ys\, z$.

7. Reasoning about Cryptographic Protocols

Cryptographic protocols are designed to let agents communicate securely over an insecure network. An obvious security goal is *secrecy*: a spy cannot read the contents of messages intended for others. Also important is *authentication*: agents can identify themselves to others, but nobody can masquerade as somebody else. Specialist applications may require other security goals, such as anonymity.

Cryptographic protocols typically employ a *server*: a trusted agent whose job is to manage everyone's encryption keys, generate fresh keys and distribute them to the appropriate agents for use in new conversations.

Without breaking the encryption method, a spy can exploit flaws in a protocol. A typical attack involves intercepting a message sent by A and replaying it at some later time, hoping to be accepted as A. Protocols try to prevent replay attacks by using *nonces*, which may be random numbers

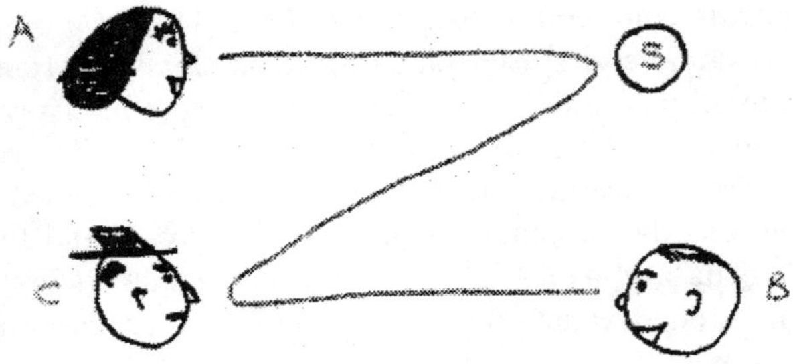

Fig. 7.1. A Private Conversation?

or counters. By including a fresh nonce in a message, and checking for its presence in the acknowledgement, an agent can reject old messages replayed by the spy.

Many cryptographic protocols have been shown to be vulnerable, often in subtle ways [3]. Rigorous correctness proofs seem essential. One popular proof method is based upon logics of belief [7]. Security protocols can also be analyzed using Isabelle.

In preliminary experiments, I have proved properties of several well-known protocols, and discovered a new attack on one of them. Inductive definitions are used to specify the elements of messages and possible traces of protocol runs. Isabelle's simplifier and classical reasoner are used heavily.

7.1 Agents and Messages

The protocol proofs rest upon a common theory of agents, messages and their analysis. Protocols are treated at a high level, rather than as strings of bits. Encryption is regarded as a primitive; we cannot detect attacks that rely on numerical idiosyncrasies of encryption methods [27]. Such attacks can be prevented by including redundancy in the body of each encrypted message.

Agents include the server, the friendly agents and the spy. We can model attacks where the spy is an insider.

```
datatype agent = Server | Friend nat | Spy
```

Messages are agent names, nonces, keys, pairs and encryptions. Keys and nonces are just natural numbers. Long messages, consisting of pairs nested to the right, have the special syntax $\{\!|X_1, \ldots, X_n|\!\}$.

```
datatype msg = Agent agent
             | Nonce nat
             | Key   key
             | MPair msg msg
             | Crypt msg key
```

We use an operator invKey to model public-key encryption. Each public key K has an inverse, written K^{-1}, which should be kept private. In shared-key cryptography, a key is its own inverse.

```
rules invKey    "invKey (invKey K) = K"
```

Three inductively defined set operators model the processing of messages. If H is a set of messages then

- parts H is the set of all components of H (going down recursively). The key is not regarded as part of an encrypted message.
- analz H is similar, but needs the key K^{-1} to analyze the body of a message encrypted with K.
- synth H is the set of all messages that can be built using elements of H as components.

Here is the definition of analz H. The set includes the whole of H as well as elements that can be obtained by taking pairs apart. It also contains the bodies of encrypted messages whose keys are available. The set parts H is defined similarly, except that it includes the bodies of all encrypted messages.

```
consts  analz   :: msg set => msg set
inductive "analz H"
  intrs
    Inj      "X ∈ H ⟹ X ∈ analz H"
    Fst      "{|X,Y|} ∈ analz H ⟹ X ∈ analz H"
    Snd      "{|X,Y|} ∈ analz H ⟹ Y ∈ analz H"
    Decrypt "[| Crypt X K ∈ analz H; Key(invKey K) ∈ analz H |]
             ⟹ X ∈ analz H"
```

The definition of synth H attempts to model the spy's capabilities. He cannot use nonces or keys except those in H: they are unguessable. But he can mention the name of any agent.

```
consts  synth   :: msg set => msg set
inductive "synth H"
  intrs
    Inj      "X ∈ H ⟹ X ∈ synth H"
    Agent    "Agent agt ∈ synth H"
    MPair    "[| X ∈ synth H; Y ∈ synth H |] ⟹ {|X,Y|} ∈ synth H"
    Crypt    "[| X ∈ synth H; Key(K) ∈ H |] ⟹ Crypt X K ∈ synth H"
```

7.2 An Algebraic Theory of Messages

More than eighty theorems are proved about parts, analz, synth and similar operators. They form a strong algebraic theory for reasoning about protocols. Thanks to Isabelle's automatic tools, the proof scripts are short: on average, each law is proved using under three commands. The proofs are mostly by rule induction, classical reasoning and rewriting.

Of the three set operators, parts is the easiest to reason about. It distributes over union, and can be evaluated symbolically. Atomic members of its argument (agents, keys and nonces) are simply extracted, becoming members of the result. Compound members are broken down and the new argument recursively evaluated. (Here $\mathsf{ins}\,X H$ denotes the set $\{X\} \cup H$.)

$$\begin{aligned}
\mathsf{parts}(\mathsf{parts}\,H) &= \mathsf{parts}\,H \\
\mathsf{parts}\,G \cup \mathsf{parts}\,H &= \mathsf{parts}(G \cup H) \\
\mathsf{parts}(\mathsf{ins}(\mathsf{Key}\,K)\,H) &= \mathsf{ins}(\mathsf{Key}\,K)(\mathsf{parts}\,H) \\
\mathsf{parts}(\mathsf{ins}\{\!|X,Y|\!\}H) &= \mathsf{ins}\{\!|X,Y|\!\}(\mathsf{parts}(\mathsf{ins}\,X(\mathsf{ins}\,Y H))) \\
\mathsf{parts}(\mathsf{ins}(\mathsf{Crypt}\,X K)H) &= \mathsf{ins}(\mathsf{Crypt}\,X K)(\mathsf{parts}(\mathsf{ins}\,X H))
\end{aligned}$$

On the other hand, analz is the hardest of the operators to reason about. During evaluation, we cannot extract a key without first showing that there are no encrypted messages that it could decrypt. Similarly, we cannot extract an encrypted message without knowing whether there is a key available to decrypt it. Here are some laws concerning analz, the last of which states that agent names can be extracted during evaluation.

$$\begin{aligned}
\mathsf{analz}(\mathsf{analz}\,H) &= \mathsf{analz}\,H \\
\mathsf{analz}\,G \cup \mathsf{analz}\,H &\subseteq \mathsf{analz}(G \cup H) \\
\mathsf{analz}(\mathsf{ins}(\mathsf{Agent}\,a)\,H) &= \mathsf{ins}(\mathsf{Agent}\,a)(\mathsf{analz}\,H)
\end{aligned}$$

Symbolic evaluation of synth H is impossible because the result is infinite. But other forms of reasoning are possible. Here are some properties of synth:

$$\begin{aligned}
\mathsf{synth}(\mathsf{synth}\,H) &= \mathsf{synth}\,H \\
\mathsf{synth}\,G \cup \mathsf{synth}\,H &\subseteq \mathsf{synth}(G \cup H) \\
\mathsf{Key}\,K \in \mathsf{synth}\,H &\Longrightarrow \mathsf{Key}\,K \in H
\end{aligned}$$

We have several laws to simplify expressions in which the operators are nested. All the operators are idempotent. Similarly we have

$$\begin{aligned}
\mathsf{parts}(\mathsf{analz}\,H) &= \mathsf{parts}\,H \\
\mathsf{analz}(\mathsf{parts}\,H) &= \mathsf{parts}\,H.
\end{aligned}$$

The following laws are more interesting. The combination of building up followed by breaking down can be separated:

$$\begin{aligned}
\mathsf{parts}(\mathsf{synth}\,H) &= \mathsf{parts}\,H \cup \mathsf{synth}\,H \\
\mathsf{analz}(\mathsf{synth}\,H) &= \mathsf{analz}\,H \cup \mathsf{synth}\,H
\end{aligned}$$

The proofs are largely automatic, but take a long time to run.

There is no law to break down the combinations synth(parts H) and synth(analz H). The latter is the set of messages that can be built from whatever can be decrypted from H. The spy might send any such message.

```
Nil  [] ∈ otway

Fake [| evs∈otway;  B≠Spy;  X∈synth (analz (sees lost Spy evs)) |]
     ⟹ Says Spy B X  # evs  ∈ otway

OR1  [| evs∈otway;  A ≠ B;  B ≠ Server |]
     ⟹ Says A B {|Nonce (newN evs), Agent A, Agent B,
                   Crypt {|Nonce (newN evs), Agent A, Agent B|}
                         (shrK A) |}
          # evs  ∈ otway

OR2  [| evs∈otway;  B ≠ Server;
        Says A' B {|Nonce NA, Agent A, Agent B, X|}∈set_of_list evs |]
     ⟹ Says B Server
             {|Nonce NA, Agent A, Agent B, X, Nonce (newN evs),
               Crypt {|Nonce NA, Agent A, Agent B|} (shrK B)|}
          # evs  ∈ otway

OR3  [| evs∈otway;  B ≠ Server;
        Says B' Server
             {|Nonce NA, Agent A, Agent B,
               Crypt {|Nonce NA, Agent A, Agent B|} (shrK A),
               Nonce NB,
               Crypt {|Nonce NA, Agent A, Agent B|} (shrK B)|}
          ∈set_of_list evs |]
     ⟹ Says Server B
             {|Nonce NA,
               Crypt {|Nonce NA, Key (newK evs)|} (shrK A),
               Crypt {|Nonce NB, Key (newK evs)|} (shrK B)|}
          # evs  ∈ otway

OR4  [| evs∈otway;  A ≠ B;
        Says S B {|Nonce NA, X, Crypt {|Nonce NB, Key K|} (shrK B)|}
          ∈set_of_list evs;
        Says B Server {|Nonce NA, Agent A, Agent B, X', Nonce NB, X''|}
          ∈set_of_list evs |]
     ⟹ Says B A {|Nonce NA, X|} # evs  ∈ otway

Oops [| evs∈otway;  B ≠ Spy;
        Says Server B {|Nonce NA, X, Crypt {|Nonce NB, Key K|} (shrK B)|}
          ∈set_of_list evs |]
     ⟹ Says B Spy {|Nonce NA, Nonce NB, Key K|} # evs  ∈ otway
```

Fig. 7.2. Specifying a Protocol

7.3 Specifying a Protocol

A protocol inductively specifies a set of possible traces. Each trace is a list of events. An *event* is something of the form $\mathsf{Says}\, A\, B\, X$, namely A says X to B. Other events could be envisaged, corresponding to internal actions of agents. We try to prove that all possible traces are safe. A typical safety guarantee states that receipt of a certain message implies that it was sent by some designated agent.

We assume that the spy already knows the keys of some agents, lost through carelessness. The protocol should still work for other agents: it must not simply collapse.

The function sees describes what an agent sees from a list of events. The spy sees all traffic; other agents see only what is intended for them. From the empty list, each agent sees his initial state, which contains only the key shared with the server. The spy holds all the "lost" keys.

Consider the Otway-Rees protocol, as simplified by Burrows et al. [7, page 247]. A protocol run starts with some agent A sending to B a message, requesting to start a secure conversation. Agent B forwards that message and further information to the server S. He replies by sending B a message containing a new key Kab and further information that B must forward to A.

1. $A \to B : Na, A, B, \{Na, A, B\}_{Ka}$
2. $B \to S : Na, A, B, \{Na, A, B\}_{Ka}, Nb, \{Na, A, B\}_{Kb}$
3. $S \to B : Na, \{Na, Kab\}_{Ka}, \{Nb, Kab\}_{Kb}$
4. $B \to A : Na, \{Na, Kab\}_{Ka}$

Figure 7.2 presents the inductive definition of this protocol. The first rule, `Nil`, allows the empty trace as the starting point. Rule `Fake` describes spy behaviour: he may say anything he is able to. (He may also use the other rules, as if he were honest.) Rules `OR1` to `OR4` describe the messages of the protocol. Each is couched in terms of what message an agent receives and what message is sent in response. Finally, `Oops` models the accidental loss of a session key to the spy; the nonces identify the protocol run.

A spy might intercept messages. We do not need to model interception because nothing forces agents to respond to messages. There are traces in which some messages are ignored and others trigger several replies. Traces allow protocol runs to take place concurrently: middle-person attacks, where the spy uses messages from one run to break another, can be detected.

The protocol shown above turns out to be faulty. A spy C intercepts message 1 and starts a new protocol run with A. He ends up connected to A, to whom he masquerades as B. Here is the attack in detail:

1. $A \to B : Na, A, B, \{Na, A, B\}_{Ka}$ (intercepted)

1′. $C \to A : Nc, C, A, \{Nc, C, A\}_{Kc}$

2′. $A \to S : Nc, C, A, \{Nc, C, A\}_{Kc}, Na', \{Nc, C, A\}_{Ka}$ (intercepted)

2″. $C \to S : Nc, C, A, \{Nc, C, A\}_{Kc}, \underline{Na}, \{Nc, C, A\}_{Ka}$
3′. $S \to A : Nc, \{Nc, Kca\}_{Kc}, \{Na, Kca\}_{Ka}$ (intercepted)
4′. $C \to A : Na, \{Na, Kca\}_{Ka}$

The attack proceeds by replacing nonce Na' by A's original nonce Na (in message 2"), thereby fooling A into accepting key K_{ca} as a key for talking with B. This attack is more serious than that discovered by Mao and Boyd [20], where the server could detect that nonces are being misused.

Unaware of Mao and Boyd's attack, I attempted to prove the protocol correct. I could not prove a subgoal containing messages 1 and 2". In the original version of Otway-Rees, where nonce Nb is encrypted, one can prove that this situation is impossible. Thus, this method of analysing protocols lets us prove correctness properties and also detect flaws.

8. Other Work and Conclusions

Isabelle has been applied to a variety of other tasks: refinement, specification languages, theoretical studies. Let us consider a few of these. (Applications of inductive definitions were presented in Sect. 6.4.)

Isabelle's logical variables support interactive refinement, where a specification is transformed step-by-step into a proven implementation. Investigators at the Max Planck Institute, Saarbrücken have worked on deriving logic programs [2], functional programs [4] and hardware [5].

Coen [8] has implemented a variant of Manna and Waldinger's approach [19] to refinement of functional programs. His Classical Computational Logic (CCL) extends first-order logic with a functional language defined by an operational semantics. He derives programs in this language, not mathematical functions. Termination arguments are expressed using well-founded recursion, not primitive recursion. Later, unpublished work extended the approach to lazy functional programs, replacing termination by reduction to weak head normal form. Coen did some extended examples, deriving functional programs for insertion sort and unification.

Rasmussen has embedded the relational hardware description language Ruby using Isabelle's ZF set theory [46]. Two separate projects aim to support the Z specification language. Kolyang et al. [15] report a promising implementation of Z schemas. The TokiZ project [17] has built a prototype including a deductive system for Z and much of Z's mathematical library.

Isabelle has been applied to studies in logic. Basin et al. [6] are applying Isabelle to study labelled deductive systems. As a first example of modular presentation of logics, they have implemented a wide variety of modal logics. Matthews is using Isabelle to implement Feferman's theory of finitary inductive definitions, FS0 [21]. Grąbczewski has mechanized the first two chapters of Rubin and Rubin's *Equivalents of the Axiom of Choice* [41].

To conclude, let us recall those features of Isabelle that have turned out to be particularly successful. Designers of new tools should bear them in mind.

- A higher-order syntax supports variable binding.
- Unknowns in goals allow refinement and automatic proof search.
- A generic framework supports a wide range of notations and methods.
- The classical reasoner and simplifier let us construct proofs of realistic size.
- Set-theoretic primitives find many uses in specifications.
- Many computational phenomena can be modelled using inductive or co-inductive definitions.

Acknowledgement. Giampaolo Bella, Rachel Cardell-Oliver, Michael Jones, Fabio Massacci, Chris Owens, Mark Staples and Myra VanInwegen commented on this article. Isabelle's simplifier is largely the work of Prof. Tobias Nipkow, now at the Technical University of Munich. Isabelle/ZF includes work by Martin Coen, Philippe de Groote and Philippe Noël. The research was funded by numerous grants including EPSRC GR/K57381 "Mechanising Temporal Reasoning" and GR/K77051 "Authentication Logics" and ESPRIT 6453 "Types." Many thanks also to the summer school organizers.

References

1. Aczel, P., An introduction to inductive definitions, In *Handbook of Mathematical Logic*, J. Barwise, Ed. North-Holland, 1977, pp. 739–782
2. Anderson, P., Basin, D., Deriving and applying logic program transformers, In *Algorithms, Concurrency and Knowledge (1995 Asian Computing Science Conference)* (Pathumthani, Thailand, December 1995), LNCS 1023, Springer, pp. 301–318
3. Anderson, R., Needham, R., Programming Satan's computer, In *Computer Science Today: Recent Trends and Developments*, J. van Leeuwen, Ed., LNCS 1000. Springer, 1995, pp. 426–440
4. Ayari, A., Basin, D., Generic system support for deductive program development, In *Second International Workshop on Tools and Algorithms for the Construction and Analysis of Systems (TACAS '96)* (1996), LNCS 1055, Springer, pp. 313–328
5. Basin, D., Friedrich, S., Modeling a hardware synthesis methodology in Isabelle, In von Wright et al. [51], pp. 33–50
6. Basin, D., Matthews, S., Viganò, L., Labelled propositional modal logics: theory and practice, Tech. Rep. MPI-I-96-2-002, Max-Planck-Institut für Informatik, Saarbrücken, 1996
7. Burrows, M., Abadi, M., Needham, R. M., A logic of authentication, *Proceedings of the Royal Society of London* **426** (1989), 233–271
8. Coen, M. D., *Interactive Program Derivation*, PhD thesis, University of Cambridge, Nov. 1992, Computer Laboratory Technical Report 272
9. Felty, A., Implementing tactics and tacticals in a higher-order logic programming language, *Journal of Automated Reasoning* **11**, 1 (1993), 43–82
10. Felty, A., Miller, D., Encoding a dependent-type λ-calculus in a logic programming language, In *10th International Conference on Automated Deduction* (1990), M. E. Stickel, Ed., LNAI 449, Springer, pp. 221–235

11. Frost, J., A case study of co-induction in Isabelle, Tech. Rep. 359, Computer Laboratory, University of Cambridge, Feb. 1995
12. Gordon, M. J. C., Melham, T. F., *Introduction to HOL: A Theorem Proving Environment for Higher Order Logic*, Cambridge University Press, 1993
13. Hennessy, M., *The Semantics of Programming Languages: An Elementary Introduction Using Structural Operational Semantics*, Wiley, 1990
14. Huet, G. P., A unification algorithm for typed λ-calculus, *Theoretical Comput. Sci.* **1** (1975), 27–57
15. Kolyang, Santen, T., Wolff, B., A structure preserving encoding of Z in Isabelle/HOL, In von Wright et al. [51], pp. 283–298
16. Korf, R. E., Depth-first iterative-deepening: an optimal admissible tree search, *Artificial Intelligence* **27** (1985), 97–109
17. Kraan, I., Baumann, P., Implementing Z in Isabelle, In *ZUM '95 : The Z Formal Specification Notation* (1995), J. P. Bowen M. G. Hinchey, Eds., LNCS 967, Springer
18. Lötzbeyer, H., Sandner, R., Proof of the equivalence of the operational and denotational semantics of IMP in Isabelle/ZF, Project report, Institut für Informatik, TU München, 1994
19. Manna, Z., Waldinger, R., Fundamantals of deductive program synthesis, *IEEE Trans. Softw. Eng.* **18**, 8 (Aug. 1992), 674–704
20. Mao, W., Boyd, C., Towards formal analysis of security protocols, In *Computer Security Foundations Workshop VI* (1993), IEEE Computer Society Press, pp. 147–158
21. Matthews, S., Implementing FS_0 in Isabelle: Adding structure at the metalevel, In *Design and Implementation of Symbolic Computation Systems. International Symposium, DISCO '96* (1996), J. Calmet C. Limongelli, Eds., LNCS 1128, Springer, pp. 228–239
22. McCune, W., OTTER 3.0 Reference Manual and Guide, Tech. Rep. ANL-94/6, Argonne National Laboratory, Argonne, IL, 1994
23. McRobbie, M., Slaney, J. K., Eds., *Automated Deduction — CADE-13 International Conference* (1996), LNAI 1104, Springer
24. Miller, D., Unification under a mixed prefix, *Journal of Symbolic Computation* **14**, 4 (1992), 321–358
25. Milner, R., *Communication and Concurrency*, Prentice-Hall, 1989
26. Milner, R., Tofte, M., Co-induction in relational semantics, *Theoretical Comput. Sci.* **87** (1991), 209–220
27. Moore, J. H., Protocol failures in cryptosystems, In *Contemporary Cryptology: The Science of Information Integrity*, G. J. Simmons, Ed. IEEE Press, 1992, pp. 541–558
28. Nipkow, T., Functional unification of higher-order patterns, In *Eighth Annual Symposium on Logic in Computer Science* (1993), M. Vardi, Ed., IEEE Computer Society Press, pp. 64–74
29. Nipkow, T., Order-sorted polymorphism in Isabelle, In *Logical Environments* (1993), G. Huet G. Plotkin, Eds., Cambridge University Press, pp. 164–188
30. Nipkow, T., More Church-Rosser proofs (in Isabelle/HOL), In McRobbie, Slaney [23], pp. 733–747
31. Nipkow, T., Winskel is (almost) right: Towards a mechanized semantics textbook, In *Foundations of Software Technology and Theoretical Computer Science* (1996), LNCS, Springer, In press
32. Paulson, L. C., *Logic and Computation: Interactive proof with Cambridge LCF*, Cambridge University Press, 1987
33. Paulson, L. C., The foundation of a generic theorem prover, *Journal of Automated Reasoning* **5**, 3 (1989), 363–397

34. Paulson, L. C., Set theory for verification: I. From foundations to functions, *Journal of Automated Reasoning* **11**, 3 (1993), 353–389
35. Paulson, L. C., A fixedpoint approach to implementing (co)inductive definitions, In *Automated Deduction — CADE-12 International Conference* (1994), A. Bundy, Ed., LNAI 814, Springer, pp. 148–161
36. Paulson, L. C., *Isabelle: A Generic Theorem Prover*, Springer, 1994, LNCS 828
37. Paulson, L. C., Set theory for verification: II. Induction and recursion, *Journal of Automated Reasoning* **15**, 2 (1995), 167–215
38. Paulson, L. C., Generic automatic proof tools, Tech. Rep. 396, Computer Laboratory, University of Cambridge, May 1996
39. Paulson, L. C., *ML for the Working Programmer*, 2nd ed., Cambridge University Press, 1996
40. Paulson, L. C., A concrete final coalgebra theorem for ZF set theory, In *Types for Proofs and Programs: International Workshop TYPES '94* (published 1995), P. Dybjer, B. Nordström, J. Smith, Eds., LNCS 996, Springer, pp. 120–139
41. Paulson, L. C., Grąbczewski, K., Mechanizing set theory: Cardinal arithmetic and the axiom of choice, *Journal of Automated Reasoning* (1996), In press
42. Pelletier, F. J., Seventy-five problems for testing automatic theorem provers, *Journal of Automated Reasoning* **2** (1986), 191–216, Errata, JAR 4 (1988), 235–236
43. Pfenning, F., The practice of logical frameworks, In *Trees in Algebra and Programming - CAAP '96. 21st International Colloquium* (1996), H. Kirchner, Ed., LNCS 1059, Springer, pp. 119–134
44. Pusch, C., Verification of compiler correctness for the WAM, In von Wright et al. [51]
45. Rasmussen, O., The Church-Rosser theorem in Isabelle: A proof porting experiment, Tech. Rep. 364, Computer Laboratory, University of Cambridge, May 1995
46. Rasmussen, O., An embedding of Ruby in Isabelle, In McRobbie, Slaney [23], pp. 186–200
47. Rich, E., Knight, K., *Artificial Intelligence*, 2nd ed., McGraw-Hill, 1991
48. Schroeder-Heister, P., A natural extension of natural deduction, *Journal of Symbolic Logic* **49**, 4 (Dec. 1984), 1284–1300
49. Slind, K., Function definition in higher-order logic, In von Wright et al. [51]
50. Suppes, P., *Axiomatic Set Theory*, Dover, 1972
51. von Wright, J., Grundy, J., Harrison, J., Eds., *Theorem Proving in Higher Order Logics* (1996), LNCS 1125
52. Winskel, G., *The Formal Semantics of Programming Languages*, MIT Press, 1993

Machine-Assisted Verification Using Theorem Proving and Model Checking

N. Shankar[1]

[1] Computer Science Laboratory, SRI International, Menlo Park, CA 94025 USA

Abstract. Theorem proving and model checking are complementary approaches to the verification of hardware designs and software algorithms. In theorem proving, the verification task is one of showing that the formal description of the program *implies* the formal statement of a putative program property, while model checking demonstrates that the program is a model that *satisfies* the putative property. Theorem proving is completely general but typically requires significant human guidance, whereas model checking though restricted to a limited range of properties of small (essentially) finite-state systems, is largely automatic. This paper is a tutorial on the combined use of theorem proving and model checking as mechanized in the PVS specification and verification environment.

Keywords. Mechanical verification, Theorem proving, Model checking

1 Introduction

There are both scientific and pragmatic reasons for pursuing the formal verification of hardware and software systems. The scientific reason is to develop the body of knowledge needed to rigorously calculate and predict the behavior of a hardware or software system from an algorithmic description of its design. The pragmatic reason is that it is increasingly the case that extremely complex systems are being designed and deployed in ways that even trivial errors can have serious consequences. A formal method is used to write precise and unambiguous specifications of the requirements, the design assumptions, and the system design, and to demonstrate by rigorous calculation that the requirements are met by the design and the assumptions.

The conventional wisdom in the engineering of complex systems has been that the most troublesome errors are typically introduced early in the design lifecycle. Validation methods such as simulation and testing are not only inefficient and labor-intensive, they contribute very little in the early design phases. An effective formal method must therefore support the early detection and elimination of high-level design errors.

As with any engineering endeavour, mathematics is the basic tool for modelling and analyzing the behavior of hardware and software systems. However, in contrast to other engineering uses of mathematics, the analysis of discrete systems requires the extremely rigorous, large-scale application of logic and mathematics. Rigorous, repeatable, large-scale formal analysis requires effective and highly efficient mechanization.

The mathematics used in the formal development of programs and proofs can range from the trivial to the abstruse. The trivial parts of verification can and should be handled automatically through the use of appropriate decision procedures. At the higher levels, that is in the selection of induction schemes, the choice and usage of lemmas, case analysis, and quantifier instantiations, an effective mechanization must provide low-level automation that can harnessed through high-level user control.

PVS (Prototype Verification System) is a specification and verification environment that provides mechanized support for developing and harnessing mathematics for the purpose of formal verification [28]. PVS carefully integrates an expressive specification language with powerful theorem proving capabilities, and also integrates low-level inference procedures with high-level proof strategies. The inference procedures available in PVS include decision procedures for equality and linear arithmetic, symbolic model checking, and conditional rewriting and simplification. The ideas and implementation underlying PVS are still undergoing development but the system has already been used in some substantial verification projects. These projects include the verification of hardware processors [14, 25], floating-point hardware [26, 31], real-time, distributed, and fault-tolerant algorithms [18,22,23,32], and in the construction of background libraries for use in verification.

This paper is a tutorial on the use of PVS for mechanized verification. Verification, even when restricted to mechanized verification using PVS, is too vast a subject to be the topic of a single paper. We will narrow our focus in this paper to the verification of transition systems, particularly finite-state ones. The main purpose of this paper is to describe how such transition systems can be specified and verified using the combination of theorem proving and model checking available in PVS.

2 Brief Overview of PVS

PVS (Prototype Verification System) is a specification and verification environment that exploits the synergy between an expressive specification language and powerful support for interactive and automated proof construction. The specification language of PVS is based on simply typed higher-order logic with several extensions such as parametric theories, predicate subtypes, dependent types, and abstract datatypes. The PVS typechecker generates proof obligations corresponding to the subtype restrictions. These proof obligations must eventually be discharged by means of the PVS proof checker before a specification can be regarded as well-formed.

A simple example of the use of predicate subtypes occurs in the case of division.[1]

```
nonzero_real: TYPE = {r: real | r /= 0}                              1
```

In the type definition in 1, the type `nonzero_real` is declared to be a subtype of the type `real` of real numbers containing all the reals distinct from `0`. The division operation can then be introduced in 2 with its denominator type constrained to range over the nonzero reals.

```
 /: [real, nonzero_real -> real]                                      2
```

Note that '/' is known to be an infix symbol in the PVS grammar. This means that when an expression of the form `1/0` is typechecked, the obviously false proof obligation `0 /= 0` is generated corresponding to the subtype constraint on the denominator of the division operator.

This kind of subtyping also extends to higher-order types, i.e., functions, relations, functions of functions, and so on. For example, the higher-order predicate `injective?` characterizes the class of injective functions from domain type `D` to range type `R`. This predicate definition is embedded in a parametric theory `functions` in 3 that takes `D` and `R` as type parameters. The predicate `injective?` can easily be used to define the subtype of injective functions of the function type `[D -> R]`.

```
functions [D, R: TYPE]: THEORY                                        3
 BEGIN
  f, g: VAR [D -> R]
  x, x1, x2: VAR D

  injective?(f): bool = (FORALL x1, x2: (f(x1) = f(x2) => (x1 = x2)))

END functions
```

PVS has an interactive proof checker that can be used to construct a proof tree by invoking proof commands. Each proof command either completes a branch of the proof, or generates subgoals that add new branches to the current proof branch. The PVS proof calculus is based on Gentzen's sequent calculus. The primitive proof commands are themselves quite powerful. Some of them employ decision procedures for quantifier-free theories of equality and linear arithmetic, conditional rewriting, and model checking. Compound proof commands can be composed from simpler ones by means of *strategies* which are programs that can control the application of proof commands with constructs for recursion, branching, and backtracking.

[1]The examples in this section are adapted from the PVS prelude library which contains a basic collection of PVS theories for logic, functions, relations, sets, and various data types.

3 Peterson's Algorithm for Mutual Exclusion

The main goal of this paper is to illustrate the use of the specification, theorem proving, and model checking capabilities of PVS in the verification of transition systems. Peterson's algorithm for 2-process mutual exclusion is a useful example of a transition system since it is simple and familiar. The version of the algorithm presented here is simplified for ease of exposition but does loosely capture the key ideas underlying Peterson's algorithm [30].

The mutual exclusion problem is to ensure that two processes are never simultaneously in their critical section. The processes are labelled p and q. The behavior of each process is modelled by control variables `pcp` and `pcq` ('`pc`' for *program control*) that determine the control states of p and q, respectively. When `pcp` (respectively, `pcq`) is set to the value `critical`, this indicates that p (respectively, q) is in its critical section. The noncritical section is divided into a `sleeping` phase and a `trying` phase. Each process is allowed to test whether the other process is sleeping but the implementation details for such a test are omitted.[2] Access to the critical section is arbitrated by means of a shared boolean variable called `turn`.

In addition to a stuttering or skip transition, each process has three transitions. In its transition from a `sleeping` state to a `trying` state, process p (respectively, q) sets the `turn` variable to `FALSE` (`TRUE`). From the `trying` state, p (correspondingly, q) enters its `critical` state when either q (p) is `sleeping`, or `turn` is `TRUE` (`FALSE`). When exiting from the `critical` state to a `sleeping` state, process p (q) once again sets the `turn` variable to `FALSE` (`TRUE`). The algorithm as described above is obviously an abstraction of a more refined system that can contain transitions within the `sleeping`, `trying`, and `critical` phases. These transitions are not observable at the level of abstraction at which the algorithm has been described. Each process has a stuttering (skip) transition to allow internal steps that have no effect on the observable state. The stuttering step allows for a refinement map to be established between the abstract and concrete transition systems [1].

We first demonstrate the formalization of the above variant of Peterson's protocol by means of predicates and relations in the higher-order logic of PVS. We then show how a proof of the mutual exclusion property can be established by means of the PVS proof checker. We then describe how model checking can be used to establish both safety and liveness properties of this simple finite-state protocol.

3.1 A Formal Description of Peterson's Algorithm

There are a large number of programming notations and program logics that can be used to formally express the algorithm that has been informally de-

[2] The status of the other process can be observed by using a shared boolean variable for each process which indicates whether the process is in its `sleeping` state.

scribed above, but algorithms of this form can just as easily be modelled in a conventional predicate calculus. The description below uses the higher-order logic of PVS.

The first step is to capture the notion of computations over a computation state using the PVS logic. This is formalized in the theory `state` in 4 which takes the type (also named) `state` as a parameter. This type parameter can be instantiated when the `state` theory is imported within another theory. A computation is modelled as an infinite sequence of computation states, i.e., of type `sequences[states]` which is just the function type `[nat -> state]`. An assertion is just a predicate on states, i.e., an expression of type `pred[state]` (which is shorthand for the type `[state -> bool]`).

```
state[state: TYPE] : THEORY                                          4
BEGIN
  IMPORTING  sequences[state]

  statepred: TYPE = PRED[state]   %assertions
  Action: TYPE =  PRED[[state, state]]
  computation : TYPE = sequence[state]

  pp: VAR statepred
  action: VAR Action
  aa, bb, cc: VAR computation

  5
END state
```

A program characterizes a set of computations and is specified as a pair of an initialization assertion and an action, a binary transition relation on states which constrains the allowable transitions of a computation. The higher-order predicate `Init` is defined so that `Init(pp)(aa)` holds of assertion `pp` and computation `aa` exactly if `pp` holds of the initial state `aa(0)`. The predicate `Inv` of an action `action` and computation `aa` is defined to hold exactly when `action` holds of each pair of adjacent states. When `Run(pp, action)(aa)` holds, a computation `aa` is a satisfying run of a given program consisting of initialization predicate `pp` and action `action`. The predicate `Inv` is overloaded to also apply to an assertion `pp` and holds exactly when `pp` holds of every state in a given computation.

```
                                                              5
  Init(pp)(aa) : bool = pp(aa(0))

  Inv(action)(aa) : bool =
    (FORALL (n : nat) : action(aa(n), aa(n+1)))

  Run(pp, action)(aa): bool =
    (Init(pp)(aa) AND Inv(action)(aa))

  Inv(pp)(aa) : bool =
   (FORALL (n : nat) : pp(aa(n)))
```

Using the theory `state`, we can now describe the Peterson algorithm by supplying the appropriate state type, the initialization predicate, and transition relation. This is done in the theory `mutex` shown in [6], [7], and [8]. The type of the global computation state is given by a record type with a field corresponding to each program variable. The two program control variables correspond to the fields `pcp` and `pcq` of the enumerated type `PC` consisting of three values `sleeping`, `trying`, and `critical`.

```
                                                              6
mutex   : THEORY
  BEGIN

   PC : TYPE = {sleeping, trying, critical}
   state : TYPE =
             [# pcp : PC,
                turn: bool,
                pcq : PC #]

   IMPORTING  state[state]
   s, s0, s1: VAR state

   [7]
```

The processes p and q are specified independently. The initialization predicate for p is given by `I_P` which asserts that p is initially in its `sleeping` state. The transition relation `G_P` for p is the disjunction (nondeterministic choice) of four actions: the stuttering action, the `sleeping-to-trying` action, the entry into the `critical` state, and the exit from the `critical` to `sleeping` state. The expression `s0 WITH [pcp := trying, turn := FALSE])` represents the result of updating the record `s0` as indicated at the fields `pcp` and `turn`.

```
I_P(s) : bool = (sleeping?(pcp(s)))                                   7

G_P(s0, s1): bool =
  (  (s1 = s0)                         %stutter
   OR
     (sleeping?(pcp(s0)) AND           %try
        s1 = s0 WITH [pcp := trying, turn := FALSE])
   OR
     (trying?(pcp(s0)) AND             %enter critical
       (turn(s0) OR sleeping?(pcq(s0))) AND
      s1 = s0 WITH [pcp := critical])
   OR (critical?(pcp(s0)) AND          %exit critical
      s1 = s0 WITH [pcp := sleeping, turn := FALSE ]))

 8
END mutex
```

The specification of q is symmetrical and is shown in 8

```
I_Q(s) : bool = (sleeping?(pcq(s)))                                   8

G_Q(s0, s1): bool =
  (  (s1 = s0)                         %stutter
   OR
     (sleeping?(pcq(s0)) AND           %try
        s1 = s0 WITH [pcq := trying, turn := TRUE])
   OR
     (trying?(pcq(s0))  AND            %enter critical
       (NOT turn(s0) OR sleeping?(pcp(s0))) AND
      s1 = s0 WITH [pcq := critical])
   OR (critical?(pcq(s0)) AND          %exit critical
      s1 = s0 WITH [pcq := sleeping, turn := TRUE]))
```

The entire system is specified in 9 in terms of the initialization predicate I which is the conjunction of I_P and I_Q, and the transition relation G which is the disjunction (i.e., interleaving) of actions G_P and G_Q.

```
I(s) : bool = (I_P(s) AND I_Q(s))                                     9

G(s0, s1) : bool = (G_P(s0, s1) OR G_Q(s0, s1))
```

3.2 Deductive Verification of Peterson's Algorithm

Having specified the algorithm in terms of an initialization predicate and a transition relation, we can now attempt to prove that the algorithm guarantees mutual exclusion. The statement is in the theory mutex_proof in 10.

(The theory connectives that is imported by the mutex_proof theory contains the lifted definitions of the boolean connectives so that, for example, AND(pp, qq) is defined as the predicate (LAMBDA s: pp(s) AND qq(s)).) The assertion safe holds of any state where at most one of p or q is critical. It is then easy to state the theorem that safe is an invariant for any computation corresponding to the above algorithm specification.

```
mutex_proof: THEORY                                                        10
 BEGIN
   IMPORTING mutex, connectives[state]
   s, s0, s1: VAR state

   safe(s) : bool = NOT (critical?(pcp(s)) AND critical?(pcq(s)))

   safety_proved: CONJECTURE
      (FORALL (aa: computation):
        Run(I, G)(aa)
        IMPLIES Inv(safe)(aa))

   19
END mutex_proof
```

We can now try the proof of the conjecture[3] safety_proved. We show a partial transcript of the proof starting with 11. The proof commands are supplied in response to the Rule? prompt following the display of the current subgoal of the proof. The skosimp command introduces an arbitrary constant aa!1 for the universally quantified parameter aa. It also "flattens" the implication across the sequent turnstile. The interpretation of a PVS sequent is as an implication between the conjunction of (*antecedent*) formulas above the sequent turnstile '|-------' and the disjunction of (*consequent*) formulas below the turnstile. The antecedent formulas are numbered -1, -2, etc., and the consequent formulas are numbered 1, 2, and so on. Since each subgoal sequent in a proof contains some new formulas and some old formulas that are inherited from the parent goal, the numbers corresponding to new formulas are in braces, e.g., {1}, whereas the numbers for the old formulas are in brackets, e.g., *[-1]*.

[3]The keyword CONJECTURE has no special semantic significance in PVS and we could just as well have used LEMMA, FACT, THEOREM, PROPOSITION, etc.

```
safety_proved :                                                    11

  |-------
{1}   (FORALL (aa: computation):
              Run(I, G)(aa) IMPLIES Inv(safe)(aa))

Rule? (skosimp)
Skolemizing and flattening,
this simplifies to:
safety_proved :

{-1}   Run(I, G)(aa!1)
  |-------
{1}   Inv(safe)(aa!1)
```

We need to establish the invariant `safe` by induction on the transitions used to construct the computation `aa!1` but the induction variable is hidden in the definition of `Inv`. Expanding `Inv` in 12 yields a universally quantified formula where we can apply induction over the parameter `n`.

```
Rule? (expand "Inv")                                               12
Expanding the definition of Inv,
this simplifies to:
safety_proved :

[-1]   Run(I, G)(aa!1)
  |-------
{1}   (FORALL (n: nat): safe(aa!1(n)))
```

Invoking induction on `n` causes PVS to instantiate the natural number induction scheme (since the type of `n` is `nat`) with a suitable induction predicate, and carry out further simplifications that yield two subgoals corresponding to the base and induction steps. In the base case, proving that the invariant holds of the initial state of the computation is straightforward and can be carried out by the `grind` command which expands out some of the definitions in the subgoal and then applies the PVS decision procedures.

13

```
Rule? (induct "n")
Inducting on n,
this yields  2 subgoals:
safety_proved.1 :

[-1]   Run(I, G)(aa!1)
  |-------
{1}   safe(aa!1(0))

Rule? (grind)
Trying repeated skolemization, instantiation, and if-lifting,

This completes the proof of safety_proved.1.
```

This brings up the induction step in 14 which requires a proof that the invariant `safe` is preserved in any transition from `aa!1(j)` to `aa!1(j + 1)`. For reasons of efficiency and succinctness of proof, we would like to prove this by case-splitting on the possible transitions allowed by the transition relation `G`. We achieve this by expanding out the definitions in the only antecedent formula using the proof command `simplify-with-rewrites`, replacing the universally quantified variable `j` in the consequent by the constant `j!1` using `skosimp*`, then instantiating the pre-state and post-state variables in the transition `G` by `aa!1(j!1)` and `aa!1(j!1 + 1)` using `inst?`, respectively. Propositional simplification using `prop` can then be used to generate the 10 subgoals corresponding to the 10 possible transitions. Since the intermediate formulas in the above sequence of proof steps are large and unreadable, they have been suppressed by using the PVS proof command `apply` to package these four steps into a single atomic proof step.

14

```
safety_proved.2 :

[-1]   Run(I, G)(aa!1)
  |-------
{1}   (FORALL (j: nat): safe(aa!1(j)) IMPLIES safe(aa!1(j + 1)))

Rule? (apply (then (simplify-with-rewrites - :defs T)
           (skosimp*)
                  (inst?)
                  (prop)))

Applying
   (THEN (SIMPLIFY-WITH-REWRITES - :DEFS T)
         (SKOSIMP*) (INST?) (PROP)),
this yields  10 subgoals:
```

The first subgoal in [15] corresponds to a stuttering transition which trivially preserves the invariant. It is proved here by `grind` but it can also be proved by other simpler proof commands.

```
safety_proved.2.1 :                                                    15

{-1}    (aa!1(1 + (j!1 + 1 - 1)) = aa!1(j!1 + 1 - 1))
{-2}    safe(aa!1(j!1))
{-3}    (sleeping?(pcp(aa!1(0))))
{-4}    (sleeping?(pcq(aa!1(0))))
  |-------
{1}   safe(aa!1(j!1 + 1))

Rule? (grind)
Trying repeated skolemization, instantiation, and if-lifting,

This completes the proof of safety_proved.2.1.
```

The second subgoal in [16] corresponds to the transition of p from a `sleeping` to `trying` state. This easily preserves the invariant at most one of p or q is `critical` since p is not critical in state `aa!1(j!1 + 1)`. This subgoal is also proved by `grind` which essentially expands the definition of `safe` and performs the required equality and enumerated datatype reasoning.

```
safety_proved.2.2 :                                                    16

{-1}    sleeping?(pcp(aa!1(j!1 + 1 - 1)))
{-2}    aa!1(1 + (j!1 + 1 - 1))
          = aa!1(j!1 + 1 - 1) WITH [pcp := trying, turn := FALSE]
{-3}    safe(aa!1(j!1))
{-4}    (sleeping?(pcp(aa!1(0))))
{-5}    (sleeping?(pcq(aa!1(0))))
  |-------
{1}   safe(aa!1(j!1 + 1))

Rule? (grind)
Trying repeated skolemization, instantiation, and if-lifting,

This completes the proof of safety_proved.2.2.
```

The third and more troublesome subgoal shown in [17] corresponds to the p transition from a `trying` to `critical` state. Here, we need to rule out the possibility that the invariant can be violated because q is already in a `critical` state.

```
safety_proved.2.3 :                                              17

{-1}   turn(aa!1(j!1 + 1 - 1))
{-2}   trying?(pcp(aa!1(j!1 + 1 - 1)))
{-3}   aa!1(1 + (j!1 + 1 - 1))
          = aa!1(j!1 + 1 - 1) WITH [pcp := critical]
{-4}   safe(aa!1(j!1))
{-5}   (sleeping?(pcp(aa!1(0))))
{-6}   (sleeping?(pcq(aa!1(0))))
  |-------
{1}   safe(aa!1(j!1 + 1))
```

Indeed, the application of the `grind` command in 18 to this subgoal fails to prove it.

```
Rule? (grind)                                                    18
Trying repeated skolemization, instantiation, and if-lifting,
this simplifies to:
safety_proved.2.3 :

{-1}   turn(aa!1(j!1))
{-2}   trying?(pcp(aa!1(j!1)))
{-3}   aa!1(1 + j!1) = aa!1(j!1) WITH [pcp := critical]
[-4]   safe(aa!1(j!1))
{-5}   critical?(pcq(aa!1(j!1)))
[-6]   (sleeping?(pcp(aa!1(0))))
[-7]   (sleeping?(pcq(aa!1(0))))
  |-------

Rule?
```

Since `grind` fails to prove this subgoal, we need to investigate whether it is in fact provable. The application of `grind` does provide a counterexample by generating a scenario where q is `critical` in state `aa!1(j!1)` and `turn(aa!1(j!1)` is `TRUE`. This cannot happen in the actual protocol since the `turn` field is checked to be `FALSE` when q becomes `critical` and p never assigns `TRUE` to `turn`. The proof fails because the invariant is not strong enough to rule out the scenario where q is `critical` but the `turn` field happens to be `TRUE`.

The above failure is instructive since it allows us to construct a strengthened invariant. The new invariant `strong_safe` in 19 is a conjunction asserting that when p or q is critical, the corresponding test in the guard for entry to the `critical` state holds. This invariant can actually be generated by simple heuristics (for related heuristics, see [4,5,24]) The test in the guard for the entry to the `critical` state for p (q) is a local invariant for p (q) that is free from interference from q (p).

```
strong_safe(s) : bool =                                              19
   ((critical?(pcp(s)) IMPLIES (turn(s) OR sleeping?(pcq(s))))
   AND
   (critical?(pcq(s)) IMPLIES (NOT turn(s) OR sleeping?(pcp(s)))))

strong_safety_proved: THEOREM
   (FORALL (aa: computation):
      Run(I, G)(aa)
     IMPLIES Inv(strong_safe)(aa))

strong_safe_implies_safe: THEOREM
   (strong_safe IMPLIES safe)(s)
```

The invariance of strong_safety stated in strong_safety_proved is proved by following the same outline as the previous proof.[4] The theorem strong_safe_implies_safe is easily proved by a single grind command. This implication is quite obvious because, by strong_safe, if p and q are simultaneously critical, then turn would have to be simultaneously both TRUE and FALSE.

To summarize the main points:

- A shared state, interleaving model of concurrent programs such as those used by Manna and Pnueli's temporal logic [24], UNITY [10], and TLA [21] can be formalized by means of predicates, relations, and sequences in a higher-order logic such as HOL [16,17] or PVS [27].
- With sufficiently powerful mechanization based on term rewriting and decision procedures, it is possible to construct mechanically verified proofs of program properties that are faithful to their informal counterparts.
- Failed invariance proofs can be used to diagnose the source of the failure.

3.3 Model Checking Mutual Exclusion

Peterson's mutual exclusion algorithm, as described above, is reused in this section to illustrate verification methodologies based on model checking. The state type state in the theory mutex in [6] in Section 3.1 is a record type with one boolean field turn and two fields, pcp and pcq, ranging over a 3-valued enumerated type PC. This yields a state space with only 18 possible states. By the invariant strong_safe, we know that 4 of these 18 possible states are unreachable from an initial state. The proof in Section 3.2 involves a

---

[4] It is possible to prove strong_safety_proved with fewer manual interactions but this typically yields a less efficient verification and a less readable proof.

ten-way case analysis which seems rather excessive for such a small reachable state space. It is clearly more expedient to carry out such verifications by means of an exhaustive exploration of the reachable state space as is done with model checking [11,12]. Model checking technology has developed to a point where it is possible to deal with quite large state spaces. It has also been extended to handle real-time constraints where the state space is actually uncountable [2,3]. The advantages of model checking over deductive verification are that model checking is largely automatic and can provide useful counterexample traces from failed verification attempts. We show in this section that the model checking verification paradigm is also formalizable within the framework of a higher-order logic such as PVS.

In the deductive approach to verification as used in the proof of `strong_safety_proved`, we are essentially proving that given a program M and a property A, that $\vdash M \supset A$ holds. In model checking, the program M is described by an automaton or a *Kripke model* and the property A as a temporal formula, and the verification establishes $M \models A$.

The basic ideas underlying model checking are quite simple. The state space consisting of 18 states in the Peterson algorithm along with the transition relation `G` which defines the set of directed transition edges between states, can be seen as defining a Kripke model on which the truth of formulas in a temporal logic can be evaluated. The most convenient temporal logic from the viewpoint of model checking is the branching time temporal logic CTL. The temporal operators of CTL can be characterized as minimal or maximal fixed points of monotone predicate transformers over the state space. The CTL operators are definable in terms of minimal and maximal fixed point operators can be defined in higher-order logic. We first examine the formalization of a fixed point calculus, the mu-calculus, in PVS before presenting the definitions of the CTL operators.

3.3.1 The Boolean Mu-Calculus

The boolean mu-calculus (BMC) [?,9] consists of boolean formulas and n-ary boolean predicates.[5]

Boolean formulas are built from the truth values `TRUE` and `FALSE` and the atomic propositions, e.g., a, b, c, by means of negation $\neg A$, disjunction $A \vee B$, boolean quantification $\exists x.\ A[x]$ or $\forall x.\ A[x]$, or by the application of an n-ary predicate r to n boolean formulas $A_1, \ldots, A_n$ as in $r(A_1, \ldots, A_n)$.

The n-ary predicates are defined either by *lambda-abstraction* so that $\lambda x_1, \ldots, x_n : A[x_1, \ldots, x_n]$ or as *least fixed points* $\mu Z.\ P[Z]$ ("the least or strongest Z such that $Z = P[Z]$") where Z is a variable ranging over n-ary boolean predicates and P is a *monotone predicate transformer*. The n-ary

[5] The boolean mu-calculus is slightly different from the modal mu-calculus [15,20] where transitions are represented by labels rather than boolean functions.

boolean predicates can be ordered so that $r \leq s$ ("r is stronger than s") when $(\forall x_1, \ldots, x_n.\ r(x_1, \ldots, x_n) \supset s(x_1, \ldots, x_n))$. A monotone predicate transformer P maps n-ary boolean functions to n-ary boolean functions so that for any n-ary predicates r and r', if $r \leq r'$, then $P[r] \leq P[r']$.[6] The greatest fixed point $\nu Z.\ P[Z]$ of a monotone predicate transformer P can be defined dually as $\neg\mu Z.\ \neg P[\neg Z]$.

Let $\perp$ represent the n-ary relation $\lambda p_1, \ldots, p_n.$ FALSE. Also let $k = 2^n$ and i range from 0 to k, then The least fixed point can be expressed directly in QBF as the disjunction $P[\perp] \vee \ldots P^i[\perp] \vee \ldots P^k[\perp]$. Typically, the computation of the fixed point will converge in a lot fewer than k iterations.

It turns out to be convenient to use reduced ordered binary decision diagrams (ROBDDs) [6–8] to represent the boolean functions defined by the boolean mu-calculus. A binary decision diagram (BDD) is a binary tree with propositional variables labelling the nodes, `TRUE/FALSE` arrows out of each node, and `TRUE/FALSE` leaves. An ordered BDD (OBDD) is BDD in which the ordering of propositional variables along each branch is consistent with some fixed total ordering on these variables. A reduced OBDD is an OBDD with a directed acyclic graph structure so that there are no duplicate occurrences of equivalent subtrees, and any redundant nodes, namely those for which both the `TRUE` and `FALSE` outgoing edges point to the same subgraph, have been eliminated. ROBDDs are a canonical representation for boolean functions. The size of an ROBDD is quite sensitive to the chosen ordering of propositional variables. The ROBDD for a circuit that checks that exactly 0 or 2 of a, b, c are high, is shown in Figure 3.3.1.

3.4 The Relational Mu-Calculus

Fixed point operators extend first-order logic by allowing the definition of operations such as transitive closure, but these operators are actually *definable* in a higher-order logic. In the higher-order logic of PVS, subsets of a given type are modelled by predicates (i.e., the characteristic predicate of the set) over the type. The PVS formalization of the relational mu-calculus is shown in [20]. The PVS theory `mucalculus` takes a parameter `T` corresponding to the state type of the computation system or Kripke model. Predicates on the state type, i.e., assertions, are of the type `pred[T]`, namely, `[bool -> T]`. The stronger-than ordering on predicates of the state type is given by the relation `p1 <= p2`. A predicate transformer is a mapping from predicates to predicates. A monotone predicate transformer is one that preserves this ordering relation as checked by `ismono`. A fixed point of a predicate transformer is characterized by the higher-order relation `isfix`.

[6] It is typical to restrict $P[Z]$ to be *syntactically monotone* so that Z only occurs within an even number of negations.

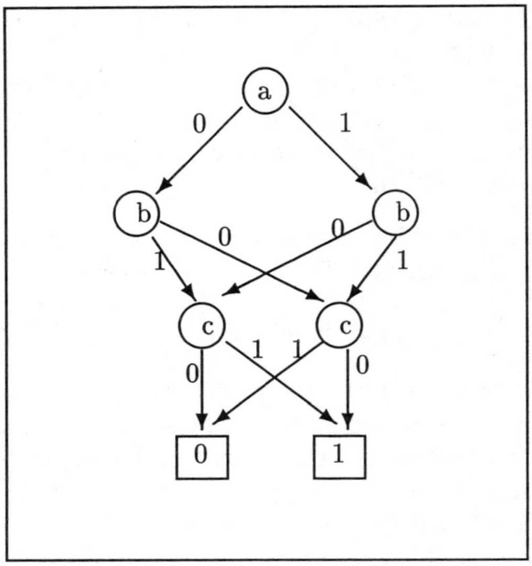

Figure 1: ROBDD for even parity on a, b, c

```
mucalculus[T:TYPE]: THEORY                                        20
BEGIN
  s: VAR T
  p,p1,p2: VAR pred[T]
  predtt: TYPE = [pred[T]->pred[T]]
  pp: VAR predtt
  setofpred: VAR pred[pred[T]]

  IMPORTING orders[pred[T]]
  <=(p1,p2):bool = FORALL s: p1(s) IMPLIES p2(s)

  ismono(pp):bool =
    FORALL p1,p2: p1 <= p2 IMPLIES pp(p1) <= pp(p2)

  isfix (pp,p):bool = (pp(p) = p)

  21
END mucalculus
```

The greatest lower bound of a set of sets (predicates) is the intersection over all these sets as defined by `glb`. The `member` operation is defined in PVS prelude library so that `member(x, p)` is just `p(x)`. The pre-fixed point of a predicate transformer `pp` is the set {`p | pp(p) <= p` }. The least fixed point is defined in 21 as the greatest lower bound of the pre-fixed points and this is used to define the μ or mu least fixed point operator.

```
glb (setofpred):pred[T] =                                              21
  LAMBDA s: (FORALL p: member(p,setofpred) IMPLIES  p(s))

lfp (pp):pred[T] = glb({p | pp(p) <= p})

mu(pp): pred[T] = lfp(pp)

22
```

Correspondingly, the greatest fixed point is defined in 22 as the least upper bound of the post-fixed points of a monotone predicate transformer which is then used to define the ν operator.

```
lub (setofpred):pred[T] =                                              22
  LAMBDA s: EXISTS p: member(p,setofpred) AND p(s)

gfp (pp):pred[T] = lub({p | p <= (pp(p))})

nu (pp):pred[T] = gfp(pp)
```

The fixed point operators can be explained intuitively in terms of infinite disjunctions as follows. Let $\top$ be $\lambda z_1 \ldots, z_n$: **true** and $\bot$ be $\lambda z_1, \ldots, z_n$: **false**, then if P is $\cup$-continuous (i.e., $P[\cup_{i \in \omega} Z_i] = \cup_{i \in \omega} P[Z_i]$, where $Z_j \leq Z_k$ for $j < k$), and Q is $\cap$-continuous (i.e., $P[\cap_{i \in \omega} Z_i] = \cap_{i \in \omega} P[Z_i]$, where $Z_j \leq Z_k$ for $k < j$), then

$$
\begin{aligned}
(\mu Z.P[Z])(z_1, \ldots, z_n) &= \bigvee_i P^i[\bot](z_1, \ldots, z_n) \\
(\nu Z.Q[Z])(z_1, \ldots, z_n) &= \bigwedge_i Q^i[\top](z_1, \ldots, z_n)
\end{aligned}
$$

When the state type is finite, all monotone predicate transformers are $\cap$-continuous and $\cup$-continuous.

Using PVS, we can prove that the `lfp` operator does actually return the least fixed point by the familiar Tarski-Knaster argument. Let l represent $\bigcap\{X | P[X] \leq X\}$:

1. l is a fixed point of P:

   (a) $P[l] \leq l$: For each l' in $\{X | P[X] \leq X\}$, we have $P[l'] \leq l'$ (since $l' \in \{X | P[X] \leq X\}$) and $l \leq l'$ (since l is a lower bound of $\{X | P[X] \leq X\}$), this yields (by monotonicity of P), $P[l] \leq P[l'] \leq l'$.

(b) $l \leq P[l]$: By 1a and monotonicity of P, $P[l]$ is itself a pre-fixed point, and hence $l \leq P[l]$.

2. If $P[Z] = Z$ then $l \leq Z$: This is obvious since if Z is a fixed point, it is a pre-fixed point.

The above description of the relational mu-calculus is general in the sense that there are no restrictions on the size of state type `T`. We say that the state type is finite when it is constructed from booleans and scalars by tuple, record, or array type constructors. When the state type `T` is finite, the least and greatest fixed points can be explicitly computed. In PVS, this is done by carrying out a binary encoding the variables of state type as an n-tuple of booleans so that a relational mu-calculus formula on a finite type is reduced to a BMC formula. The resulting BMC formula can then be processed by a package due to Janssen [19] which employs binary decision diagrams (BDDs) to check the validity of mu-calculus formulas.

3.5 Using the Mu-Calculus to Define CTL

Formulas in a branching-time temporal logic such as CTL characterize sets of states in a Kripke model, whereas formulas in a linear-time temporal logic (LTL) characterizes sets of computation traces in a Kripke model. CTL and LTL are incomparable in their expressiveness. CTL formulas consist of:

- atomic propositions: a, b, ...
- propositional combinations: $A \wedge B$, $A \vee B$, $\neg A$, $A \supset B$.
- universal path quantified formulas: $\mathbf{AG}A$ ("along every computation path, always A"), $\mathbf{AF}A$ ("along every path, eventually A"), A $\mathbf{EU}$ B (" .
- existential path quantified formulas: $\mathbf{EG}A$ ("along some path always A"), $\mathbf{EF}A$ ("along some path eventually A")

We give just three simple instances of the definitions of CTL operators using the relational mu-calculus as formalized in PVS. As a prelude to this, we first *lift* the boolean connectives to the domain of predicates in a pointwise manner. The `connectives` theory overloads the prefix/infix operators `AND`, `OR`, `NOT` to operate on predicates rather than booleans. As an example, the expression `f AND g` for predicates `f` and `g` returns a predicate of the form `(LAMBDA u:  f(u) AND g(u))`.

```
connectives [ T: TYPE ] : THEORY                                    23
  BEGIN

  f, g: VAR pred[T]
  u,v,w: VAR T

  AND(f,g)(u):bool =  f(u) AND g(u);
  OR(f,g)(u):bool =  f(u) OR g(u)
  NOT(f)(u):bool =  NOT f(u);
  IMPLIES(f,g)(u):bool = f(u) IMPLIES g(u);
  IFF(f,g)(u):bool = f(u) IFF g(u)

  TRUE(u) : bool = TRUE;
  FALSE(u) : bool = FALSE;

  END connectives
```

The CTL operators as defined below are indexed by the next-state relation N of the computation. The temporal operator EX is defined so that EX(N, p) holds of a state if p holds of some next state. The temporal formula EG(N, p) holds of a state if there is an infinite computation path leading out where p always holds. The temporal formula EU(N, p, q) holds of a state if there is computation path where q eventually holds and upto which point p holds.

```
ctlops[state : TYPE]: THEORY                                        24
 BEGIN
  IMPORTING mucalculus[state], connectives[state]
  u,v,w: VAR state
  f,g,Q,P,p1,p2: VAR pred[state]
  Z: VAR pred[[state, state]]
  N: VAR [state, state->bool]

  EX(N,f)(u):bool =  (EXISTS v: (f(v) AND N(u, v)))

  EG(N,f):pred[state] = nu (LAMBDA Q: (f AND EX(N,Q)))

  EU(N,f,g):pred[state] = mu (LAMBDA Q: (g OR (f AND EX(N,Q))))
  :
END ctlops
```

It is not hard to establish that the above definitions capture the intended meaning of the CTL operators. In the case of EX, the correspondence is obvious. The definition of EG is an infinite conjunction of the form

```
f AND EX(N, f) AND EX(N, (EX(N, f))) ....
```

This is easily seen to be equivalent to the existence of an infinite computation path on which f always holds.

Similarly, the definition of EU(N, f, g) expands out to the infinite disjunction

```
g OR (f AND EX(N, g)) OR (f AND EX(N, (f AND EX(N, g)))) ....
```

Since this is true exactly when at least one of the disjuncts holds, and each disjunct is equivalent to the existence of a finite path of f states terminating in a g state.

Invariance properties of transition systems can be captured in CTL as AG formulas. First, EF(N, f) is defined to hold on a state exactly when there is a computation path leading from that state to a state where f holds. Then AG(N, f) is defined to hold of any state from which no NOT f state is reachable.

```
EF(N,f):pred[state] = EU(N, TRUE, f)                                    25

AG(N,f):pred[state] = NOT EF(N, NOT f)
```

We can then, in 26, state the main invariant for Peterson's algorithm to assert that any state reachable by a computation path from a state s where the initialization predicate I holds, satisfies the assertion safe defined in 10.

```
safety: LEMMA                                                           26
   I(s) IMPLIES
        AG(G, safe)(s)
```

This property can be automatically proved by means of symbolic model checking. The model-check proof command expands out the initialization predicate I, the transition relation G, and also rewrites the CTL operators into their mu-calculus definitions. The type state is then checked to be finite and the formula is translated into the boolean mu-calculus and processed by a BDD-based validity checker.

```
safety :                                                                27

  |-------
[1]   (FORALL (s: state): I(s) IMPLIES AG(G, safe)(s))

Rule? (model-check)
   :
By rewriting and mu-simplifying,

Q.E.D.
```

The fixed point in the above case is computed in just two iterations by the BDD package and the BMC validity checking only takes milliseconds. With more complicated transition systems, the above definition of `AG` might be less efficient since it computes the set of states that can reach a state violating the invariant, and there could be many such states. In such cases, there is an alternative definition of `AG` that computes the set of states reachable from an initial state. The set of states reachable by a computation path of `N` transitions from a state in `P` can be defined by the predicate `Reachable(N, P)`.

```
Reachable(N, P)(s): bool =                                              28
  mu(LAMBDA Q:
      P OR
      (LAMBDA s2: (EXISTS s1: Q(s1) AND N(s1, s2))))(s)
```

The predicate `AGI(N, P, f)` then checks if the set of states reachable by a computation path from a state in `P` all satisfy the predicate `f`.

```
AGI(N, P, f): bool                                                      29
 = (FORALL u: Reachable(N, P)(u) IMPLIES f(u))
```

The safety property of Peterson's algorithm can then be expressed using `AGI`.

```
safety: LEMMA AGI(G, I, safe)                                           30
```

The mu-calculus model checker is similarly efficient on this "forward" version of `AG`. The latter definition turns out to be significantly more efficient on certain examples such as a finite-state version of the Philips bounded retransmission protocol.

BDDs are used as a compact symbolic representation for sets of states, the transition predicate, and the iterates of the predicate transformer used in computing the fixed points. The size of examples that can be handled by symbolic model checking is typically bounded by the size of the BDDs generated during the fixed point computation. As noted earlier, BDD size is sensitive to the variable ordering. Unlike other model checkers, in PVS there is no way for a user to control the variable ordering. Janssen's BDD package, like many other BDD packages, uses a heuristic for dynamically reordering the variables during BDD construction.

CTL as defined is in the alternation-free fragment of the mu-calculus since even though the temporal modalities of CTL are nested, there is no nesting of predicate variables bound by μ or ν in the expanded mu-calculus representation of a CTL formula.

3.6 Liveness

The safety property of mutual exclusion could be trivially satisfied by a program that denies entry into the critical section to both processes. As a desirable correctness property, one would like to demonstrate that if a process is in its `trying` state then it eventually makes the transition to its `critical` state. This *progress* or *liveness* property is not automatically guaranteed by the mutual exclusion algorithm as we modelled above since a valid computation can stutter indefinitely. In order to guarantee progress, we need to rule out computations with indefinite stuttering on states where a non-stuttering transition is possible. This can be done by means of fairness conditions. There is a large variety of notions of fairness. The typical ones are *strong fairness* (if a transition is infinitely often enabled, then it is infinitely often taken), and *weak fairness* (if a transition is indefinitely enabled, then it is eventually taken). These notions of fairness are not expressible in CTL, but CTL can be extended with a weaker notion of fairness that is often adequate.

A *fairness condition* `Ff` is just an assertion, i.e., a predicate on the `state` type of the computation. A fair computation path with respect to a fairness condition `Ff` is one where `Ff` occurs infinitely often. CTL can be extended so that the path quantifiers `A` and `E` are restricted to ranging over fair paths. For example, a fair version of the `EG` operator labelled `fairEG` can be defined using an alternation of a ν and μ operator.

31

```
fairEG(N, f)(Ff): pred[state] =
  nu(LAMBDA P: EU(N, f, f AND Ff AND EX(N, P)))
```

The `fairEG` operator defines a predicate that holds of states `P` such that there exists a path from any `P` state where `f` holds until both the `f` and `Ff` hold and the next state is a `P` state. If `Fflist` is a list of fairness conditions, then the `fairEG` operator can be extended to a list of fairness conditions by means of an operator `FairEG` where `FairEG(N, f)(Ff)` defines the set of set of states that lead off an infinite computation path along which `f` always holds and each `Ff` in `Fflist` holds infinitely often. We omit the definition of the CTL operators constrained with multiple fairness conditions.

The theorem `liveness_p` in 32 uses `AG` and `FairAF` to assert that in any state reachable from an initial state, if p is `trying`, then is always possible to reach along a fair path, a state where p is eventually `critical`. The restriction of the `AF` to fair paths with the accompanying fairness conditions ignores paths where either p or q is stuck in its `trying` state, or where q never exits the `critical` state.

```
                                                              32
  trying_p?(s): bool = trying?(pcp(s))
  trying_q?(s): bool = trying?(pcq(s))
  critical_p?(s): bool = critical?(pcp(s))
  critical_q?(s): bool = critical?(pcq(s))

 enabled_critical_p?(s) : bool =
    (trying?(pcp(s)) AND
          (turn(s) OR sleeping?(pcq(s))))

  ltl_liveness_1: CONJECTURE
    valid(G, Holds(I) IMPLIES (F(G(Holds(enabled_critical_p?)))
                               OR F(G(Holds(critical_q?)))
                               OR G(Holds(trying_p?) IMPLIES
                                    F(Holds(critical_p?)))))

 enabled_critical_q?(s) : bool =
    (trying?(pcq(s)) AND
          (NOT turn(s) OR sleeping?(pcp(s))))

 ctl_liveness: LEMMA
    I(s) IMPLIES
     AG(G, trying_p? IMPLIES
              FairAF(G, critical_p?)(
                  (: NOT enabled_critical_q?,
                     NOT enabled_critical_p?,
                     NOT critical_q? :)))(s)
```

Thus, the extension of CTL model checking with even a limited notion of fairness turns out to be useful in verifying liveness properties. CTL model checking is also useful in checking out a variety of small but useful properties. For example, the lemma `liveness_p` since it uses only universal path quantification could be trivially satisfied by an implementation that was deadlocked in its initial state. The lemma `sanity` provides a sanity check that there is a computation path along which p is eventually `critical`.

```
                                                              33
  sanity: LEMMA I(s) IMPLIES EF(G, critical_p?)(s)
```

In summary, model checking is an effective automatic technique for verifying temporal properties of finite-state systems. Model checking can be unified with theorem proving. This is done in PVS by first defining a relational mu-calculus using higher-order logic. The temporal operators of the branching-time temporal logic CTL can then be defined using the relational mu-calculus. When the computation state is of a finite type, relational mu-calculus formulas can be reduced to equivalent formulas in the boolean mu-calculus which can then be checked for validity using ROBDDs.

4 Model Checking Linear-Time Temporal Logic

We already noted the differences in the expressiveness of CTL and LTL. Unlike the branching-time temporal logic CTL, where formulas characterize sets of computation states, a linear-time temporal logic (LTL) formula characterize sets of computation traces. LTL formulas consist of

- atomic propositions: a, b, ...
- propositional combinations: $A \wedge B$, $A \vee B$, $\neg A$, $A \supset B$
- Next formulas : $\mathbf{X}A$ ("next A" or $\circ A$)
- Until formulas : $A \mathbf{U} B$

We can then define the more familar modalities:

- Eventuality formulas: $\mathbf{F}A$ ("eventually A" or $\Diamond A$) as `TRUE` $\mathbf{U}$ A.
- Stability formulas: $\mathbf{G}A$ ("henceforth A" or $\Box A$) as $\neg\mathbf{F}\neg A$.

Though the model checking problem for LTL formulas is PSPACE-complete and the best known algorithms are exponential in the size of the given formula, LTL is especially useful in capturing useful notions of fairness and in stating and verifying progress properties that require fairness constraints.

For example, we can then state and prove the LTL variant of the property `ctl_liveness` in 32 as `ltl_liveness` in 34. The formula asserts that for traces defined using the transition relation `G`, if the initial state satisfies `I`, then whenever p is `trying`, it is eventually `critical`, unless either p or q's entry into the critical section is enabled indefinitely but not taken, or q is stays indefinitely in its critical section. The `Holds(r)` construct essentially represents an assertion `r` as an atomic LTL proposition. The formula `valid(G, A)` asserts that every G computation path satisfies the LTL formula `A`.

34

```
ltl_liveness: LEMMA
   valid(G, Holds(I) IMPLIES (F(G(Holds(enabled_critical_p?)))
                              OR F(G(Holds(critical_q?)))
                              OR F(G(Holds(enabled_critical_q?)))
                              OR G(Holds(trying_p?) IMPLIES
                                    F(Holds(critical_p?)))))
```

The proof of `ltl_liveness` is by means of a strategy that converts an LTL validity assertion into an model checking problem in CTL augmented with fairness which is then discharged by means of the BDD-based mu-calculus validity checker. The details of this transformation are omitted. The key idea in model checking an LTL formula A is to extend each computation state s with labels corresponding to those subformulas of A that are satisfied by some computation path leading out of s. It is sufficient to label the states with formulas of the form $\mathbf{X}B$ since this is enough for computing the satisfiability of the other subformulas. Since subformulas of the form $B \mathbf{U} C$ are treated as being equivalent to $C \vee \mathbf{X}B \mathbf{U} C$, this raises the possibility that the satisfiability of $B \mathbf{U} C$ is given by a computation path along which $\mathbf{X}B \mathbf{U} C$ holds indefinitely but C never holds. Fairness conditions can be used to rule out such false computation paths by insisting that $C \vee \neg \mathbf{X}B \mathbf{U} C$ should hold infinitely often along a satisfying computation path.

5 Combining Theorem Proving and Model Checking

The previous sections have shown that:

- The same logical representation of the verification problem can be used for deductive verification or for model checking.
- Model checking can be viewed as a decision procedure for a finite-state fixed point fragment of the PVS logic.

The combination of the deductive and algorithmic (model checking) approaches has certain advantages:

1. Many significant properties of transition systems often depend only on the finite-state control part of the system where model checking can be quite effective.
2. Theorem proving can be used to *abstract* out the relevant finite-state control part of a larger, possibly infinite-state system in a *property-preserving* manner.
3. Model checking can be used to verify properties of system components that can then be composed using theorem proving.
4. Model checking can be combined with induction in the verification of parametric systems consisting of N similar processes.

The combination of theorem proving and model checking in PVS has been used in exactly this manner to verify an N-process mutual exclusion protocol that is an iterated version of Peterson's 2-process protocol. Here theorem proving is used to set up the induction and discharge the base case. The induction hypothesis is used to derive an abstraction from an $n+1$ process algorithm to a 2-process algorithm that is sufficient for verifying the mutual exclusion property. The 2-process abstraction is easily verified by model checking.

The combination of induction and model checking can also be used to verify progress properties. Here the combination of theorem proving and model checking is particularly useful since deductive proofs of liveness are usually quite tedious.

6 Conclusions

In this brief tutorial on the mechanized verification of transition systems, we have applied the PVS specification language, proof checker, and model checker to the verification of a simple instance of Peterson's mutual exclusion algorithm. The main conclusions to be drawn from this exercise are:

- The key concepts and logics used in the formalization of simple transition systems can be captured in a higher-order logic such as that of HOL [16] or PVS [28].
- Model checking and automated theorem proving are complementary but compatible verification techniques.
- Symbolic model checking using BDDs can be viewed as a decision procedure for a well-defined fragment of the PVS logic.
- The same representation of transition systems can be used as a basis for verification using either theorem proving or model checking or both in combination.
- Effective mechanized verification can be achieved with the combination of low-level decision procedures such as model checking, BDDs, and linear arithmetic, and high-level user interaction.

There are a number of open research questions in the mechanized verification of transition systems that merit further investigation. These include:

- The automated generation of invariants and invariant strengthenings.
- The discovery of property-preserving abstractions of transition systems.
- Experimentation with decision procedures for various data domains such as monadic second-order logic and recursive datatypes.
- Domain-specific techniques for application areas such as security, fault tolerance, real-time and hybrid systems, control systems, and communication protocols.

Mechanized formal verification is rapidly gaining acceptance particularly within the hardware design community as a useful technique for detecting design errors. Model checking has been especially popular in this regard because it is mostly automatic and yields intelligible counterexamples. Formal specification has also been gaining in popularity as a methodology for capturing requirements at varying levels of abstraction. Mechanized theorem proving usually requires some degree of manual guidance and is therefore not yet widely popular but deductive proof is clearly necessary to tie together high-level specification with low-level verification techniques. Such a unification can only be achieved by integrating various verification paradigms and techniques within a single coherent semantic framework.

References

[1] Martín Abadi and Leslie Lamport. The existence of refinement mappings. *Theoretical Computer Science*, 82(2):253–284, May 1991.

[2] R. Alur, C. Courcoubetis, N. Halbwachs, T. A. Henzinger, P.-H. Ho, X. Nicollin, A. Olivero, J. Sifakis, and S. Yovine. The algorithmic analysis of hybrid systems. *Theoretical Computer Science*, 138(1):3–34, 6 February 1995.

[3] Rajeev Alur, Costas Courcoubetis, and David Dill. Model-checking in dense real-time. *Information and Computation*, 104(1):2–34, May 1993.

[4] S. Bensalem, Y. Lakhnech, and H. Saidi. Powerful techniques for the automatic generation of invariants. In Rajeev Alur and Tom Henzinger, editors, *Computer-Aided Verification96*, Lecture Notes in Computer Science, New Brunswick, NJ, July 1996. Springer Verlag. To appear.

[5] Nikolaj Bjørner, Anca Browne, and Zohar Manna. Automatic generation of invariants and intermediate assertions. In Ugo Montanari and Francesca Rossi, editors, *Principles and Practice of Constraint Programming – CP '95*, volume 976 of *Lecture Notes in Computer Science*, Cassis, France, 1995. Springer Verlag.

[6] K. S. Brace, R. L. Rudell, and R. E. Bryant. Efficient implementation of a BDD package. In *Proc. of the 27th ACM/IEEE Design Automation Conference*, pages 40–45, 1990.

[7] R. E. Bryant. Graph-based algorithms for Boolean function manipulation. *IEEE Transactions on Computers*, C-35(8):677–691, August 1986.

[8] Randal E. Bryant. Symbolic boolean manipulation with ordered binary-decision diagrams. *ACM Computing Surveys*, 24(3):293–318, September 1992.

[9] J. R. Burch, E. M. Clarke, K. L. McMillan, D. L. Dill, and L. J. Hwang. Symbolic model checking: 10^{20} states and beyond. *Information and Computation*, 98(2):142–170, June 1992.

[10] K. Mani Chandy and Jayadev Misra. *Parallel Program Design: A Foundation.* Addison-Wesley, Reading, MA, 1988.

[11] E. Clarke, O. Grumberg, and D. Long. Verification tools for finite-state concurrent systems. pages 124–175.

[12] E. M. Clarke and O. Grumberg. Research on automatic verication of finite state concurrent systems. In *Annual Review of Computer Science*, pages 269–290. Annual Reviews, Inc., 1987.

[13] Costas Courcoubetis, editor. *Computer-Aided Verification, CAV '93*, volume 697 of *Lecture Notes in Computer Science*, Elounda, Greece, June/July 1993. Springer-Verlag.

[14] David Cyrluk. Inverting the abstraction mapping: A methodology for hardware verification. In Srivas and Camilleri [33], pages 172–186.

[15] E.A. Emerson and C.L Lei. Efficient model checking in fragments of the propositional mu-calculus. In *Proceedings of the 10th Symposium on Principles of Programming Languages*, pages 84–96, New Orleans, LA, January 1985. Association for Computing Machinery.

[16] M. J. C. Gordon and T. F. Melham, editors. *Introduction to HOL: A Theorem Proving Environment for Higher-Order Logic.* Cambridge University Press, Cambridge, UK, 1993.

[17] Michael J. C. Gordon. Mechanizing programming logics in higher-order logic. In G. Birtwistle and P. A. Subrahmanyam, editors, *Current Trends in Hardware Verification and Theorem Proving*, pages 387–439, New York, NY, 1989. Springer-Verlag.

[18] Jozef Hooman. Correctness of real time systems by construction. In H. Langmaack, W.-P. de Roever, and J. Vytopil, editors, *Formal Techniques in Real-Time and Fault-Tolerant Systems*, volume 863 of *Lecture Notes in Computer Science*, pages 19–40, Lübeck, Germany, September 1994. Springer-Verlag.

[19] G. L. J. M. Janssen. *ROBDD Software.* Department of Electrical Engineering, Eindhoven University of Technology, October 1993.

[20] Dexter Kozen. Results on the propositional μ-calculus. *Theoretical Computer Science*, 27:333–354, 1983.

[21] Leslie Lamport. The temporal logic of actions. *ACM TOPLAS*, 16(3):872–923, May 1994.

[22] Patrick Lincoln and John Rushby. Formal verification of an algorithm for interactive consistency under a hybrid fault model. In Courcoubetis [13], pages 292–304.

[23] Patrick Lincoln and John Rushby. Formal verification of an interactive consistency algorithm for the Draper FTP architecture under a hybrid fault model. In *COMPASS '94 (Proceedings of the Ninth Annual Conference on Computer Assurance)*, pages 107–120, Gaithersburg, MD, June 1994. IEEE Washington Section.

[24] Zohar Manna and Amir Pnueli. *Temporal Verification of Reactive Systems, Volume 2: Safety.* Springer Verlag, New York, NY, 1995.

[25] Steven P. Miller and Mandayam Srivas. Formal verification of the AAMP5 microprocessor: A case study in the industrial use of formal methods. In *WIFT '95: Workshop on Industrial-Strength Formal Specification Techniques*, pages 2–16, Boca Raton, FL, 1995. IEEE Computer Society.

[26] Paul S. Miner and James F. Leathrum, Jr. Verification of IEEE compliant subtractive division algorithms. In Srivas and Camilleri [33], pages 64–78.

[27] S. Owre, J. M. Rushby, and N. Shankar. PVS: A prototype verification system. In Deepak Kapur, editor, *11th International Conference on Automated Deduction (CADE)*, volume 607 of *Lecture Notes in Artificial Intelligence*, pages 748–752, Saratoga, NY, June 1992. Springer-Verlag.

[28] Sam Owre, John Rushby, Natarajan Shankar, and Friedrich von Henke. Formal verification for fault-tolerant architectures: Prolegomena to the design of PVS. *IEEE Transactions on Software Engineering*, 21(2):107–125, February 1995.

[29] David Park. Finiteness is mu-ineffable. *Theoretical Computer Science*, 3:173–181, 19.

[30] G. L. Peterson. Myths about the mutual exclusion problem. *Information Processing Letters*, 12(3):115–116, June 1981.

[31] H. Rueß, N. Shankar, and M. K. Srivas. Modular verification of SRT division. In Rajeev Alur and Thomas A. Henzinger, editors, *Computer-Aided Verification, CAV '96*, volume 1102 of *Lecture Notes in Computer Science*, pages 123–134, New Brunswick, NJ, July/August 1996. Springer-Verlag.

[32] Natarajan Shankar. Verification of real-time systems using PVS. In Courcoubetis [13], pages 280–291.

[33] Mandayam Srivas and Albert Camilleri, editors. *Formal Methods in Computer-Aided Design (FMCAD '96)*, volume 1166 of *Lecture Notes in Computer Science*, Palo Alto, CA, November 1996. Springer-Verlag.

NATO ASI Series F

Including Special Programmes on Sensory Systems for Robotic Control (ROB) and on Advanced Educational Technology (AET)

Vol. 1: Issues in Acoustic Signal - Image Processing and Recognition. Edited by C. H. Chen. VIII, 333 pages. 1983. *(out of print)*

Vol. 2: Image Sequence Processing and Dynamic Scene Analysis. Edited by T. S. Huang. IX, 749 pages. 1983.

Vol. 3: Electronic Systems Effectiveness and Life Cycle Costing. Edited by J. K. Skwirzynski. XVII, 732 pages. 1983. *(out of print)*

Vol. 4: Pictorial Data Analysis. Edited by R. M. Haralick. VIII, 468 pages. 1983.

Vol. 5: International Calibration Study of Traffic Conflict Techniques. Edited by E. Asmussen. VII, 229 pages. 1984.

Vol. 6: Information Technology and the Computer Network. Edited by K. G. Beauchamp. VIII, 271 pages. 1984.

Vol. 7: High-Speed Computation. Edited by J. S. Kowalik. IX, 441 pages. 1984.

Vol. 8: Program Transformation and Programming Environments. Report on a Workshop directed by F. L. Bauer and H. Remus. Edited by P. Pepper. XIV, 378 pages. 1984.

Vol. 9: Computer Aided Analysis and Optimization of Mechanical System Dynamics. Edited by E. J. Haug. XXII, 700 pages. 1984. *(out of print)*

Vol. 10: Simulation and Model-Based Methodologies: An Integrative View. Edited by T. I. Ören, B. P. Zeigler, M. S. Elzas. XIII, 651 pages. 1984.

Vol. 11: Robotics and Artificial Intelligence. Edited by M. Brady, L. A. Gerhardt, H. F. Davidson. XVII, 693 pages. 1984.

Vol. 12: Combinatorial Algorithms on Words. Edited by A. Apostolico, Z. Galil. VIII, 361 pages. 1985.

Vol. 13: Logics and Models of Concurrent Systems. Edited by K. R. Apt. VIII, 498 pages. 1985.

Vol. 14: Control Flow and Data Flow: Concepts of Distributed Programming. Edited by M. Broy. VIII, 525 pages. 1985. *Reprinted as Springer Study Edition 1986.*

Vol. 15: Computational Mathematical Programming. Edited by K. Schittkowski. VIII, 451 pages. 1985.

Vol. 16: New Systems and Architectures for Automatic Speech Recognition and Synthesis. Edited by R. De Mori, C.Y. Suen. XIII, 630 pages. 1985. *(out of print)*

Vol. 17: Fundamental Algorithms for Computer Graphics. Edited by R. A. Earnshaw. XVI, 1042 pages. 1985. *Reprinted as Springer Study Edition 1991.*

Vol. 18: Computer Architectures for Spatially Distributed Data. Edited by H. Freeman and G. G. Pieroni. VIII, 391 pages. 1985. *(out of print)*

Vol. 19: Pictorial Information Systems in Medicine. Edited by K. H. Höhne. XII, 525 pages. 1986.

Vol. 20: Disordered Systems and Biological Organization. Edited by E. Bienenstock, F. Fogelman Soulié, G. Weisbuch. XXI, 405 pages.1986

Vol. 21: Intelligent Decision Support in Process Environments. Edited by E. Hollnagel, G. Mancini, D. D. Woods. XV, 524 pages. 1986.

Vol. 22: Software System Design Methods. The Challenge of Advanced Computing Technology. Edited by J. K. Skwirzynski. XIII, 747 pages. 1986. *(out of print)*

Vol. 23: Designing Computer-Based Learning Materials. Edited by H. Weinstock and A. Bork. IX, 285 pages. 1986. *(out of print)*

Vol. 24: Database Machines. Modern Trends and Applications. Edited by A. K. Sood and A. H. Qureshi. VIII, 570 pages. 1986.

NATO ASI Series F

Including Special Programmes on Sensory Systems for Robotic Control (ROB) and on Advanced Educational Technology (AET)

Vol. 25: Pyramidal Systems for Computer Vision. Edited by V. Cantoni and S. Levialdi. VIII, 392 pages. 1986. *(ROB)*

Vol. 26: Modelling and Analysis in Arms Control. Edited by R. Avenhaus, R. K. Huber and J. D. Kettelle. VIII, 488 pages. 1986. *(out of print)*

Vol. 27: Computer Aided Optimal Design: Structural and Mechanical Systems. Edited by C. A. Mota Soares. XIII, 1029 pages. 1987.

Vol. 28: Distributed Operating Systems. Theory und Practice. Edited by Y. Paker, J.-P. Banatre and M. Bozyiğit. X, 379 pages. 1987.

Vol. 29: Languages for Sensor-Based Control in Robotics. Edited by U. Rembold and K. Hörmann. IX, 625 pages. 1987. *(ROB)*

Vol. 30: Pattern Recognition Theory and Applications. Edited by P. A. Devijver and J. Kittler. XI, 543 pages. 1987.

Vol. 31: Decision Support Systems: Theory and Application. Edited by C. W. Holsapple and A. B. Whinston. X, 500 pages. 1987.

Vol. 32: Information Systems: Failure Analysis. Edited by J. A. Wise and A. Debons. XV, 338 pages. 1987.

Vol. 33: Machine Intelligence and Knowledge Engineering for Robotic Applications. Edited by A. K. C. Wong and A. Pugh. XIV, 486 pages. 1987. *(ROB)*

Vol. 34: Modelling, Robustness and Sensitivity Reduction in Control Systems. Edited by R.F. Curtain. IX, 492 pages. 1987.

Vol. 35: Expert Judgment and Expert Systems. Edited by J. L. Mumpower, L. D. Phillips, O. Renn and V. R. R. Uppuluri. VIII, 361 pages. 1987.

Vol. 36: Logic of Programming and Calculi of Discrete Design. Edited by M. Broy. VII, 415 pages. 1987.

Vol. 37: Dynamics of Infinite Dimensional Systems. Edited by S.-N. Chow and J. K. Hale. IX. 514 pages. 1987.

Vol. 38: Flow Control of Congested Networks. Edited by A. R. Odoni, L. Bianco and G. Szegö. XII, 355 pages. 1987.

Vol. 39: Mathematics and Computer Science in Medical Imaging. Edited by M. A. Viergever and A. Todd-Pokropek. VIII, 546 pages. 1988.

Vol. 40: Theoretical Foundations of Computer Graphics and CAD. Edited by R. A. Earnshaw. XX, 1246 pages. 1988. *(out of print)*

Vol. 41: Neural Computers. Edited by R. Eckmiller and Ch. v. d. Malsburg. XIII, 566 pages. 1988. *Reprinted as Springer Study Edition 1989, 1990.*

Vol. 42: Real-Time Object Measurement and Classification. Edited by A. K. Jain. VIII, 407 pages. 1988. *(ROB)*

Vol. 43: Sensors and Sensory Systems for Advanced Robots. Edited by P. Dario. XI, 597 pages. 1988. *(ROB)*

Vol. 44: Signal Processing and Pattern Recognition in Nondestructive Evaluation of Materials. Edited by C. H. Chen. VIII, 344 pages. 1988. *(ROB)*

Vol. 45: Syntactic and Structural Pattern Recognition. Edited by G. Ferraté, T. Pavlidis, A. Sanfeliu and H. Bunke. XVI, 467 pages. 1988. *(ROB)*

NATO ASI Series F

Including Special Programmes on Sensory Systems for Robotic Control (ROB) and on Advanced Educational Technology (AET)

Vol. 46: Recent Advances in Speech Understanding and Dialog Systems. Edited by H. Niemann, M. Lang and G. Sagerer. X, 521 pages. 1988.

Vol. 47: Advanced Computing Concepts and Techniques in Control Engineering. Edited by M. J. Denham and A. J. Laub. XI, 518 pages. 1988. *(out of print)*

Vol. 48: Mathematical Models for Decision Support. Edited by G. Mitra. IX, 762 pages. 1988.

Vol. 49: Computer Integrated Manufacturing. Edited by I. B. Turksen. VIII, 568 pages. 1988.

Vol. 50: CAD Based Programming for Sensory Robots. Edited by B. Ravani. IX, 565 pages. 1988. *(ROB)*

Vol. 51: Algorithms and Model Formulations in Mathematical Programming. Edited by S. W. Wallace. IX, 190 pages. 1989.

Vol. 52: Sensor Devices and Systems for Robotics. Edited by A. Casals. IX, 362 pages. 1989. *(ROB)*

Vol. 53: Advanced Information Technologies for Industrial Material Flow Systems. Edited by S. Y. Nof and C. L. Moodie. IX, 710 pages. 1989.

Vol. 54: A Reappraisal of the Efficiency of Financial Markets. Edited by R. M. C. Guimar ães, B. G. Kingsman and S. J. Taylor. X, 804 pages. 1989.

Vol. 55: Constructive Methods in Computing Science. Edited by M. Broy. VII, 478 pages. 1989.

Vol. 56: Multiple Criteria Decision Making and Risk Analysis Using Microcomputers. Edited by B. Karpak and S. Zionts. VII, 399 pages. 1989.

Vol. 57: Kinematics and Dynamic Issues in Sensor Based Control. Edited by G. E. Taylor. XI, 456 pages. 1990. *(ROB)*

Vol. 58: Highly Redundant Sensing in Robotic Systems. Edited by J. T. Tou and J. G. Balchen. X, 322 pages. 1990. *(ROB)*

Vol. 59: Superconducting Electronics. Edited by H. Weinstock and M. Nisenoff. X, 441 pages. 1989.

Vol. 60: 3D Imaging in Medicine. Algorithms, Systems, Applications. Edited by K. H. Höhne, H. Fuchs and S. M. Pizer. IX, 460 pages. 1990. *(out of print)*

Vol. 61: Knowledge, Data and Computer-Assisted Decisions. Edited by M. Schader and W. Gaul. VIII, 421 pages. 1990.

Vol. 62: Supercomputing. Edited by J. S. Kowalik. X, 425 pages. 1990.

Vol. 63: Traditional and Non-Traditional Robotic Sensors. Edited by T. C. Henderson. VIII, 468 pages. 1990. *(ROB)*

Vol. 64: Sensory Robotics for the Handling of Limp Materials. Edited by P. M. Taylor. IX, 343 pages. 1990. *(ROB)*

Vol. 65: Mapping and Spatial Modelling for Navigation. Edited by L. F. Pau. VIII, 357 pages. 1990. *(ROB)*

Vol. 66: Sensor-Based Robots: Algorithms and Architectures. Edited by C. S. G. Lee. X, 285 pages. 1991. *(ROB)*

Vol. 67: Designing Hypermedia for Learning. Edited by D. H. Jonassen and H. Mandl. XXV, 457 pages. 1990. *(AET)*

Vol. 68: Neurocomputing. Algorithms, Architectures and Applications. Edited by F. Fogelman Soulié and J. Hérault. XI, 455 pages. 1990.

Vol. 69: Real-Time Integration Methods for Mechanical System Simulation. Edited by E. J. Haug and R. C. Deyo. VIII, 352 pages. 1991.

NATO ASI Series F

Including Special Programmes on Sensory Systems for Robotic Control (ROB) and on Advanced Educational Technology (AET)

Vol. 70: Numerical Linear Algebra, Digital Signal Processing and Parallel Algorithms. Edited by G. H. Golub and P. Van Dooren. XIII, 729 pages. 1991.

Vol. 71: Expert Systems and Robotics. Edited by T. Jordanides and B.Torby. XII, 744 pages. 1991.

Vol. 72: High-Capacity Local and Metropolitan Area Networks. Architecture and Performance Issues. Edited by G. Pujolle. X, 536 pages. 1991.

Vol. 73: Automation and Systems Issues in Air Traffic Control. Edited by J. A. Wise, V. D. Hopkin and M. L. Smith. XIX, 594 pages. 1991.

Vol. 74: Picture Archiving and Communication Systems (PACS) in Medicine. Edited by H. K. Huang, O. Ratib, A. R. Bakker and G. Witte. XI, 438 pages. 1991.

Vol. 75: Speech Recognition and Understanding. Recent Advances, Trends and Applications. Edited by P. Laface and Renato De Mori. XI, 559 pages. 1991.

Vol. 76: Multimedia Interface Design in Education. Edited by A. D. N. Edwards and S. Holland. XIV, 216 pages. 1992. *(AET)*

Vol. 77: Computer Algorithms for Solving Linear Algebraic Equations. The State of the Art. Edited by E. Spedicato. VIII, 352 pages. 1991.

Vol. 78: Integrating Advanced Technology into Technology Education. Edited by M. Hacker, A. Gordon and M. de Vries. VIII, 185 pages. 1991. *(AET)*

Vol. 79: Logic, Algebra, and Computation. Edited by F. L. Bauer. VII, 485 pages. 1991.

Vol. 80: Intelligent Tutoring Systems for Foreign Language Learning. Edited by M. L. Swartz and M. Yazdani. IX, 347 pages. 1992. *(AET)*

Vol. 81: Cognitive Tools for Learning. Edited by P. A. M. Kommers, D. H. Jonassen, and J. T. Mayes. X, 278 pages. 1992. *(AET)*

Vol. 82: Combinatorial Optimization. New Frontiers in Theory and Practice. Edited by M. Akgül, H. W. Hamacher, and S. Tüfekçi. XI, 334 pages. 1992.

Vol. 83: Active Perception and Robot Vision. Edited by A. K. Sood and H. Wechsler. IX, 756 pages. 1992.

Vol. 84: Computer-Based Learning Environments and Problem Solving. Edited by E. De Corte, M. C. Linn, H. Mandl, and L. Verschaffel. XVI, 488 pages. 1992. *(AET)*

Vol. 85: Adaptive Learning Environments. Foundations and Frontiers. Edited by M. Jones and P. H. Winne. VIII, 408 pages. 1992. *(AET)*

Vol. 86: Intelligent Learning Environments and Knowledge Acquisition in Physics. Edited by A. Tiberghien and H. Mandl. VIII, 285 pages. 1992. *(AET)*

Vol. 87: Cognitive Modelling and Interactive Environments. With demo diskettes (Apple and IBM compatible). Edited by F. L. Engel, D. G. Bouwhuis, T. Bösser, and G. d'Ydewalle. IX, 311 pages. 1992. *(AET)*

Vol. 88: Programming and Mathematical Method. Edited by M. Broy. VIII, 428 pages. 1992.

Vol. 89: Mathematical Problem Solving and New Information Technologies. Edited by J. P. Ponte, J. F. Matos, J. M. Matos, and D. Fernandes. XV, 346 pages. 1992. *(AET)*

Vol. 90: Collaborative Learning Through Computer Conferencing. Edited by A. R. Kaye. X, 260 pages. 1992. *(AET)*

Vol. 91: New Directions for Intelligent Tutoring Systems. Edited by E. Costa. X, 296 pages. 1992. *(AET)*

NATO ASI Series F

Including Special Programmes on Sensory Systems for Robotic Control (ROB) and on Advanced Educational Technology (AET)

Vol. 92: Hypermedia Courseware: Structures of Communication and Intelligent Help. Edited by A. Oliveira. X, 241 pages. 1992. *(AET)*

Vol. 93: Interactive Multimedia Learning Environments. Human Factors and Technical Considerations on Design Issues. Edited by M. Giardina. VIII, 254 pages. 1992. *(AET)*

Vol. 94: Logic and Algebra of Specification. Edited by F. L. Bauer, W. Brauer, and H. Schwichtenberg. VII, 442 pages. 1993.

Vol. 95: Comprehensive Systems Design: A New Educational Technology. Edited by C. M. Reigeluth, B. H. Banathy, and J. R. Olson. IX, 437 pages. 1993. *(AET)*

Vol. 96: New Directions in Educational Technology. Edited by E. Scanlon and T. O'Shea. VIII, 251 pages. 1992. *(AET)*

Vol. 97: Advanced Models of Cognition for Medical Training and Practice. Edited by D. A. Evans and V. L. Patel. XI, 372 pages. 1992. *(AET)*

Vol. 98: Medical Images: Formation, Handling and Evaluation. Edited by A. E. Todd-Pokropek and M. A. Viergever. IX, 700 pages. 1992.

Vol. 99: Multisensor Fusion for Computer Vision. Edited by J. K. Aggarwal. XI, 456 pages. 1993. *(ROB)*

Vol. 100: Communication from an Artificial Intelligence Perspective. Theoretical and Applied Issues. Edited by A. Ortony, J. Slack and O. Stock. XII, 260 pages. 1992.

Vol. 101: Recent Developments in Decision Support Systems. Edited by C. W. Holsapple and A. B. Whinston. XI, 618 pages. 1993.

Vol. 102: Robots and Biological Systems: Towards a New Bionics? Edited by P. Dario, G. Sandini and P. Aebischer. XII, 786 pages. 1993.

Vol. 103: Parallel Computing on Distributed Memory Multiprocessors. Edited by F. Özgüner and F. Erçal. VIII, 332 pages. 1993.

Vol. 104: Instructional Models in Computer-Based Learning Environments. Edited by S. Dijkstra, H. P. M. Krammer and J. J. G. van Merriënboer. X, 510 pages. 1993. *(AET)*

Vol. 105: Designing Environments for Constructive Learning. Edited by T. M. Duffy, J. Lowyck and D. H. Jonassen. VIII, 374 pages. 1993. *(AET)*

Vol. 106: Software for Parallel Computation. Edited by J. S. Kowalik and L. Grandinetti. IX, 363 pages. 1993.

Vol. 107: Advanced Educational Technologies for Mathematics and Science. Edited by D. L. Ferguson. XII, 749 pages. 1993. *(AET)*

Vol. 108: Concurrent Engineering: Tools and Technologies for Mechanical System Design. Edited by E. J. Haug. XIII, 998 pages. 1993.

Vol. 109: Advanced Educational Technology in Technology Education. Edited by A. Gordon, M. Hacker and M. de Vries. VIII, 253 pages. 1993. *(AET)*

Vol. 110: Verification and Validation of Complex Systems: Human Factors Issues. Edited by J. A. Wise, V. D. Hopkin and P. Stager. XIII, 704 pages. 1993.

Vol. 111: Cognitive Models and Intelligent Environments for Learning Programming. Edited by E. Lemut, B. du Boulay and G. Dettori. VIII, 305 pages. 1993. *(AET)*

Vol. 112: Item Banking: Interactive Testing and Self-Assessment. Edited by D. A. Leclercq and J. E. Bruno. VIII, 261 pages. 1993. *(AET)*

Vol. 113: Interactive Learning Technology for the Deaf. Edited by B. A. G. Elsendoorn and F. Coninx. XIII, 285 pages. 1993. *(AET)*

NATO ASI Series F

Including Special Programmes on Sensory Systems for Robotic Control (ROB) and on Advanced Educational Technology (AET)

Vol. 114: Intelligent Systems: Safety, Reliability and Maintainability Issues. Edited by O. Kaynak, G. Honderd and E. Grant. XI, 340 pages. 1993.

Vol. 115: Learning Electricity and Electronics with Advanced Educational Technology. Edited by M. Caillot. VII, 329 pages. 1993. *(AET)*

Vol. 116: Control Technology in Elementary Education. Edited by B. Denis. IX, 311 pages. 1993. *(AET)*

Vol. 117: Intelligent Learning Environments: The Case of Geometry. Edited by J.-M. Laborde. VIII, 267 pages. 1996. *(AET)*

Vol. 118: Program Design Calculi. Edited by M. Broy. VIII, 409 pages. 1993.

Vol. 119: Automating Instructional Design, Development, and Delivery. Edited by. R. D. Tennyson. VIII, 266 pages. 1994. *(AET)*

Vol. 120: Reliability and Safety Assessment of Dynamic Process Systems. Edited by T. Aldemir, N. O. Siu, A. Mosleh, P. C. Cacciabue and B. G. Göktepe. X, 242 pages. 1994.

Vol. 121: Learning from Computers: Mathematics Education and Technology. Edited by C. Keitel and K. Ruthven. XIII, 332 pages. 1993. *(AET)*

Vol. 122: Simulation-Based Experiential Learning. Edited by D. M. Towne, T. de Jong and H. Spada. XIV, 274 pages. 1993. *(AET)*

Vol. 123: User-Centred Requirements for Software Engineering Environments. Edited by D. J. Gilmore, R. L. Winder and F. Détienne. VII, 377 pages. 1994.

Vol. 124: Fundamentals in Handwriting Recognition. Edited by S. Impedovo. IX, 496 pages. 1994.

Vol. 125: Student Modelling: The Key to Individualized Knowledge-Based Instruction. Edited by J. E. Greer and G. I. McCalla. X, 383 pages. 1994. *(AET)*

Vol. 126: Shape in Picture. Mathematical Description of Shape in Grey-level Images. Edited by Y.-L. O, A. Toet, D. Foster, H. J. A. M. Heijmans and P. Meer. XI, 676 pages. 1994.

Vol. 127: Real Time Computing. Edited by W. A. Halang and A. D. Stoyenko. XXII, 762 pages. 1994.

Vol. 128: Computer Supported Collaborative Learning. Edited by C. O'Malley. X, 303 pages. 1994. *(AET)*

Vol. 129: Human-Machine Communication for Educational Systems Design. Edited by M. D. Brouwer-Janse and T. L. Harrington. X, 342 pages. 1994. *(AET)*

Vol. 130: Advances in Object-Oriented Database Systems. Edited by A. Dogac, M. T. Özsu, A. Biliris and T. Sellis. XI, 515 pages. 1994.

Vol. 131: Constraint Programming. Edited by B. Mayoh, E. Tyugu and J. Penjam. VII, 452 pages. 1994.

Vol. 132: Mathematical Modelling Courses for Engineering Education. Edited by Y. Ersoy and A. O. Moscardini. X, 246 pages. 1994. *(AET)*

Vol. 133: Collaborative Dialogue Technologies in Distance Learning. Edited by M. F. Verdejo and S. A. Cerri. XIV, 296 pages. 1994. *(AET)*

Vol. 134: Computer Integrated Production Systems and Organizations. The Human-Centred Approach. Edited by F. Schmid, S. Evans, A. W. S. Ainger and R. J. Grieve. X, 347 pages. 1994.

Vol. 135: Technology Education in School and Industry. Emerging Didactics for Human Resource Development. Edited by D. Blandow and M. J. Dyrenfurth. XI, 367 pages. 1994. *(AET)*

Vol. 136: From Statistics to Neural Networks. Theory and Pattern Recognition Applications. Edited by V. Cherkassky, J. H. Friedman and H. Wechsler. XII, 394 pages. 1994.

NATO ASI Series F

Including Special Programmes on Sensory Systems for Robotic Control (ROB) and on Advanced Educational Technology (AET)

Vol. 137: Technology-Based Learning Environments. Psychological and Educational Foundations. Edited by S. Vosniadou, E. De Corte and H. Mandl. X, 302 pages. 1994. *(AET)*

Vol. 138: Exploiting Mental Imagery with Computers in Mathematics Education. Edited by R. Sutherland and J. Mason. VIII, 326 pages. 1995. *(AET)*

Vol. 139: Proof and Computation. Edited by H. Schwichtenberg. VII, 470 pages. 1995.

Vol. 140: Automating Instructional Design: Computer-Based Development and Delivery Tools. Edited by R. D. Tennyson and A. E. Barron. IX, 618 pages. 1995. *(AET)*

Vol. 141: Organizational Learning and Technological Change. Edited by C. Zucchermaglio, S. Bagnara and S. U. Stucky. X, 368 pages. 1995. *(AET)*

Vol. 142: Dialogue and Instruction. Modeling Interaction in Intelligent Tutoring Systems. Edited by R.-J. Beun, M. Baker and M. Reiner. IX, 368 pages. 1995. *(AET)*

Vol. 143: Batch Processing Systems Engineering. Fundamentals of Chemical Engineering. Edited by G. V. Reklaitis, A. K. Sunol, D. W. T. Rippin, and Ö. Hortaçsu. XIV, 868 pages. 1996.

Vol. 144: The Biology and Technology of Intelligent Autonomous Agents. Edited by Luc Steels. VIII, 517 pages. 1995.

Vol. 145: Advanced Educational Technology: Research Issues and Future Potential. Edited by T. T. Liao. VIII, 219 pages. 1996. *(AET)*

Vol. 146: Computers and Exploratory Learning. Edited by A. A. diSessa, C. Hoyles and R. Noss. VIII, 482 pages. 1995. *(AET)*

Vol. 147: Speech Recognition and Coding. New Advances and Trends. Edited by A. J. Rubio Ayuso and J. M. López Soler. XI, 505 pages. 1995.

Vol. 148: Knowledge Acquisition, Organization, and Use in Biology. Edited by K. M. Fisher and M. R. Kibby. X, 246 pages. 1996. *(AET)*

Vol. 149: Emergent Computing Methods in Engineering Design. Applications of Genetic Algorithms and Neural Networks. Edited by D.E. Grierson and P. Hajela. VIII, 350 pages. 1996.

Vol. 150: Speechreading by Humans and Machines. Edited by D. G. Stork and M. E. Hennecke. XV, 686 pages. 1996.

Vol. 151: Computational and Conversational Discourse. Burning Issues – An Interdisciplinary Account. Edited by E. H. Hovy and D. R. Scott. XII, 202 pages. 1996.

Vol. 152: Deductive Program Design. Edited by M. Broy. IX, 467 pages. 1996.

Vol. 153: Identification, Adaptation, Learning. Edited by S. Bittanti and G. Picci. XIV, 553 pages. 1996.

Vol. 154: Reliability and Maintenance of Complex Systems. Edited by S. Özekici. XI, 589 pages. 1996.

Vol. 155: Cooperation: Game-Theoretic Approaches. Edited by S. Hart and A. Mas-Colell. VIII, 328 pages. 1997.

Vol. 156: Microcomputer-Based Labs: Educational Research and Standards. Edited by R.F. Tinker. XIV, 398 pages. 1996. *(AET)*

Vol. 157: Logic of Computation. Edited by H. Schwichtenberg. VII, 396 pages. 1997.

Vol. 158: Mathematical Methods in Program Development. Edited by M. Broy and B. Schieder. VIII, 528 pages. 1997.

Springer and the environment

At Springer we firmly believe that an international science publisher has a special obligation to the environment, and our corporate policies consistently reflect this conviction.
We also expect our business partners – paper mills, printers, packaging manufacturers, etc. – to commit themselves to using materials and production processes that do not harm the environment. The paper in this book is made from low- or no-chlorine pulp and is acid free, in conformance with international standards for paper permanency.

Springer